# The Receptors

VOLUME II

# Contributors

I. M. Adcock

Mario Ascoli

J. J. Blum

Hector F. DeLuca

Boris Draznin

William F. Greenlee

B. D. Greenstein

L. E. Limbird

Rebecca Link

Allen P. Minton

Robert A. Neal

James C. Paulson

M. R. Sairam

Yoram Shechter

J. D. Sweatt

J. Craig Venter

# The Receptors

## VOLUME II

Edited by

### P. MICHAEL CONN

*Department of Pharmacology*
*University of Iowa*
*College of Medicine*
*Iowa City, Iowa*

**1985**

ACADEMIC PRESS, INC.
*(Harcourt Brace Jovanovich, Publishers)*

Orlando   San Diego   New York   London
Toronto   Montreal   Sydney   Tokyo

ACADEMIC PRESS, INC.
Orlando, Florida 32887


*United Kingdom Edition published by*
ACADEMIC PRESS INC. (LONDON) LTD.
24–28 Oval Road, London NW1 7DX


Library of Congress Cataloging in Publication Data
(Revised for vol. II)
Main entry under title:

The Receptors.

  Includes indexes.
    1. Cell receptors–Collected works.   2. Drug
receptors–Collected works.   I. Conn, P. Michael.
[DNLM: 1. Receptors, Drug.   2. Receptors, Endogenous
substances.   3. Receptors, Immunologic.   WK 102 R2955]
QH603.C43R428   1984        615'.7        84-6378
ISBN  0–12–185202–4 (v. 2: alk. paper)

PRINTED IN THE UNITED STATES OF AMERICA

85 86 87 88        9 8 7 6 5 4 3 2 1

# Contents

CHAPTER 1

**The Vitamin D Receptor**

*Rebecca Link and Hector F. DeLuca*

CHAPTER 2

**On Experimental Discrimination between Alternative Mechanistic Models for the Receptor-Mediated Stimulation of Adenylate Cyclase**

*Allen P. Minton*

CHAPTER 6

**Studies on Insulin Receptors: Implications for Insulin Action**

*Yoram Shechter*

CHAPTER 7

**Size of Neurotransmitter Receptors as Determined by Radiation Inactivation–Target Size Analysis**

*J. Craig Venter*

CHAPTER 8

**$\alpha_2$-Adrenergic Receptors: Apparent Interaction with Multiple Effector Systems**

*L. E. Limbird and J. D. Sweatt*

CHAPTER 9

**Protein Glycosylation and Receptor–Ligand Interactions**

*M. R. Sairam*

CHAPTER 10

**Role of Steroid Hormone Receptors in Development
and Puberty**

*B. D. Greenstein and I. M. Adcock*

CHAPTER 11

**Functions and Regulation of Cell Surface Receptors in Cultured Leydig Tumor Cells**

*Mario Ascoli*

CHAPTER 12

**Somatostatin Receptors in Endocrine Cells**

*Boris Draznin*

# Contributors

*Numbers in parentheses indicate the pages on which the authors' contributions begin.*

I. M. ADCOCK (341), Department of Pharmacology, St. Thomas's Hospital Medical School, London SE1 7EH, United Kingdom

MARIO ASCOLI[1] (367), Division of Endocrinology, Departments of Medicine and Biochemistry, School of Medicine, Vanderbilt University, Nashville, Tennessee 37232

J. J. BLUM (57), Department of Physiology, Duke University Medical Center, Durham, North Carolina 27710

HECTOR F. DELUCA (1), Department of Biochemistry, University of Wisconsin, Madison, Madison, Wisconsin 53706

BORIS DRAZNIN (401), Research Service, Veterans Administration Medical Center and Department of Medicine, University of Colorado Health Sciences Center, Denver, Colorado 80220

WILLIAM F. GREENLEE (89), Department of Cell Biology, Chemical Industry Institute of Toxicology, Research Triangle Park, North Carolina 27709

B. D. GREENSTEIN (341), Department of Pharmacology, St. Thomas's Hospital Medical School, London SE1 7EH, United Kingdom

L. E. LIMBIRD (281), Department of Pharmacology, School of Medicine, Vanderbilt University, Nashville, Tennessee 37232

REBECCA LINK (1), Department of Biochemistry, University of Wisconsin, Madison, Madison, Wisconsin 53706

ALLEN P. MINTON (37), Laboratory of Biochemical Pharmacology, National Institute of Arthritis, Diabetes, and Digestive and Kidney Diseases, National Institutes of Health, Bethesda, Maryland 20205

[1]*Present address: Center for Biomedical Research, The Population Council, 1230 York Avenue, New York, New York 10021*

ROBERT A. NEAL (89), Department of Biochemical Toxicology and Pathobiology, Chemical Industry Institute of Toxicology, Research Triangle Park, North Carolina 27709

JAMES C. PAULSON (131), Department of Biological Chemistry, UCLA School of Medicine, Los Angeles, California 90024

M. R. SAIRAM (307), Reproduction Research Laboratory, Clinical Research Institute of Montreal, Montreal, Quebec H2W 1R7, Canada

YORAM SHECHTER (221), Department of Hormone Research, The Weizmann Institute of Science, 76100 Rehovot, Israel

J. D. SWEATT (281), Department of Pharmacology, School of Medicine, Vanderbilt University, Nashville, Tennessee 37232

J. CRAIG VENTER (245), Section on Receptor Biochemistry, Laboratory of Neurophysiology, NINCDS, National Institutes of Health, Bethesda, Maryland 20205

# Preface

It has been noted that those scientific discoveries that form the basis of our central dogmas appear to pass through three stages. Initially most scientists are skeptical and believe the discovery is "new, but not true." If many laboratories are able to repeat the finding and skeptics cannot find the means to disprove it, the discovery is elevated: "true, but not important." Time passes. If the work continues to survive scrutiny and establishes significance, the discovery reaches the ultimate accolade: "true *and* important, but not new." Usually, by the time an idea arrives at the third stage, the concept is shown to pervade many systems and becomes formative in the design of testable experimental models.

This is the rationale which encourages scientists to read widely, often in areas which have only little direct association to one's own restricted scientific interest. Often a finding in one area is seminal in another. For this reason, I hope this treatise will be helpful. Its intent is to provide a cross-disciplinary forum for those interested in biological recognition and signals.

The chapters continue to be of three types: (1) those that cover particularly well understood receptor systems, (2) those that reexamine old ideas about receptors in light of new data or that describe groups of recognition systems and their relation to a single theme, and (3) those that include techniques and advances with unusual potential for receptor study.

With these themes in mind, this volume contains chapters that deal with our current understanding of receptors for somatostatin, vitamin D, insulin, and animal viruses, as well as for the $\alpha_2$-adrenergic and *Ah* systems. Information is presented on the significance of translational modifications of receptor ligands, and current thoughts on the mechanisms of receptor–ligand interactions are examined. Two additional chapters assess the role of receptors in development and their regulation by tumors. Finally, a recently appreciated technology, target size analysis, is described since it offers a unique opportunity to obtain considerable physical information about receptors while they are still membrane components.

*P. Michael Conn*

# Contents of Previous Volume

# The Receptors

VOLUME II

# 1

# The Vitamin D Receptor

**REBECCA LINK AND HECTOR F. DELUCA**
Department of Biochemistry
University of Wisconsin, Madison
Madison, Wisconsin

## I. INTRODUCTION

Vitamin D is a secosteroid and is the only vitamin known to be converted to a hormone (DeLuca, 1983; DeLuca and Schnoes, 1983; Norman *et al.*, 1982). In actual fact, vitamin D is not required in the diet when animals are exposed to sufficient amounts of sunlight. It is now well known that ultraviolet light in the range of 280–300 nm induces a chemical photolysis of the sterol 7-dehydrocholesterol that is found in abundant quantities in the epidermis (DeLuca, 1983). The immediate product of the photolysis reaction is previtamin $D_3$, as illustrated in Fig. 1. Previtamin D then undergoes a thermally dependent isomerization into an equilibrium mixture of vitamin D and previtamin D. Achievement of this equilibrium requires 36 hr at body temperature (Holick, 1981). Vitamin D (but not previtamin $D_3$) is then recognized by the vitamin D transport protein that transports it to the liver where it begins a series of metabolic modifications

**1**

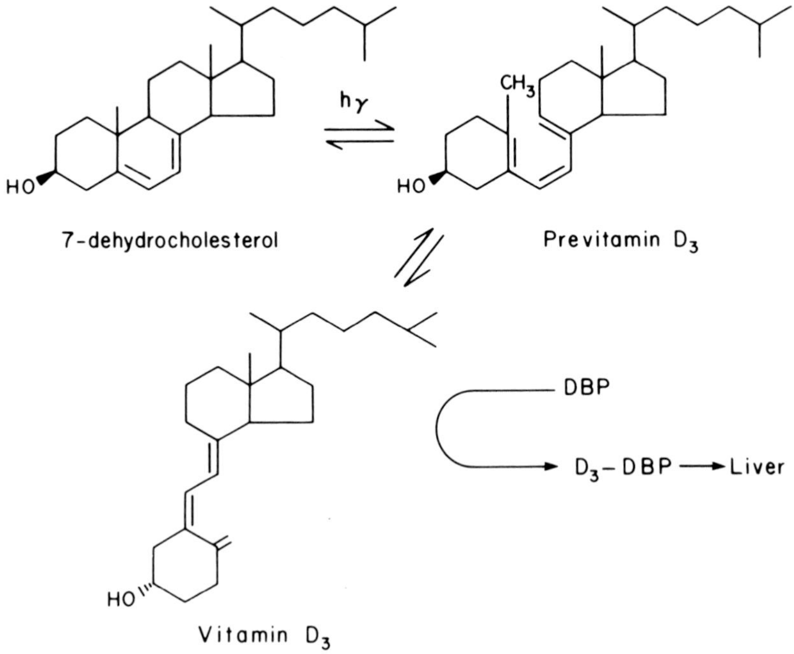

**Fig. 1.** Biogenesis of vitamin $D_3$.

required for function. Vitamin D can also be absorbed from the diet in two forms, vitamin $D_2$ and vitamin $D_3$. These compounds are absorbed in the distal small intestine with fats into the lacteal system, ultimately reaching the liver (DeLuca, 1983; DeLuca and Schnoes, 1983; Norman *et al.*, 1982; Dueland *et al.*, 1982). Thus, vitamin $D_3$ is in fact not a vitamin, since it is normally produced in skin under ultraviolet light and is then converted in a series of reactions to a hormone to carry out its functions.

Figure 2 illustrates the metabolism of vitamin $D_3$ required for function (De-Luca, 1983; DeLuca and Schnoes, 1983; Norman *et al.*, 1982). In the endoplasmic reticulum of the liver parenchymal cells (DeLuca, 1983; DeLuca and Schnoes, 1983; Norman *et al.*, 1982; Dueland *et al.*, 1981), vitamin D is converted to 25-hydroxyvitamin $D_3$ (25-OH-$D_3$) (DeLuca and Schnoes, 1983), although some of this compound can be produced in mitochondria by a system not specific for vitamin D (Björkhem *et al.*, 1980). 25-Hydroxyvitamin D is the major circulating form of the vitamin and is also transported on the vitamin D-transport $\alpha$-globulin (DeLuca, 1983; DeLuca and Schnoes, 1983; Norman *et al.*, 1982). The kidney modifies 25-OH-D further before it can function. In the kidney, 25-OH-D is $1\alpha$-hydroxylated to form the final vitamin D hormone, $1\alpha,25$-dihydroxyvitamin $D_3$ [$1\alpha,25$-(OH)$_2$D$_3$]. This reaction occurs in the mito-

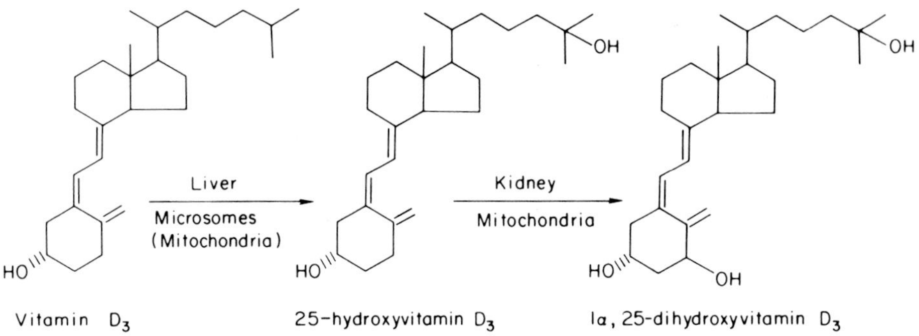

Vitamin D$_3$        25-hydroxyvitamin D$_3$        1α, 25-dihydroxyvitamin D$_3$

**Fig. 2.** Metabolism of vitamin D$_3$ required for function.

chondria as a result of the action of a three-component mixed-function monooxygenase which is dependent upon a specific cytochrome *P*-450 (DeLuca, 1983; DeLuca and Schnoes, 1983; Ghazarian *et al.*, 1974). The 1,25-(OH)$_2$D$_3$ that is formed in the kidney is then transported by the plasma-transport protein to the target tissues for function.

The kidney also carries out several other metabolic reactions on vitamin D. The most prominent at physiologic concentrations is 24-hydroxylation to produce 24*R*,25-dihydroxyvitamin D$_3$ [24*R*,25-(OH)$_2$D$_3$] (DeLuca, 1983; DeLuca and Schnoes, 1983; Norman *et al.*, 1982). Other physiologic hydroxylations include 26-hydroxylation, 23-hydroxylation, and the formation of the 26,23-lactone (DeLuca, 1981, 1983). Several other metabolic modifications have been reported to occur in liver preparations either *in vitro* or *in vivo* under circumstances of large doses or concentrations of 25-OH-D (DeLuca, 1983, 1985; DeLuca and Schnoes, 1983; Norman *et al.*, 1982). It is unclear what the physiological significance of these reactions might be. Interested readers are directed elsewhere for reviews on the metabolism of vitamin D (DeLuca, 1981, 1983; DeLuca and Schnoes, 1983; Norman *et al.*, 1982).

1,25-Dihydroxyvitamin D$_3$ has three well-defined functions and several other postulated functions. The best known of the functions is the responsibility for initiating intestinal transport of calcium in the villus cells from the lumen of the small intestine to the plasma (DeLuca, 1983; DeLuca and Schnoes, 1983; Norman *et al.*, 1982). Similarly, it initiates a phosphate-transport system in the intestinal villus cells as well. Both reactions are metabolically active, requiring metabolic energy for the transfer. The molecular mechanism of the transcellular transport of either calcium or phosphorus is poorly understood and will be discussed later.

The 1,25-(OH)$_2$D$_3$ also plays an important role in the mobilization of calcium from bone (DeLuca, 1983; DeLuca and Schnoes, 1983; Norman *et al.*, 1982). *In vivo* this is best observed by measuring the rise in serum calcium of a vitamin D-deficient animal on a low-calcium diet in response to the hormone. It is believed that calcium is transferred from the bone-fluid compartment, across the osteoblastic and bone-lining cell membranes, to the plasma compartment. This process requires the presence of not only 1,25-$(OH)_2D_3$ but also the parathyroid hormone (DeLuca, 1983; DeLuca and Schnoes, 1983; Norman *et al.*, 1982). In addition to this reaction, 1,25-$(OH)_2D_3$ can be added to embryonic cultures of bone to bring about a resorption of bone (Raisz *et al.*, 1972). It is not known how this system is connected to the bone-calcium mobilization system or whether it is related to the bone remodeling sequence, which involves resorption of old bone and formation of new bone (Frost, 1966). Thus, the responses to 1,25-$(OH)_2D_3$ in bone are complex and have yet to be studied in molecular detail.

Another proven site of action of 1,25-$(OH)_2D_3$ is the nephron of the kidney. It is believed to operate primarily in the distal renal tubule to facilitate renal reabsorption of calcium (DeLuca, 1983; DeLuca and Schnoes, 1983). It is also very possible that the renal reabsorption of calcium in the distal tubule is under control of the vitamin D hormone with the parathyroid hormone working in concert. There have been other reports that 1,25-$(OH)_2D_3$ may function in the proximal convoluted tubule cells where it in some way brings about regulation of the enzymes that metabolize 25-$OH$-$D_3$ (DeLuca, 1983; DeLuca and Schnoes, 1983; Norman *et al.*, 1982). By initiating intestinal calcium transport, intestinal phosphate transport, the mobilization of calcium from bone, and renal reabsorption of calcium, vitamin D brings about an elevation of plasma calcium and phosphorus to levels that are required for normal neuromuscular function and for the mineralization of bone (DeLuca, 1983; DeLuca and Schnoes, 1983; Norman *et al.*, 1982). It is now quite clear that 1,25-$(OH)_2D_3$ or in fact any form of vitamin D is not required for the actual transfer of mineral from the plasma compartment to bone (Underwood and DeLuca, 1984; Weinstein *et al.*, 1984).

Of all of the target sites of action of 1,25-$(OH)_2D_3$, studies on the molecular mechanism of action have only been carried out on the intestinal villus cell and the bone-calcium mobilization system. The mobilization of calcium from bone in response to 1,25-$(OH)_2D_3$ is blocked by the prior administration of actinomycin D but not by the postadministration of the antibiotic (Tanaka and DeLuca, 1971). Thus, that action appears to involve nuclear events including transcription. The nature of the gene products expressed in response to 1,25-$(OH)_2D_3$ in bone has not been determined. Calcium binding protein, a gene product in chick intestine, has been identified in small amounts in bone (Christakos and Norman, 1978). It is unclear, however, if it is this protein or some other protein that is involved in the mobilization of calcium.

There have been reports that 1,25-$(OH)_2D_3$ may play a role in osteocalcin synthesis, especially in osteosarcoma cell lines (Price and Baukol, 1980, 1981). The role of osteocalcin in bone is unknown, and, furthermore, a lack of osteocalcin production in vitamin D deficiency has not been satisfactorily demonstrated.

By far, the most extensive information is available on the intestinal calcium transport system of the small intestine. The response of the intestinal calcium transport system to 1,25-$(OH)_2D_3$ is a complex biphasic one, as will be discussed in Section IV (Halloran and DeLuca, 1981a).

## II. DISCOVERY OF THE RECEPTOR PROTEINS FOR 1,25-$(OH)_2D_3$

Following the clear demonstration of nuclear localization of 1,25-$(OH)_2D_3$ prior to a target organ response (Stumpf et al., 1979; Zile et al., 1978), it was logical to assume that a receptor protein might exist for 1,25-$(OH)_2D_3$ primarily because of its steroidal nature. Early work resulted in the conclusion that vitamin D itself was active on the nucleus (Haussler and Norman, 1967) and was later revised to a metabolite of vitamin D (Haussler et al., 1968). This metabolite was then identified as 1,25-$(OH)_2D_3$ (Holick et al., 1971). In 1973, Brumbaugh and Haussler (1973a) reported the existence of a specific binding protein for 1,25-$(OH)_2D_3$ in the cytoplasm prepared from rachitic chick intestine. This protein was reported to be responsible for the transfer of 1,25-$(OH)_2D_3$ to chromatin. However, the selectivity of this transfer was not clear from these early reports. Attempts in other laboratories to find the receptor protein were frustrated because of the relatively great instability of the molecule, especially to proteolytic cleavage. By washing intestinal mucosa at least three times prior to homogenization, much of the extracellular proteolytic activity could be eliminated. The use of high-salt extraction together with sulfhydryl reagents, such as dithiothreitol or mercaptoethanol, permitted a clear demonstration of the 1,25-$(OH)_2D_3$ receptor in chick intestine (Kream et al., 1976). This was possible not only by sucrose density gradient sedimentation analysis but also by Scatchard plot analysis. The stabilization of this receptor protein to that degree by high-salt extraction and dithiothreitol together with elimination of proteolytic activity by washing permitted its use as a specific agent for a radioreceptor assay for 1,25-$(OH)_2D_3$ (Eisman et al., 1976). It also permitted the demonstration of this receptor in rat intestine (Kream and DeLuca, 1977), rat and chicken bone (Chen et al., 1979; Kream et al., 1977a; Mellon and DeLuca, 1980), kidney (Colston and Feldman, 1980; Simpson et al., 1980), skin (Simpson and DeLuca, 1980), and elsewhere (Colston et al., 1980; Norman et al., 1982). It also permitted the first physical constants to be measured for this receptor molecule. The details of these characteristics are found in a later section.

With the chemical synthesis of tritiated 1,25-$(OH)_2D_3$ having a specific activity of 160 Ci/mmole (Napoli et al., 1980) came the possibility of frozen-section autoradiography using freeze–thaw techniques originally used for other steroid hormones (Stumpf et al., 1979; Zile et al., 1978). These techniques quickly revealed specific nuclear localization of $1,25\text{-}(OH)_2D_3$ not only in the target tissues of intestine, bone, and distal renal tubule cells (Stumpf *et al.*, 1979, 1980, 1981; Narbaitz *et al.*, 1983; Zile *et al.*, 1978) but also in a number of tissues not previously appreciated as target sites of vitamin D action (Stumpf *et al.*, 1979, 1980, 1981; Narbaitz *et al.*, 1983). For example, specific localization could be demonstrated in the parathyroid glands, in the endocrine cells of the stomach, in the epidermis, in cells of the pituitary, certain brain cells, odontoblasts of teeth, and epithelial cells of mammary tissues. This specific nuclear localization raises the significant question of whether $1,25\text{-}(OH)_2D_3$ may function in many tissues not previously appreciated as targets of vitamin D action.

Of considerable importance is the presence of the $1,25\text{-}(OH)_2D_3$ receptor in a number of cancerous cell lines. For example, in malignant melanoma cells (Frampton *et al.*, 1983a), in mammary carcinoma cells (Eisman *et al.*, 1980a), in osteogenic sarcoma cells (Partridge *et al.*, 1980), and so forth (Eisman *et al.*, 1980b), the existence of receptors and nuclear localization of $1,25\text{-}(OH)_2D_3$ can be shown. Furthermore, stimulation of growth or suppression of growth depending upon the concentration of $1,25\text{-}(OH)_2D_3$ in these cells can be demonstrated (Frampton *et al.*, 1983b; Freake *et al.* 1981; Haussler *et al.*, 1981). Work from Haussler, Koeffler, and others has shown that cancer cell lines that respond to $1,25\text{-}(OH)_2D_3$ can be rendered unresponsive in mutants that lack the receptor (Haussler *et al.*, 1983). There is, therefore, excitement that vitamin D compounds may be useful in controlling the growth of malignant cells.

Of considerable interest is the recent discovery by Abe *et al.* (1979, 1981) and others (Miyaura *et al.*, 1981; Tanaka *et al.*, 1982) that $1,25\text{-}(OH)_2D_3$, when added to myeloid leukemia cells in culture, will bring about a differentiation into monocytes. This basic discovery has been made with murine M-1 and human HL-60 cell lines and confirmed (McCarthy *et al.*, 1983). Further work by Suda and colleagues has also shown that monocytes can be made to accumulate into multinuclear cells in culture, perhaps representing precursors of osteoclasts or cells that are responsible for bone resorption (Abe *et al.*, 1983; Bar-Shavit *et al.*, 1983). It is unclear from these and other studies whether the multinuclear cells produced in culture are in fact osteoclastic in nature. Nevertheless, these interesting and exciting new developments illustrate that $1,25\text{-}(OH)_2D_3$ and its receptor molecule may have important basic functions in a large number of cells.

## III. BIOCHEMICAL AND PHYSICAL PROPERTIES OF THE $1,25\text{-}(OH)_2D_3$ RECEPTOR

Partial characterization of the $1,25\text{-}(OH)_2D_3$ receptor has been performed using nuclear and cytosolic tissue extracts. The receptor–hormone complex has been shown to sediment between 3.2 and 3.7 S in a 5–20% sucrose density gradient that contains 0.3–0.5 $M$ KCl (Brumbaugh and Haussler, 1974b; Chen *et al.*, 1979; Colston *et al.*, 1980; Kream *et al.*, 1976, 1977b; Wecksler *et al.*, 1980a). Reports of the apparent molecular weight have varied from 37,000– 99,700 for the receptor–hormone complex, as determined by gel filtration (J. E. Bishop *et al.*, 1982; Brumbaugh *et al.*, 1975; Franceschi and DeLuca, 1979; Hughes and Haussler, 1978; Wecksler *et al.*, 1979, 1980a). The corresponding Stokes molecular radii ($R_s$) varied from 29–36 Å. Most recent reports indicate an apparent molecular weight of 60,000–70,000, including determinations with partially purified receptor (Pike and Haussler, 1979; Pike *et al.*, 1983; Simpson *et al.*, 1983).The molecular weight estimations that have been observed from gel filtration experiments are usually larger than the molecular weight values obtained from sucrose density gradient analysis of the same cytosol. The observed molecular weight differences may indicate that the receptor is an asymmetric molecule. J. E. Bishop *et al.* (1982) reported evidence for multiple molecular weight forms of the chick intestinal receptor. These authors suggest a higher molecular weight species for the hormone-occupied receptor than for the hormone-unoccupied receptor, reporting molecular weights of 99,700 and 51,400, respectively.

Equilibrium and kinetic studies have been reported for 1,25-$(OH)_2D_3$ binding to its receptor. Equilibrium constants of 5–50 × $10^{-11}$ $M$ at 25°C have been obtained for the hormone–receptor complex by Scatchard analysis (Brumbaugh *et al.*, 1975; Colston *et al.*, 1980; Mellon and DeLuca, 1979, 1980; Wecksler *et al.*, 1980a). Wecksler and Norman (1980) determined $K_d$ values at three different temperatures in order to calculate thermodynamic constants for the binding process. Their $K_d$ values ranged from 2–5 × $10^{-10}$ $M$, providing a $\Delta H°$ of 6.7 ± 1.2 kcal/mole, a $\Delta G°$ of 12.2 ± 0.1 kcal/mole, and a $\Delta S°$ of −19 ± 3.2 cal/degree mole at 4°C for the dissociation process. An entropy-driven process for the binding of hormone to receptor is suggested by these thermodynamic values. Kinetic analyses of 1,25-$(OH)_2D_3$ binding have been reported with chick (Mellon and DeLuca, 1979, 1980; Simpson *et al.*, 1980; Wecksler and Norman, 1980; Wecksler *et al.*, 1980a,b), rat (Wecksler *et al.*, 1979), and human (Wecksler *et al.*, 1980a,b) cytosolic receptors. The association and dissociation constants that were determined are presented in Table I, along with the equilibrium constants. There are inconsistencies in the $K_d$ values obtained from Scatchard analysis and the $K_d$ values derived from the ratio of $K_d/K_a$, which may be a reflection of the impure receptor preparations.

There has been evidence to suggest that a cysteine residue is in or near the 1,25-$(OH)_2D_3$ binding site of the receptor. Preincubation of receptor cytosols with sulfhydryl reagents, N-ethylmaleimide or iodoacetamide, greatly reduces hormone binding activity (Coty, 1980; Wecksler *et al.*, 1979, 1980a,b). The effects of these reagents were minimal with the hormone-protected site of the hormone–receptor complex. Sulfhydryl protecting reagents, dithiothreitol and

**TABLE I**

**Equilibrium and Kinetic Binding Constants for 1,25-(OH)$_3$D$_3$ Receptor**

| Cytosol | Equilibrium $K_d$ (M) | Kinetic analysis | | | Reference |
|---|---|---|---|---|---|
| | | $k_a$ $(M^{-1}min^{-1})$ | $k_d$ $(min^{-1})$ | $k_d/k_a$ (M) | |
| Chick intestine | $7.1 \times 10^{-11a}$ | $9.5 \times 10^8$ | $7.1 \times 10^{-3}$ | $7.5 \times 10^{-10}$ | Mellon and DeLuca (1979) |
| | $2 \times 10^{-10b}$ | $0.5 \times 10^7$ | $3.6 \times 10^{-5}$ | $7.2 \times 10^{-12}$ | Wecksler et al. (1980b) |
| Chick bone | $7.6 \times 10^{-11a}$ | $9.5 \times 10^8$ | $2.3 \times 10^{-2}$ | $2.4 \times 10^{-11}$ | Mellon and DeLuca (1980) |
| Cultured chick kidney cells | $5.6 \times 10^{-11a}$ | $9.2 \times 10^7$ | $5.0 \times 10^{-3}$ | $5.4 \times 10^{-11}$ | Simpson et al. (1980) |
| Chick parathyroid glands | $5.4 \times 10^{-10b}$ | $1.0 \times 10^7$ | $1.3 \times 10^{-5}$ | $1.3 \times 10^{-12}$ | Wecksler et al. (1980a) |
| Rat intestine | $7.4 \times 10^{-10b}$ | $1.7 \times 10^7$ | $7.2 \times 10^{-4}$ | $4.2 \times 10^{-11}$ | Wecksler et al. (1979) |
| Human intestine | $1.8 \times 10^{-10b}$ | $2.5 \times 10^7$ | $6.4 \times 10^{-4}$ | $2.56 \times 10^{-11}$ | Wecksler et al. (1980b) |
| Human parathyroid gland | $2.1 \times 10^{-10b}$ | $1.2 \times 10^7$ | $5.3 \times 10^{-4}$ | $4.4 \times 10^{-11}$ | Wecksler et al. (1980a) |

[a] Values were obtained at 25°C.
[b] $K_d$ and $k_d$ values were obtained at 4°C and $k_a$ values were obtained at 0°C.

monothioglycerol, have been shown to stabilize the hormone binding site (Mellon *et al.,* 1980; Walters *et al.,* 1980a, 1982).

The aggregation properties of the 1,25-$(OH)_2D_3$ receptor have been investigated in cytosolic and nuclear receptor preparations. Ion-dependent aggregation of the receptor was shown to occur by sucrose density gradients in chick (Franceschi and DeLuca, 1979) and rat (Feldman *et al.,* 1979) intestinal cytosols. The 3.2–3.7 S sedimenting species observed in 0.3 $M$ KCl was shown to aggregate to a 6–8 S sedimenting species in hypotonic buffers. Partially purified receptor from chick intestinal cytosols did not retain the aggregation properties observed in the crude cytosols (Franceschi and DeLuca, 1979; Pike, 1982a). A comparison of the sedimentation properties of cytoplasmic and nuclear preparations of mouse kidney receptors showed that the nuclear preparations did not display the ion-dependent aggregation seen in the cytosolic preparations (Colston and Feldman, 1980). Cytoplasmic components, which contaminate crude cytosolic receptor preparations, appear to contribute to the observed aggregation. Franceschi and DeLuca (1979) reported that incubation of chick intestinal cytosols at 25°C produced a loss of receptor aggregating ability. The temperature-dependent effect on receptor aggregation was irreversible as opposed to the ion-dependent effect which was reversible. The loss of aggregating ability at 25°C was shown to follow a similar time course to the loss of DNA-binding ability of receptor (Franceschi *et al.,* 1983). The authors demonstrated that RNase treatment inactivated or removed cytosolic aggregating factors and that addition of RNA induced receptor aggregation (Franceschi *et al.,* 1983). The authors proposed that the aggregated receptor may be a ribonucleoprotein and that DNA and RNA may interact with a common receptor site that is temperature sensitive.

The stability of the hormone binding site of the receptor in cytosolic preparations has been of interest. A decrease in the receptor's hormone binding ability has been observed to follow increases in temperature (Mellon *et al.,* 1980). Receptor–hormone binding was also observed to decrease outside the pH range of 6–10 (Mellon *et al.,* 1980). The presence of 1,25-$(OH)_2D_3$ appears to improve the stability of the receptor–hormone binding region (McCain *et al.,* 1978). The importance of sulfhydryl groups on the receptor has been demonstrated for maintenance of hormone binding ability (Walters *et al.,* 1980a; Wecksler *et al.,* 1979, 1980a,b). Reagents that maintain a reducing environment, such as dithiothretiol and monothioglycerol, reduce the lability of the receptor (Mellon *et al.,* 1980; Walters *et al.,* 1982). Although several protease inhibitors have been applied in studies on receptor stability, there has been no direct evidence of a loss of hormone binding ability that results from proteolytic modification of the receptor. Several protease inhibitors, L-1-tosylamido-2-phenylethyl chloromethyl ketone (TPCK), $N^\alpha$-p-tosyl-L-lysine chloromethyl ketone (TLCK), and p-hydroxymercuribenzoate (PHMB) were shown to decrease hormone binding to the receptor, which suggested that these inhibitors interact with the hormone binding

site (Mellon et al., 1980). J. E. Bishop et al. (1982) demonstrated that phenylmethylsulfonyl fluoride (PMSF) could imitate $1,25\text{-}(OH)_2D_3$ in its ability to
alter the gel filtration properties of its receptor, which suggested that PMSF also
interacts with the receptor at the hormone binding site. The interaction of TPCK
with receptor was applied by Hunziker et al. (1980) to develop an assay to measure
the endogenously $1,25\text{-}(OH)_2D_3$ bound receptor, and Simpson and DeLuca
(1980) showed that the addition of PMSF to homogenization buffers improved rat
skin receptor preparations.

The specificity of the receptor for $1,25\text{-}(OH)_2D_3$ has been studied by competitive binding assays with analogues and metabolites of $1,25\text{-}(OH)_2D_3$. Chick
intestinal cytosol and cytosol–chromatin preparations have been the primary
sources of receptor for these studies. The competitive binding studies employed a
cytosol–chromatin assay (Brumbaugh and Haussler, 1974a,b; Procsal et al.,
1975; Siebert et al., 1979; Wecksler et al., 1978), cytosol–polyethylene glycol
precipitation assay (Eisman and DeLuca, 1977), and sucrose density gradients
(Kream et al., 1977c). The specificity of the receptor for $1,25\text{-}(OH)_2D_3$ is
presented diagrammatically in Fig. 3. The $1\alpha$-hydroxyl and the 25-hydroxyls are
extremely important for the binding of ligand to receptor. A hydroxyl group in
the $24(R)$ position appears to be an effective substitute for the 25-hydroxyl
(Eisman and DeLuca, 1977; Kream et al., 1976; Napoli et al., 1979b; Siebert et
al., 1979). However, the $24(S)$ hydroxyl is much less effective as is an additional
hydroxyl in the $24(R)$ position, as in $1,24(R),25\text{-}(OH)_3D_3$. Equivalent binding
was observed for $1,25\text{-}(OH)_2D_2$, indicating a tolerance for the $24(R)$-methyl
group at the hormone–receptor binding site (Eisman and DeLuca, 1977; Kream
et al., 1976). The hydroxyl group does not appear to be as important in the $3\beta$-
position as it is in the $1\alpha$- and 25-positions. The data suggests that all three
hydroxyl groups of $1,25\text{-}(OH)_2D_3$ contribute to the hormone–receptor interaction. The high selectivity of the receptor for $1,25\text{-}(OH)_2D_3$ has also been observed in receptor from other chick tissues (Brumbaugh et al., 1975; Mellon and
DeLuca, 1980; Pike, 1982b; Simpson et al., 1980) and from mammalian tissues
(Chen et al., 1979; Feldman et al., 1979; Simpson and DeLuca, 1980).

Of particular interest are a number of metabolites of vitamin D and their
interaction with the receptor. Some of the metabolites reported are only observed
under pharmacological circumstances; nevertheless they are useful in understanding the structure–function relationship in regard to the receptor. Virtually
all of the receptor binding studies in these circumstances have been with the
chick intestinal receptor obtained by high-salt extraction of the mucosa. So far
there has not been a metabolite or analogue reported that is superior to
$1,25\text{-}(OH)_2D_3$ itself in interacting with the receptor (DeLuca and Schnoes, 1983;
Franceschi et al., 1981; Norman et al., 1982). Of the physiologically important
metabolites found, it is quite clear that the introduction of a hydroxyl on the 24-
position in addition to the 25-hydroxyl grouping diminishes receptor binding by

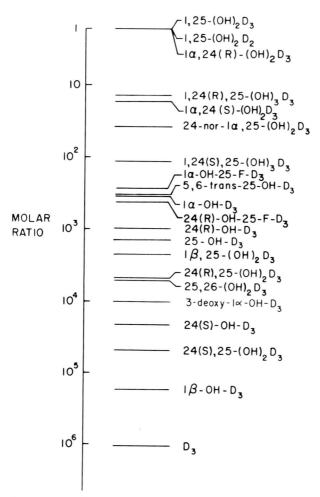

**Fig. 3.** A diagrammatic illustration of the specificity of 1,25-(OH)₂D₃ binding to its receptor in chick intestinal cytosolic preparations. The molar ratio indicates metabolite or analogue required for 50% displacement from the receptor.

one order of magnitude. A similar effect is observed with the introduction of the 26-hydroxyl (Tanaka *et al.*, 1981) or the 24-oxo derivative (Takasaki *et al.*, 1982; Yamada *et al.*, 1983). Furthermore, the introduction of the 23-hydroxyl or 23-oxo groupings reduces receptor binding markedly, as does the formation of 26,23-lactone (Y. Tanaka and H. F. DeLuca, unpublished observations). Of these metabolites, only 1,24(*R*),25-(OH)₃D₃ can be considered to be of physiological importance. Thus, it is clear that the receptor shows poor tolerance for metabolic alterations in the side chain of 1,25-(OH)₂D₃.

Another interesting area has been the introduction of fluoro groups into the $1,25\text{-}(OH)_2D_3$ molecule. Substitution of a 25-fluoro group or a 1-fluoro group reduces receptor binding to the level of no hydroxyl or hydrogen, again illustrating the importance of a hydroxyl on the 1- and 25-positions for binding (Napoli *et al.*, 1978, 1979a). $24,24\text{-Difluoro-}1,25\text{-}(OH)_2D_3$ $[24,24\text{-}F_2\text{-}1,25\text{-}(OH)_2D_3]$ is equal to $1,25\text{-}(OH)_2D_3$ in its ability to bind to the chick intestinal receptor (Okamoto *et al.*, 1983; Tanaka *et al.*, 1983) whereas the presence of the $26,26,26,27,27,27\text{-}F_6$ group on the $1,25\text{-}(OH)_2D_3$ molecule reduces receptor binding to a factor of approximately 1.8 (Tanaka *et al.*, 1984). It is of some interest, however, that both the $24,24\text{-}F_2\text{-}1,25\text{-}(OH)_2D_3$ and the $26,26,26,27,27,27\text{-}F_6\text{-}1,25\text{-}(OH)_2D_3$ are approximately 5–10 times more biologically active *in vivo* than $1,25\text{-}(OH)_2D_3$ (Tanaka *et al.*, 1983, 1984). Although the reason for this is not entirely clear, it seems likely that the presence of the fluoro groups at these positions markedly reduces metabolic degradation of the hormone, increasing its lifetime in plasma and tissues.

The presence of a fluoro group in the C-25 position when a $1\alpha$-hydroxyl group is present results in receptor binding equal to $1\alpha\text{-OH-}D_3$ (Napoli *et al.*, 1978). However, *in vivo* this compound becomes 24-hydroxylated, and thus the $1,24\text{-}(OH)_2\text{-}25$-fluoro derivative is a potent binder to the receptor and a highly biologically active compound (Napoli *et al.*, 1979b). These results confirm the fact that a hydroxyl on C-24 can substitute for a hydroxyl on C-25, but the presence of both 24- and 25-hydroxyls reduces biological activity.

Although the above results illustrate that the 1- and 25-hydroxyls of $1,25\text{-}(OH)_2D_3$ are extremely important to binding, one cannot eliminate the importance of the $3\beta$-hydroxyl or the intact side chain. Many substitutions are possible, and also the *cis*-triene structure undoubtedly plays an essential role in placing the hydroxyls in the correct position for binding. In short, much remains to be learned concerning the molecular requirements for interaction with the receptor, although much more useful information will result from a characterization of the binding site of the receptor. Thus far, no information is available on that aspect.

## IV. MECHANISM OF ACTION OF THE $1,25\text{-}(OH)_2D_3$ RECEPTOR

The mechanism by which $1,25\text{-}(OH)_2D_3$ performs its biological functions is still largely unknown, as is the role of the $1,25\text{-}(OH)_2D_3$ receptor. There is evidence that $1,25\text{-}(OH)_2D_3$ behaves as a classic steroid hormone through a nuclear mechanism that results in the production of proteins *de novo* and that its receptor is involved in this process. A nuclear mechanism for the hormone was suggested by early *in vivo* studies (Chen and DeLuca, 1973a; Haussler *et al.*,

1968; Lawson and Wilson, 1974). Subcellular fractionation studies also suggested an association of 1,25-$(OH)_2D_3$ with nuclear material (Tsai *et al.*, 1972), although a firm conclusion could not be reached because of a possible artifactual distribution of radiolabeled 1,25-$(OH)_2D_3$ during fractionation (Chen and De-Luca, 1973b). The localization of 1,25-$(OH)_2D_3$ in the nucleus of target cells was firmly established by freeze–thaw mount autoradiography (Stumpf *et al.*, 1979, 1980; Zile *et al.*, 1978). A time course of accumulation of the hormone in the crude chromatin fraction of the intestinal mucosa was reported by Tsai *et al.* (1972) and Brumbaugh and Haussler (1974a), and the presence of 1,25-$(OH)_2D_3$ in the chromatin fraction was shown to precede the 1,25-$(OH)_2D_3$-dependent calcium transport. Evidence for hormonal regulation of protein synthesis was suggested by the stimulation of RNA synthesis in rat kidney (Chen and DeLuca, 1973b) and chick intestine (Tsai and Norman, 1973a), the enhancement of RNA polymerase II activity (Zerwehk *et al.*, 1974), and an increase in chromatin template activity in response to 1,25-$(OH)_2D_3$ (Zerwekh *et al.*, 1976). However, investigations studying the effect of protein and RNA synthesis inhibitors on intestinal calcium transport produced unclear results (Bikle *et al.*, 1979; Tanaka *et al.*, 1971; Tsai *et al.*, 1973). While actinomycin D could block the calcium-binding protein induction, it could not block the calcium transport response (Bikle *et al.*, 1978, 1979). Unfortunately, 30% of the RNA synthesis in the intestine could be blocked by that dose, while higher doses resulted in mortality. Thus, interpretation of these results is compromised. Corradino (1973a, 1978) developed an embryonic chick intestinal organ culture system that responds to 1,25-$(OH)_2D_3$ by increasing calcium uptake and calcium binding protein (CaBP) synthesis. Long-term inhibition of the hormone action by actinomycin D and α-amanitin was observed in this culture system (Corradino, 1973b). Franceschi and DeLuca (1981b) further characterized the embryonic chick duodenal system. The calcium uptake system was found to be similar to the calcium transport system of the young chick duodenum except for the energy dependence of the calcium uptake process. A 1,25-$(OH)_2D_3$ receptor was also shown in this tissue. Cyclo-heximide, an inhibitor of polypeptide chain elongation, and actinomycin D, an inhibitor of RNA synthesis, completely blocked the hormone-dependent calcium uptake (Franceschi and DeLuca, 1981a). The effects of cycloheximide were totally reversible, while the effects of actinomycin D were partially reversible by removal of the inhibitor, showing that cell necrosis was not responsible for the block in responsiveness to 1,25-$(OH)_2D_3$. Anisomycin, another polypeptide chain elongation inhibitor, and α-amanitin, an RNA polymerase II inhibitor, were also found to block hormone-dependent calcium uptake. In this system, therefore, the calcium transport response to 1,25-$(OH)_2D_3$ clearly involves a nuclear mechanism.

In the rat, the intestinal calcium transport (Halloran and DeLuca, 1981a) and the intestinal phosphate transport responses (Kabakoff *et al.*, 1982) show complex biphasic responses, suggesting two independent but related mechanisms. The first 6-hr response is by existent villus cells showing nuclear localization of 1,25-$(OH)_2D_3$. The second or 24-hr response is probably by nuclear action in the crypt cells that migrate onto the villus. Nuclear localization for 1,25-$(OH)_2D_3$ also has been demonstrated in these cells as well (Stumpf et al., 1979). Rasmussen and colleagues (1981) have suggested that 1,25-$(OH)_2D_3$ functions directly on the brush border membranes to bring about the 6-hr transport response. To support these observations, changes in lipid composition of the brush border membranes and in calcium uptake by isolated vesicles in response to 1,25-$(OH)_2D_3$ have been reported (Goodman et al., 1972; Max et al., 1978; Rasmussen et al., 1979). However, Kendrick et al. (1981) have shown that, under appropriate conditions, actinomycin D will block the intestinal calcium transport response to 1,25-$(OH)_2D_3$ in the rat. The best available evidence, therefore, dictates that the intestinal calcium transport response to 1,25-$(OH)_2D_3$ involves nuclear activity.

The only gene product that has been clearly characterized is calcium binding protein in the chick (Taylor and Wasserman, 1967) and a much smaller molecular weight version in the rat and other mammalian species (Wasserman and Feher, 1977). In the case of the chicken, there is no doubt that this protein is absent in vitamin D deficiency and appears in response to 1,25-$(OH)_2D_3$. There is considerable debate as to whether this protein appears in sufficient time to carry out the intestinal calcium transport response (Spencer et al., 1976, 1978). Very recently, using highly sensitive computer analyzed fluorographs and radiographs of two-dimensional gel preparations of intestine, it has been shown that the calcium binding protein in the case of embryonic chick intestine is present prior to initiation of the intestinal calcium transport response (C. W. Bishop et al., 1982, 1984). It seems abundantly clear, therefore, that in the case of the chick, the calcium binding protein plays some role in the transport phenomenon, but clearly other factors that are made in response to 1,25-$(OH)_2D_3$ must be present to bring about the transport response. Sodium is also required for the calcium transport phenomenon in response to 1,25-$(OH)_2D_3$. Figure 4 provides a rough illustration of the probable molecular mechanism of action of 1,25-$(OH)_2D_3$ in eliciting the intestinal calcium transport response. As depicted, 1,25-$(OH)_2D_3$ enters the nucleus of a target cell and binds to its receptor, producing a structural change in the receptor. The "activated" hormone–receptor complex interacts with chromatin, which initiates events that result in the translation of specific calcium and phosphorus transport proteins. A major missing link, however, is the identification of the gene products made in response to 1,25-$(OH)_2D_3$ and information regarding the entry of 1,25-$(OH)_2D_3$ into the target cells, the molecular mechanism involved in gene expression, and the transcellular events involved in the calcium transport process.

In eliciting a target organ response, the 1,25-$(OH)_2D_3$ receptor was thought to

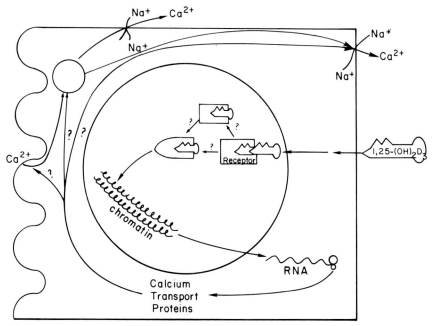

**Fig. 4.** A representation of a probable molecular mechanism of action of 1,25-(OH)$_2$D$_3$.

behave as other steroid hormone receptors. Thus, 1,25-(OH)$_2$D$_3$ binding to the receptor in target cells induces an "activation" or an alteration of the receptor that is required for binding to the chromatin and eliciting a target response. An activated receptor–hormone complex was believed to be required for interaction with specific nuclear acceptor sites that initiate the hormonal response. Therefore, areas of interest in receptor research have been: (1) the receptor's involvement in transport of hormone from cytoplasm to nucleus, (2) the identification of an "activated" receptor–hormone complex, and (3) the interaction of the receptor–hormone complex with DNA and chromatin.

Evidence for the 1,25-(OH)$_2$D$_3$ receptor in the cytoplasmic fraction of intestinal mucosa preparations from rachitic chicks in a hormone-free state was first obtained by Brumbaugh and Haussler (1974b) and Tsai and Norman (1973b) and in the nuclear fraction by Brumbaugh and Haussler (1974a) and Lawson and Wilson (1974). Early evidence suggested the involvement of the receptor in the translocation of hormone from cytoplasm to nucleus. Brumbaugh and Haussler (1973b, 1974b, 1975) suggested that the nuclear localization of the hormone–receptor complex was temperature dependent. Employing a reconstituted cytosol–chromatin system, they demonstrated a temperature-dependent transfer of the receptor–hormone complex to the chromatin fraction. However, selectivity of this binding to nuclei of target cells could not be demonstrated. More recently, Pike (1982a) has shown that the binding of chick intestinal 1,25-$(OH)_2D_3$ receptor to the nuclear chromatin fraction was temperature independent. Pike's reconstruction experiments differed from early experiments in that the cytosolic receptor–hormone complex was prepared prior to incubation with the chromatin fraction. This study did demonstrate a hormone dependence for nuclear binding of the receptor. In intermediate ionic strength buffers, the receptor–hormone complex bound to nuclei while the hormone-free receptor did not in the reconstituted system. This study also demonstrated that hormone–receptor complex interaction with chromatin was sensitive to KCl concentration, as might be expected since 0.3 $M$ KCl is known to extract nuclear receptors (Walters et al., 1980b). The ionic-strength effect on hormone–receptor interaction with chromatin has been investigated by Walters et al. (1980b). They were interested in the employment of intermediate or high-ionic-strength buffers in the preparation of cytosolic receptors. They found that in vitamin D-deficient intestinal mucosal homogenates prepared in low-salt buffers 90% of the receptors were in the chromatin fraction. Increasing the ionic strength of the buffers resulted in an increase in receptor concentration in the "cytosolic" fraction. The ionic strength effects on receptor–chromatin interaction observed by Walters et al. (1981) were observed while investigating hormone-free receptor interactions with nuclei. Therefore, unoccupied receptor was found to associate with nuclear material in low-salt buffers. In order to avoid the ionic-strength effects, Walters et al. (1981) utilized a hexylene glycol method to purify nuclei from vitamin D-deficient chick intestinal mucosa. A high nuclear/cytosolic receptor ratio was obtained, indicating the presence of hormone-free receptor in the nucleus. The in vitro experiments which have been performed may not be reflecting in vivo conditions. However, the presence of hormone-free receptor in the nucleus in vivo cannot be firmly established with presently available techniques. Likewise, in vitro nuclear translocation studies may be a reflection of the different ionic-strength effects on the hormone–receptor complex and the hormone-free receptor rather than the ability of the hormone–receptor complex to translocate from cytosol to nucleus.

In this chapter, a conformational change in the receptor which is believed to follow the formation of the hormone–receptor complex will be termed "activation." In other steroid-hormone systems, this has been identified by the ability of the "activated" hormone–receptor complex to bind to DNA-cellulose or chromatin. However, hormone-free, 1,25-$(OH)_2D_3$ receptor has been shown to bind to chromatin and DNA-cellulose (Hunziker et al., 1983; Pike, 1981; Walters et al., 1980b, 1981). Therefore, the presumed structural conformational change that occurs in the receptor with hormone binding may be more subtle in the case of the 1,25-$(OH)_2D_3$ receptor. The reconstitution studies of Pike (1982a) with chick intestinal receptor, in which he found that the hormone–receptor complex bound to nuclei and that the hormone-free receptor remained in the cytosol, may have been an indication of receptor activation upon hormone binding. In that study he did not observe a difference between hormone–receptor complex and hormone-free receptor in their interactions with DNA-cellulose. Pike and Haussler (1983), working with receptor in cultured 3T6 mouse fibroblasts, demonstrated a difference in hormone–receptor and hormone-free receptor interactions with DNA-cellulose. Although both receptor forms bound to DNA-cellulose, the hormone-free receptor eluted in 0.165 $M$ KCl while the hormone–receptor complex eluted in 0.22 M KCl. The differences in behavior of the hormone–receptor and hormone-free receptor from chick intestinal mucosa toward DNA-cellulose have been investigated by Hunziker *et al.* (1983). These authors also found that both forms of receptor would bind to DNA-cellulose. The differences in elution of the two forms of receptor were found to be more subtle in these experiments, with the hormone-free receptor having lower affinity for DNA-cellulose than the hormone–receptor complex. When a mixture of both forms was applied to a DNA-cellulose column, a difference was observed in DNA binding affinities. The subtle difference was that observed on DNA-cellulose could not be reproduced on DEAE-Sepharose, suggesting that ligand-induced changes in overall surface charge of the receptor were not major. These authors also showed a difference in salt extractability from chromatin of the hormone–receptor and hormone-free receptor. The hormone-free receptor was found to associate with intestinal chromatin but could be extracted from the chromatin fraction with a lower salt concentration than the hormone–receptor complex. These authors reported a third form of receptor, which was the hormone–receptor complex formed *in vivo,* suggesting a difference between the *in vitro-* and *in vivo*-formed hormone–receptor complex. It is difficult to interpret and compare the material available in this area because of the crude receptor preparations and different experimental conditions. However, available data suggests a structural change in the receptor molecule with binding of the hormone.

Recently, the interaction with DNA has been explored in greater detail. Radparvar and Mellon (1982) performed a series of experiments to characterize the interaction of the hormone–receptor complex with DNA. Chick intestinal cytoplasmic receptor was studied using a competitive DNA-cellulose binding assay. These studies supported the findings that the binding of the hormone–receptor to DNA-cellulose does not require a thermal- or salt-induced activation. A preference for double-stranded DNA over single-stranded DNA was observed, as well as a distinction between DNA and RNA. The hormone–receptor complex also appeared to prefer AT-rich segments of DNA. Pike (1981) investigated the possible presence of a reactive sulfhydryl in the receptor's DNA binding region. Mercurials, $HgCl_2$ and *p*-chloromercuribenzene sulfonate (pCMBS), were found to inhibit hormone-free receptor binding to DNA-cellulose in a concentration-dependent manner. These reagents were also capable of eluting receptor from DNA-cellulose. The mercurial effect on the receptor was found to be reversible in excess thiol reagents. Iodoacetamide was found to partially block hormone–receptor binding to DNA (37% at 25 n$M$), but it was unable to dissociate the hormone–receptor–DNA complex at the same concentration. Franceschi *et al.* (1983) reported that the receptor's DNA binding ability is temperature sensitive. These studies employed intestinal mucosa cytosols from vitamin D-deficient chicks. The authors found that 40–60% of the hormone–receptor complexes bound to DNA-cellulose while practically all would bind to DEAE-cellulose. Furthermore, incubation of the hormone–receptor complex at 25°C for 1 hr resulted in the totally irreversible loss of DNA binding ability with retention of DEAE binding ability. A series of protease inhibitors did not stabilize the temperature sensitivity, suggesting that the loss of DNA binding was not the result of proteolytic modification of the receptor. Several divalent cations were found to stabilize the receptor to the temperature sensitivity. Magnesium chloride and $MnCl_2$ were shown to totally protect the receptor's DNA binding ability from the effect of 25°C incubation. Also, $Na_2MoO_4$ was found to partially protect the receptor's DNA binding site, but other phosphatase inhibitors were ineffective, suggesting that the $Na_2MoO_4$ stabilization was not the result of phosphoprotein phosphatase inhibition. Furthermore, the inclusion of 5 m$M$ $MgCl_2$ in buffers for the preparation of the cytosol provided a greater quantity of receptor that retained its DNA binding ability.

## V. REGULATION OF RECEPTOR NUMBER

By means of Scatchard plot analysis, receptor numbers can be easily determined under circumstances in which the receptors are in the hormone-free form. In the case of $1,25\text{-}(OH)_2D_3$, this occurs only in absolutely vitamin D-deficient animals. Under these circumstances, receptor numbers vary from 50 fmoles/mg protein to values as high as 2000 fmoles/mg, depending on species, tissue, and preparation of sample. To understand the relationship of receptor numbers to vitamin D function, it is obvious that a determination of both hormone-free and hormone-occupied receptors is required inasmuch as the regulation undoubtedly occurs in animals receiving normal amounts of vitamin D. This determination is not trivial. The major reason for this is that in the absence of ligand, the receptor for $1,25\text{-}(OH)_2D_3$ has been reported to be quite labile, becoming destroyed during any incubation without ligand being present. In addition, in the case of mammalian receptor in the hormone-occupied form, dissociation rates are quite slow, prohibiting measurement of the hormone–receptor complex by an exchange reaction. Coty (1980) has introduced the technique of mersalyl to facilitate the dissociation of ligand from receptor; then excess thiol reagent displaces bound mersalyl and permits subsequent reformation of the hormone–receptor complex. If rebinding is done in the presence of radiolabeled hormone, then total receptor numbers can be obtained. This technique appears to be at least partially successful in the case of chicken receptor, but it is less successful in the case of mammalian receptor, since up to 90 min or more is required for complete dissociation of ligand from receptor in the presence of mersalyl in the case of rat small intestinal receptor (Massaro et al., 1983b). Furthermore, rebinding of labeled ligand is not completely successful. Nevertheless, a rough approximation of hormone-occupied and hormone-free receptors can be obtained (Massaro et al., 1983b).

Hunziker et al. (1980) have introduced another technique in which hormone-free receptors can be measured in the usual fashion and in an accompanying sample, the unoccupied hormone sites are blocked with TPCK which then allows subsequent quantitation of the hormone-occupied receptors by exchange with $1,25\text{-}(OH)_2[^3H]D_3$ at 37°C for 30 min. This technique has also been applied and is only partially successful. More recently, with the use of monoclonal antibodies, total receptor levels can be measured, allowing determinations of hormone-free and hormone-occupied receptor levels, but so far this technique has not been widely applied (Dokoh et al., 1983). Under normal circumstances, it appears that with circulating levels of $1,25\text{-}(OH)_2D_3$ at 2–3 times normal, only 20% of the total receptors found in intestine are located in the hormone-occupied form (Hunziker et al., 1980; Massaro et al., 1983b). By increasing the dosage of $1,25\text{-}(OH)_2D_3$ from external sources, Hunziker et al. (1982) have reported that up to 70% of the receptors become hormone occupied in a transient fashion. This high level of hormone–receptor complex very quickly decays, within 4 to 8 hr postadministration. The exact meaning for the excess of receptor numbers to hormone occupancy is unknown. Furthermore, measurements of hormone-occupied, hormone-free, and inactive receptors have not reached a state of perfection to allow confidence in the data obtained. It is, therefore, quite evident that clearly proven methods for measurement of these forms of receptors are required before any conclusion can be made.

There have been studies reporting regulation of the $1,25\text{-}(OH)_2D_3$ receptor numbers. Perhaps one of the clearest examples is the level of the rat intestinal receptor during neonatal development. During studies on the role of vitamin D and reproduction, it was discovered that intestinal calcium transport and responsiveness to vitamin D does not appear until 14–16 days postpartum in the neonatal rat (Halloran and DeLuca, 1980). This lack of responsiveness was not a failure of conversion of vitamin D to its active hormonal form, since administration of $1,25\text{-}(OH)_2D_3$ to the neonatal rat prior to this time did not elicit a calcium transport response. The lack of responsiveness of the neonatal rat intestine to $1,25\text{-}(OH)_2D_3$ is the result of a lack of receptor (Halloran and DeLuca, 1981b). Receptor does not appear in the normal neonatal rat until 14–16 days postpartum and from 16 to 18 days in the case of the vitamin D-deficient rat (Halloran and DeLuca, 1981b; Massaro et al., 1983a). This is precisely when intestinal calcium transport makes its appearance, as does the responsiveness to $1,25\text{-}(OH)_2D_3$. Adrenalectomy delays the appearance of receptor whereas injections of hydrocortisone stimulate the appearance of receptor in the neonatal rat intestine (Massaro et al., 1983a). Intestinal cultures explanted from neonatal rats at 14 days postpartum do not develop receptor in culture for 48 hr, but if they are cultured with hydrocortisone within 24 hr, the receptor makes its appearance (Massaro et al., 1982). Thus, hydrocortisone, in playing its well-known role in the maturation of intestine, also brings about the appearance of the $1,25\text{-}(OH)_2D_3$ receptor. The presence of receptor, therefore, is required for the neonatal rat intestine to elicit a response to the vitamin D hormone.

Another example of regulation of the $1,25\text{-}(OH)_2D_3$ receptor is that found in the genetic defect called vitamin D-dependency rickets Type II (Bell et al., 1978). This is an experiment of nature in which an autosomal recessive trait appears in which the target organs are not responsive to $1,25\text{-}(OH)_2D_3$. Marx and colleagues (Eil et al., 1981; Liberman et al., 1983) were the first to culture skin fibroblasts taken from such subjects as compared to normal subjects, and they demonstrated that in at least one subpopulation, no apparent receptor for $1,25\text{-}(OH)_2D_3$ could be found in clear contrast to the finding of 50 fmoles/mg protein of receptor in the normal cells. Similar results have been obtained by others (Feldman et al., 1982). This, however, should be classified as a genetic defect in the receptor and not a matter of regulation.

Only one example of hormonal regulation of the $1,25\text{-}(OH)_2D_3$ receptor has been published. It is well known that glucocorticoids bring about a reduction in $1,25\text{-}(OH)_2D_3$-induced intestinal calcium transport (Kimberg, 1969; Lindgren and DeLuca, 1983). Since glucocorticoid-inhibited intestinal calcium transport does not appear very responsive to $1,25\text{-}(OH)_2D_3$, some end-organ resistance is envisioned. It is, therefore, logical to determine whether glucocorticoids given at high doses might reduce the receptor number. In the case of the mouse, Hirst and Feldman (1982a) have reported that glucocorticoid administration markedly reduces $1,25\text{-}(OH)_2D_3$ receptor number. However, in the rat in which most of the intestinal calcium transport inhibition has been reported, glucocorticoids actually cause a 50% increase in unoccupied $1,25\text{-}(OH)_2D_3$ receptor levels (Hirst and Feldman, 1982b). Glucocorticoids added to cultures of osteoblast-like cells also increase receptor number (Chen et al., 1983). Since the results do not correlate with the physiological responses, it is unclear what the glucocorticoid-induced changes in receptor number in these experiments mean, if they have significance at all.

One final regulation has been observed by Feldman and co-workers in which cultures of fibroblasts during the rapid phase of growth show much larger numbers of $1,25\text{-}(OH)_2D_3$ receptors than when the same cells are not in the log phase of growth (Chen and Feldman, 1981). It is not clear whether the receptor number for $1,25\text{-}(OH)_2D_3$ in these cells has any relationship to the growth of cells or whether the log phase of growth simply results in the existence of a larger number of receptors. It may be significant that $1,25\text{-}(OH)_2D_3$ in malignant cells will suppress growth at high concentrations in those cells that possess the $1,25\text{-}(OH)_2D_3$ receptor (Haussler *et al.*, 1983).

Much can be expected in the area of regulation of $1,25\text{-}(OH)_2D_3$ receptor numbers and populations. However, the technology is such that this is not likely to be available in the near future.

## VI. PURIFICATION OF CHICK INTESTINAL 1,25-(OH)₂D₃ RECEPTOR

Purification of the cytosolic receptor for $1,25\text{-}(OH)_2D_3$ has been a difficult undertaking, largely because of the low quantity of this protein. Receptor levels found in intestinal mucosa, the major target tissue of $1,25\text{-}(OH)_2D_3$, are less than 0.001% of the total mucosa protein. The use of radiolabeled $1,25\text{-}(OH)_2D_3$ to trace the receptor through the isolation procedures has made the task possible. An early step in all receptor purifications has been the labeling of the receptor with $1,25\text{-}(OH)_2[^3H]D_3$. The formation of the hormone–receptor complex not only provides a means of tracking the receptor during the purification procedures but also helps stabilize the receptor during these procedures (McCain *et al.*, 1978).

Partial purification of the chicken intestinal receptor was first reported by Haussler and Norman (1969). A 0.3 $M$ KCl extraction of the chromatin fraction followed by selective precipitation with ammonium sulfate and protamine sulfate provided a 167-fold purification relative to crude mucosa homogenate. It was estimated that a $10^6$-fold purification would be required to isolate the receptor in homogeneous form.

In 1979, an approximately 50% pure form of the receptor was reported by Pike and Haussler (1979). The authors observed an 86,000-fold purification from the cytosolic fraction of intestinal mucosa obtained from vitamin D-deficient chicks. The isolation procedures included selective precipitation with Polymin P (polyethyleneimine), followed by sequential chromatography on DNA-cellulose, Sephacryl S-200, Blue dextran-Sepharose, DNA-cellulose, and heparin-Sepharose. One major and three minor bands of molecular weights 50,000–65,000 were observed on sodium dodecyl sulfate–polyacrylamide gel electrophoresis. A 14% yield was reported for this purification based on recovery of hormone-receptor complex containing radiolabeled hormone. Following these procedures, 26 μg of semipure receptor was obtained from the intestinal mucosa of 350 rachitic chicks (800 g of tissue).

A modification of this procedure was described by Pike *et al.* (1983) for large scale preparations of semipure receptor from vitamin $D_3$-deficient chicks. The initial preparation of cytosol was omitted. The receptor fraction was collected from a 0.06–0.15% Polymin P fractionation of whole rinsed gut homogenate. The supernatant of the 0.06% Polymin P precipitation, termed polysol, was found to contain the same specific activity of receptor which they observed in the cytosol. Sequential chromatography in preparative DNA-cellulose and Blue dextran-Sepharose was followed by analytical applications on a gel filtration column, Ultragel AcA-44 and DNA-cellulose. The above purification scheme was used to isolate 412 µg of receptor from 10 kg of chick intestine. The average overall purity of receptor was approximately 13% in 37% yield.

Recently, Simpson and DeLuca (1982) reported that the chicken intestinal receptor was purified to apparent homogeneity. These studies showed that adult vitamin D-replete chicken duodena could be used as a source of receptor. The use of these animals creates a problem in the determination of receptor recovery, since endogenous $1,25-(OH)_2D_3$ is present. However, the advantages of obtaining intestines of mature vitamin D-replete adult chickens over raising vitamin D-deficient chicks for large scale preparations is evident. A crude nuclear extract of chicken intestinal mucosa was used as starting material for these preparations. This crude nuclear high-salt extract was shown to yield a receptor preparation 4- to 5-fold enriched with receptor, as compared to the corresponding high-salt whole-cell extract.

A five-step purification scheme shown in Table II was applied to obtain the highly purified receptor (Simpson et al., 1983). The crude nuclear extract was prepared from approximately 600 g of mucosa and brought to 38% saturation of ammonium sulfate. This precipitation provided a means of concentrating the starting material and a 3.3-fold decrease in total protein. DNA-cellulose chromatography, which had been previously shown to be effective in receptor purification (Pike and Haussler, 1979), was employed using batch adsorption and column elution for a 50- to 100-fold purification with a 50–75% yield. Hydroxylapatite was used to give a twofold purification and a fourfold concentration of

**TABLE II**

**Receptor Purification Scheme**

| Step | Protein (mg) | Receptor (dpm × $10^{-6}$) | Specific activity (dpm × $10^{-3}$/mg) | Purification (x-fold) | Yield (%) |
|---|---|---|---|---|---|
| Nuclear extract | 2100 | 9.00 | 4.3 | 1 | 100 |
| $(NH_4)_2SO_4$ | 640 | | | | |
| DNA-cellulose | 6.8 | 3.00 | 441.2 | 103 | 33 |
| Hydroxylapatite | 2.0 | 1.80 | 900.1 | 214 | 20 |
| Sephacryl S-200 | 0.34 | 1.22 | 3614.0 | 861 | 14 |
| DEAE-cellulose | 0.029 | 0.72 | 24828.4 | 5774 | 8 |

the post-DNA-cellulose receptor. Gel exclusion chromatography on Sephacryl S-200, which was run in a low-salt buffer, provided an additional sixfold purification. The final step of the purification employed the ion exchange resin DEAE-Sepharose, which gave approximately 10-fold purification with 75% recovery. Recoveries from Sephacryl S-200 and DEAE-Sepharose were improved when 0.5% Triton X-100 was included in the buffers. From a crude extract containing 2.1 g of protein, 29 μg of purified receptor was obtained. The five-step purification provided an approximately 6000-fold purification of the crude nuclear extract with a 5–9% yield, based on the recovery of labeled hormone–receptor complex. A 30,000-fold purification was estimated necessary to isolate a homogeneous form of the receptor from the crude nuclear extract. As determined by specific activity, the receptor was approximately 20% pure. The purity estimation determined by specific activity would not consider receptor bound to endogenous unlabeled hormone or hormone-free receptor.

The authors believe higher purification was achieved based on gel electrophoresis and an iodination experiment of the purified receptor. A single band was seen on a sodium dodecyl sulfate–polyacrylamide gel, with a molecular weight of approximately 62,000–65,000 (Fig. 5). Two-dimensional electrophoresis of the purified receptor showed a single major spot, elongated along

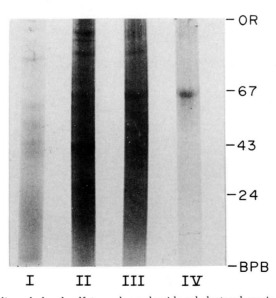

**Fig. 5.**   Sodium dodecyl sulfate–polyacrylamide gel electrophoresis of fractions from the chick intestinal receptor purification scheme of Simpson *et al.* (1983). Samples are I, $(NH_4)_2SO_4$ precipitate (40 μg); II, hydroxylapatite (20 μg); III, Sephacryl S-200 (10 μg), and IV, DEAE-cellulose (4 μg). The molecular weight standards are bovine serum albumin (67,000), ovalbumin (43,000), trypsinogen (24,000), and bromphenol blue (BPB; tracking dye).

the isoelectric focus axis (Fig. 6), with a molecular weight of 63,000 ± 3,900 and a p*I* of 6.0–6.3. The purified receptor showed a single 3.7 S peak on a 4–20% linear sucrose gradient and retained its ability to bind to DNA-cellulose. When the receptor preparation was iodinated and run on a 4–20% linear sucrose gradient, approximately 90% of the [125]I label was recovered with a 3.7 S sedimentation coefficient.

The purification procedures described demonstrate great advances toward the isolation of the 1,25-$(OH)_2D_3$ receptor. However, the low yields which accompany the chromatography procedures provide very small amounts of the purified

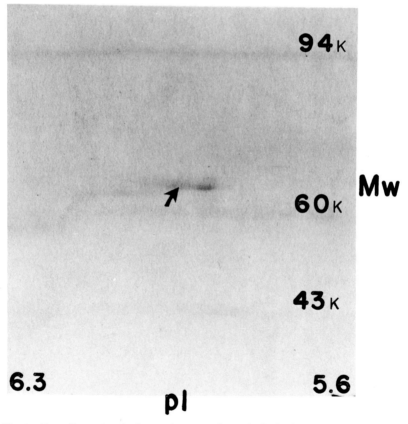

**Fig. 6.** Two-dimensional electrophoresis of purified chick intestinal receptor of Simpson *et al.* (1983). The protein was isoelectric focused horizontally and electrophoresed in sodium dodecyl sulfate along the vertical axis. The indicated major polypeptide has a molecular weight of 64,000 and a p*I* of approximately 6.0. The internal standards in the second dimension of this gel are phosphorylase a (94,000), catalase (60,000), and actin (43,000).

receptor. To obtain sufficient quantities of purified receptor to advance the study of receptor–hormone interaction and hormone mechanism may require the application of other purification techniques.

## VII. ANTIBODIES TO CHICK INTESTINAL 1,25-(OH)$_2$D$_3$ RECEPTOR

Serum and monoclonal antibodies have been prepared against partially purified chick intestinal 1,25-(OH)$_2$D$_3$ receptor by Pike *et al.* (1982, 1983). The partial purification provided an approximately 13% pure receptor fraction for antigen. The presence of antireceptor antibodies in rat sera and hybridoma media was determined by the difference in mobility of radiolabeled hormone–receptor complex versus hormone–receptor–antibody complex on sucrose gradients or gel filtration columns and by immunoprecipitation of the labeled complex with antirat immunoglobulins.

Polyclonal serum antibodies were obtained from two Lewis rats that were given three intradermal injections over a period of two months. Significant antireceptor antibody titers were obtained in both Lewis rats. However, rat C, whose antigen contained 6.6 pmoles of receptor binding per milligram of protein, possessed an approximately threefold higher titer than rat F, whose antigen contained 2.3 pmoles of receptor binding per milligram of protein. Hybridomas were produced by the fusion of spleen cells of the two rats, which had received an intravenous injection of antigen prior to fusion, with the nonsecreting mouse myeloma cell lines SP2/0-Ag14 and P3-X63-Ag8.653. The spleen cell–myeloma cell fusions produced four positive antibody secretors: three were derived from the SP2/0-Ag14 fusions (SP2/0-4A5, SP2/0-9A7, SP2/0-8D3), and one was derived from the P3-X63-Ag8.653 fusion (P3-8C8). A shift in the sedimentation of hypertonic 10–30% sucrose gradients from 3.3 to the 7–8 S region indicated a single receptor–immunoglobulin complex for SP2/0-4A5, SP2/0-9A7, and P3-8C8 media. The SP2/0-8D3 media indicated a higher molecular weight complex, which suggested an IgM antibody.

The monoclonal antibodies were found to react with both the hormone–receptor complex and the hormone-free receptor. The specificity and cross-reactivity of the antisera to rat C and the monoclonal antibodies were determined and are presented in Table III. The antibodies appear specific for the 1,25-(OH)$_2$D$_3$ receptor, since they would not react with either glucocorticoid or estrogen receptors. Different patterns of tissue and species recognition were observed for the individual monoclonal antibodies. Antibody SP2/0-8D3 appeared to be the most specific, reacting with only the chick receptor while P3-8C8 displayed variable reactivity and SP2/0-4A5 and SP2/0-9A7 reacted with all 1,25-(OH)$_2$D$_3$ receptors.

**TABLE III**

**Specificity and Cross-Reactivity of Antibodies to Chick 1,25-Dihydroxyvitamin D$_3$ Receptors[d]**

| Receptor/binding protein | Antibody | | | | |
|---|---|---|---|---|---|
| | Antiserum rat C | Monoclonal | | | |
| | | 4A5γ | 9A7γ | 8C8γ | 8D3λ |
| 1,25(OH)$_2$D$_3$ receptor | | | | | |
| Intestinal tissue | | | | | |
| Fish (*Micropterus salmoides*) | + | ±[a] | ± | ND[b] | — |
| Frog (*Rana catesbeiana*) | + | + | + | ND | — |
| Chick | + | + | + | + | + |
| Rat | + | + | + | ND | − |
| Kidney tissue | | | | | |
| Rat | + | + | + | ND | − |
| Cultured cells[c] | | | | | |
| Mouse fibroblasts (3T6) | + | ± | + | − | − |
| Rat pituitary (GH$_3$) | + | + | ND | ND | ND |
| Rat bone (ROS 17/2.8) | + | + | + | − | − |
| Hamster ovary (CHO) | + | + | ND | ND | ND |
| Pig kidney (LLC-PK$_1$) | + | + | + | ± | − |
| Human breast (MCF-7) | + | + | + | ± | − |
| Human intestine (IN-407) | + | + | + | ± | − |
| Vitamin D binding protein (4.1 and 5.8 S) | − | ND | ND | ND | ND |
| Glucocorticoid receptor (rat liver) | − | − | − | ND | − |
| Estrogen receptor (MCF-7) | − | − | − | ND | − |

[a] Signifies limited reactivity when assessed in comparison to chick receptor.
[b] ND, not determined.
[c] Nuclear extraction following intact cell exposure to 1,25-(OH)$_2$D$_3$.
[d] Data from Pike *et al.*, 1983.

## VIII. CONCLUSIONS

The hormonal form of vitamin D acts, at least in part, through a nuclear-mediated mechanism. A specific receptor protein has been found for this molecule in intestine, bone, and kidney, where it is known to carry out specific actions in stimulating calcium and phosphorus transport reactions. This receptor molecule has been quite thoroughly characterized in its crude form. It has a $K_d$ of $5 \times 10^{-11}$ $M$ for 1,25-(OH)$_2$D$_3$ and shows remarkable selectivity for this molecule. Upon interaction of the receptor with 1,25-(OH)$_2$D$_3$, a subtle change occurs in the receptor molecule that permits it to bind to specific sites in nuclear chromatin. It is believed but not proved that this initiates transcription of specific genes that code for calcium and phosphorus transport proteins. Thus far, the calcium and phosphorus transport proteins have not been identified except for two calcium binding proteins found in the soluble fraction. It also is unclear whether these proteins play a role in the calcium transport process. Virtually nothing is known concerning the molecular mechanism of phosphate transport in response to $1,25\text{-}(OH)_2D_3$.

In addition to the classic actions of vitamin D, it has now been demonstrated that a number of tissues possess not only the receptor for $1,25\text{-}(OH)_2D_3$ but also show specific nuclear localization of radiolabeled $1,25\text{-}(OH)_2D_3$. These include the parathyroid gland cells, islet cells of the pancreas, certain cell types of the pituitary, brain cells, odonoblasts, endocrine cells of the stomach, epidermal cells of the skin, and epithelial cells of mammary tissue. These results suggest that more subtle actions of $1,25\text{-}(OH)_2D_3$ will be found in those organs not previously appreciated to be targets of vitamin D action.

A more recent development has been the finding of $1,25\text{-}(OH)_2D_3$ receptors in a number of cancer cell lines. 1,25-Dihydroxyvitamin $D_3$ functioning through a receptor mechanism can stimulate growth of certain cancer cells at low concentrations and inhibit growth at high concentrations. In addition, recent developments have shown that $1,25\text{-}(OH)_2D_3$ will cause differentiation of myeloid leukemia cells into monocytes. There is also additional evidence that these monocytes may accumulate into multinuclear cells perhaps as an osteoclast precursor. Again, the molecular mechanism of action of $1,25\text{-}(OH)_2D_3$ in these responses remains unknown.

The receptor for $1,25\text{-}(OH)_2D_3$ in chick intestinal epithelial cells has been purified to homogeneity and monoclonal antibodies prepared to it. These developments together with modern technological advances in the receptor field and in molecular biology show great promise in contributing new advances in our understanding of how $1,25\text{-}(OH)_2D_3$ interacts with its target cell.

## ACKNOWLEDGMENTS

This work was supported by program project Grant No. AM-14881 and postdoctoral research fellowship award (R.L.) No. 1 F32 AM07194 from the National Institutes of Health and by the Harry Steenbock Research Fund of the Wisconsin Alumni Research Foundation.

## REFERENCES

Abe, E., Tanabe, R., Suda, T., Yoshiki, S., Horikawa, H., Masumura, T., and Sugahara, M. (1979). Circadian rhythm of 1α,25-dihydroxyvitamin $D_3$ production in egg-laying hens. *Biochem. Biophys. Res. Commun.* **88**, 500–507.

Abe, E., Miyaura, C., Sakagami, H., Takeda, M., Konno, K., Yamazaki, T., Yoshiki, S., and Suda, T. (1981). Differentiation of mouse myeloid leukemia cells induced by $1\alpha,25$-dihydroxyvitamin $D_3$. *Proc. Natl. Acad. Sci. U.S.A.* **78**, 4990–4994.

Abe, E., Miyaura, C., Tanaka, H., Shiina, Y., Kuribayashi, T., Suda, S., Nishii, Y., DeLuca, H. F., and Suda, T. (1983). $1\alpha,25$-Dihydroxyvitamin $D_3$ promotes fusion of mouse alveolar macrophages both by a direct mechanism and by a spleen cell-mediated indirect mechanism. *Proc. Natl. Acad. Sci. U.S.A.* **80**, 5583–5587.

Bar-Shavit, Z., Teitelbaum, S., Reitsma, P., Hall, A., Pegg, L. E., Trial, J., and Kahn, A. J. (1983). Induction of monocytic differentiation and bone resorption by 1,25-dihydroxyvitamin $D_3$. *Proc. Natl. Acad. Sci. U.S.A.* **80**, 5907–5911.

Bell, N. H., Hamstra, A. J., and DeLuca, H. F. (1978). Vitamin D-dependent rickets Type II: Resistance of target organs to 1,25-dihydroxyvitamin D. *N. Engl. J. Med.* **298**, 996–999.

Bikle, D. D., Zolock, D. T., Morrissey, R. L., and Herman, R. H. (1978). Independence of 1,25-dihydroxyvitamin $D_3$-mediated calcium transport from *de novo* RNA and protein synthesis. *J. Biol. Chem.* **253**, 484–488.

Bikle, D. D., Morrissey, R. L., Zolock, D. T., and Herman, R. H. (1979). Stimulation of chick gut alkaline phosphatase activity by actinomycin D and 1,25-dihydroxyvitamin $D_3$: Evidence for independent mechanisms. *J. Lab. Clin. Med.* **94**, 88–94.

Bishop, C. W., Kendrick, N. C., and DeLuca, H. F. (1982). Induction of calcium binding protein before 1,25-dihydroxyvitamin $D_3$ stimulation of duodenal calcium uptake. *J. Biol. Chem.* **258**, 1305–1310.

Bishop, C. W., Kendrick, N. C., and DeLuca, H. F. (1984). The early time course of calcium-binding protein induction by 1,25-dihydroxyvitamin $D_3$ as determined by computer analysis of two-dimensional electrophoresis gels. *J. Biol. Chem.* **259**, 3355–3360.

Bishop, J. E., Hunziker, W., and Norman, A. W. (1982). Evidence for multiple molecular weight forms of the chicken intestinal 1,25-dihydroxyvitamin $D_3$ receptor. *Biochem. Biophys. Res. Commun.* **108**, 140–145.

Björkhem, I., Holmberg, I., Oftebro, H., and Pedersen, J. I. (1980). Properties of a reconstituted vitamin $D_3$ 25-hydroxylase from rat liver mitochondria. *J. Biol. Chem.* **255**, 5244–5249.

Brumbaugh, P. F., and Haussler, M. R. (1973a). $1\alpha,25$-Dihydroxyvitamin $D_3$ receptor: Competitive binding of vitamin D analogs. *Life Sci.* **13**, 1737–1746.

Brumbaugh, P. F., and Haussler, M. R. (1973b). Nuclear and cytoplasmic receptors for 1,25-dihydroxycholecalciferol in intestinal mucosa. *Biochem. Biophys. Res. Commun.* **51**, 74–80.

Brumbaugh, P. F., and Haussler, M. R. (1974a). $1\alpha,25$-Dihydroxycholecalciferol receptors in intestine. I. Association of $1\alpha,25$-dihydroxycholecalciferol with intestinal mucosa chromatin. *J. Biol. Chem.* **249**, 1251–1257.

Brumbaugh, P. F., and Haussler, M. R. (1974b). 1,25-Dihydroxycholecalciferol receptors in chicken intestine. II. Temperature dependent transfer of the hormone to chromatin and a specific cytosol receptor. *J. Biol. Chem.* **249**, 1258–1262.

Brumbaugh, P. F., and Haussler, M. R. (1975). Specific binding of $1\alpha,25$-dihydroxycholecalciferol to nuclear components of chick intestine. *J. Biol. Chem.* **250**, 1588–1594.

Brumbaugh, P. F., Hughes, M. R., and Haussler, M. R. (1975). Cytoplasmic and nuclear binding components for $1\alpha,25$-dihydroxyvitamin $D_3$ in chick parathyroid glands. *Proc. Natl. Acad. Sci. U.S.A.* **72**, 4871–4875.

Chen, T. C., and DeLuca, H. F. (1973a). Receptors of 1,25-dihydroxycholecalciferol in rat intestine. *J. Biol. Chem.* **248**, 4890–4895.

Chen, T. C., and DeLuca, H. F. (1973b). Stimulation of [$^3$H] Uridine incorporation in nuclear RNA of rat kidney by vitamin D metabolites. *Arch. Biochem. Biophys.* **156**, 321–327.

Chen, T. L., and Feldman, D. (1981). Regulation of 1,25-dihydroxyvitamin $D_3$ receptors in cultured mouse bone cells. Correlation of receptor concentration with the rate of cell division. *J. Biol. Chem.* **256**, 5561–5665.

Chen, T. L., Hirst, M. A., and Feldman, D. (1979). A receptor-like binding macromolecule for $1\alpha,25$-dihydroxycholecalciferol in cultured mouse bone cells. *J. Biol. Chem.* **254**, 7491–7494.

Chen, T. L., Cone, C. M., Morey-Holton, E., and Feldman, D. (1983). $1\alpha,25$-Dihydroxyvitamin $D_3$ receptors in cultured rat osteoblast-like cells. Glucocorticoid treatment increases receptor content. *J. Biol. Chem.* **258**, 4350–4355.

Christakos, S., and Norman, A. W. (1978). Vitamin $D_3$-induced calcium binding protein in bone tissue. *Science (Washington, D.C.)* **202**, 70–71.

Colston, K., and Feldman, D. (1980). Nuclear translocation of the 1,25-dihydroxycholecalciferol receptor in mouse kidney. *J. Biol. Chem.* **255**, 7510–7513.

Colston, K., Hirst, M., and Feldman, D. (1980). Organ distribution of the cytoplasmic 1,25-dihydroxycholecalciferol receptor in various mouse tissues. *Endocrinology (Baltimore)* **107**, 1916–1922.

Corradino, R. A. (1973a). Embryonic chick intestine in organ culture: Response to vitamin $D_3$ and its metabolites. *Science (Washington, D.C.)* **179**, 402–404.

Corradino, R. A. (1973b). 1,25-Dihydroxycholecalciferol: Inhibition of action in organ-cultured intestine by actinomycin D and $\alpha$-amanitin. *Nature (London)* **243**, 41–42.

Corradino, R. A. (1978). Calcium-binding protein of intestine: Induction by biologically significant cholecalciferol-like steroid *in vitro*. *J. Steroid Biochem.* **9**, 1183–1187.

Coty, W. A. (1980). Reversible dissociation of steroid hormone receptor complexes by mercurial reagents. *J. Biol. Chem.* **255**, 8035–8037.

DeLuca, H. F. (1981). Recent advances in the metabolism of vitamin D. *Annu. Rev. Physiol.* **43**, 199–209.

DeLuca, H. F. (1983). Metabolism and mechanism of action of vitamin D—1982. *In* "Annual Advances in Bone and Mineral Research" (W. A. Peck, ed.), pp. 7–73. Excerpta Med. Found., Amsterdam.

DeLuca, H. F. (1985). The vitamin D–calcium axis: 1983. *In* "Calcium in Biological Systems," 3M Award Lecture, pp. 491–511. Plenum, New York.

DeLuca, H. F., and Schnoes, H. K. (1983). Vitamin D: Recent advances. *Annu. Rev. Biochem.* **52**, 411–439.

Dokoh, S., Donaldson, C. A., Marion, S. L., and Haussler, M. R. (1983). A radioimunoassay of receptor for 1,25-dihydroxyvitamin $D_3$ in its target organs. *Int. Conf. Calcium Regul. Horm.*, *7th*, Abstr., p. 80 (E-7).

Dueland, S., Holmberg, I., Berg, T., and Pedersen, J. I. (1981). Uptake and 25-hydroxylation of vitamin $D_3$ by isolated rat liver cells. *J. Biol. Chem.* **256**, 10430–10434.

Dueland, S., Pedersen, J. I., Helgerud, P., and Drevon, C. A. (1982). Transport of vitamin $D_3$ from chylomicrons to $\alpha$-globulins. *J. Biol. Chem.* **257**, 146–152.

Eil, C., Liberman, U. A., Rosen, J. F., and Marx, S. J. (1981). A cellular defect in hereditary vitamin D-dependent rickets Type II: Defective nuclear uptake of 1,25-dihydroxyvitamin D in cultured skin fibroblasts. *N. Engl. J. Med.* **304**, 1588–1591.

Eisman, J. A., and DeLuca, H. F. (1977). Intestinal 1,25-dihydroxyvitamin $D_3$ binding protein: Specificity of binding. *Steroids* **30**, 245–257.

Eisman, J. A., Hamstra, A. J., Kream, B. E., and DeLuca, H. F. (1976). 1,25-Dihydroxyvitamin D in biological fluids: A simplified and sensitive assay. *Science (Washington, D.C.)* **193**, 1021–1023.

Eisman, J. A., Martin, T. J., and MacIntyre, I. (1980a). Presence of 1,25-dihydroxy vitamin D receptor in normal and banormal breast tissue. *In* "Hormones and the Kidney: Progress in Biochemical Pharmacology" (G. S. Stokes and J. F. Mahoneys, eds.), Vol. 17, pp. 143–150. Karger, Basel.

Eisman, J. A., Martin, T. J., and MacIntyre, I. (1980b). 1,25-Dihydroxyvitamin $D_3$ receptors in cancer. *Lancet* **1**, 1188–1190.

Feldman, D., McCain, T. A., Hirst, M. A., Chen, T. L., and Colston, K. W. (1979). Characterization of a cytoplasmic receptor-like binder for $1\alpha,25$-dihydroxycholecalciferol in rat intestinal mucosa. *J. Biol. Chem.* **254,** 10378–10384.

Feldman, D., Chen, T., Cone, C., Hirst, M., Shani, S., Benderli, A., and Hochberg, Z. (1982). Vitamin D resistant rickets with alopecia: Cultured skin fibroblasts exhibit defective cytoplasmic receptors and unresponsiveness to $1,25\text{-}(OH)_2D_3$. *J. Clin. Endocrinol. Metab.* **55,** 1020–1024.

Frampton, R. J., Omond, S., and Eisman, J. A. (1983a). Effects of 1,25-dihydroxyvitamin $D_3$ on cultured human cancer cells. *Clin. Exp. Pharmacol. Physiol.* **10,** 439–441.

Frampton, R. J., Omond, S. A., and Eisman, J. A. (1983b). Inhibition of human cancer cell growth by 1,25-dihydroxyvitamin $D_3$ metabolites. *Cancer Res.* **43,** 4443–4446.

Franceschi, R. T., and DeLuca, H. F. (1979). Aggregation properties of the 1,25-dihydroxyvitamin $D_3$ receptor from chick intestinal cytosol. *J. Biol. Chem.* **254,** 11629–11635.

Franceschi, R. T., and DeLuca, H. F. (1981a). The effect of inhibitors of protein and RNA synthesis on 1,25-dihydroxyvitamin $D_3$-dependent calcium uptake in cultured embryonic chick duodenum. *J. Biol. Chem.* **256,** 3848–3852.

Franceschi, R. T., and DeLuca, H. F. (1981b). Characterization of 1,25-dihydroxyvitamin $D_3$-dependent calcium uptake in cultured embryonic chick duodenum. *J. Biol. Chem.* **256,** 3840–3847.

Franceschi, R. T., Simpson, R. U., and DeLuca, H. F. (1981). Binding proteins for vitamin D metabolites: Serum carriers and intracellular receptor. *Arch. Biochem. Biophys.* **210,** 1–13.

Franceschi, R. T., DeLuca, H. F., and Mercado, D. L. (1983). Temperature-dependent inactivation of nucleic acid binding and aggregation of the 1,25-dihydroxyvitamin $D_3$ receptor. *Arch. Biochem. Biophys.* **222,** 504–517.

Freake, H. C., Marcocci, C., Iwasaki, J., and MacIntyre, I. (1981). 1,25-Dihydroxyvitamin $D_3$ specifically binds to a human breast cancer cell line (T47D) and stimulates growth. *Biochem. Biophys. Res. Commun.* **101,** 1131–1138.

Frost, H. M. (1966). "Bone Dynamics in Osteoporosis and Osteomalacia," Henry Ford Hospital Surgical Monograph Series. Thomas, Springfield, Illinois.

Gelbard, H. A., Stern, P. H., and Prichard, D. C. (1980). $1\alpha$-25-Dihydroxyvitamin $D_3$ nuclear receptors in the pituitary. *Science (Washington, D.C.)* **209,** 1247–1249.

Ghazarian, J. G., Jefcoate, C. R., Knutson, J. C., Orme-Johnson, W. H., and DeLuca, H. F. (1974). Mitochondrial cytochrome $P_{450}$: A component of chick kidney 25-hydroxycholecalciferol-$1\alpha$-hydroxylase. *J. Biol. Chem.* **249,** 3026–3033.

Goodman, D. B. P., Haussler, M. R., and Rasmussen, H. (1972). Vitamin $D_3$ induced alteration of microvillar membrane lipid composition. *Biochem. Biophys. Res. Commun.* **46,** 80–86.

Halloran, B. P., and DeLuca, H. F. (1980). Calcium transport in the small intestine during early development: The role of vitamin D. *Am. J. Physiol.* **239,** G473–G479.

Halloran, B. P., and DeLuca, H. F. (1981a). Intestinal calcium transport: Evidence for two distinct mechanisms of action of 1,25-dihydroxyvitamin $D_3$. *Arch. Biochem. Biophys.* **208,** 477–486.

Halloran, B. P., and DeLuca, H. F. (1981b). Appearance of the intestinal cytosolic receptor for 1,25-dihydroxyvitamin $D_3$ during neonatal development in the rat. *J. Biol. Chem.* **256,** 7338–7342.

Haussler, M. R., and Norman, A. W. (1967). The subcellular distribution of physiological doses of vitamin $D_3$. *Arch. Biochem. Biophys.* **118,** 145–153.

Haussler, M. R., and Norman, A. W. (1969). Chromosomal receptor for a vitamin D metabolite. *Proc. Natl. Acad. Sci. U.S.A.* **62,** 155–158.

Haussler, M. R., Myrtle, J. F., and Norman, A. W. (1968). The association of a metabolite of vitamin $D_3$ with intestinal mucosa chromatin *in vivo*. *J. Biol. Chem.* **243,** 4055–4064.

Haussler, M. R., Pike, J. W., Chandler, J. S., Manolagas, S. C., and Deftos, L. J. (1981). Molecular action of 1,25-dihydroxyvitamin $D_3$: New cultured cell models. *Ann. N.Y. Acad. Sci.* **372,** 502–505.

Haussler, M. R., Mangelsdorf, D. J., Koeffler, H. P., Donaldson, C. A., and Pike, J. W. (1983). 1,25-Dihydroxyvitamin D induces differentiation of human leukemia (HL-60) cells to monocytes/macrophages via its specific intracellular receptor. *Int. Conf. Calcium Regul. Horm., 8th,* Abstr. p. 82 (E-11).

Hirst, M., and Feldman, D. (1982a). Glucocorticoids down-regulate the number of 1,25-dihydroxyvitamin $D_3$ receptors in mouse intestine. *Biochem. Biophys. Res. Commun.* **105,** 1590–1596.

Hirst, M., and Feldman, D. (1982b). Glucocorticoid regulation of 1,25(OH)$_2$vitamin $D_3$ receptors: Divergent effects on mouse and rat intestine. *Endocrinology (Baltimore)* **111,** 1400–1402.

Holick, M. F. (1981). The cutaneous photosynthesis of previtamin $D_3$: A unique photoendocrine system. *J. Invest. Dermatol.* **76,** 51–58.

Holick, M. F., Schnoes, H. K., DeLuca, H. F., Suda, T., and Cousins, R. J. (1971). Isolation and identification of 1,25-dihydroxycholecalciferol. A metabolite of vitamin D active in intestine. *Biochemistry* **10,** 2799–2804.

Hughes, M. R., and Haussler, M. R. (1978). 1,25-Dihydroxyvitamin $D_3$ receptors in parathyroid glands: Preliminary characterization of cytoplasmic and nuclear binding components. *J. Biol. Chem.* **253,** 1065–1073.

Hunziker, W., Walters, M. R., and Norman, A. W. (1980). 1,25-Dihydroxyvitamin $D_3$ receptors: Differential quantitation of endogenously occupied and unoccupied sites. *J. Biol. Chem.* **255,** 9534–9537.

Hunziker, W., Walters, M. R., Bishop, J. E., and Norman, A. W. (1982). Effect of vitamin D status on the equilibrium between occupied and unoccupied 1,25-dihydroxyvitamin D intestinal receptors in the chick. *J. Clin. Invest.* **69,** 826–834.

Hunziker, W., Walters, M. R., Bishop, J. E., and Norman, A. W. (1983). Unoccupied and *in vitro* and *in vivo* occupied 1,25-dihydroxyvitamin $D_3$ intestinal receptors. Multiple biochemical forms and evidence for transformation. *J. Biol. Chem.* **258,** 8642–8648.

Kabakoff, B., Kendrick, N. C., and DeLuca, H. F. (1982). 1,25-Dihydroxyvitamin $D_3$-stimulated active uptake of phosphate by rat jejunum. *Am. J. Physiol.* **6,** E470–E475.

Kendrick, N. C., Kabakoff, B., and DeLuca, H. F. (1981). Oxygen-dependent 1,25-dihydroxycholecalciferol-induced calcium ion transport in rat intestine. *Biochem. J.* **194,** 178–186.

Kimberg, D. V. (1969). Effects of vitamin D and steroid hormones on the active transport of calcium by the intestine. *N. Engl. J. Med.* **280,** 1396–1405.

Kream, B. E., and DeLuca, H. F. (1977). A specific binding protein for 1,25-dihydroxyvitamin $D_3$ in rat intestinal cytosol. *Biochem. Biophys. Res. Commun.* **76,** 735–738.

Kream, B. E., Reynolds, R. D., Knutson, J. C., Eisman, J. A., and DeLuca, H. F. (1976). Intestinal cytosol binders of 1,25-dihydroxyvitamin $D_3$ and 25-hydroxyvitamin $D_3$. *Arch. Biochem. Biophys.* **176,** 779–787.

Kream, B. E., Jose, M., Yamada, S., and DeLuca, H. F. (1977a). A specific high-affinity binding macromolecule for 1,25-dihydroxyvitamin $D_3$ in fetal bone. *Science (Washington, D.C.)* **197,** 1086–1088.

Kream, B. E., Yamada, S., Schnoes, H. K., and DeLuca, H. F. (1977b). Specific cytosol-binding protein for 1,25-dihydroxyvitamin $D_3$ in rat intestine. *J. Biol. Chem.* **252,** 4501–4506.

Kream, B. E., Jose, M. J. L., and DeLuca, H. F. (1977c). The chick intestinal cytosol binding protein for 1,25-dihydroxyvitamin $D_3$: A study of analog binding. *Arch. Biochem. Biophys.* **179,** 462–468.

Lawson, D. E. M., and Wilson, P. W. (1974). Intranuclear localization and receptor proteins for 1,25-dihydroxycholecalciferol in chick intestine. *Biochem. J.* **144,** 573–583.

Liberman, U. A., Eil, C., Holst, P., Rosen, J. F., and Marx, S. J. (1983). Hereditary resistance to 1,25-dihydroxyvitamin D: Defective function of receptors for 1,25-dihydroxyvitamin D in cell cultured from bone. *J. Clin. Endocrinol. Metab.* **57,** 958–960.

Lindgren, J. U., and DeLuca, H. F. (1983). Oral 1,25(OH)$_2$D$_3$: An effective prophylactic treatment for glucocorticoid osteopenia in rats. *Calcif. Tissue Int.* **35,** 107–110.

McCain, R. A., Haussler, M. R., Okrent, D., and Hughes, M. R. (1978). Partial purification of the chick intestinal receptor for 1,25-dihydroxyvitamin D by ion-exchange and blue dextran-sepharose chromatography. *FEBS Lett.* **86**, 65–68.

McCarthy, D. M., San Miguel, J. F., Freake, H. C., Green, P. M., Zola, H., Catovsky, D., and Goldman, J. M. (1983). 1,25-Dihydroxyvitamin $D_3$ inhibits proliferation of human pro-myelocytic leukaemia (HL60) cells and induces monocyte-macrophage differentiation in HL60 and normal human bone marrow cells. *Leuk. Res.* **7**, 51–56.

Massaro, E., Simpson, R. U., and DeLuca, H. F. (1982). Stimulation of specific 1,25-dihydrox-yvitamin $D_3$ binding protein in cultured postnatal rat intestine by hydrocortisone. *J. Biol. Chem.* **257**, 13736–13739.

Massaro, E. R., Simpson, R. U., and DeLuca, H. F. (1983a). Glucocorticoids and appearance of 1,25-dihydroxyvitamin $D_3$ receptor in rat intestine. *Am. J. Physiol.* **244**, E230–E235.

Massaro, E. R., Simpson, R. U., and DeLuca, H. F. (1983b). Quantitation of endogenously occupied and unoccupied binding site for 1,25-dihydroxyvitamin $D_3$ in rat intestine. *Proc. Natl. Acad. Sci. U.S.A.* **80**, 2549–2553.

Max, E. E., Goodman, D. B. P., and Rasmussen, H. (1978). Purification and characterization of chick intestine brush border membrane: Effects of 1$\alpha$(OH) vitamin $D_3$ treatment. *Biochim. Biophys. Acta* **511**, 224–239.

Mellon, W., and DeLuca, H. F. (1979). An equilibrium and kinetic study of 1,25-dihydroxyvitamin $D_3$ binding to chicken intestinal cytosol employing high specific activity 1,25-dihydroxy-[$^3$H-26,27]vitamin $D_3$. *Arch. Biochem. Biophys.* **197**, 90–95.

Mellon, W. S., and DeLuca, H. F. (1980). A specific 1,25-dihydroxyvitamin $D_3$ binding mac-romolecule in chicken bone. *J. Biol. Chem.* **255**, 4081–4086.

Mellon, W. S., Franceschi, R. T., and DeLuca, H. F. (1980). An *in vitro* study of the stability of the chicken intestinal cytosol 1,25-dihydroxyvitamin $D_3$-specific receptor. *Arch. Biochem. Biophys.* **202**, 83–92.

Miyaura, C., Abe, E., Kuribayashi, T., Tanaka, H., Konno, K., Nishii, Y., and Suda, T. (1981). 1$\alpha$,25-Dihydroxyvitamin $D_3$ induces differentiation of human myeloid leukemia cells. *Biochem. Biophys. Res. Commun.* **102**, 937–943.

Napoli, J. L., Fivizzani, M. A., Schnoes, H. K., and DeLuca, H.F. (1978). 1$\alpha$-hydroxy-25-fluorovitamin $D_3$: A potent analogue of 1$\alpha$,25-dihydroxyvitamin $D_3$. *Biochemistry* **17**, 2387–2392.

Napoli, J. L., Fivizzani, M. A., Schnoes, H. K., and DeLuca, H. F. (1979a). 1-Fluorovitamin $D_3$: A vitamin $D_3$ analog more active on bone-calcium mobilization than intestinal-calcium transport. *Biochemistry* **18**, 1641–1646.

Napoli, J. L., Mellon, W. S., Schnoes, H. K., and DeLuca, H. F. (1979b). Evidence for the metabolism of 24R-hydroxy-25-fluorovitamin $D_3$ and 1$\alpha$-hydroxy-25-fluorovitamin $D_3$ to 1$\alpha$,24R-dihydroxy-25-fluorovitamin $D_3$. *Arch. Biochem. Biophys.* **197**, 193–198.

Napoli, J. L., Mellon, W. S., Fivizzani, M. A., Schnoes, H. K., and DeLuca, H. F. (1980). Direct chemical synthesis of 1$\alpha$,25-dihydroxy[26,27-$^3$H]vitamin $D_3$ with specific activity: Its use in receptor studies. *Biochemistry* **19**, 2515–2521.

Narbaitz, R., Stumpf, W. E., Sar, M., Huang, S., and DeLuca, H. F. (1983). Autoradiographic localization of target cells for 1$\alpha$,25-dihydroxyvitamin $D_3$ in bones from fetal rats. *Calcif. Tissue Int.* **35**, 177–182.

Norman, A. W., Roth, J., and Orci, L. (1982). The vitamin D endocrine system: Steroid metabo-lism, hormone receptors, and biological response (calcium binding proteins). *Endocr. Rev.* **3**, 331–366.

Okamoto, S., Tanaka, Y., DeLuca, H. F., Kobayashi, Y., and Ikekawa, N. (1983). Biological activity of 24,24-difluoro-1,25-dihydroxyvitamin $D_3$. *Am. J. Physiol.* **7**, E159–E163.

Partridge, N. C., Frampton, R. J., Eisman, J. A., Michelangeli, V. P., Elms, E., Bradley, T. R., and Martin, T. J. (1980). Receptors for 1,25(OH)$_2$-vitamin D$_3$ enriched in cloned osteoblast-like rat osteogenic sarcoma cells. *FEBS Lett.* **115**, 139–142.

Pike, J. W. (1981). Evidence for a reactive sulfhydryl in the DNA binding domain of the 1,25-dihydroxyvitamin D$_3$ receptor. *Biochem. Biophys. Res. Commun.* **100**, 1713–1719.

Pike, J. W. (1982a). Interaction between 1,25-dihydroxyvitamin D$_3$ receptors and intestinal nuclei. Binding to nuclear constituents *in vitro*. *J. Biol. Chem.* **257**, 6766–6775.

Pike, J. W. (1982b). Receptors for 1,25-dihydroxyvitamin D$_3$ in chick pancreas: A partial physical and functional characterization. *J. Steroid Biochem.* **16**, 385–395.

Pike, J.W., and Haussler, M. R. (1979). Purification of chicken intestinal receptor for 1,25-dihydroxyvitamin D. *Proc. Natl. Acad. Sci. U.S.A.* **76**, 5485–5489.

Pike, J. W., and Haussler, M. R. (1983). Association of 1,25-dihydroxyvitamin D$_3$ with cultured 3T6 mouse fibroblasts. *J. Biol. Chem.* **258**, 8554–8560.

Pike, J. W., Donaldson, C. A., Marion, S. L., and Haussler, M. R. (1982). Development of hybridomas secreting monoclonal antibodies to the chicken intestinal 1α,25-dihydroxyvitamin D$_3$ receptor. *Proc. Natl. Acad. Sci. U.S.A.* **79**, 7719–7723.

Pike, J.W., Marion, S. L., Donald, C. A., and Haussler, M. R. (1983). Serum and monoclonal antibodies against the chick intestinal receptor for 1,25-dihydroxyvitamin D$_3$. Generation by a preparation enriched in a 64,000-dalton protein. *J. Biol. Chem.* **258**, 1289–1296.

Price, P. A., and Baukol, S. A. (1980). 1,25-Dihydroxyvitamin D$_3$ increases synthesis of the vitamin K-dependent bone protein by osteosarcoma cells. *J. Biol. Chem.* **255**, 11660–11666.

Price, P. A., and Baukol, S. A. (1981). 1,25-Dihydroxyvitamin D$_3$ increases serum levels of the vitamin K-dependent bone protein. *Biochem. Biophys. Res. Commun.* **99**, 928–935.

Procsal, D. A., Okamura, W. H., and Norman, A. W. (1975). Structural requirements for the interaction of 1α,25(OH)$_2$vitamin D$_3$ with its chick intestinal receptor system. *J. Biol. Chem.* **250**, 8382–8388.

Radparvar, S., and Mellon, W. S. (1982). Characterization of 1,25-dihydroxyvitamin D$_3$-receptor complex interactions with DNA by a competitive assay. *Arch. Biochem. Biophys.* **217**, 552–563.

Raisz, L. G., Trummel, C. L., Holick, M. F., and DeLuca, H. F. (1972). 1,25-Dihydroxycholecalciferol: A potent stimulator of bone resorption in tissue culture. *Science (Washington, D.C.)* **175**, 768–769.

Rasmussen, H., Fontaine, O., Max, E., and Goodman, D. B. P. (1979). Lhe effect of 1α-hydroxyvitamin D$_3$ administration on calcium transport in chick intestine brush border membrane vesicles. *J. Biol. Chem.* **254**, 2993–2999.

Rasmussen, H., Fontaine, O., and Matsumoto, T. (1981). Liponomic regulation of calcium transport by 1,25-(OH)$_2$D$_3$. *Ann. N.Y. Acad. Sci.* **372**, 518–522.

Siebert, P. D., Ohnuma, N., and Norman, A. W. (1979). Studies on the mode of action of calciferol. XIX. A 24R-hydroxy-group can replace the 25-hydroxy group of 1α,25-dihydroxyvitamin D$_3$ for optimal binding to the chick intestinal receptor. *Biochem. Biophys. Res. Commun.* **91**, 827–834.

Simpson, R. U., and DeLuca, H. F. (1980). Characterization of a receptor-like protein for 1,25-dihydroxyvitamin D$_3$ in rat skin. *Proc. Natl. Acad. Sci. U.S.A.* **77**, 5822–5826.

Simpson, R. U., and DeLuca, H. F. (1982). Purification of chicken intestinal receptor for 1α25-dihydroxyvitamin D$_3$ to apparent homogeneity. *Proc. Natl. Acad. Sci. U.S.A.* **79**, 16–20.

Simpson, R. U., Franceschi, R. T., and DeLuca, H. F. (1980). 1α,25-dihydroxyvitamin D$_3$ in cultured chick kidney cells. *J. Biol. Chem.* **255**, 10160–10166.

Simpson, R. U., Hamstra, A., Kendrick, N. C., and DeLuca, H. F. (1983). Purification of the receptor for 1α,25-dihydroxyvitamin D$_3$ from chicken intestine. *Biochemistry* **22**, 2586–2594.

Spencer, R., Charman, M., Wilson, P., and Lawson, E. (1976). Vitamin D-stimulated intestinal calcium absorption may not involve calcium binding protein directly. *Nature (London)* **263**, 161–163.

Spencer, R., Chapman, M., Wilson, P. W., and Lawson, D. E. M. (1978). The relationship between vitamin D-stimulated calcium transport and intestinal calcium-binding protein in the chicken. *Biochem. J.* **170**, 93–102.

Stumpf, W. E., Sar, M., Reid, F. A., Tanaka, Y., and DeLuca, H. F. (1979). Target cells for 1,25-dihydroxyvitamin $D_3$ in intestinal tract, stomach, kidney, skin, pituitary and parathyroid. *Science (Washington, D.C.)* **206**, 1188–1190.

Stumpf, W. E., Sar, M., Narbaitz, R., Reid, F. A., DeLuca, H. F., and Tanaka, Y. (1980). Cellular and subcellular localization of 1,25-$(OH)_2$-vitamin $D_3$ in rat kidney: Comparison and localization of parathyroid hormone and estradiol. *Proc. Natl. Acad. Sci. U.S.A.* **77**, 1149–1153.

Stumpf, W. E., Sar, M., and DeLuca, H. F. (1981). Site of action of 1,25 $(OH)_2$ vitamin $D_3$ identified by thaw-mount autoradiography. *In* "Hormonal Control of Calcium Metabolism" (D. V. Cohn, R. V. Talmage, and J. L. Matthews, eds.), pp. 222–229. Excerpta Med. Found., Amsterdam.

Takasaki, Y., Suda, T., Yamada, S., Ohmori, M., Takayama, H., and Nishii, Y. (1982). Chemical synthesis, biological activity, and metabolism of 25-hydroxy-24-oxovitamin $D_3$. *J. Biol. Chem.* **257**, 3732–3738.

Tanaka, H., Abe, H., Miyaura, C., Kuribayashi, T., Konno, K., Nishii, Y., and Suda, T. (1982). 1α,25-Dihydroxycholecalciferol and a human myeloid leukaemia cell line (HL-60). The presence of a cytosol receptor and induction of differentiation. *Biochem. J.* **204**, 713–719.

Tanaka, Y., and DeLuca, H. F. (1971). Bone mineral mobilization activity of 1,25-dihydroxycholecalciferol, a metabolite of vitamin D. *Arch. Biochem. Biophys.* **146**, 574–578.

Tanaka, Y., DeLuca, H. F., Omdahl, J., and Holick, M. F. (1971). Mechanism of action of 1,25-dihydroxycholecalciferol on intestinal calcium transport. *Proc. Natl. Acad. Sci. U.S.A.* **68**, 1286–1288.

Tanaka, Y., Schnoes, H. K., Smith, C. M., and DeLuca, H. F. (1981). 1,25,26-Trihydroxyvitamin $D_3$: Isolation, identification, and biological activity. *Arch. Biochem. Biophys.* **210**, 104–109.

Tanaka, Y., Wichmann, J. K., DeLuca, H. F., Kobayashi, Y., and Ikekawa, N. (1983). Metabolism and binding properties of 24,24-difluoro-25-hydroxyvitamin $D_3$. *Arch. Biochem. Biophys.* **225**, 649–655.

Tanaka, Y., DeLuca, H. F., Kobayashi, Y., and Ikekawa, N. (1984). 26,26,26,27,27,27-Hexafluoro-1,25-dihydroxyvitamin $D_3$: A highly potent, long-lasting analog of 1,25-dihydroxyvitamin $D_3$. *Arch. Biochem. Biophys.* **229**, 348–354.

Taylor, A. N., and Wasserman, R. H. (1967). Vitamin $D_3$-induced calcium-binding protein: Partial purification, electrophoretic visualization, and tissue distribution. *Arch. Biochem. Biophys.* **119**, 536–540.

Tsai, H. C., and Norman, A. W. (1973a). Studies on the mode of action of calciferol. VI. Effect of 1,25-dihydroxyvitamin $D_3$ on RNA synthesis in the intestinal mucosa. *Biochem. Biophys. Res. Commun.* **54**, 622–627.

Tsai, H. C., and Norman, A. W. (1973b). Studies on calciferol metabolism. VIII. Evidence for a cytoplasmic receptor for 1,25-dihydroxyvitamin $D_3$ in the intestinal mucosa. *J. Biol. Chem.* **248**, 5967–5975.

Tsai, H. C., Wong, R. G., and Norman, A. W. (1972). Studies on calciferol metabolism. IV. Subcellular localization of 1,25-dihydroxyvitamin $D_3$ in intestinal mucosa and correlation with increased calcium transport. *J. Biol. Chem.* **247**, 5511–5519.

Tsai, H. C., Midgett, R. J., and Norman, A. W. (1973). Studies on calciferol metabolism. VII. The effects of actinomycin D and cycloheximide on the metabolism tissue and subcellular localization, and action of vitamin $D_3$. *Arch. Biochem. Biophys.* **157**, 339–347.

Underwood, J. L., and DeLuca, H. F. (1984). Vitamin D is not directly necessary for growth and mineralization. *Am. J. Physiol.* **246**, E493–E498.

Walters, M. R., Hunziker, W., and Norman, A. W. (1980a). Cytosol preparations are inadequate for quantitating unoccupied receptors for 1,25-dihydroxyvitamin $D_3$. *J. Recept. Res.* **1**, 313–317.

Walters, M. R., Hunziker, W., and Norman, A. W. (1980b). Unoccupied 1,25-dihydroxyvitamin $D_3$ receptors nuclear/cytosol rates depends on ionic strength. *J. Biol. Chem.* **255**, 6799–6805.

Walters, M. R., Hunziker, W., and Norman, A. W. (1981). Apparent nuclear localization of unoccupied receptors for 1,25-dihydroxyvitamin $D_3$. *Biochem. Biophys. Res. Commun.* **98**, 990–995.

Walters, M. R., Hunziker, W., Konami, D., and Norman, A. W. (1982). Factors affecting the distribution and stability of unoccupied 1,25-dihydroxyvitamin $D_3$ receptors. *J. Recept. Res.* **2**, 331–346.

Wasserman, R. H., and Feher, J. J. (1977). Vitamin D-dependent calcium-binding proteins. *In* "Calcium Binding Proteins and Calcium Function" (R. H. Wasserman, R. A. Corradino, E. Carafoli, R. H. Kretsinger, D. H. MacLennan, and S. L. Siegel, eds.), pp. 292–302. Elsevier/North-Holland, New York.

Wecksler, W. R., and Norman, A. W. (1980). A kinetic and equilibrium binding study of 1$\alpha$,25-dihydroxyvitamin $D_3$ with its cytosol receptor from chick intestinal mucosa. *J. Biol. Chem.* **255**, 3571–3574.

Wecksler, W. R., Okamura, W. H., and Norman, A. W. (1978). Studies on the mode of action of vitamin D: XIV. Quantitative assessment of the structural requirements for the interaction of 1$\alpha$,25-dihydroxyvitamin $D_3$ with its chick intestinal mucosa receptor system. *J. Steroid Biochem.* **9**, 929–932.

Wecksler, W. R., Ross, F. P., and Norman, A. W. (1979). Characterization of the 1$\alpha$,25-dihydroxyvitamin $D_3$ receptors from rat intestinal cytosol. *J. Biol. Chem.* **254**, 9488–9491.

Wecksler, W. R., Ross, F. P., Mason, R. S., and Norman, A. W. (1980a). Biochemical properties of the 1$\alpha$,25-dihydroxyvitamin $D_3$ cytosol receptors from human and chicken intestinal mucosa. *J. Clin. Endocrinol. Metab.* **50**, 152–157.

Wecksler, W. R., Ross, F. P., Mason, R. S., Posen, S., and Norman, A. W. (1980b). Biochemical properties of the 1$\alpha$,25-dihydroxyvitamin $D_3$ cytoplasmic receptors from human and chick parathyroid glands. *Arch. Biochem. Biophys.* **201**, 95–103.

Weinstein, R. S., Underwood, J. L., Hutson, M. S., and DeLuca, H. F. (1984). Bone histomorphometry in vitamin D-deficient rats infused with calcium and phosphorus. *Am. J. Physiol.* **246**, E499–E505.

Yamada, S., Ohmori, M., Takayama, H., Takasaki, Y., and Suda, T. (1983). Isolation and identification of 1$\alpha$- and 23-hydroxylated metabolites of 25-hydroxy-24-oxovitamin $D_3$ from *in vitro* incubates of chick kidney homogenates. *J. Biol. Chem.* **258**, 457–463.

Zerwekh, J. E., Haussler, M. R., and Lindell, T. J. (1974). Rapid enhancement of chick intestinal DNA-dependent RNA polymerase II activity by 1$\alpha$,25-dihydroxyvitamin $D_3$, *in vivo. Proc. Natl. Acad. Sci. U.S.A.* **71**, 2337–2341.

Zerwekh, J. E., Lindell, T. J., and Haussler, M. R. (1976). Increased intestinal chromatin template activity. Influence of 1$\alpha$,25-dihydroxyvitamin $D_3$ and hormone-receptor complexes. *J. Biol. Chem.* **251**, 2388–2394.

Zile, M., Bunge, E. C., Barsness, L., Yamada, S., Schnoes, H. K., and DeLuca, H. F. (1978). Localization of 1,25-dihydroxyvitamin $D_3$ in intestinal nuclei *in vivo. Arch. Biochem. Biophys.* **186**, 15–24.

# 2

# On Experimental Discrimination between Alternative Mechanistic Models for the Receptor-Mediated Stimulation of Adenylate Cyclase

**ALLEN P. MINTON**
Laboratory of Biochemical Pharmacology
National Institute of Arthritis, Diabetes, and Digestive and Kidney Diseases
National Institutes of Health
Bethesda, Maryland

## I. INTRODUCTION

A variety of extracellular hormones and drugs (collectively termed *agonists*) are known to stimulate the intracellular production of cyclic AMP by interacting with specific agonist binding sites, called receptors, on the outer surface of the

THE RECEPTORS, VOL. II

ISBN 0-12-185202-4

plasma membrane. The mechanism by which occupancy of the receptor by agonist molecules results in stimulation of cyclase activity has been intensively studied for many years in numerous laboratories (for recent reviews, see Ross *et al.*, 1983; Lefkowitz *et al.*, 1983; Aurbach, 1982; Levitzki, 1981). In spite of the very large quantity of published data characterizing the receptor-mediated stimulation of cyclic AMP by agonists, an unambiguous mechanistic description of this process, even at the qualitative level, has not emerged, primarily because of the complexity of the system studied. At least three separable membrane protein components and a guanine nucleotide cofactor are known to be necessary for receptor-mediated stimulation of adenylate cyclase. A receptor component, denoted R, contains the specific binding site for the agonist. A regulatory component, denoted N, contains a binding site for the guanine nucleotide and, under at least some conditions, appears to catalyze the hydrolysis of guanosine triphosphate (GTP) to guanosine diphosphate (GDP). A catalytic component, denoted C, contains the active site for conversion of ATP to cyclic AMP.

Several mechanistic models for the receptor-mediated stimulation of cyclic AMP production have been proposed. Most of these take the three macromolecular components R, N, and C, and the guanine nucleotide into account either explicitly or implicitly. There is, however, a significant divergence of views as to the way the three protein moieties interact and how the presence of agonist and guanine nucleotide affect their interactions. The purpose of this chapter is to explore the extent to which currently available data permit a choice to be made between alternate hypotheses regarding the role of the agonist–receptor complex in stimulating adenylate cyclase and regarding the functional relationships between the R, N, and C components. More specifically, we shall pose and attempt to answer the following two questions: (1) Does the agonist–receptor complex act as a reactant or as a catalyst with respect to stimulation of cyclase activity? (2) Mechanistic models for receptor-mediated stimulation of adenylate cyclase have been proposed wherein the N component is postulated to be associated with R throughout the process of stimulation (Rodbell, 1980), associated with C throughout the process of stimulation (Tolkovsky *et al.*, 1982), and transferred between R and C during the process of stimulation (Lefkowitz *et al.*, 1983). Do currently existing data permit one or more of these three alternatives to be excluded?

## II. GENERAL ASSUMPTIONS, DEFINITIONS, AND NOMENCLATURE

At this point we shall explicitly state the fundamental assumptions underlying the analysis presented in this chapter: (1) It is assumed that the experimental results on which the analysis is based are at least semiquantitatively reliable (i.e., not experimental artifacts). (2) It is assumed that differences between receptor-mediated stimulation of adenylate cyclase observed in different cell types and with different agonist–receptor combinations reflect quantitative rather than qualitative differences in the underlying mechanism. (3) During the last few years attention has been directed toward the inhibition of cyclase activity by the interaction of a different set of pharmacologically active substances (inhibitory ligands) with their own receptors, which are distinct from the receptors involved in the stimulation of cyclase by agonists (Rodbell, 1980). This inhibitory process is thought to involve an inhibitory regulatory component denoted $N_I$. It will be assumed that inhibitory ligands, acting through $N_I$, modulate the interaction between N and catalytic unit C in a competitive fashion. It follows that even though we do not explicitly include inhibitory ligands, receptors, and $N_I$ in the reaction schemes to be presented and discussed later, the effects of these moieties will be manifest in the magnitude of apparent rate and equilibrium constants characterizing reactions between N and C. (4) It is assumed for simplicity that a particular catalytic unit may exist in one of two catalytic states (inactive or fully active) and that variation in cyclase activity is obtained by altering the fraction of active C units. Clearly, the possibility of a variable degree of catalytic activity within a particular C unit should not be excluded from consideration *a priori*. Should models based upon the two-state hypothesis prove untenable, permitting the activity of an individual C unit to vary provides an obvious route to generalization.

Throughout the chapter we shall employ the terms, conventions, and notation introduced below.

1. The term "amount" as applied to a chemical species refers to the amount per unit volume and is hence synonymous with the term concentration. The concentration of species X will be denoted by [X].
2. Fractional saturation of binding, denoted by $y_B$, is the ratio of the amount of ligand bound to the specific binding capacity of the substrate (membrane, solubilized receptor, etc.).
3. Fractional reponse, denoted by $y_R$, is the ratio of the observed cyclase activity to the theoretical maximum cyclase activity.
4. Binding isotherm refers to the functional dependence of steady-state ligand binding upon ligand concentration.
5. Response isotherm refers to the functional dependence of the level of steady-state response upon agonist concentration.
6. A reference type isotherm is an isotherm characteristic of a single class of independent sites. A reference type binding isotherm may be represented by the Langmuir relation

$$y_B = K[L]/(1 + K[L]) \qquad (1)$$

in which [L] is the concentration of free (unbound) ligand, $y_B$ is the equilibrium average number of molecules of L bound per binding site, and $K$ is the equilibrium constant for association of L to a site. The titration plot ($y$ versus log[L]) of a reference isotherm is sigmoid, symmetrical about the point of half-saturation, and spans the range of $y$ between 0.09 and 0.91 in two log units of [L] (Fig. 1). A Scatchard plot ($y/$[L] versus $y$) of this isotherm is a straight line, and the Hill coefficient at half-saturation is equal to one.

7. Apparent multisite-class type isotherm refers to a phenomenological description of any isotherm which, when plotted in the form of a titration plot, exhibits a shallower dependence upon free-ligand concentration than does the reference type isotherm, i.e., more than two log units in [L] are required to span the 0.09–0.91 range of $y$ (Fig. 1). A Scatchard plot of an apparent multisite-class isotherm is concave upwards, and the Hill coefficient at half-saturation is less than one. A multisite-class type binding isotherm may be analytically represented by the function

$$y_B = \sum_{i=1}^{n} f_i \left[ K_i[L]/(1 + K_i[L]) \right] \qquad (2)$$

for which $n$ represents the number of apparent site classes, $f_i$ the mole fraction of sites of class i, and $K_i$ the equilibrium constant for the association of L to sites of class i. It is important to keep in mind that different

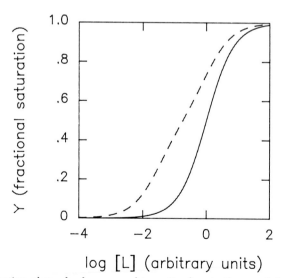

**Fig. 1.** Titration plots of reference and two-site-class isotherms. Solid curve is the reference isotherm calculated according to eq. (1) with $K = 1$. The dashed curve is the two-site-class isotherm calculated according to eq. (2) with $n = 2, f_1 = f_2 = 0.5, K_1 = 30$ and $K_2 = 1$.

apparent classes of sites may arise from interactions between various components of the system and that they do not necessarily represent multiple classes of independent sites possessing intrinsically different affinities for L.

8. The concentration of free-ligand L corresponding to half-maximal specific binding of L is a useful rough measure of overall binding site affinity and is denoted by $[L]_B^{50}$.

9. The concentration of free-ligand L corresponding to half-maximal stimulation of cyclase activity ($y_R = 0.5$) is denoted by $[L]_R^{50}$.

10. Rate constants will be denoted by lower case letters. Equilibrium constants (or pseudo-equilibrium constants) will be denoted by upper case letters and will be association constants unless otherwise indicated.

11. A catalytically active C unit will be denoted by C*.

## III. IS THE RECEPTOR A CATALYST OR A REACTANT?

The earliest and simplest possible mechanistic model for stimulation of intracellular cyclase activity by extracellular agonist postulates a single transmembrane protein containing a receptor site on the outside of the membrane and a catalytic binding site for ATP on the inside of the membrane. Upon binding of agonist to the receptor site, the conformation of the protein is altered so as to promote the conversion of bound ATP to cyclic AMP. This model does not account for the well-documented "spare receptor" effect in cyclase systems, i.e., the observation that receptor-mediated cyclase activity reaches its maximum value when only a small fraction of the total number of receptors are occupied by agonist (Rodbell et al., 1974; Ross et al., 1977).

The first model to offer a reasonable explanation of the spare receptor effect was the nonstoichiometric floating-receptor model, proposed in somewhat different forms at about the same time by DeHaën (1976) and Jacobs and Cuatrecasas (1976). According to this model, cyclase activity resides on a catalytic unit C, which is separate from receptor R in the absence of agonist L and inactive when separated from R. C is presumed to be present in a stoichiometric ratio to R of significantly less than one. It is postulated that conformational changes in R associated with the binding of L increase the affinity of R for C, which becomes catalytically active upon association with receptor.

An alternate mechanism was proposed by Tolkovsky and Levitzki (1978) to account for certain kinetic data which they felt were not well accommodated by the floating-receptor model. According to their collision-coupling model, the binding of agonist to receptor serves to activate the receptor as a catalyst for the conversion of inactive to active C units. The notion that cyclase, once activated, no longer requires the presence of either agonist or receptor has since been incorporated into models other than the original collision-coupling model (Ross *et al.*, 1983; Lefkowitz *et al.*, 1983; Aurbach, 1982).

The question of whether or not the receptor forms part of an active cyclase complex would seem at first glance to be amenable to an unambiguous resolution. In fact, it has not been resolved, and in the remainder of this section we shall demonstrate that it cannot be resolved without additional data.

In order to distinguish more clearly between the concepts of the receptor as reactant and the receptor as catalyst, we shall employ two archetypical models for the stimulation of cyclase by agonist. The models are highly simplified and identical except for the postulated functional role of the receptor.

Model I: Receptor as reactant (Boeynaems and Dumont, 1975)

$$L + R \underset{k_{12}}{\overset{k_{11}}{\rightleftharpoons}} LR$$

$$LR + C \underset{k_{22}}{\overset{k_{21}}{\rightleftharpoons}} LRC^*$$

Model II: Receptor as catalyst (Tolkovsky and Levitzki, 1978)

$$L + R \underset{k_{12}}{\overset{k_{11}}{\rightleftharpoons}} LR$$

$$LR + C \underset{k_{22}}{\overset{k_{21}}{\rightleftharpoons}} LRC$$

$$LRC \overset{k_3}{\rightarrow} LR + C^*$$

$$C^* \overset{k_4}{\rightarrow} C$$

We shall attempt to distinguish between the two models on the basis of three different types of experimental data: (1) steady-state agonist binding and response isotherms, (2) kinetics of appearance of enhanced cyclase activity following addition of agonist, and (3) retention of enhanced cyclase activity following extensive washing of membranes to remove agonist.

## A. Steady-State Binding and Response Isotherms

Tables I and II summarize the various types of binding and response isotherms which may be exhibited by models I and II, respectively. Comparison of the two

**TABLE I**

**Steady-State Binding and Response Isotherms Generated by Model I[a]**

| Condition | Binding isotherm | Response isotherm | Comment |
|---|---|---|---|
| $C_{TOT}/R_{TOT} > 0.95$ | Reference type: $[L]_B^{50} \sim 1/2K_1K_2$ | Reference type: $[L]_R^{50} \sim [L]_B^{50}$ | |
| $C_{TOT}/R_{TOT} < 0.05$ | Reference type: $[L]_B^{50} \sim 1/K_1$ | Reference type: $[L]_R^{50} \sim 1/K_2K_2$ | "Spare receptors": $[L]_R^{50} < [L]_B^{50}$ |
| $0.05 < C_{TOT}/R_{TOT} < 0.95$ | Two-site-class type: High-affinity sites $[L]_B^{50} \sim 1/K_1K_2(1 + C_{TOT}/R_{TOT})$ Low-affinity sites $[L]_B^{50} \sim 1/K_1$ Fraction of high-affinity sites $\sim C_{TOT}/R_{TOT}$ | Reference type: $[L]_R^{50} \sim [L]_B^{50}$ of high-affinity sites | "Spare receptors": response associated with binding to high-affinity sites |

[a] $K_1 \equiv k_{11}/k_{12}$; $K_2 \equiv k_{21}/k_{22} > 1$.

**TABLE II**

**Steady-State Binding and Response Isotherms Generated by Model II**[a]

| Condition | Binding isotherm | Response isotherm | Comment |
|---|---|---|---|
| $J > 0.95$ | Reference type: $[L]_B^{50} \sim 1/[K_1K_2(1+K_3)(1+C_{TOT}/R_{TOT})]$ | Reference type: $[L]_R^{50} \sim [L]_B^{50}$ | |
| $J < 0.05$ | Reference type: $[L]_B^{50} \sim 1/K_1$ | Reference type: $[L]_R^{50} \sim 1/[K_1K_2(1+K_3)(1+C_{TOT}/R_{TOT})]$ | "Spare receptors": $[L]_R^{50} < [L]_B^{50}$ |
| $0.05 < J < 0.95$ | Two-site-class type: <br> High-affinity sites <br> $[L]_B^{50} \sim 1/[K_1K_2(1+K_3)(1+C_{TOT}/R_{TOT})]$ <br> Low-affinity sites <br> $[L]_B^{50} \sim 1/K_1$ <br> Fraction of high-affinity sites $\sim J$ | Reference type: <br> $[L]_R^{50} \sim [L]_B^{50}$ of high-affinity sites | "Spare receptors"; response associated with binding to high-affinity sites |

[a] $K_1 \equiv k_{11}/k_{12}$; $K_2 \equiv k_{21}/(k_{22}+k_3) > 1$; $K_3 \equiv k_3/k_4$; $J \equiv (C_{TOT}/R_{TOT})[k_4/(k_3+k_4)]$.

tables reveals that both models can exhibit the same qualitative types of behavior and that the difference between them resides in the conditions under which the various types of behavior may be exhibited. In particular, model I may exhibit a spare-receptor effect only when $C_{TOT}/R_{TOT}$ is significantly less than unity. If $k_3 \gg k_4$, model II may exhibit a spare-receptor effect even if $C_{TOT}/R_{TOT}$ is equal to or greater than unity. Thus, the two models may be distinguished on the basis of steady-state binding and activity isotherms if and only if independent measurements can establish that the amount of C is stoichiometrically comparable to or greater than the amount of R in the system.

## B. Kinetics of Activation of C

Tolkovsky and Levitzki (1978) measured the rate of activation of cyclase in turkey erythrocyte membranes following addition of the β-adrenergic agonist epinephrine together with the nonhydrolyzable guanine nucleotide analogue guanyl-5'-yl imidodiphosphate [Gpp(NH)p]. They found that the appearance of active cyclase was first order in C (to within the precision of their experiments) and that the dependence of the apparent first-order rate constant upon agonist concentration was describable by a Langmuir isotherm [i.e., proportional to the right hand side of Eq. (1)]. They argued that such behavior was consistent with the collision-coupling model but inconsistent with the floating-receptor model and similar models. In fact, the observed behavior is fully consistent with model I provided that $C_{TOT}/R_{TOT} < 0.2$. Under such conditions, the concentration of free-LR is diminished by not more than 20% as the active complex LRC* is formed, and the reaction appears to be first order in C to within experimental precision. Thus, the kinetic experiments of Tolkovsky and Levitzki (1978) can distinguish between models I and II if and only if it can be established that the amount of C is stoichiometrically comparable to the amount of R.

## C. Persistence of Activation

It has been observed that adenylate cyclase which has been activated by agonist in the presence of Gpp(NH)p retains its activity after extensive washing to remove agonist (Schramm and Rodbell, 1975; Sevilla et al., 1976). It has been argued that this finding indicates that the agonist–receptor complex functions only as a catalyst for the production of active cyclase and not as a component of the active cyclase (Sevilla et al., 1976). This argument fails to take into account the following two possibilities inherent in model I and similar models: (1) If $k_{22}$ is much smaller than $k_{21}$—a possibility consistent with the large extent of receptor spareness observed in the presence of Gpp(NH)p (Ross et al., 1977; Tolkovsky and Levitzki, 1978)—then the rate of dissociation and inactivation of LRC* could be extremely slow on the time scale of the experiment, and hence the formation of LRC* would appear to be quasi-irreversible (Minton, 1982). (2) If $C_{TOT}$ is much less than $R_{TOT}$, then most of the agonist bound to the membrane is bound to R only and is not a part of the active complex LRC*. According to model I, L is more tightly bound in LRC* than it is in LR. Thus, most of the agonist bound to the membrane could be washed off without substantially affecting the amount of preformed LRC*. Quantitative removal of all agonist and receptor has not been demonstrated for any preparation exhibiting quasi-irreversible activation by agonist.

In summary, it appears that a clear-cut determination of the stoichiometric ratio of total C to total R is necessary before the question of whether the receptor functions as a catalyst for the activation of cyclase or as a component of active cyclase can be resolved.

## IV. WHAT ARE THE FUNCTIONAL RELATIONSHIPS BETWEEN R, N, AND C, AND HOW ARE THEY AFFECTED BY AGONIST AND GUANINE NUCLEOTIDE?

In order to simplify the analysis of this complex system, we shall not explicitly take into account the role of hydrolysis of guanine nucleotide (denoted by G), except as a possible mechanism for promoting the dissociation of G from its binding site. In any event, most of the data to be considered were obtained in systems utilizing the nonhydrolyzable analogue Gpp(NH)p in lieu of GTP. Even with this restriction, a system exhibiting receptor-mediated stimulation of adenylate cyclase activity might conceivably contain as many as 16 distinct macromolecular species, to wit:

Possible unitary species: R, LR, N, NG, C
Possible binary species: RN, LRN, RNG, RC, LRC, NC, NGC
Possible ternary species: RNC, LRNC, RNGC, LRNGC

In principle, a complete description of the time-dependent and steady-state behavior of such a system requires that the rate constants for each constituent reaction in both directions be specified for a given set of experimental conditions (temperature, pH, ionic strength, etc.). It should be clear from the above that any attempt to formulate a quantitative mechanistic model which correctly describes the major time-dependent and steady-state properties of the system, while remaining simple enough to be comprehensible, is necessarily based upon two extremely optimistic assumptions: (1) A large fraction of the 16 possible macromolecular species are either nonexistent or otherwise nonessential to the process of receptor-mediated stimulation of cyclase and may thus be neglected; and (2) The time-dependent behavior is dominated by a small number of rate-limiting processes; at any given instant most of the species present may be assumed to be in equilibrium (or quasiequilibrium) with their respective reaction partners.

Our initial objective in this section is to classify the various qualitatively different types of behavior which may be exhibited by a receptor-mediated cyclase system. As previously stated, we assume that the different types of behavior exhibited by different systems are due to quantitative rather than qualitative differences in the underlying mechanism. If this assumption is true, then any mechanistic model which is minimally acceptable ought to be able to simulate any of the observed types of behavior through variation in the values of appropriate rate or equilibrium constants. Conversely, if a model is qualitatively incapable of simulating one or more of the observed types of behavior, it is either incomplete or based upon one or more incorrect underlying assumptions, or both. If two or more models are capable of accommodating all of the data, then any preference between them will rest upon subjective considerations, such as Occam's Razor (preference for the simplest acceptable model).

Our second objective is to investigate whether specific models based upon general mechanistic hypotheses are capable of accounting for the accumulated data. If a given model representative of a general class of models is incapable of accounting for a particular observation, we shall attempt to determine whether the model may be modified to accommodate the observation without altering the hypothesis characterizing the general class of models.

The various types of behavior which we believe should be accommodated by a minimally acceptable general model for receptor-mediated stimulation of adenylate cyclase are summarized in Table III. For convenience in referring to these observations, we classify each table entry according to the type of property (indexed by letter) and the system studied (indexed by number). For example, the effect of increasing guanine nucleotide concentration upon affinity for agonist (property C) in turkey erythrocyte membranes (system 3) will be referred to as observation C3.

Three reaction schemes will be introduced as models embodying alternate mechanistic hypotheses. The steady-state and kinetic properties of each of these models (as well as a number of variations of each model) have been extensively explored via numerical simulation using a broad spectrum of parameter values and combinations thereof. Our approach is to be contrasted with that of Tolkovsky and Levitzki (1978, 1981; Tolkovsky et al., 1982) who have characterized various kinetic models on the basis of analytic solutions obtained in certain limiting cases. We feel that by restricting their analysis to these limiting cases, Tolkovsky and Levitzki have failed to appreciate the full range of kinetic and steady-state behaviors which a given model is capable of exhibiting.

## A. N Is Associated with R throughout the Process of Receptor-Mediated Stimulation of Cyclase

A model based upon this hypothesis will be referred to as an RN model. Prior to the resolution of the N subunit, early studies of the role of guanine nucleotide in receptor-mediated cyclase stimulation led to the hypothesis that the guanine

# TABLE III

**Summary of Experimental Observations on Different Systems for Receptor-Mediated Stimulation of Adenylate Cyclase**

| Property \ System | (1) β-Adrenergic receptors: S-49 lymphoma cell membranes | (2) β-Adrenergic receptors: frog erythrocyte membranes | (3) β-Adrenergic receptors: turkey erythrocyte membranes | (4) Glucagon receptors: rat liver membranes |
|---|---|---|---|---|
| A. Agonist binding isotherm in the absence of G | a. Multisite-class[a] | a. Reference[b] <br> b. Multisite-class[b] | a. Reference[c] <br> b. Multisite-class[b] | Reference[g] |
| B. Agonist binding isotherm in the presence of high concentrations of G | Reference[a] | Reference[b] | Reference[c] | Two-site[g] |
| C. Effect of increasing G upon overall affinity for agonist | Decrease[a] | Decrease[b] | a. None[c] <br> b. Decrease (membrane pretreated with agonist + GMP)[d] | Decrease coupled with appearance of small amount of high-affinity site[g] |
| D. Kinetics of appearance of cyclase | Apparent first order[a] | | Apparent first order[c] | |

| | | |
|---|---|---|
| activity following addition of GppNHp | | |
| E. Effect of increasing agonist on cyclase activation | Increases apparent first-order rate constant for appearance of activity; little or no effect upon final extent of activation in presence of GppNHp[a] | Increases apparent first-order rate constant for appearance of activity[c]; increases final extent of activation in the presence of GppNHp[e] |
| F. Effect of increasing GppNHp on cyclase activation | | Increases final extent of activation; little or no effect upon apparent first-order rate constant for appearance of activity[f] |

[a] Ross et al. (1977).
[b] De Lean et al. (1980).
[c] Tolkovsky and Levitzki (1978).
[d] Lad et al. (1980).
[e] Sevilla et al. (1976).
[f] Tolkovsky et al. (1982).
[g] Lin et al. (1977).

nucleotide binding site was on the receptor. Although the early model of Ross *et al.* (1977) did not formally take into account the R, N, and C components, it may in retrospect be seen to be an implicit RN model, functionally similar to the explicit RN model introduced later by Rodbell (1980). The basic features of these models are included in reaction scheme (3)

$$L + \underline{RN} \overset{K_L}{\rightleftharpoons} \underline{LRN}$$

$$\underline{RN} + G \overset{K_G}{\rightleftharpoons} \underline{RNG}$$

$$L + \underline{RNG} \overset{\alpha K_L}{\rightleftharpoons} \underline{LRNG} \qquad\qquad (3)$$

$$\underline{RNG} + C \underset{k_{12}}{\overset{k_{11}}{\rightleftharpoons}} \underline{RNGC}^*$$

$$\underline{LRNG} + C \underset{k_{22}}{\overset{k_{21}}{\rightleftharpoons}} \underline{LRNGC}^*$$

The types of behavior which may be exhibited by this model are listed below, together with model conditions which are sufficient (but may not be necessary) to produce the specified behavior: (1) reference type agonist binding curve in absence of added G: intrinsic property of model; (2) reference type agonist binding isotherm in presence of large amounts of G: $C_{TOT}/R_{TOT} < 0.2$; (3) two-site-class type agonist binding isotherm in presence of large amounts of G: $0.1 < C_{TOT}/R_{TOT} < 0.9$; $k_{21}/k_{22} \gg k_{11}/k_{12}$; (4) addition of G decreases overall affinity for agonist; $\alpha \ll 1$; (5) addition of G has no effect on affinity for agonist: $\alpha \cong 1$; (6) apparent first-order kinetics for appearance of cyclase activity following addition of Gpp(NH)p: $C_{TOT}/R_{TOT} < 0.2$; (7) increasing agonist concentration increases rate constant for appearance of cyclase activity while not affecting final extent: $k_{11} < k_{21}$; $k_{12} < k_{22}$; $k_{11}/k_{12} \cong k_{21}/k_{22}$; (8) increasing agonist concentration increases the rate constant for the appearance of cyclase activity and final extent of activation: $k_{11} < k_{21}$; $k_{12} < k_{22}$; $k_{11}/k_{12} < k_{21}/k_{22}$.

As presented earlier, scheme (3) cannot account for the following types of observed behavior: (1) two-site-class type agonist binding isotherm in the absence of G (observations A1, A2b, A3b); and (2) increasing concentrations of G do not significantly affect the apparent first-order rate constant for the appearance of cyclase activity (observation F3).

We have found that both of these observations may be accounted for by a simple extension of scheme (3), namely the addition of the following reaction

$$\underline{LRN} + C \underset{k_{32}}{\overset{k_{31}}{\rightleftharpoons}} \underline{LRNC}$$

in which $\underline{LRNC}$ is a catalytically inactive complex. If $0.1 < C_{TOT}/R_{TOT} < 0.9$ and $k_{31} \ll k_{32}$, then the agonist binding isotherm will be a two-site-class type isotherm. If $C_{TOT}/R_{TOT} < 0.2$, $k_{31} \cong k_{21}$, and $k_{32} \cong k_{22}$, then addition of G (in the presence of a saturating amount of agonist) will affect the steady-state concentration of $\underline{LRNGC}^*$ without affecting the apparent first-order rate constant for the formation of $\underline{LRNGC}^*$. Thus, the extended scheme (3) presented above can account for all of the various types of behavior summarized in Table III.

## B. N Is Associated with C throughout the Process of Receptor-Mediated Stimulation of Cyclase

Models based upon this hypothesis will be generically referred to as NC models. The first such model was an extension of the collision-coupling model, proposed by Tolkovsky *et al.* (1982), in order to account for the observed effect of increasing Gpp(NH)p upon the kinetics of the appearance of cyclase activity (observation F3). The particular reaction scheme proposed by Tolkovsky *et al.* (1982) was based upon the assumption that Gpp(NH)p binds tightly and irreversibly to NC and that only NC which binds Gpp(NH)p may be activated through transient interaction with the hormone–receptor (HR) complex. According to this hypothesis, the level of steady-state activity should increase in direct proportion to the total amount of Gpp(NH)p added until all of the Gpp(NH)p binding sites on the NC complex are saturated. Gpp(NH)p added in excess of the saturating amount should not affect the level of steady-state activity. If, however, Gpp(NH)p binds ''reversibly'' to the NC complex, then the steady-state activity should be proportional to the amount of NC–Gpp(NH)p complex existing at equilibrium, calculated according to Eq. (1). In Fig. 2, we have plotted the dependence of steady-state cyclase activity upon GPP(NH)p, as reported by Tolkovsky *et al.* (1982), together with best-fit functions based upon the above two hypotheses. Since the calculation based upon the reversible binding hypothesis fits the data as well as (actually slightly better than) the calculation based upon the irreversible binding hypothesis, we believe that the possible dissociation of Gpp(NH)p from its binding site(s) over the time-course of an experiment should not be ruled out *a priori*. The NC version of the collision-coupling model, modified to take this possibility into account, is presented below as reaction scheme (4).

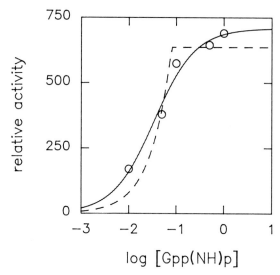

**Fig. 2.** Dependence of steady-state cyclase activity upon the concentration of added Gpp(NH)p. Data points are taken from Fig. 1 of Tolkovsky *et al.* (1982) in the long-time limit. The solid and dashed curves are best-fit functions calculated according to the respective hypotheses of reversible and irreversible Gpp(NH)p binding.

$$L + R \overset{k_{11}}{\underset{k_{12}}{\rightleftharpoons}} LR$$

$$G + \underline{NC} \overset{k_{21}}{\underset{k_{22}}{\rightleftharpoons}} \underline{GNC}$$

$$LR + \underline{GNC} \overset{k_{31}}{\underset{k_{32}}{\rightleftharpoons}} LR\underline{GNC}$$

(4)

$$LR\underline{GNC} \overset{k_4}{\rightarrow} LR + \underline{GNC}^*$$

$$\underline{GNC}^* \overset{k_5}{\rightarrow} G + \underline{NC}$$

The types of behavior which may be exhibited by reaction scheme (4) are listed, together with model conditions sufficient to produce the specified behavior: (1) reference type agonist binding isotherm in the absence of added G: intrinsic property of model; (2) reference type agonist binding isotherm in the presence of large amounts of G: $C_{TOT} \ll R_{TOT}$; (3) two-site-class type agonist binding isotherm in the presence of large amounts of G: $0.1 < C_{TOT}/R_{TOT} < 0.9$; $k_{22} \gg$

$k_4$; (4) no effect of G on overall affinity for agonist: $C_{TOT} \ll R_{TOT}$; (5) appearance of cyclase activity is pseudo-first order following addition of Gpp(NH)p: the reaction between LR and GNC is rate limiting; LR $\gg$ GNC; (6) addition of G does not significantly affect the apparent first-order rate constant for appearance of cyclase activity: $k_{21}[G] > k_{22}$ and $k_{31}$; $C_{TOT} \ll R_{TOT}$; (7) addition of agonist increases the rate of formation of active cyclase but not the final extent of activation: $k_5 \cong 0$ (all C is eventually converted to GC*; the decay of GC* does not occur on the time scale of the experiment); and (8) addition of agonist increases both the rate of formation of active cyclase and the final extent of activation: $k_5 > 0$ (the final steady-state level of GC* reflects a balance between the rates of production and decay).

Reaction scheme (4) cannot account for the following observations: (1) a two-site-class type agonist binding isotherm in the absence of G (observations A1, A2b, A3b) and (2) increasing the concentration of G reduces the overall affinity for agonist (observations C1, C2, C3b, C4).

The first of these two observations may be accommodated within the context of an NC model by permitting the HR and NC moieties to form a catalytically inactive HRNC complex in the absence of G. The second observation is more difficult to explain. It is not apparent how scheme (4) could be modified to account for the depression of affinity for agonist observed in several systems upon addition of G, short of introducing a second binding site for G on the R unit which modulates the affinity of R for L. However, a model so extended would no longer be an NC model as we have defined it (i.e., the binding site for G residing on the catalytic component as opposed to the receptor component.

## C. Receptor-Mediated Activation of Cyclase Requires the Transfer of N from R to C

Models based upon this hypothesis, which we shall generically term shuttle models, have been proposed by Lefkowitz and co-workers (Lefkowitz *et al.*, 1983) and Swillens and Dumont (1980). These models are far too complex to be quantitatively analyzed in their entirety. The simplified reaction scheme (5)

$$L + R \underset{k_{12}}{\overset{k_{11}}{\rightleftharpoons}} LR$$

$$LR + N \underset{k_{22}}{\overset{k_{21}}{\rightleftharpoons}} LRN$$

$$LRN + G \overset{k_3}{\rightleftharpoons} LR + NG \tag{5}$$

$$NG + C \xrightarrow{k_4} NGC^*$$

$$NGC^* \xrightarrow{k_5} N + C$$

presented above appears to retain the basic features common to shuttle models The types of behavior which may be exhibited by reaction scheme (5) are listed below, together with model conditions sufficient to produce the specified behavior: (1) reference type agonist binding isotherm in the absence of added G: $N_{TOT}/R_{TOT} < 0.05$ or $> 0.95$; (2) two-site-class type agonist binding isotherm in the absence of added G: $0.05 < N_{TOT}/R_{TOT} < 0.95$; $k_{21} \ll k_{22}$; (3) reference type agonist binding isotherm in the presence of large amounts of G: an intrinsic property of the model; (4) increasing G decreases overall affinity for agonist: $k_{21} \gg k_{22}$; (5) increasing G has little or no effect on affinity for agonist: $k_{21} \leq k_{22}$; (6) appearance of cyclase activity is apparent first order following the addition of G: the reaction of NG with C is rate limiting; $C_{TOT} \ll NG$; and (7) increasing agonist concentration increases both the rate constant for the appearance of activation and the extent of activation: $k_5 > 0$ (the steady-state level of NGC* reflects a balance of the rates of production and decay).

Reaction scheme (5) cannot account for the following observations: (1) two-site-class type agonist binds isotherm in the presence of excess G (observation B4), and (2) increasing concentrations of G increase the steady-state level of activation but not the apparent first-order rate constant for the appearance of activation (observation E3).

The first of these observations may be accommodated within the context of a shuttle model by invoking intrinsic heterogeneity of agonist binding sites (an inelegant but permissible solution). The second of the two observations appears, however, to be fundamentally inconsistent with the shuttle hypothesis, as was pointed out by Tolkovsky *et al.* (1982). If cyclase is activated via a reaction between C and an excess of NG, then any change (such as an increase or decrease in the concentration of G) which alters the concentration of NG must concomitantly alter the apparent first-order rate constant for the formation of NGC*. We therefore conclude that shuttle models in general, and not just scheme (5), are unable to provide a unified interpretation of all of the data presented in Table III.


## V. SUMMARY

In this chapter we have specifically focused on the role of agonist and receptor in receptor-mediated activation of cyclase. While the presence of guanine nucleotide (or an analogue thereof) is recognized as a *sine qua non* for the activation of cyclase, its role has not been analyzed in detail.

The major conclusions drawn from consideration of the data reviewed in this chapter are twofold: (1) In the absence of independent information indicating that the stoichiometric ratio of total C to total R in a particular membrane type exceeds about 0.2, experimental observations cited as evidence for a catalytic role of the agonist–receptor complex in the stimulation of cyclase activity are also compatible with the hypothesis that the receptor forms part of the active cyclase complex. (2) According to the simplest model capable of accounting for all of the data summarized in Tables II and III, the N component exists as a subunit of the receptor and remains associated with R throughout the process of receptor-mediated activation of adenylate cyclase.

## ACKNOWLEDGMENTS

I thank Drs. L. D. Kohn, M. Rodbell, and K. Seamon of the National Institutes of Health for helpful discussions and for critically reviewing the initial draft of this work.

## REFERENCES

Aurbach, G. D. (1982). Polypeptide and amine hormone regulation of adenylate cyclase. *Annu. Rev. Physiol.* **44,** 653–666.

Boeynaems, J. M., and Dumont, J. E. (1975). Quantitative analysis of the binding of ligands to their receptors. *J. Cyclic Nucleotide Res.* **1,** 123–142.

DeHaën, C. (1976). The non-stoichiometric floating receptor model for hormone-sensitive adenylyl cyclase. *J. Theor. Biol.* **58,** 383–400.

De Lean, A., Stadel, J. M., and Lefkowitz, R. J. (1980). A ternary complex model explains the agonist-specific binding properties of the adenylate cyclase-coupled β-adrenergic receptor. *J. Biol. Chem.* **255,** 7108–7117.

Jacobs, S., and Cuatrecasas, P. (1976). The mobile receptor hypothesis and 'cooperativity' of hormone binding. *Biochim. Biophys. Acta* **433,** 482–495.

Lad, P. M., Nielsen, T. B., Preston, M. S., and Rodbell, M. (1980). The role of the guanine nucleotide exchange reaction in the regulation of the β-adrenergic receptor and in the actions of catecholamines and cholera toxin on adenylate cyclase in turkey erythrocyte membranes. *J. Biol. Chem.* **255,** 988–995.

Lefkowitz, R. J., Stadel, J. M., and Caron, M. G. (1983). Adenylate cyclase-coupled β-adrenergic receptors. *Annu. Rev. Biochem.* **52,** 159–186.

Levitzki, A. (1981). The β-adrenergic receptor and its mode of coupling to adenylate cyclase. *CRC Crit. Rev. Biochem.* **10,** 81–112.

Lin, M. C., Nicosia, S., Lad, P. M., and Rodbell, M. (1977). Effects of GTP on binding of [³H]glucagon to receptors in rat hepatic plasma membranes. *J. Biol. Chem.* **252,** 2790–2792.

Minton, A. P. (1982). Steady-state relations between hormone binding and elicited response: quantitative mechanistic models. *In* "Hormone Receptors" (L. D. Kohn, ed.), pp. 43–65. Wiley, New York.

Rodbell, M. (1980). The role of hormone receptors and GTP-regulatory proteins in membrane transduction. *Nature (London)* **284,** 17–22.

Rodbell, M., Lin, M. C., and Salomon, Y. (1974). Evidence for interdependent action of glucagon and nucleotides on the hepatic adenylate cyclase system. *J. Biol. Chem.* **249**, 59–65.

Ross, E. M., Maguire, M. E., Sturgill, T. W., Biltonen, R. L., and Gilman, A. G. (1977). Relationship between the β-adrenergic receptor and adenylate cyclase. *J. Biol. Chem.* **252**, 5761–5775.

Ross, E. M., Pedersen, S. E., and Florio, V. A. (1983). Hormone-sensitive adenylate cyclase: Identity, function and regulation of the protein components. *Curr. Top. Membr. Transp.* **18**, 109–142.

Schramm, M. E., and Rodbell, M. (1975). A persistent active state of the adenylate cyclase system produced by the combined actions of isoproterenol and guanylyl imidodiphosphate in frog erythrocyte membranes. *J. Biol. Chem.* **250**, 2232–2237.

Sevilla, N., Steer, M. L., and Levitzki, A. (1976). Synergistic activation of adenylate cyclase by guanylyl imidodiphosphate and epinephrine. *Biochemistry* 15, 3493–3499.

Swillens, S., and Dumont, J. E. (1980). A unifying model of current concepts and data on adenylate cyclase activation by β-adrenergic agonists. *Life Sci.* **27**, 1013–1028.

Tolkovsky, A. M., and Levitzki, A. (1978). Model of coupling between the β-adrenergic receptor and adenylate cyclase in turkey erythrocytes. *Biochemistry* **17**, 3795–3810.

Tolkovsky, A. M., and Levitzki, A. (1981). Theories and predictions of models describing sequential interactions between the receptor, the GTP regulatory unit, and the catalytic unit of hormone-dependent adenylate cyclases. *J. Cyclic Nucleotide Res.* **7**, 139–150.

Tolkovsky, A. M., Braun, S., and Levitzki, A. (1982). Kinetics of interaction between β-receptors, GTP protein, and the catalytic unit of turkey erythrocyte adenylate cyclase. *Proc. Natl. Acad. Sci. U.S.A.* **79**, 213–217.

# 3

# The Role of Microaggregation
# in Hormone–Receptor–Effector
# Interactions

**J. J. BLUM**
Department of Physiology
Duke University Medical Center
Durham, North Carolina

## I. INTRODUCTION

It has been appreciated for several decades that the binding of a hormone, neurotransmitter, or other ligand that is active at very low concentrations to

**57**

Copyright © 1985 by Academic Press, Inc.
All rights of reproduction in any form reserved.
ISBN 0-12-185202-4

highly specific receptors is an interaction of fundamental importance for the
regulation of cell function. It was initially thought that the interaction of ligand
with receptor was, in and of itself, the event that led to the production of the
signal (second messenger) that influenced cell function. This viewpoint seemed
justified by the ability to describe hormone response curves in terms of the
binding of the hormone to the receptor, as evidenced by linear Scatchard plots
which yield values for the affinity constant that are close to the value obtained
from dose–response curves. More recently, however, it has become evident that
formation of a ligand–receptor complex is only the first step in a complex series
of events that lead to production of one or more second messengers. The intermediate steps may include interaction with coupling factors, such as the activating $(N_s)$ and inhibitory $(N_i)$ coupling proteins that mediate the interaction between various receptors and adenylate cyclase or dimerization of the receptor–
hormone complex. Other types of intermediate steps occur in the case of steroid
hormones with nuclear receptors, but this chapter is concerned only with receptors localized in the plasma membrane. The formal analytical methods that will
be used, however, may also be applicable for analysis of the interaction of
steroid hormones with their receptors. Section II examines the interaction of go-
nadatropin releasing hormone (GnRH) with its receptors in the membranes of
pituitary cells capable of releasing luteinizing hormones (LH), using evidence
obtained from studies with antibodies to introduce the concept that receptor
dimerization may be an obligatory step between the binding of GnRH to its
receptor and the entry of $Ca^{2+}$ into the pituitary cell. A discussion of the
mathematical procedure used to analyze the model is, however, postponed to
Section IV. Section III reviews the evidence that receptor dimerization is involved in the ability of antireceptor antibodies to mimic the biological actions of
a number of hormones including some, such as insulin, for which the second
messenger has not yet been identified. Section IV then analyzes the adenylate
cyclase system in which a regulatory protein mediates the interactions between
the receptor (e.g., a β-receptor) and the effector (adenylate cyclase). Because
this is a fairly complex kinetic scheme, the mathematical procedures for analysis
of ligand–receptor–effector models are developed in some detail in this section.
Section V then briefly reexamines the evidence concerning receptor–effector
interactions and the possible role of membrane properties on these interactions.

## II. THE GnRH RECEPTOR

When GnRH is added to pituitary cells in culture, luteinizing hormone is
released via a calcium-dependent mechanism (Conn, 1984; Conn *et al.*, 1984),
according to the dose–response curve shown in Fig. 1. The dose–response curve
is unusually broad, spanning almost four log units of [GnRH]. Scatchard analy-

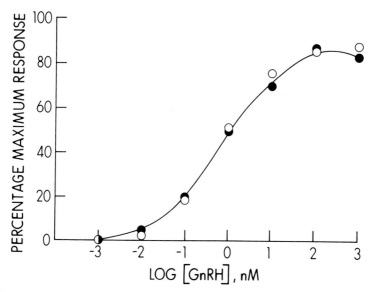

**Fig. 1.**  Dose–response curve for LH release by pituitary glands in culture after GnRH addition ●——●, unpublished data of Dr. P. M. Conn; ○——○, predicted response computed acccording to Eq. (15)–(20), as discussed in the text.

ses of binding data for GnRH and for GnRH-A, a "super agonist" analogue of GnRH (Clayton and Catt, 1981), indicate a single class of saturable high-affinity receptors. The $ED_{50}$ values for GnRH and GnRH-A-stimulation of LH release by cultured pituitary cells are 0.5 and 0.01 n$M$, respectively, whereas the respective $K_d$ values are 50 and 0.25 n$M$ (Clayton and Catt, 1981), suggesting that occupation of only a fraction of the available receptor is sufficient for a maximal release of LH.

Other analogues of GnRH have also been prepared. One of these, [D-Glu$^1$-D-Phe$^2$-D-Trp$^3$-D-Lys$^6$]GnRH, is a competitive antagonist that inhibits GnRH-stimulated release (Conn et al., 1982a). This antagonist (GnRH-Ant) can be dimerized with glycol bissuccinimidyl succinate (EGS); the dimer is also an antagonist with properties closely similar to those of the monomer. Incubation of the dimer with a cross-reacting antibody yields a "conjugate" with one GnRH-Ant dimer attached to each arm of the antibody. One monomeric unit of each EGS–GnRH-Ant dimer is tightly bound to the antigen binding site and inaccessible to the GnRH receptor, but the two "outward" facing monomers of each GnRH-Ant molecule are separated by about 150 Å and are available to the receptor.

In contrast to the monomer or dimer, the divalent antibody–dimer conjugate is a potent agonist. The monovalent reduced-pepsin fragment of the conjugate, however, is not an agonist (Conn et al., 1982a). This suggests that the formation of a receptor dimer is sufficient to initiate signal transduction. It is further found that when the conjugate level is raised above an optimally effective level, LH release declines, with practically no LH release at sufficiently high concentrations of conjugate (Fig. 2). To explain these data, it has been postulated (Blum and Conn, 1982) that the divalent conjugate interacts initially with one receptor and then with another receptor, the second interaction being favored by the increase in effective concentration of the second ligand of the conjugate resulting from the binding of the first ligand of the conjugate. Dembo and Goldstein (1980) have proposed a similar model to account for the IgE-mediated release of histamine from human basophils. When the two receptors have been positioned an appropriate distance apart, they may then interact with an effector, in this case considered to be a $Ca^{2+}$-conducting channel (see Fig. 3).

Formally, the model can be written as follows

$$A + R \rightleftharpoons A \cdot R \qquad\qquad K_1 = [A \cdot R]/[A][R] \qquad\qquad (1)$$

$$A \cdot R + R \rightleftharpoons A \cdot R_2 \qquad K_2 = [A \cdot R_2]/[A \cdot R][R] \qquad (2)$$

$$A \cdot R_2 + E \rightleftharpoons A \cdot R_2 \cdot E \qquad K_3 = [A \cdot R_2 \cdot E]/[A \cdot R_2][E] \qquad (3)$$

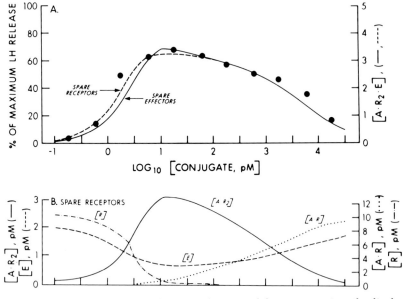

**Fig. 2.** Luteinizing hormone release as a function of the concentration of a divalent antibody conjugate, AB-[(GnRH-Ant₂)-EGS]₂. (A) The solid symbols represent the measured release of LH relative to the amount released by $10^{-7}$ M GnRH. Two theoretical fits to these data are shown. Parameters for the fit labeled "spare receptors" (....) were: $R_T = 10$ pM; $E_T = 2$ pM; $K_1 = 0.18$ pM$^{-1}$; $K_2 = 10$ pM$^{-1}$; $K_3 = 0.6$ pM$^{-1}$. Parameters for the fit labeled "spare effectors" (- - -) were: $R_T = 10$ pM; $E_T = 10$ pM; $K_1 = 0.2$ pM$^{-1}$; $K_2 = 10$ p M$^{-1}$; $K_3 = 0.4$ pM$^{-1}$. (B) The computed values of [E], [R], [A·R], and [A·R₂] for the spare-receptor fit shown in (A). (For further details, see Blum and Conn, 1982.)

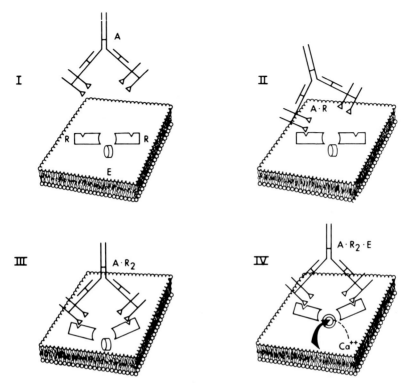

**Fig. 3.** Schematic diagram of possible molecular species in a system containing a divalent antibody conjugate, (A), a receptor (R), and an effector (E), which is represented here as a channel for the passage of $Ca^{2+}$ ions. (From Blum and Conn, 1982.)

for which [A] is the concentration of free-divalent antibody conjugate; [R], the concentration of free-receptor; [A·R], the concentration of antibody–conjugate molecules complexed to a single receptor; [A·R$_2$], the concentration of antibody–conjugate molecules with a receptor bound to each of the two available GnRH-Ant monomers; [E], the concentration of free-effectors (e.g., closed $Ca^{2+}$-conducting channels); and [A·R$_2$·E], the concentration of active effector molecules. The response is assumed to be directly proportional to [A·R$_2$·E], and the system is assumed to be at equilibrium. Conservation of mass requires that

$$A_T = [A] + [A\cdot R] + [A\cdot R_2] + [A\cdot R_2\cdot E] \qquad (4)$$

$$R_T = [R] + [A\cdot R] + 2[A\cdot R_2\cdot E] + 2[A\cdot R_2] \qquad (5)$$

$$E_T = [E] + [A\cdot R_2\cdot E] \qquad (6)$$

for which $A_T$, $R_T$, and $E_T$ are the total concentrations of GnRH-Ant antibody–conjugate, receptor, and effector molecules present, respectively.

The qualitative response characteristics of this model are compatible with the observed responses. At low levels of bivalent agonist, the response will increase as increasing numbers of receptor dimers positioned the correct distance apart to interact with the effector are formed (Fig. 3). However, with a higher concentration of uncomplexed conjugate, there is an increasing probability of the formation of antibody complexes with only one receptor bound, and the response will decline. The solution to this set of equations can be obtained by algebraic manipulation (Blum and Conn, 1982) or, much more easily, by the matrix procedure to be described later. With suitable values for the affinity constants $K_1$, $K_2$, and $K_3$ and receptor ($R_T$) and effector ($E_T$) concentrations of 10 and 2 p$M$, respectively, the experimental data could be fit over the entire range of conjugate concentrations studied. These values for $E_T$ and $R_T$ correspond to 60% spare receptors, a value close to the experimentally determined number of spare receptors in pituitary cells (Naor *et al.*, 1980). It was also possible to fit these data with $R_T = 10$ p$M$ and $E_T = 10$ p$M$, corresponding to 50% spare effectors (Blum and Conn, 1982). This illustrates the importance of receptor-binding studies in limiting the choice of parameter values in constructing models of ligand–receptor–effector interaction.

In addition to the stimulation of LH release, GnRH also causes an initial decrease in receptor number (down-regulation) followed (after about 5 hr of exposure) by an increase in receptor number (up-regulation) (Conn *et al.*, 1983). The bivalent antibody–conjugate that acts as an agonist also evokes this biphasic response, although the antagonist molecules by themselves do not evoke either up- or down-regulation. Thus, dimerization of the GnRH receptors appears to be sufficient to cause both the release of LH and the subsequent down- and up-regulation processes.

Further evidence that dimerization of the GnRH receptor may be a step in the hormonally induced release of LH from pituitary cells comes from studies of the interaction of antibodies against GnRH, or analogues thereof, and GnRH-like peptides with either agonist or antagonist activity. Gregory *et al.* (1982) prepared an analogue of GnRH that was a good antagonist of LH release by GnRH but, when added together with an antibody that reacted with its N-terminal region, behaved as an agonist. They concluded that the COOH-terminal portion of GnRH or of its agonist or antagonist analogues plays the dominant role in binding to the receptor and that the Glu-His-Trp-Ser sequence at the N-terminal end of the peptide is involved in the dimerization of occupied receptors. Conn and co-workers (Conn, 1984) prepared an EGS dimer of D-Lys[6]-GnRH. This dimer stimulated LH release with the same efficacy and only slightly less potency than the monomeric form. At very low concentrations of dimer, in which LH release was only slightly above basal levels, addition of an antibody to D-Lys[6]-GnRH caused considerable enhancement of potency. Addition of the monovalent antibody, however, was ineffective.

More recently, it has been shown (P. M. Conn, personal communication) that the polycation polylysine, but not monomeric lysine or spermidine, can evoke

LH release in pituitary cell cultures. Whereas the antibody studies described earlier indicate that LH release can be evoked by divalent antibodies with bound agonist or antagonist molecules, the polylysine data indicate that merely bringing the GnRH receptors the correct distance apart is sufficient to evoke LH release.

In view of these results, it seems likely that GnRH itself may cause LH release by a mechanism involving receptor dimerization, and Zolman (1983) has interpreted complex kinetic data on the rate of interactions of GnRH receptors to immobilized GnRH as indicating that receptor aggregation is an early consequence of the binding of GnRH. As mentioned earlier, however, the dose–response curve for GnRH-stimulated release of LH is unusually broad (see Fig. 1). Such broadening could occur because of events occurring before or after receptor dimerization. The observations that LH release caused by calcium ionophores (Conn *et al.*, 1979), veratridine (Conn and Rogers, 1980), or polylysines of various chain lengths (P. M. Conn, personal communication) occurs over a relatively narrow ligand concentration range suggests that the broadening of the dose–response curve observed with GnRH is not due to events occurring immediately after the formation of the receptor dimer. It is also unlikely that events occurring well after the initial interaction of GnRH with its receptor, such as up- or down-regulation or desensitization (Clayton and Catt, 1981; Marion *et al.*, 1981; Smith *et al.*, 1983; Keri *et al.*, 1983), are responsible for the broadening of the GnRH dose–response curve, since the shape of the curve is unaltered if the measurements are made at 15-min (the shortest time compatible with reliable measurements) instead of the 3- to 5-hr times usually employed (P. M. Conn, personal communication).

It therefore seemed likely that processes preceding the formation of the putative (GnRH·GnRH receptor)$_2$ dimer might be responsible for the broadening of the dose–response curve. To investigate this, a number of models were tested in which various processes that seemed likely to occur following the binding of GnRH to its receptor were investigated. One such model is represented in the following equations

$$H + R \rightleftharpoons H \cdot R \tag{7}$$
$$H \cdot R + E \rightleftharpoons H \cdot R \cdot E \tag{8}$$
$$H \cdot R + H \cdot R \rightleftharpoons H_2R_2 \tag{9}$$
$$H_2R_2 + E \rightleftharpoons H_2R_2E \tag{10}$$
$$H \cdot R + H \cdot R \cdot E \rightleftharpoons H_2R_2E \tag{11}$$
$$H \cdot R + H_2R_2 \rightleftharpoons H_3R_3 \tag{12}$$
$$H_2R_2 + H_2R_2 \rightleftharpoons H_4R_4 \tag{13}$$
$$H_3R_3 + H \cdot R \rightleftharpoons H_4R_4 \tag{14}$$

The corresponding steady-state equations, along with the appropriate conservation relations, were solved for a wide range of values for the equilibrium constants. With suitable values and the assumption that the response was proportional to [H$_2$·R$_2$·E], it was possible to simulate full- or partial-agonist behavior, with or without a decrease in response at very high-hormone concentrations. However, no combination of $K_a$ values was found which yielded the broad dose–response curve characteristic of GnRH. The possibility that both H·R·E and $H_2$·$R_2$·E were active was also examined, but no combination of such active species was found that could account for dose–response curve broadening.

Avissar *et al.* (1982, 1983) have suggested that muscarinic receptors in rat pituitary cells, which are coupled to $Ca^{2+}$ channels in several systems, may exist as a mixture of forms with differing states of aggregation and of binding affinity. The possibility that the GnRH receptor existed as a mixture of tetrameric ($R_4$) and dimeric ($R_2$) species even in the absence of GnRH was investigated with the aid of the following model

$$H + R_4 \rightleftharpoons H \cdot R_2 + R_2 \qquad K_1 = [R_2][H \cdot R_2]/[H][R_4] \qquad (15)$$
$$H \cdot R_2 + H \rightleftharpoons H \cdot R + H \cdot R \qquad K_2 = [H \cdot R]^2/[H \cdot R_2][H] \qquad (16)$$
$$R_2 + H \rightleftharpoons H \cdot R + R \qquad K_3 = [H \cdot R][R]/[R_2][H] \qquad (17)$$
$$R + H \rightleftharpoons H \cdot R \qquad K_4 = [H \cdot R]/[H][R] \qquad (18)$$
$$H \cdot R + E \rightleftharpoons H \cdot R \cdot E \qquad K_5 = [H \cdot R \cdot E]/[H \cdot R][E] \qquad (19)$$
$$H \cdot R + H \cdot R \cdot E \rightleftharpoons H_2R_2E \qquad K_6 = [H_2 \cdot R_2 \cdot E]/[H \cdot R][H \cdot R \cdot E] \qquad (20)$$

In this, as in the preceding model, it was assumed that the response was proportional to $[H_2 \cdot R_2 \cdot E]$ The curve drawn through the data in Fig. 1 derives from this model, with the following parameter values: $K_1 = 10^{-3}$; $K_2 = 2.8 \times 10^{-4}$; $K_3 = 1$; $K_4 = 10^3$ n$M^{-1}$; $K_5 = 2 \times 10^3$ n$M^{-1}$; $K_6 = 2 \times 10^3$ n$M^{-1}$; $E_o = 3$ p$M$; $R_o = 10$ p$M$. Despite the good fit to the data, this model does *not* describe the behavior of the GnRH receptor system, for a plot of $B/F$ versus $B$ (in which $B$ is the sum of all bound forms of the receptor and $F$ is the concentration of the free-receptor) does not match the experimentally observed Scatchard plot (Clayton and Catt, 1981). This again emphasizes the value of a Scatchard plot as a constraint upon models which might otherwise seem satisfactory.

At present, therefore, there is no satisfactory explanation for the shape of the dose–response curve for GnRH itself. From the evidence presented, though, it seems likely that the process(es) responsible for the breadth of the dose–response curve lie between the formation of the hormone–receptor complex, H·R, and the entrance of $Ca^{2+}$ into the cytosol.

## III. EVIDENCE THAT MICROAGGREGATION PLAYS A KEY ROLE IN OTHER ANTIBODY–RECEPTOR–EFFECTOR SYSTEMS

### A. Insulin

Autoantibodies to the insulin receptor occur in certain forms of diabetes and can also be prepared by immunizing animals with purified insulin receptor preparations (Jacobs *et al.*, 1978; Kahn *et al.*, 1978; Le Marchand-Brustel *et al.*, 1978; Schechter *et al.*, 1979a, 1982; Baldwin *et al.*, 1980; King and Cuatrecasas, 1981). Each of these antibodies can mimic one or more of the biological effects of insulin. Some interfere with the binding of insulin to the native receptors *in situ* (Kahn *et al.*, 1978; Baldwin *et al.*, 1980; Schechter *et al.*, 1982) while others (Jacobs *et al.*, 1978) do not compete with $^{125}$I-insulin for binding to the insulin receptor but do precipitate solubilized receptors labeled with $^{125}$I-insulin, indicating an interaction with the receptor at a location other than the insulin binding site. In the earlier group, both monovalent and divalent Fab′ fragments inhibit $^{125}$I-insulin binding, but only the divalent fragments retain the ability to mimic the biological effects of insulin (Kahn *et al.*, 1978; Baldwin *et al.*, 1980). However, addition of anti-Fab′ serum, which cross-links the monovalent Fab fragment–receptor complexes, restores the insulin-like activity to the Fab′ fragments studied by Kahn *et al.* (1978) but not to those studied by Baldwin *et al.* (1980).

Le Marchand-Brustel *et al.* (1978) examined the insulin-like response to an antireceptor serum over a wide range of concentrations. As the concentration of antiserum increased, uptake of 2-deoxyglucose rose to a maximum and then declined. Since these experiments were performed with a crude serum, it is possible that this biphasic action was not due to a particular anti-insulin receptor (IgG) but to other antibodies in the serum. Studies with purified monoclonal antibodies would be required to remove this uncertainty. Studies performed over a wide range of concentrations with a purified anti-insulin IgG and a suboptimal concentration of insulin also yielded a biphasic response of DNA synthesis in fibroblasts (Schechter *et al.*, 1979a). Thus, studies with antibodies to insulin and with antibodies to the insulin receptor are consistent with the model [Eq. (1)–(6)] used to interpret the biphasic response observed for the interaction of antibody conjugate with GnRH receptors.

The report (Schechter *et al.*, 1979a) that a bivalent antibody to insulin caused a 10-fold increase in the binding of added insulin to hepatocyte membranes and to cultured fibroblasts but not to adipocytes is of interest in view of the finding (Jarrett *et al.*, 1980) that two thirds of the receptors occupied by ferritin–insulin on adipocytes occur in groups of two or more, whereas up to two thirds of the receptors on hepatocytes are single. Ferritin–insulin does not, however, cause aggregation of receptors in either tissue. Jarrett *et al.* (1980) have suggested that the higher degree of insulin-receptor aggregation in adipocyte membranes could in part account for the greater sensitivity of adipocytes to insulin, but it is not yet clear whether receptor dimerization is a required step in the initiation of the response to insulin itself.

It has been known for some time that the plant lectin concanavalin A (αCon A), which consists at physiological pH largely of tetramers of identical (intact) subunits, can mimic the effect of insulin on isolated adipocytes (Czech and Lynn, 1973; Cuatrecasas and Tell, 1973). Suya *et al.* (1982) have prepared a monovalent concanavalin A (mCon A) which does not mimic the action of insulin. However, exposure of adipocytes to mCon A in the presence of an antibody raised to mCon A causes stimulation of glucose oxidation and inhibition of epinephrine-mediated lipolysis. These results demonstrate that antibody-induced aggregation of receptors for mCon A (presumably not the insulin receptor itself since neither αCon A or mCon A interfere appreciably with insulin binding to adipocytes) can induce an insulin-like action in rat adipocytes. A similar conclusion, i.e., that membrane events elicited by multivalent agents which bind to cell surface sites other than insulin receptors can activate transport, was reached by Czech (1980) and Minton (1981). Such behavior is susceptible to explanation by the hormone–receptor–effector model. If receptors that do not normally interact with certain effectors (e.g., ion-conducting channels or transmembrane glucose transporting channels), acquired, upon binding of their natural ligand (e.g., mCon A), a high affinity for those effectors when spaced the correct distance apart by interaction with a divalent antibody, the "nonspecific" stimulation observed by Suya *et al.* (1982) and Czech (1980) would result.

Recently, a monoclonal antibody that binds to the insulin receptor and blocks insulin binding to it (and therefore blocks the metabolic actions of insulin) has been prepared (Roth *et al.*, 1982). This monoclonal anti-insulin receptor antibody is approximately 100 times more potent than insulin at down-regulating the receptor (Roth *et al.*, 1983), suggesting that for the insulin receptor, at least, dimerization per se may be a crucial event in the initiation of the sequence of steps leading to down-regulation. Experiments with insulin dimers, however, are consistent with the possibility that receptor occupancy is sufficient for down-regulation (Roth *et al.*, 1984). It would be of interest to ascertain whether monovalent fragments of this antibody could also cause down-regulation.

## B. Epidermal Growth Factor

Schechter *et al.* (1979a,b) first showed that the stimulation of DNA synthesis normally evoked by epidermal growth factor (EGF) could be elicited from a relatively inert CNBr fragment of EGF by the addition of an antibody raised against EGF. As observed for GnRH (Conn *et al.*, 1982b), the effectiveness of very low concentrations of EGF itself was markedly enhanced by addition of this antibody. For both EGF and CNBr–EGF, there was an optimum concentration of antibody above which the stimulatory effects were progressively reduced. Furthermore, monoclonal antibodies to EGF receptors have been prepared which not only mimic many of the actions of EGF but which also show a decline in activity above an optimal value (Schreiber *et al.*, 1981). This behavior was simulated with the aid of a bivalent ligand hypothesis (Minton, 1981), but is also simulable by the model described in equations (1)–(3). It is noteworthy (see Section III, G) that among the early effects of EGF (but presumably subsequent to microaggregation) are increases in $Rb^+$ and $Ca^{2+}$ fluxes (reviewed in O'Connor-Mc-Court and Hollenberg, 1983).

## C. Prolactin

Prolactin is a major regulator of growth and differentiation in the rabbit mammary gland, causing both up- and down-regulation of prolactin receptors, DNA synthesis, and the synthesis of casein messenger RNA and casein (Djiane et al., 1982). Antibodies prepared against partially purified prolactin receptors not only interfered with many of the actions of added prolactin on mammary gland explants but also had the capacity to mimic the effects of prolactin. High concentrations of these (polyclonal) antibodies inhibited their own actions (Djiane et al., 1981). Similar results were observed in rat mammary tumor explants (Edery et al., 1983).

Rat hepatocytes also have prolactin receptors. These decline markedly in numbers during 48 hr in culture but this can be prevented by the addition of prolactin to the medium. Addition of antiprolactin receptor serum can block this up-regulatory action of prolactin, yet these same antibodies, added alone, can instead mimic this action. Thus, both in rat liver and in rabbit mammary gland, most of the effects of prolactin could be mimicked by antibodies to prolactin and, for cases tested, the effects declined above an optimal antibody concentration.

## D. Thyrotropin Receptor

Sera from patients with Graves' disease generally have antibodies ($T_sAb$; also called LATS) that stimulate the thyroid gland. These antibodies stimulate adenylate cyclase yet inhibit the adenylate cyclase stimulating activity of thyrotropin stimulating hormone (TSH) in isolated thryoid plasma membranes (Yamashita and Field, 1972; Mehdi and Kriss, 1978; Endo et al., 1981; Vitti et al., 1983). This activity is localized on the Fab portion of the antibody (Rees-Smith et al., 1977). Several workers (Mehdi and Kris, 1978; Mutoh et al., 1980; Zakarija and McKenzie, 1983) report, for IgGs either obtained from patients' serum or raised against a thyroid membrane preparation, a biphasic effect on the adenylate cyclase activity and the $^{125}$-I-TSH binding capacity of thyroid membranes.

While these results are clearly consistent with the hormone–receptor–effector model, they are far from conclusive. Thus, Zakarija and McKenzie (1983) believe that the decline in adenylate cyclase activity at high-IgG levels is due to the presence of an antibody that inhibits the stimulatory action of $T_sAb$ (see also Mutoh et al., 1980). There is also evidence that monovalent as well as divalent Fab fragments can activate thyroid adenylate cyclase (Lindstrom, 1977). Mutoh et al. (1980) interpret their finding that antithyroid membrane IgG can partially inhibit the stimulatory effect of glucagon on dog liver membrane adenylate cyclase as indicating that the IgGs raised against the thyroid membranes may act via conformational changes in membranes rather than interacting specifically with TSH (and glucagon) receptors. However, their data is also consistent with the presence of a mixed population of antibodies, those against the TSH receptor acting according to the model shown in Fig. 2, others acting directly on some component(s) of the adenylate cyclase–effector system which may be common to both thyroid and liver membranes. Clearly, further work is necessary to clarify the mechanism of the biphasic response of thyroid membranes to $T_sAb$.

## E. Enkephalin Receptors

Shimohigashi et al. (1982) synthesized a series of dimeric analogues of [D-Ala$^2$,Leu$^5$]enkephalin, by using $\alpha,\omega$-diaminoalkanes of various lengths to cross-link at the COOH termini. These dimeric pentapeptide enkephalins had increased affinity and selectivity for the "$\delta$"-type opiate receptor of rat brain membranes. At the present time, no other functional consequences of dimerization of these receptors are known.

## F. Luteinizing Hormone Receptors

Monoclonal antibodies ($M_{ab}$) against the LH receptor have recently been prepared. Antibodies from two of the five clones isolated were capable of stimulating Leydig cells, indicating that suitable dimers were formed. Antibodies from the other three clones behaved as competitive antagonists to LH binding. Their antagonist behavior could be converted into agonist behavior by the addition of a cross-linking antimouse IgG. A possible interpretation of these observations is that the dimers found by the antagonist $M_{ab}$s were improperly spaced and/or in the wrong orientation for interaction with the effectors. Interaction with IgG may have changed the conformation of the $M_{ab}$ so that the dimers were then spaced and oriented properly for interaction with the effector.

## G. Acetylcholine Receptors

Although the nicotinic receptor has been extensively characterized (Popot and Changeux, 1984) and over 160 monoclonal antibodies to it have been prepared (Lindstrom et al., 1983), I am not aware of any report that antibodies against this receptor act as agonists. A few monoclonal antibodies to the muscarinic receptor have been prepared (Lieber et al., 1984). One of these mimics agonist action on intact guinea pig myometrium, causing contraction, a rise in cGMP content, and inhibition of cAMP accumulation due to prostacyclin. Another monoclonal antibody reduces contraction but does not have agonist-like activity with respect to the cyclic nucleotides. As pointed out by Lieber et al. (1984), their data imply that the muscarinic receptors of the myometrium are coupled to multiple effector systems.

## H. Immunoglobulin Interactions with Fc Receptors

The binding of IgE to its receptor in the mast cell membrane, which occurs via the Fc portion of the antibody, does not cause histamine release. If a divalent—but not a monovalent—antigen to which that IgE has specific binding sites is then added, activation of secretion occurs (Foreman, 1980). Further evidence that the binding of adjacent IgE molecules activates the secretory process in mast cells or basophilic leukocytes includes (Foreman, 1980; Metzger and Ishizaka, 1982) (1) anti-IgE antibodies (directed against the Fc region of IgE) can induce histamine release; (2) concanavalin A, which binds to carbohydrates associated with the Fc region, can induce histamine secretion; (3) chemically dimerized IgE can cause histamine secretion; and (4) a divalent antibody (IgG) prepared against the receptor for IgE can also stimulate histamine release.

The receptor for IgE on basophilic leukemia cells has recently been shown to consist of one $\alpha$-, one $\beta$-, and two $\gamma$-chains, none of which appeared to change their molecular weights or relative proportions when aggregated by a suitable IgG (Perez-Montfort et al., 1983). However, this cross-linking of receptors by the anti-receptor antibody caused an increase in membrane permeability to $Ca^{2+}$, which in turn triggered the release of histamine (Ishizaka et al., 1979, 1984). Foreman (1980) had earlier suggested that the opening of the $Ca^{2+}$ channel was mediated by the formation of lysophosphatidylcholine, but direct interaction of the dimerized receptors with the $Ca^{2+}$ channel (cf. Fig. 3) is an alternative mechanism.

The changes in permeability noted above have also been shown to be associated with a rapid and reversible depolarization of the basophilic leukemia membrane. This suggested to Kanner and Metzger (1983) that the microaggregated receptors themselves might form the ion-conducting channels. Evidence supporting this concept comes from recent studies on the binding of IgG–Fc fragments to a macrophage cell line (Young et al., 1983a). When these fragments (bound to their receptors on the cell surface) interacted with monovalent ligands, no appreciable change in membrane potential occurred; but when they interacted with divalent or multivalent ligands, depolarization resulted. Extracellular $Ca^{2+}$ did not appear to play a significant role in the depolarization. A hyperpolarization followed the depolarization and was partially blocked by ouabain and quinine. It was suggested that this hyperpolarization resulted from a combination of $Na^+,K^+$-ATPase activity and $Ca^{2+}$-activated $K^+$ channels. In plasma membrane vesicles prepared from these macrophages (Young et al., 1983b), both soluble and immobilized immunocomplexes caused a cation flux; similar results were obtained when the purified receptors were reconstituted in liposomes or in planar bilayer membranes. Since this Fc receptor appears to consist of a single polypeptide chain, Young *et al.* (1983b) suggest that it functions as a ligand-dependent ionophore that will carry monovalent cations when (1) occupied by the Fc portion of an IgG molecule and (2) in a dimeric (or higher) state of aggregation. Recently, Pfefferkorn (1984) has shown that $Na^+/K^+$ fluxes are not necessary for activation of phagocytosis by the ionophoretic Fc receptor of a macrophage cell line, indicating that transmembrane signaling by the Fc receptor occurs via a monovalent cation-flux-*independent* mechanism. Since Fc-receptor interaction with multivalent ligands led to an increase in intracellular free $Ca^{2+}$ that was only partially blocked in the absence of external $Ca^{2+}$, Young *et al.* (1984) suggest that the multivalent ligands cause an influx of external $Ca^{2+}$ in addition to a release from internal compartments.

The possibility that the receptor is an ionophore could also be applicable to the IgE–Fc receptors of mast cells and basophils, in which an influx of $Ca^{2+}$ is associated with receptor dimerization, and to other systems (e.g., the GnRH receptor; see Section II) involving a $Ca^{2+}$-dependent stimulus–secretion coupling. If indeed the IgE or GnRH receptors are themselves $Ca^{2+}$-conducting channels, then the notion of a separate effector, a (closed) $Ca^{2+}$ channel that opens upon interaction with a suitably spaced pair of (occupied) receptors (see Fig. 3), may be erroneous. The Fc receptors in various cell lines could be cation-conducting channels which do not "open" unless brought into suitably close juxtaposition with one another by divalent ligands. Any number of variations and combinations of these themes of intrinsic and separate ionophore activities may occur, and it is not excluded that events taking place in the lipid phase (e.g., involving phosphatidylinositol) may play a key role, especially in the $Ca^{2+}$-dependent systems. The possibility that certain classes of receptors, upon dimerization, either manifest intrinsic ion-conducting properties or become associated with effector molecules that then act as ion-conducting channels may be a fruitful guide for further research in this field.

Extensive theoretical treatments of receptor cross-linking on cell surfaces have been developed to explain the large number of observations on IgE-induced cellular responses (Perelson and DeLisi, 1980; Dembo and Goldstein, 1978; DeLisi and Siraganian, 1979a,b; Wofsy, 1979). Although these treatments provide a comprehensive analysis of many of the factors involved in the formation of receptor dimers, they implicitly assume that dimerization per se is the effective signal leading to cellular response. In so far as the receptor dimers must interact with an effector (such as an ion-conducting channel) or with a coupling protein (see Section IV) and then an effector, further theoretical analysis may be required before these treatments may fully account for IgE-mediated responses and for application to some of the other systems described earlier.

Minton (1981) has examined in detail the properties of a model in which the ligand is assumed to be bivalent, having high affinity and high specificity for the receptor but low affinity and specificity toward the effector. This model has many attractive features, although it is unlikely to be applicable to the analysis of equilibrium responses to divalent antibodies binding relatively small ligands such as GnRH or catecholamines (see Section IV). A variety of dose–response curves, including the biphasic ones under discussion, may, however, occur in systems far from equilibrium (Minton, 1982).

## IV. THE ADENYLATE CYCLASE SYSTEM

### A. Properties of a Hormone–Receptor–Coupling Protein–Effector System

In the adenylate cyclase system, the receptor, after binding a suitable ligand, interacts with the effector, adenylate cyclase, via a coupling protein, N, which is also embedded in the membrane. The coupling protein is often referred to as guanine nucleotide regulatory protein because the complex effects that guanine nucleotides have upon adenylate cyclase activity involve their binding to this protein.

A scheme commonly used to account for the interaction of ligands such as β-adrenergic agonists with the β-receptor (Stadel et al., 1982) is shown in Fig. 4. In this scheme, the ligand, H, interacts with a low-affinity (for H) form of the receptor, R, forming H·R. The latter can interact with N·GDP (hereafter written as N·D) to form H·R·N·D, which now has a low affinity for GDP. Dissociation of GDP results in formation of H·R·N, which has a high affinity both for the receptor and for GTP. H·R·N·T is then formed and it subsequently dissociates into H·R, which in turn dissociates relatively easily into receptor and ligand and N·T, which can interact with the inactive catalytic unit of adenylate cyclase, C, to form the active adenylate cyclase, C·N·T. The rate of cAMP production is assumed proportional to the steady-state level of C·N·T, this being determined by the balance between the rate of formation of C·N·T and its intrinsic GTPase activity.

Although this scheme, or its equivalent, has been used as a model for thinking about how ligands and guanine nucleotides could act to stimulate adenylate cyclase activity, it was only recently that the expected adenylate cyclase activity as a function of increasing ligand concentration for any assumed set of rate constants has been examined (Blum, 1984). In the model of Stadel et al. (1982), no provision was made for the dissociation of ligand from the complexes H·R·N·T, H·R·N, or H·R·N·D nor for the reaction of guanine nucleotides with the regulatory protein in the absence of hormone. However, in many cells (e.g., hepatocytes), the cyclase may be stimulated by guanine nucleotides alone. This requires that N·T and N·D be formed as indicated in Fig. 4 (via rate constants

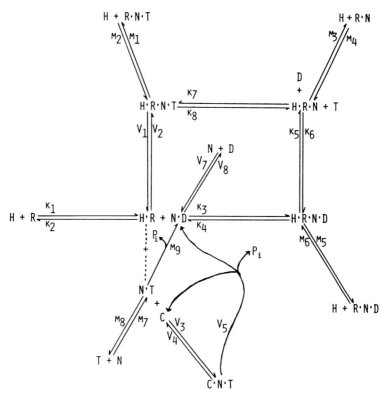

**Fig. 4.** The model of β-receptor (R), interaction with coupling factor (N), to activate adenylate cyclase (C), according to Stadel *et al.* (1982) has been modified to include the reactions described by rate constants $M_7$, $M_8$, $M_9$, $V_7$, and $V_8$ [which permit adenylate cyclase activation by guanine nucleotides in the absence of catecholamine or other ligand (H)] and the dissociation reactions indicated by rate constants $M_1...M_6$. Abbreviations: D and T, the concentrations of GDP and GTP, respectively; $P_i$, orthophosphate.

$M_7$, $M_8$, and $V_7$, $V_8$) and that N·T have a GTPase activity (rate constant $M_9$), even in the absence of the adenylate cyclase moiety. The following discussion will describe some of the properties of this more complete model and compare them to those of the simpler model.

The rate equations governing this model are

$$dx_1/dt = -k_1x_1y_1 + k_2y_2 - M_6x_1x_5 - M_2x_1x_6 + M_5x_2 \tag{21}$$
$$+ M_1x_3 - M_4x_1x_7 + M_3y_3$$
$$dy_1/dt = -k_1x_1y_1 + k_2y_2 \tag{22}$$
$$dy_2/dt = -k_2y_2y_4 + k_4x_2 - V_2y_2y_5 + V_1x_3 + k_1x_1y_1 - k_2y_2 \tag{23}$$
$$dy_4/dt = -k_3y_2y_4 + V_5x_4 + M_9y_5 + V_7x_8D_0 - V_8y_4 + k_4x_2 \tag{24}$$
$$dx_2/dt = +k_3y_2y_4 - (k_4 + k_5 + M_5)x_2 + k_6y_3D_0 + M_6x_1x_5 \tag{25}$$

$$dy_3/dt = k_5x_2 - (k_6D_0 + M_3 + k_7T_0)y_3 + k_8x_3 + M_4x_1x_7 \tag{26}$$
$$dx_3/dt = k_7T_0y_3 - (k_8 + V_1 + M_1)x_3 + V_2y_2y_5 + M_2x_1x_6 \tag{27}$$
$$dy_5/dt = V_1x_3 - V_2y_2y_5 - V_3y_5y_6 + V_4x_4 + M_7T_0x_8 - (M_8 + M_9)y_5 \tag{28}$$
$$dy_6/dt = -V_3y_5y_6 + (V_4 + V_5)x_4 \tag{29}$$
$$dx_4/dt = +V_3y_5y_6 - (V_4 + V_5)x_4 \tag{30}$$
$$dx_5/dt = M_5x_2 - M_6x_1x_5 \tag{31}$$
$$dx_6/dt = M_1x_3 - M_2x_1x_6 \tag{32}$$
$$dx_7/dt = M_3y_3 - M_4x_1x_7 \tag{33}$$
$$dx_8/dt = -(M_7T_0 + V_7D_0)x_8 + M_8y_5 + V_8y_4 \tag{34}$$

for which the concentration of each species is defined as follows: $x_1$ = H; $x_2$ = H·R·N·D: $x_3$ = H·R·N·T; $x_4$ = C·N·T: $x_5$ = R·N·D; $x_6$ = R·N·T; $x_7$ = R·N; $X_8$ = N; $y_1$ = R; $y_2$ = H·R; $y_3$ = H·R·N; $y_4$ = N·D; $y_5$ = N·T; and $y_6$ = C. Since the concentrations of GDP and GTP are assumed constant, their concentrations have been written as $D_o$ and $T_o$, respectively. The mass balance equations for this system are

$$H_o = x_1 + x_2 + x_3 + y_2 + y_3 \tag{35}$$
$$R_o = x_2 + x_3 + x_5 + x_6 + x_7 + y_1 + y_2 + y_3 \tag{36}$$
$$C_o = x_4 + y_6 \tag{37}$$
$$N_o = x_2 + x_3 + x_4 + x_5 + x_6 + x_7 + x_8 + y_3 + y_4 + y_5 \tag{38}$$

for which $H_o$, $R_o$, $C_o$, and $N_o$ are the total concentrations of ligand, receptor, adenylate cyclase, and guanine nucleotide regulatory protein, respectively.

At this point there are 18 algebraic equations (21)–(38) but only 14 independent variables. To eliminate the four redundant equations, one first notes that $dy_6/dt + dx_4/dt = 0$. Therefore, Eq. (30) can be replaced by Eq. (37) Since

$$dR_o/dt = 0 = dy_1/dt + dy_2/dt + dy_3/dt + dx_2/dt + dx_3/dt + dx_5/dt + dx_6/dt + dx_7/dt \tag{39}$$

one of these variables is determined if the other seven are specified. Choosing $y_2$ as the dependent variable, one may then replace Eq. (23) by Eq. (36). Similarly, Eq. (27) may be replaced by Eq. (38) and Eq. (21) may be replaced by Eq. (35). Thus Eqs. (22), (24), (25), (26), (28), (29), (31), (32), (33), (34), (35), (36), (37), and (38) form a set of 14 independent equations which can be solved for the 14 independent variables. Nevertheless, even in the steady state, for which all the derivatives equal zero, these equations are nonlinear (because of product terms such as $k_1x_1y_1$). A convenient way to solve this system is to linearize the product terms by a Taylor series expansion and then use the resulting system of linear equations in an iterative matrix-inversion procedure to obtain an exact solution, as described in detail elsewhere (Blum, 1984). Then for any assumed set of rate constants and any assumed values of $H_o$, $R_o$, $C_o$, and $N_o$, one can compute the concentrations of all of the species in the model. It should be noted that the principle of detailed balance requires that $k_7V_1k_3k_5 = k_8k_6k_4V_2$ in Fig. 4;

specification of any seven of these rate constants thus determines the eighth. In the discussion that follows, only the steady-state properties of this model will be considered; extension of the analysis to nonsteady-state conditions is simple but further information on the values of some of the rate constants is desirable before such studies will be worthwhile.

Because of the GTPase activities of N·T and C·N·T, one cannot assume that the various reactions shown in Fig. 4 are at equilibrium even after steady state has been attained. Indeed, inspection of Eq. (21)–(24) when all derivatives have been set equal to zero shows that only the ratios of $k_1/k_2$, $M_1/M_2$, $M_3/M_4$, and $M_5/M_6$ need be specified; all of the other rate constants must be specified individually if the equations are to be solved. In particular cases some of the remaining reactions may also, for all practical purposes, be at equilibrium. This will not, however, be true in general, and theoretical treatments that assume that all the reactions are at equilibrium should be reexamined from this point of view (see also Minton, 1982).

## B.  Adenylate Cyclase Activity as a Function of Agonist Concentration

For certain sets of rate constants, increasing concentrations of ligand produce the straightforward increase in adenylate cyclase activity characteristic of full agonist. For other values of the rate constants, partial agonists are mimicked, as discussed in detail elsewhere (Blum, 1984). Of particular interest in the simpler model was the finding that, for certain sets of rate constants, it predicted that as the concentration of the ligand increased, the adenylate cyclase activity would rise to a maximum and then decline. Since such behavior has been observed for many systems (see, e.g., Blum, 1984; England et al., 1983; Gardner and Jensen, 1982, for references to such observations), it was of interest to examine the effect of including the reactions characterized by rate constants $M_7$, $M_8$, $M_9$, $V_7$, and $V_8$ on such biphasic behavior.

The heavy line in each of the four graphs of Fig. 5B depicts the predicted adenylate cyclase activity as a function of ligand concentration for the parameters listed in Fig. 5A. For a ligand–receptor system with this set of rate constants, peak adenylate cyclase activity would be obtained at 1 n$M$ ligand.* At concentrations above 1 n$M$, the adenylate cyclase activity declines. The effects of changing rate parameters other than $M_7$, $M_8$, $M_9$, $V_7$, and $V_8$ have been described (Blum, 1984) and will not be repeated here. Inclusion of these additional parameters has very little effect on the basic shapes of the curves of adenylate cyclase activity versus ligand concentration for a variety of configurations, including full

---

*The units in which the rate constants and concentrations are expressed must be consistent. For the sake of definiteness, concentrations are expressed in nmoles/liter and time in seconds.

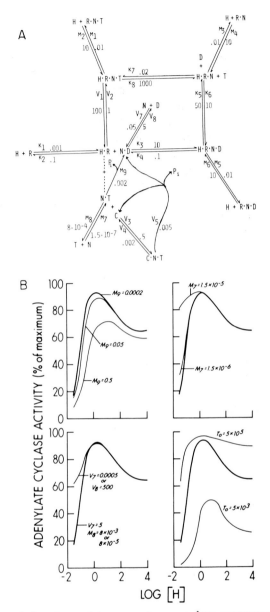

**Fig. 5.** Effect of changing certain parameters on a dose–response curve that shows partial inhibition of adenylate cyclase activity at high concentrations of ligand H. In each of the four panels of (B) the curve drawn with a heavy solid line was computed for the parameter values shown in (A) plus the following parameter values: $R_o = 2$; $C_o = 1$; $N_o = 1.4$; $D_o = 10^4$; $T_o = 5 \times 10^4$. (Units of concentration and time are taken as nM and sec, respectively.) Each of the four panels in (B) show the effect of varying one (or, in the lower left panel, two) parameters to the indicated values, keeping all of the other parameters constant.

agonists, partial agonists, and ligands which show the biphasic response, as shown in Fig. 5 and described as follows.

The upper left panel shows that reducing the rate of the GTPase reaction in the N·T complex ($M_9$) causes little change in the shape of the biphasic response curve. Raising $M_9$ 25-fold (from 0.002–0.05 sec$^{-1}$) also has very little effect on the shape of the response curve but raising $M_9$ to 0.5 sec$^{-1}$ reduces the maximum response and moves the curve to the right. Increasing $M_7$, the rate constant for formation of the N·T complex, from $1.5 \times 10^{-7}$–$1.5 \times 10^{-6}$ has scarcely any effect on the shape of the response curve, except that one now observes appreciable adenylate cyclase activity even in the absence of hormone (upper right panel). A further 10-fold increase in $M_7$ increases the "basal" adenylate cyclase activity almost to the level observed with an optimum concentration of agonist. Similar effects (again without influencing the shape of the descending portion of the response curve) can be obtained by decreasing $V_7$ or increasing $V_8$, i.e., by increasing the concentration of regulatory protein available to interact with GTP (lower left panel). The effect of increasing the GTP concentration (lower right panel) is as expected, i.e., cyclase activity in the absence of hormone becomes large, the maximum activity at optimal [H] increases slightly, and the activity decreases very little as [H] increases further.

Figure 4 also provides for the dissociation of H from H·R·N·T, H·R·N, and H·R·N·D. The effects of increasing the dissociation rate constants $M_1$, $M_3$, and $M_5$ on the response curve are shown in Fig. 6. As dissociation becomes appreciable (which requires much larger values of $M_1$, $M_3$, or $M_5$ than were chosen for the reference curve), there is a marked shift to the right in the response curve and

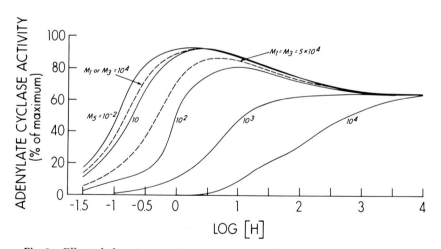

**Fig. 6.** Effect of changing certain parameters on a dose–response curve that shows partial inhibition of adenylate cyclase values at high concentrations of ligand. The parameter values are as shown in the legend to Fig. 5 except for the indicated individual changes.

a small decrease in the amount of adenylate cyclase activity expressed at optimal [H]. At sufficiently high values of $M_5$ (i.e., dissociation of H from H·R·N·D: see Korner *et al.*, 1982), the biphasic nature of the response curve is lost and the ligand then behaves as a partial agonist.

## C. Inhibitory and Excitatory Coupling Factors

Although the model shown in Fig. 4 was developed for the (stimulatory) reaction of catecholamines with the β-receptor, many other types of receptors are now known to interact—via stimulatory regulatory proteins ($N_s$) or inhibitory regulatory proteins ($N_i$)—with adenylate cyclase, causing an increase or decrease, respectively, in adenylate cyclase activity (see, e.g., Rodbell, 1980). With minor modifications, the scheme shown in Fig. 4 could serve as a model for the inhibitory effects of a ligand in the presence of GTP. To model the simultaneously stimulatory and inhibitory effects of two ligands, one would need not only to include a scheme for the inhibitory ligand but, also, one would need to know whether the $N_i$ and $N_s$ proteins or their subunits compete for the same site on adenylate cyclase or whether they alter its activity by binding to separate sites (Hildebrand *et al.*, 1984). A more detailed discussion of the possible interrelations between the $N_i$ and $N_s$ systems is available (Cooper and Londos, 1982).

Levitzki (1982) has proposed a somewhat different model for the activation of adenylate cyclase than that shown in Fig. 4. In the "collision-coupling" model, N is always associated with either R or C. A number of properties of this model have been examined in detail by Macfarlane (1982), Minton (1982, see also this volume Chapter 2), and Blum (1984). The collision-coupling model shows a variety of full and partial agonist behaviors but does not yield a biphasic dose–response curve.

Rodbell (1980, 1983) has drawn attention to evidence suggesting that R·N exists as an oligomeric complex which dissociates into monomeric R·N units upon binding of ligand and that these units would interact with C to form the active adenylate cyclase. This scheme also does not yield a biphasic dose–response curve, although it is possible to modify the scheme so that such a response pattern can occur.

Perhaps the most extensively studied example of a biphasic dose–response curve is that of the glucagon modulation of adenylate cyclase activity in hepatocyte (England *et al.*, 1983) and adipocyte membranes (Jean-Baptiste *et al.*, 1982). In the study on hepatocyte membranes, 12 ligands (glucagon and a variety of analogues) were examined. Each bound to the membranes with a single equilibrium constant and there was no indication of cooperativity. For many of the compounds, including glucagon, a biphasic dose–response curve was observed. In some cases, the maximal adenylate cyclase activity achieved was equal to that observed at the optimal concentration of glucagon. For others, the response at optimal concentrations was less (partial agonism). England *et al.* (1983) could fit the data for all of the ligands with an equation of the form

$$A = A_{min} + (A_{max} - A_{min})r/[1 + (K_A r/H) + (Hr/K_I)] \qquad (40)$$

for which $A$ is the adenylate cyclase activity observed at ligand concentration $H$, $A_{max}$ is the theoretical maximum adenylate cyclase activity, $A_{min}$ is the basal activity in the absence of $H$ (but, in these experiments, in the presence of 1 $\mu M$ GTP), $K_A$ and $K_I$ are the dissociation constants for binding of H that lead to activation and inhibition, respectively, and $r = K'_A/(K_A + K'_A)$, with $K'_A$ representing a nonproductive binding introduced by DeLean et al. (1979) as a mechanism to account for partial agonism. England et al. (1983) discussed several other mechanisms that might give rise to the biphasic dose–response curves, such as receptor cross-linking and desensitization or exhaustion of the response system, but do not find these to be likely explanations for their observations.

While the formalism embodied in equation (40) provides a straightforward quantitative description of the interaction of glucagon and its analogues with the hepatocyte membrane adenylate cyclase system, it necessarily ignores the mechanistic features of the $H/R/N_s/C$ interactions schematized in Fig. 4. Since biphasic dose–response curves may be observed for certain ranges of the rate constants (e.g., Fig. 5), the data of England et al. (1983) and of Jean-Baptiste et al. (1982) might also be fit according to the scheme of Fig. 4. If such fits were obtained, this would provide a basis for beginning to understand the differences between these ligands in terms of specific molecular complexes.

Rodbell (1980) summarized evidence which suggests that GTP also mediates interactions between a number of ligands (e.g., dopamine, angiotensin, and opiates) and effectors other than adenylate cyclase, presumably via a guanine nucleotide regulatory protein $N_x$. Recent evidence indicates that insulin activates hepatocyte cAMP phosphodiesterase and a membrane kinase (Heyworth et al., 1983) and inhibits adenylate cyclase activity (Heyworth and Houslay, 1983) by a process that involves a guanine nucleotide regulatory protein(s). Similarly, introduction of nonhydrolyzable analogues of GTP into the cytosol of mast cells cause exocytosis in response to the addition of extracellular $Ca^{2+}$, suggesting that "$N_x$" may serve to couple receptors for $Ca^{2+}$-mobilizing hormones with their effectors, $Ca^{2+}$-conducting channels (Gomperts, 1983). A number of other systems also appear to use guanine nucleotide regulatory proteins as links between receptors and effectors (Chang et al., 1983). It remains to be seen whether these $H/R/N_x/E$ systems can also be described in terms of a model such as that of Fig. 4.

## V. RECEPTOR–EFFECTOR INTERACTIONS REVISITED

Evidence that receptor dimerization plays a key role in the process by which antibodies to a variety of hormones and/or receptors can mimic the effect of the "normal" hormone–receptor interaction was summarized in Sections II and III.

Although Rodbell (1980, 1983) has presented evidence suggesting that disaggregation of an R·N oligomer may occur upon binding of H in the glucagon–adenylate cyclase system, there is no indication that aggregation (e.g., dimerization) should play a role in any adenylate cyclase system, and more specifically, the β-adrenergic receptor system, whether characterized by the model shown in Fig. 4 or by any other model I have encountered. Strosberg and others (see Strosberg, 1983), however, have prepared antibodies to alprenolol in rabbits and used these antibodies to prepare anti-idiotypic antibodies that bind to the β-adrenergic receptors of various types of cells. Binding of the anti-idiotypic antibodies resulted in the activation or the inhibition of basal or catecholamine-stimulated adenylate cyclase activity, depending on the system used. There was a synergistic effect between hormone and antibody, but no synergistic effect on adenylate cyclase activation was observed between the antibodies and either GTP or, guanyl-5'-yl imidodiphosphate [Gpp(NH)p]. This suggests that dimerization of the β-adrenergic receptor may have bypassed the coupling protein and permitted the receptor to interact directly with the adenylate cyclase. Alternatively, it is possible that dimerization sufficiently perturbs membrane structure so that adenylate cyclase molecules located near the dimers assume an active conformation.

Some further indication that microaggregation of catecholamine receptors can occur comes from studies on β-adrenergic antagonists with multiple pharmacophores (Pitha *et al.*, 1983). Whereas the monovalent antagonists bound and dissociated as expected, compounds containing two or three pharmacophores had somewhat lower binding affinities but dissociated very slowly. Unfortunately, possible physiological responses to the slowly desorbing divalent or trivalent compounds were not examined.

Although this chapter has focused on receptor dimerization as a key step in evoking agonist effects in response to certain ligands, it should be clear from several of the studies mentioned previously that receptors may form larger clusters in the presence of multivalent ligands. A number of theoretical treatments of clustering were mentioned (e.g., Goldstein and Perelson, 1984). MacGlashan *et al.* (1983) have studied the release of histamine from human basophils in response to dimeric, trimeric, and tetrameric cross-linked IgE. They found that histamine release by trimer was enhanced by indomethacin, whereas release induced by dimeric IgE was not. Fewtrell and Metzger (1980) showed that larger oligomers of IgE were more effective (per monomer unit bound) in releasing serotonin from rat basophilic leukemia cells. Thus, both the qualitative and quantitative behavior of receptor-mediated secretion varies with the number of receptors per cluster and, presumably, their precise spacing and orientation.

That polylysine can mimic the action of GnRH has already been mentioned and similar observations have been reported for beef thyroid membrane basal and TSH-, prostaglandin E- (PGE-), and Gpp(NH)p-activated adenylate cyclase activity (Wolff and Cook, 1975). It is noteworthy that in this system the polycations caused a biphasic response. Wolff and Cook (1975) suggest that the polycation effect results from an electrostatic effect on membrane conformation not restricted to any particular receptor domain. Lockwood and East (1974) reported that certain polyamines could mimic the effect of insulin on adipocyte membranes at a locus subsequent to the insulin receptor site. These observations raise the possibility that the dimers formed by interaction of antibodies with receptors may act in two ways: (1) by forming an appropriate receptor configuration so that the dimer may interact in a specific way with the effector, as illustrated in Fig. 3, or (2) by merely perturbing the membrane in the vicinity of the effector in such a way as to mimic the effect of the occupied receptor on the effector. Of course, both mechanisms may contribute to varying degrees for any particular system. For example, the insulin–ricin B hybrid molecule, acting via binding to the ricin receptor, "may induce a cell-surface perturbation such as receptor clustering which in turn triggers some or all of the pleiotropic effects associated with the hormone" (Hofmann *et al.*, 1983). Evidence has also been presented that both surface charge and hydrophobic interactions influence the behavior of the nucleotide regulatory protein that couples the glucagon receptor to adenylate cyclase in hepatocyte membranes (Rubalcava *et al.*, 1983).

In addition to the effects of antibodies, polycations, lectins, etc., on receptor spacing (both perpendicular to and in the plane of the membrane), further consequences of the fact that the receptor–effector system is embedded in a membrane are that the density of receptors per unit area of membrane and that the diffusion coefficient of the various complexes (and hence the fluidity of the membrane) must be considered in attempting to understand any particular system.

Incardona (1983) has reviewed the literature on the effects of receptor density on the binding of ligands (including viruses) on cells. Generally, the forward rate constant for binding H to R depends on receptor density, $\bar{R}$, as $\bar{R}/(\bar{R} + S)$, for which $S$ is a function of the effective radius of the ligand–receptor complex in the plane of the membrane. Thus at a sufficiently high receptor density, the rate of formation of H·R becomes independent of density. Since the rate constant for the dissociation of the H·R complex is inversely proportional to $(\bar{R} + S)$, the equilibrium association is directly proportional to receptor density. A reduction in receptor number due to, e.g., internalization will not only cause a loss in response because of the loss of receptors per se but may also reduce the response to low concentrations of ligand via a reduction in the effective affinity constant if $\bar{R}$ is sufficiently low.

Changes in membrane fluidity may affect the dose–response curve for ligands not only by moving the receptors perpendicularly to the plane of the membrane and hence affecting their accessibility to the ligand and/or the interaction with effector on the cytosolic face of the membrane. This has been suggested to account for modulation of transferrin receptors in bone marrow cells by alterations in membrane fluidity (Muller and Shinitzky, 1979). Changes in membrane fluidity will also change the diffusion coefficient of macromolecules in the plane of the membrane. Theoretical considerations pertaining to the possible role of membrane fluidity on initiation of the IgG-mediated complement fixation cascade suggest that if diffusion in the plane of the membrane does limit multiple C1q binding, there may be an optimal range for the diffusion coefficient, $D$, and that outside this range one would observe a sharp drop in probability of activation (DeLisi and Weigel, 1983). The effect of variations in $D$ on the "mean trapping time, $\tau$" (i.e., the time required for a diffusing species such as a receptor to be captured by another species such as an effector) has been analyzed in some detail (see Weaver, 1983, and references therein). Applications of such theoretical treatments to various receptor–effector systems may increase our understanding of the roles that changes in receptor density and membrane fluidity have on hormonal response. Thus, Swillens (1982) has suggested that restricted diffusion may account for the observation that, for a given number of occupied β-receptors, the adenylate cyclase activity in S49 lymphoma cells depends on the total number of receptors present. A direct measurement of the possible role of diffusion in the interaction of IgE with its receptor in a cultured cell line, however, demonstrated that for this ligand–receptor system the reaction kinetics were not diffusion controlled (Wank *et al.*, 1983).

It should be stressed that receptors can interact not only with one another, but with other membrane proteins as well. Such interactions may have important consequences. Henis (1984), for example, has shown that above a certain threshold of local coverage of the Con A receptors on lymphocytes there was an inhibition not only of Con A receptor mobility, but also of the mobility of surface immunoglobulins. These results indicate specificity in the interactions between different membrane proteins and the cytoskeleton. Interaction between major histocompatibility complex antigens and the receptors for glucagon (Lafuse and Edidin, 1980) and epidermal growth factor (Schreiber *et al.*, 1984) have also been demonstrated. Further evidence for bivalent ligand–induced effects that are not necessarily restricted to direct interaction with receptors comes from experiments with antibodies raised against rat myoblast cell membranes. Lo and Duronio (1984) discovered that several such antibodies could activate the transport system for hexoses and for certain amino acids. It is unlikely that each of these antibodies (raised in both sheep and rabbits, one of them monoclonal) interacts only with the proteins of the transport systems. The possibility must therefore be considered that such effects are manifestations of the ability of divalent ligands to alter the function of membrane proteins by "nonspecific" interactions. The nature of these interactions is even less well understood than the nature of homologous receptor–receptor interactions.

It seems clear that a receptor–coupling protein–effector model is a useful concept with which to examine data on ligand-induced cellular responses. The various rate constants, however, will not only be determined by the specific chemical interactions between the groups responsible for binding and catalysis but may also depend on membrane properties, e.g., fluidity, diffusion constants, receptor density, perturbation by antibodies or polycations, and interactions between receptors and other membrane proteins.

## ACKNOWLEDGMENT

I am grateful to: Dr. A. P. Minton and J. Leiser for a careful reading of this manuscript; Dr. P. M. Conn for many helpful discussions and for allowing me to use unpublished data from his laboratory; Dr. M. Hines for invaluable aid with the methodology for computer analysis of nonlinear steady-state systems; and to R. Hougom for patiently and expertly typing this manuscript.

## REFERENCES

Avissar, S., Moscona-Amir, E., and Sokolovsky, M. (1982). Photoaffinity labeling reveals two muscarinic receptor macromolecules associated with the presence of $Ca^{++}$ in rat adenohypophysis. *FEBS Lett.* **150**, 343–346.

Avissar, S., Amitai, G., and Sokolovsky, M. (1983). Oligomeric structure of muscarinic receptors is shown by photoaffinity labeling: subunit assembly may explain high- and low-affinity agonist states. *Proc. Natl. Acad. Sci. U.S.A.* **80**, 156–159.

Baldwin, D., Terris, S., and Steiner, D. (1980). Characterization of insulin-like actions of anti-insulin receptor antibodies. *J. Biol. Chem.* **255**, 4028–4034.

Blum, J. J. (1984). Hormone receptor-coupling factor-adenylate cyclase interactions: theoretical considerations. *J. Theor. Biol.* **111**, 589–608.

Blum, J. J., and Conn, P. M. (1982). Gonadotropin-releasing hormone stimulation of luteinizing hormone release: A ligand–receptor–effector model. *Proc. Natl. Acad. Sci. U.S.A.* **79**, 7307–7311.

Chang, K. J., Blanchard, S. G., and Cuatrecasas, P. (1983). Unmasking of magnesium-dependent high-affinity sites for [DALA$^2$,DLEU$^5$] enkephalin after pretreatment of brain membranes with guanine nucleotides. *Proc. Natl. Acad. Sci. U.S.A.* **80**, 940–944.

Clayton, R. N., and Catt, K. J. (1981). Gonadotropin-releasing hormone receptors: Characterization, physiological regulation, and relationship to reproductive function. *Endocr. Rev.* **2**, 186–209.

Conn, P. M. (1984). Molecular mechanism of gonadotropin releasing hormone action. *In* "Biochemical Actions of Hormones" (G. Litwack, ed.), Vol. 11, pp. 67–92. Academic Press, Orlando.

Conn, P. M., and Rogers, D. C. (1980). Gonadotropin release from pituitary cultures following activation of endogenous ion channels. *Endocrinology (Baltimore)* **107**, 2133–2134.

Conn, P. M., Rogers, D. C., and Sandhu, F. S. (1979). Alteration of the intracellular calcium level stimulates gonadotropin release from cultured rat anterior pituitary cells. *Endocrinology (Baltimore)* **105**, 1122–1127.

Conn, P. M., Rogers, D. C., Stewart, J. M., Niedel, J., and Sheffield, T. (1982a). Conversion of a GnRH antagonist to an agonist: implication for a receptor microaggregate as the functional unit for signal transduction. *Nature (London)* **296**, 653–655.

Conn, P. M., Rogers, D. C., and McNeil, R. (1982b). Potency enhancement of a GnRH agonist: GnRH-receptor microaggregation stimulates gonadotropin release. *Endocrinology (Baltimore)* **111**, 335–337.

Conn, P. M., Rogers, D. C., and Seay, S. G. (1983). Biphasic regulation of the GnRH receptor by receptor microaggregation and intracellular $Ca^{++}$ levels. *Mol. Pharmacol.* **25**, 51–55.

Conn, P. M., Hsueh, A. J. W., and Crowley, W. F., Jr. (1984). Gonadotropin-releasing hormone: molecular and cell biology, physiology, and clinical applications. *Fed. Proc., Fed. Am. Soc. Exp. Biol.* **43**, 2351–2361.

Cooper, D. M. F., and Londos, C. (1982). GTP-dependent stimulation and inhibition of adenylate cyclase. *In* "Hormone Receptors" (L. D. Kohn, ed.), Vol. 6, pp. 309–333. Wiley, New York.

Cuatrecasas, P., and Tell, G. P. E. (1973). Insulin-like activity of concanavalin A and wheat germ agglutinin—direct interactions with insulin receptors. *Proc. Natl. Acad. Sci. U.S.A.* **70**, 485–489.

Czech, M. P. (1980). Insulin action and the regulation of hexose transport. *Diabetes* **29**, 399–409.

Czech, M. P., and Lynn, W. C. (1973). Stimulation of glucose metabolism by lectins in isolated white fat cells. *Biochim. Biophys. Acta* **297**, 386–397.

DeLean, A., Munson, P. J., and Rodbard, D. (1979). Multi-subsite receptors for multivalent ligands. Application to drugs, hormones and neurotransmitters. *Mol. Pharmacol.* **15**, 60–70.

DeLisi, C., and Siraganian, R. P. (1979a). Receptor cross-linking and histamine release. I. The quantitative dependence of basophil degranulation on the number of receptor doublets. *J. Immunol.* **122**, 2286–2292.

DeLisi, C., and Siraganian, R. P. (1979b). Receptor cross-linking and histamine release. II. Interpretation and analysis of anomalous dose response patterns. *J. Immunol.* **122**, 2293–2299.

DeLisi, C., and Weigel, F. W. (1983). Membrane fluidity and the probability of complement fixation. *J. Theor. Biol.* **102**, 307–322.

Dembo, M., and Goldstein, B. (1978). Theory of equilibrium binding of symmetric bivalent haptens to cell surface antibody: Application to histamine release from basophils. *J. Immunol.* **121**, 345–353.

Dembo, M., and Goldstein, B. (1980). A model of cell activation and desensitization by surface immunoglobulin: the case of histamine release from human basophils. *Cell* **22**, 59–67.

Djiane, J., Houdebine, L.-M., and Kelly, P. A. (1981). Prolactin-like activity of prolactin receptor antibodies on casein and DNA synthesis in the mammary gland. *Proc. Natl. Acad. Sci. U.S.A.* **78**, 7445–7448.

Djiane, J., Houdebine, L.-M., and Kelly, P. A. (1982). Correlation between prolactin-receptor interaction, down regulation of receptors, and stimulation of casein and deoxyribonucleic acid biosynthesis in rabbit mammary gland explants. *Endocrinology (Baltimore)* **110**, 791–795.

Dower, S. K., Titus, J. A., and Segal, D. M. (1984). The binding of multivalent ligands to cell surface receptors. *In* "Cell Surface Dynamics" (A. S. Perelson, C. DeLisi, and F. W. Wiegel, eds.), Vol. 3, pp. 277–328. Dekker, New York.

Edery, M., Djiane, J., Houdebine, L.-M., and Kelly, P. A. (1983). Prolactinlike activity of anti-prolactin receptor antibodies in rat mammary tumor explants. *Cancer Res.* **43**, 3170–3174.

Endo, K., Amir, S. M., and Ingbar, S. H. (1981). Development and evaluation of a method for the partial purification of immunoglobulins specific for Graves' disease. *J. Clin. Endocrinol. Metab.* **52**, 1113–1123.

England, R. D., Jenkins, W. T., Flanders, K. C., and Gurd, R. S. (1983). Noncooperative receptor interactions of glucagon and eleven analogues: Inhibition of adenylate cyclase. *Biochemistry* **22**, 1722–1728.

Fewtrell, C., and Metzger, H. (1980). Larger oligomers of IgE are more effective than dimers in stimulating rat basophilic leukemia cells. *J. Immunol.* **125**, 701–710.

Foreman, J. (1980). Receptor–secretion coupling in mast cells. *Trends Pharmacol. Sci.* **2**, 460–462.

Gardner, J. D., and Jensen, R. T. (1982). Receptor modulation of calcium and cyclic AMP: differences in coupling to specific hormone-receptor complexes in cells responsive to several hormones. *In* "Hormone Receptors" (L. D. Kohn, ed.), Vol. 6, pp. 277–308. Wiley, New York.

Gilman, A. G. (1984). G proteins and dual control of adenylate cyclase. *Cell* **36**, 577–579.

Gomperts, B. D. (1983). Involvement of guanine nucleotide-binding protein in the gating of $Ca^{++}$ by receptors. *Nature (London)* **306**, 64–66.

Gregory. H., Taylor, C. L., and Hopkins, C. R. (1982). Luteinizing hormone release from dissociated pituitary cells by dimerization of occupied LHRH receptors. *Nature (London)* **300**, 269–271.

Henis, Y. (1984). Mobility modulation by local concanavalin A binding. *J. Biol. Chem.* **259**, 1515–1519.

Heyworth, C. M., and Houslay, M. D. (1983). Insulin exerts actions through a distinct species of guanine nucleotide regulatory protein: inhibition of adenylate cyclase. *Biochem. J.* **214**, 547–552.

Heyworth, C. M., Wallace, A. V., and Houslay, M. D. (1983). Insulin and glucagon regulate the activation of two distinct membrane-bound cyclic AMP phosphodiesterases in hepatocytes. *Biochem. J.* **214**, 99–110.

Hildebrand, J. D., Codina, J., and Birnbaumer, L. (1984). Interaction of the stimulatory and inhibitory regulatory proteins of the adenyl cyclase system with the catalytic component of cyc-S49 cell membranes. *J. Biol. Chem.* **259**, 13178–13185.

Hofmann, C. A., Lotan, R. M., Roth, G. D., and Oeltmann, T. N. (1983). Insulin-ricin B hybrid molecules: receptor binding and biological activity in a minimal deviation hepatoma cell line. *Arch. Biochem. Biophys.* **227**, 448–456.

Incardona, N. L. (1983). Binding of viruses and ligands to cells: Effect of receptor density. *J. Theor. Biol.* **104**, 693–699.

Ishizaka, T., Foreman, J. C., Sterk, A. R., and Ishizaka, K. (1979). Induction of calcium flux across the rat mast cell membrane by bridging IgE receptors. *Proc. Natl. Acad. Sci. U.S.A.* **76**, 5858–5862.

Ishizaka, T., Conrad, D. H., Schulman, E. S., Sterk, A. R., Ko, C. G. L., and Ishizaka, K. (1984). IgE-mediated triggering signals for mediator release from human mast cells and basophils. *Fed. Proc., Fed. Am. Soc. Exp. Biol.* **43**, 2840–2845.

Jacobs, S., Chang, K. J., and Cuatrecasas, P. (1978). Antibodies to purified insulin receptor have insulin like activity. *Science (Washington, D.C.)* **200**, 1283–1284.

Jarrett, L., Schwitzer, J. B., and Smith, R. M. (1980). Insulin receptors: Differences in structural organization on adipocyte and liver plasma membranes. *Science (Washington, D.C.)* **210**, 1127–1128.

Jean-Baptiste, E., Rizack, M. A., and Epand, R. M. (1982). Lipolytic and adenyl-cyclase stimulating activity of glucagon$_{1-6}$: Comparison with glucagon derivatives chemically modified in the 7-29 sequence. *Biosci. Rep.* **2**, 819–824.

Kahn, C. R., Baird, K. L., Jarrett, D. B., and Flier, J. S. (1978). Direct demonstration that receptor crosslinking or aggregation is important in insulin action. *Proc. Natl. Acad. Sci. U.S.A.* **75**, 4209–4213.

Kanner, B. I., and Metzger, H. (1983). Cross linking of receptors for immunoglobulin E depolarizes the plasma membrane of rat basophilic leukemia cells. *Proc. Natl. Acad. Sci. U.S.A.* **80**, 5744–5748.

Keri, G., Nikolics, K., Teplan, I., and Molnar, J. (1983). Desensitization of luteinizing hormone release in cultured pituitary cells by gonadotropin-releasing hormone. *Mol. Cell. Endocrinol.* **30**, 109–120.

King, A. C., and Cuatrecasas, P. (1981). Peptide hormone-induced receptor mobility, aggregation, and internalization. *N. Engl. J. Med.* **305**, 77–78.

Kohn, L. D., Valente, W. A., Laccetti, P., Cohen, J. L., Aloj, S. M., and Grollman, E. F. (1983).

Multicomponent structure of the thyrotropin receptor: Relationship to Graves' disease. *Life Sci.* **32,** 15–30.

Korner, M., Gilon, C., and Schramm, M. (1982). Locking of hormone in the β-adrenergic receptor by attack on a sulfhydryl in an associated component. *J. Biol. Chem.* **257,** 3389–3396.

Lafuse, W., and Edidin, M. (1980). Influence of mouse major histocompatibility complex, H-2, on liver adenylate cyclase activity and on glucagon binding to liver cell membranes. *Biochemistry* **19,** 49–54.

Le Marchand-Brustel, Y., Gorden, P., Flier, J. S., Kahn, C. R., and Freychet, P. (1978). Insulin receptor antibodies inhibit insulin binding and stimulate glucose metabolism in skeletal muscle. *Diabetologia* **14,** 311–317.

Levitzki, A. (1982). Activation and inhibition of adenylate cyclase by hormones: mechanistic aspects. *Trends Pharmacol. Sci.* **4,** 203–208.

Lieber, D., Harbon, S., Guillet, J.-G., Andre, C., and Strosberg, A. D. (1984). Monoclonal antibodies to purified muscarinic receptor display agonist-like-activity. *Proc. Natl. Acad. Sci. U.S.A.* **81,** 4331–4334.

Lindstrom, J. (1977). Antibodies to receptors for acetylcholine and other hormones. *In* "Receptors and Recognition" (P. Cuatrecasas and M. F. Greaves, eds.), Vol. 3, Series A, pp. 1–44. Wiley, New York.

Lindstrom, J., Tzartos, S., Gullick, W., Hochschwender, S., Swanson, L., Sargent, P., Jacob, M., and Montal, M. (1983). Use of monoclonal antibodies to study acetylcholine receptors from electric organs, muscle, and brain and the autoimmune response to receptor in *Myasthenia gravis. Cold Spring Harbor Symp. Quant. Biol.* **48,** 89–99.

Lo, T. C. Y., and Duronio, V. (1984). Activation of hexose transport by antibody. *Can. J. Biochem. Cell Biol.* **62,** 245–254.

Lockwood, D. H., and East, L. E. (1974). Studies on the insulin-like actions of polyamines on lipid and glucose metabolism in adipose tissue cells. *J. Biol. Chem.* **249,** 7717–7722.

Macfarlane, D. E. (1982). Bidirectional collision coupling in the regulation of adenylate cyclase. *Mol. Pharmacol.* **22,** 580–588.

MacGlashan, D. W., Jr., Schliemer, R. P., and Lichtenstein, L. M. (1983). Qualitative differences between dimeric and trimeric stimulation of human basophils. *J. Immunol.* **130,** 4–6.

Marion, J., Cooper, R. L., and Conn, P. M. (1981). Regulation of the rat pituitary gonadotropin-releasing hormone receptor. *Mol. Pharmacol.* **19,** 339–405.

Mehdi, S. Q., and Kriss, J. P. (1978). Preparation of radiolabeled thyroid-stimulating immunoglobulins (TSI) by recombining TSI heavy chains with $^{125}$I-labeled light chains: Direct evidence that the product binds to the membrane thyrotropin receptor and stimulates adenylate cyclase. *Endocrinology (Baltimore)* **103,** 296–301.

Mellman, I., and Plutner, H. (1984). Internalization and degradation of macrophage $F_c$ receptors bound to polyvalent immune complexes. *J. Cell Biol.* **98,** 1170–1177.

Metzger, H., and Ishizaka, T. (1982). Transmembrane signaling by receptor aggregation: the mast cell receptor for IgE as a case study. *Fed. Proc., Fed. Am. Soc. Exp. Biol.* **41,** 7.

Minton, A. P. (1981). The bivalent ligand hypothesis. A quantitative model for hormone action. *Mol. Pharmacol.* **19,** 1–14.

Minton, A. P. (1982). Steady-state relations between hormone binding and elicited response: quantitative mechanistic models. *In* "Hormone Receptors" (L. D. Kohn, ed.), Vol. 6, pp. 43–65. Wiley, New York.

Muller, C., and Shinitzky, M. (1979). Modulations of transferrin receptors in bone marrow cells by changes in lipid fluidity. *Br. J. Haematol.* **42,** 355–362.

Mutoh, H., Totsuka, Y., Chou, M. C. Y., and Field, J. B. (1980). Effects of antibodies to bovine thyroid plasma membranes on *in vitro* basal and thyroid stimulating hormone stimulation of bovine thyroid adenylate cyclase. *Endocrinology (Baltimore)* **107,** 707–713.

Naor, Z., Clayton, R. N., and Catt, K. J. (1980). Characterization of gonadotropin-releasing hormone receptors in cultured rat pituitary cells. *Endocrinology (Baltimore)* **107**, 1144–1152.

O'Connor-McCourt, M., and Hollenberg, M. D. (1982). Receptors acceptors, and the action of polypeptide hormones: illustrative studies with epidermal growth factor (urogastrone). *Can. J. Biochem. Cell Biol.* **61**, 670–682.

Perelson, A. S. (1984). Some mathematical models of receptor clustering by multivalent ligands. *In* "Cell Surface Dynamics" (A. S. Perelson, C. DeLisi, and F. W. Wiegel, eds.), pp. 223–276. Dekker, New York.

Perelson, A. S., and DeLisi, C. (1980). Receptor clustering on a cell surface. I. Theory of receptor cross-linking by ligands bearing two chemically identical functional groups. *Math. Biosci.* **48**, 71–110.

Perez-Montfort, R., Fewtrell, C., and Metzger, H. (1983). Changes in the receptor for immunoglobulin E coincident with receptor-mediated stimulation of basophilic leukemia cells. *Biochemistry* **22**, 5733–5737.

Pffeferkorn, L. C. (1984). Transmembrane signaling: an ion-flux model for signal transduction by complexed Fc receptors. *J. Cell Biol.* **99**, 2231–2240.

Pitha, J., Milecki, J., Czajkowska, T., and Kusiak, J. W. (1983). β-Adrenergic antagonists with multiple pharmacophores: Persistent blockage of receptors. *J. Med. Chem.* **26**, 7–11.

Podesta, E. J., Solano, A. R., Attar, R., Sanchez, M. L., and Molina y Vedia, L. (1983). Receptor aggregation induced by antilutropin receptor antibody and biological response in rat testis Leydig cells. *Proc. Natl. Acad. Sci. U.S.A.* **80**, 3986–3990.

Popot, J.-L., and Changeux, J.-P. (1984). Nicotinic receptor of acetylcholine: structure of an oligomeric integral membrane protein. *Physiol. Rev.* **64**, 1162–1239.

Rees-Smith, B., Pyle, G. A., Petersen, V. B., and Hall, R. (1977). Interaction of thyroid-stimulating antibodies with the human thyrotopin receptor. *J. Endocrinol.* **75**, 401–407.

Rodbell, M. (1980). The role of hormone receptors and GTP-regulatory proteins in membrane transduction. *Nature (London)* **284**, 17–22.

Rodbell, M. (1983). The actions of glucagon at its receptor: regulation of adenylate cyclase. *In* "Handbook of Experimental Pharmacology" (P. J. Lefebvre, ed.), Vol. 66/I, pp. 263–290. Springer-Verlag, Berlin and New York.

Roth, R. A., Cassell, D. J., Morgan, D. O., Tatnell, M. A., Jones, R. H., Schüttler, A., and antibodies to the human insulin receptor block insulin binding and inhibit insulin action. *Proc. Natl. Acad. Sci. U.S.A.* **79**, 7312–7316.

Roth, R. A., Maddux, B. A., Cassell, D. J., and Goldfine, I. D. (1983). Regulation of the insulin receptor by a monoclonal anti-receptor antibody. *J. Biol. Chem.* **258**, 12094–12097.

Roth, R. A., Cassell, D. J., Morgan, D. O., Tatnell, M. A., Jones, R. H., Schttler, A., and Brandenberg, D. (1984). Effects of covalently linked insulin dimers on receptor kinase activity and receptor down regulation. *FEBS Lett* **170**, 360–364.

Rubalcava, B., Grajales, M. O., Cerbon, J., and Pliego, J. A. (1983). The role of surface charge and hydrophobic interaction in the activation of rat liver adenyl cyclase. *Biochim. Biophys. Acta* **759**, 243–249.

Schechter, Y., Chang, K., Jacobs, S., and Cuatrecasas, P. (1979a). Modulation of binding and bioactivity of insulin by anti-insulin antibody: Relation to possible role of receptor self-aggregation in hormone action. *Proc. Natl. Acad. Sci. U.S.A.* **76**, 2720–2724.

Schechter, Y., Hernaez, L., Schlessinger, J., and Cuatrecasas, P. (1979b). Local aggregation of hormone-receptor complexes is required for activation by epidermal growth factor. *Nature (London)* **278**, 835–838.

Schechter, Y., Maron, R., Elias, D., and Cohen, I. R. (1982). Autoantibodies to insulin receptor spontaneously develop as anti-idiotypes in mice immunized with insulin. *Science (Washington, D.C.)* **216**, 542–545.

Schreiber, A. B., Lax, I., Yarden, Y., Eshlar, Z., and Schlessinger, J. (1981). Monoclonal antibodies against receptor for epidermal growth factor induce early and delayed effects of epidermal growth factor. *Proc. Natl. Acad. Sci. U.S.A.* **78**, 7535–7539.

Schreiber, A. B., Schlessinger, J., and Edidin, M. (1984). Interaction between major histocompatibility complex antigens and epidermal growth factor receptors on human cells. *J. Cell Biol.* **98**, 725–731.

Shimohigashi, Y., Costa, T., Matsumura, S., Chen, H., and Rodbard, D. (1982). Dimeric enkephalins display enhanced affinity and selectivity for the *Delta* opiate receptor. *Mol. Pharmacol.* **21**, 558–563.

Smith, M. A., Perrin, M. H., and Vale, W. W. (1983). Desensitization of cultured pituitary cells to gonadotropin-releasing hormone: Evidence for a post-receptor mechanism. *Mol. Cell. Endocrinol.* **30**, 85–96.

Stadel, J. M., DeLean, A., and Lefkowitz, R. J. (1982). Molecular mechanisms of coupling in hormone receptor–adenylate cyclase systems. *Adv. Enzymol. Relat. Areas Mol. Biol.* **53**, 1–43.

Strosberg, A. D. (1983). Anti-idiotype and anti-hormone receptor antibodies. *Springer Semin. Immunopathol.* **6**, 67–78.

Suya, H., Abe, Y., Tanaka, I., Ishii, S., and Itaya, K. (1982). Insulin-like activity of photochemically obtained monovalent monomeric concanavalin A in the presence of anti-concanavalin A antibodies: Dependence on multivalency for stimulation of glucose oxidation of rat fat cells. *J. Biochem. (Tokyo)* **92**, 1251–1257.

Swillens, S. (1982). Modulation of catecholamine activation by the number of active β-adrenergic receptors: Theoretical considerations on the role of receptor diffusion in the cell membrane. *J. Cyclic Nucleotide Res.* **8**, 71–82.

Wank, S.A., DeLisi, C., and Metzger, H. (1983). Analysis of the rate limiting step in a ligand–cell receptor interaction: the immunoglobulin E system. *Biochemistry* **22**, 954–959.

Weaver, D. L. (1983). Diffusion-mediated localization on membrane surfaces. *Biophys. J.* **41**, 81–86.

Vitti, P., Rotella, C., Valente, W. A., Cohen, J., Aloj, S., Laccetti, P., Ambesi-Impiombato, F. S., Grollman, E. F., Pinchera, A., Toccafondi, R., and Kohn, L. (1983). Characterization of the optimal stimulatory effects of Graves' monoclonal and serum immunoglobulin G on adenosine 3', 5'-monophosphate production in FRTL-5 thyroid cells: a potential clinical assay. *J. Clin. Endocrinol. Metab.* **57**, 782–791.

Wofsy, C. (1979). Some mathematical problems related to allergic reactions. *In* "Lectures on Mathematics in the Life Sciences," Vol. 12, pp. 135–172. Am. Math. Soc., Providence, Rhode Island.

Wolff, J., and Cook, H. G. (1975). Charge effects in the activation of adenylate cyclase. *J. Biol. Chem.* **250**, 6897–6903.

Yamashita, K., and Field, J. B. (1972). Effects of long acting thyroid stimulator on thyrotropin stimulation of adenyl cyclase activity in thyroid plasma membranes. *J. Clin. Invest.* **51**, 463–472.

Young, J. D.-E., Unkeless, J. C., Kaback, H. R., and Cohn, Z. A. (1983a). Macrophage membrane potential changes associated with γ2b/γ1Fc receptor–ligand binding. *Proc. Natl. Acad. Sci. U.S.A.* **80**, 1357–1361.

Young, J. D.-E., Unkeless, J. C., Kaback, H. R., and Cohn, Z. A. (1983b). Mouse macrophage Fc receptor for IgGγ2b/γ1 in artificial and plasma membrane vesicles functions as a ligand-dependent ionophore. *Proc. Natl. Acad. Sci. U.S.A.* **80**, 1636–1640.

Zakarija, M., and McKenzie, J. M. (1983). Thyroid-stimulating antibody (TSAb) of Graves' disease. *Life Sci.* **32**, 31–44.

Zolman, J. C. (1983). Studies on the temporal relationship between receptor binding and cellular response, especially concerning the luteinizing hormone release rate. *Biochim. Biophys. Acta* **755**, 474–480.

4

# The *Ah* Receptor: A Biochemical and Biologic Perspective

**WILLIAM F. GREENLEE\* and ROBERT A. NEAL†**

Departments of \*Cell Biology and
†Biochemical Toxicology and Pathobiology
Chemical Industry Institute of Toxicology
Research Triangle Park, North Carolina

## I. INTRODUCTION

The identification of the induction receptor (*Ah* receptor)\* for the polycyclic aromatic hydrocarbon-inducible microsomal monooxygenases [cytochrome(s)

---

\*The receptor protein for TCDD and halogenated analogues has been variously termed the "TCDD receptor," the "dioxin receptor," or the "*Ah* receptor." In this chapter, we will use the more generalized designation, *Ah* receptor.

**89**

$P_1$-450] by Poland *et al.* (1976a) was a significant event in the maturing of toxicology as a scientific discipline. Subsequent studies on the role of the *Ah* receptor in mediating the toxicity of 2,3,7,8-tetrachlorodibenzo-*p*-dioxin (TCDD),* the prototype ligand, reveal that this protein regulates the expression of a gene battery which includes the structural genes for cytochrome(s) $P_1$-450 and in certain tissues, interacts with the product of at least one other regulatory locus to control a broader pleiotropic response, resulting in altered patterns of cell proliferation (hyperplasias and/or atrophy) and differentiation (hyper-keratinization and metaplasias) (reviewed in Poland and Knutson, 1982; see also Section III). Thus, the disciplines of toxicology and receptor biology find a common focus. Intensive study of the cell and molecular biology of the *Ah* receptor is providing fundamental insight into understanding the mechanisms of toxicity of TCDD and related halogenated aromatic compounds. Conversely, the use of TCDD as a molecular probe in defining the biology of the *Ah* receptor as a regulatory protein is adding to the knowledge of basic cellular and biochemical processes in eukaryotic cells which function, at least in part, to control pro-grammed patterns of cell proliferation and differentiation. In this chapter, we detail the available knowledge on the biochemistry and biological function of the *Ah* receptor, with particular emphasis on the cell biology of the receptor, as studied using cultured human target cells, and discuss the potential value pro-vided by the understanding of cellular mechanisms to broader issues relevant to TCDD as a major public health problem.

## Evidence for the *Ah* Receptor

### 1. Genetically Determined Differences in the Induction of Cytochrome(s) $P_1$-450

The administration of 3-methylcholanthrene (MC) and structurally related polycyclic aromatic hydrocarbons (PAH) to some (but not all) inbred strains of mice results in the induction of cytochrome $P_1$-450 and associated monoox-ygenase activities, including aryl hydrocarbon hydroxylase (AHH) and 7-ethoxy-coumarin *O*-deethylase (ECOD) (reviewed in Nebert, 1979; Nebert and Jensen, 1979; Poland *et al.*, 1979). In the prototype strains C57BL/6 (responsive) and DBA/2 (nonresponsive) PAH responsiveness, as measured by the induction of hepatic AHH activity, is inherited as a simple autosomal dominant trait (Nebert

---

*Abbreviations include: AHH, aryl hydrocarbon hydroxylase; cAMP, cyclic AMP; BP, ben-zo[*a*]pyrene; Con A, concanavalin A; DMSO, dimethyl sulfoxide; ECOD, 7-ethoxycoumarin *O*-deethylase; EGF, epidermal growth factor; MC, 3-methylcholanthrene; MNNG, *N*-methyl-*N'*-nitro-*N*-nitrosoguanidine; PAH, polycyclic aromatic hydrocarbons; PB, phenobarbital; PHA, phy-tohemagglutinin; SCC, squamous cell carcinoma; TCDD, 2,3,7,8-tetrachlorodibenzo-*p*-dioxin; TPA, 12-*O*-tetradecanoylphorbol 13-acetate.

TABLE I

**Comparison of the Sensitivity of Inbred Strains of Mice to TCDD**[a]

| Mouse strain | $ED_{50}$ (nmoles/kg) |
| --- | --- |
| PAH-responsive | |
| C57BL/6J | 1.2 |
| BALB/cJ | 1.0 |
| A/J | 1.2 |
| PAH-nonresponsive | |
| DBA/2J | 13.8 |
| AKR/J | 11.0 |
| SJL/J | 25.0 |

[a] Mice were injected with p-dioxane or various concentrations of TCDD and hepatic AHH activity was assayed 24 hr later. The $ED_{50}$ values were calculated from the dose–response curves. Data from Poland and Glover (1975).

*et al.*, 1972; Thomas *et al.*, 1972). The gene locus controlling this trait is designated the *Ah* locus.*

TCDD is 30,000 times more potent than MC as an inducer of hepatic AHH activity in the rat (Poland and Glover, 1974) and induces AHH activity in both PAH-responsive and nonresponsive murine strains (Poland *et al.*, 1975). Strains nonresponsive to PAH inducers are less sensitive to TCDD, as judged by comparison of $ED_{50}$ values (dose required to elicit 50% of the maximal response) (Table I). Poland *et al.* (1976a) identified and characterized a receptor protein for TCDD, MC, and other inducers in the hepatic cytosol fractions from C57BL/6J mice. The available data indicate that this protein is the product of the *Ah* locus (Poland *et al.*, 1976a, 1979; Nebert, 1979). In the nonresponsive murine strains, a mutation in the *Ah* locus results in a receptor with a reduced affinity for TCDD (Poland *et al.*, 1976a, 1979). The PAH inducers are more readily metabolized than TCDD and presumably do not achieve a sufficient concentration in target cells to bind to the altered receptor.

The presence of the *Ah* locus (or regulatory gene loci equivalent to the murine *Ah* locus) in human cells is suggested by the findings of several studies. The inducibility of AHH activity in cultured mitogen-activated lymphocytes from

*Studies by Thomas and Hutton (1973) and Robinson *et al.* (1974) in other strains of mice have shown that the inheritance of PAH responsiveness is more complex, involving two or more nonlinked loci. In this chapter we will only consider the dominant inheritance pattern found for C57BL/6 and DBA/2 mice.

several individuals was found to be trimodal and the Hardy–Weinberg equilibrium data indicated that AHH induction was regulated by a single gene locus with two alleles (Kellerman *et al.*, 1973). Other investigators confirmed a heritable component in the inducibility of AHH activity in cultured lymphocytes (Atlas *et al.*, 1976) and monocytes (Okuda *et al.*, 1977) but were not able to distinguish between a monogenic or polygenic mode of inheritance. Chromosome mapping studies in mouse–human cell hybrids have shown that either the structural or the regulatory genes for the induction of AHH activity are located on human chromosome 2 (Brown *et al.*, 1976; Wiebel *et al.*, 1981); however, the participation of murine genes in the observed response cannot be ruled out (Wiebel *et al.*, 1981). Direct demonstration of a receptor protein for TCDD has been reported in human lymphocytes (Carlstedt-Duke *et al.*, 1982) and human squamous cell carcinoma (SCC) lines (Table II) (Hudson *et al.*, 1983). In the latter study, it was observed that the human SCC lines show differential sensitivity to TCDD analogous to the differential responsiveness of murine strains regulated by the *Ah* locus (see Section IV,A).

## 2. Ligand Specificity

The relative binding affinities of several PAH inducers and halogenated dibenzo-*p*-dioxin, dibenzofurans, azo- and azoxybenzenes, and biphenyl isomers for the hepatic receptor from C57BL/6J mice have been determined (Poland *et al.*, 1976a,b; Poland and Glover, 1977). Isomers which compete with [$^3$H]TCDD for specific binding sites *in vitro* are biologically active *in vivo* (Poland *et al.*, 1979). Halogenated aromatic compounds which bind the *Ah* receptor can be visualized as fitting into a rectangle (3 × 10 Å) (Fig. 1). Halogen atoms must occupy at least

TABLE II

Specific Binding of [$^3$H]-TCDD to Cytosol Fractions from Human SCC Lines

| Cell line | Specific binding[a] (fmoles/mg cytosol protein) | Relative response | |
| --- | --- | --- | --- |
| | | Specific binding[b] | ECOD activity[c] |
| SCC-9 | 9.1 (0.6) | 5.3 | 5.2 |
| SCC-15 | 6.1 (1.8) | 3.5 | 2.2 |
| SCC-12F | 1.8 (0.3) | 1.0 | 1.0 |

[a] Specific binding was measured by sucrose density gradient analysis. Values shown represent the average from two separate experiments. The range is given in parentheses. From Hudson *et al.*, (1983).

[b] Calculated from the average values for the specific binding in each line and normalized with respect to the value obtained for the line SCC-12F.

[c] Determined from the maximally induced activities and normalized with respect to the value obtained for line SCC-12F.

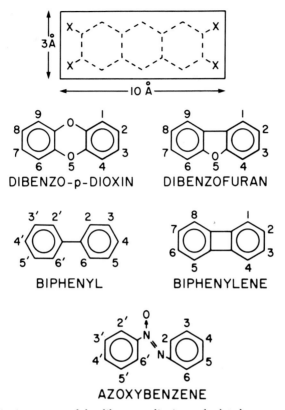

**Fig. 1.** Basic ring system of the dibenzo-p-dioxins and related compounds. Planar or nearly planar isomers, with chlorine atoms occupying at least three of the four corners of a rectangle 3 × 10 Å, bind to the TCDD receptor (Poland et al., 1979).

three of the four corners and planarity or near planarity seems to be essential for optimal binding activity (Poland *et al.*, 1979). The stereospecificity of the receptor binding site is discussed in greater detail in Section II,D.

### 3. Subcellular Distribution of the Ah Receptor

The results of initial studies on the subcellular distribution of receptor indicated that the *Ah* receptor, in the absence of ligand binding, was predominantly distributed within the cytosolic compartment (Poland *et al.*, 1976a). These conclusions were based on subcellular fractionation of [$^3$H]TCDD, following incubation of the radioligand with liver homogenates from C57BL/6J mice pretreated with solvent vehicle, unlabeled TCDD, or active and inactive isomers. The rationale for this approach was that unlabeled active isomers administered *in vivo* would bind to the receptor, blocking the specific binding sites for

[³H]TCDD during the *in vitro* incubation. The hypothesis that the TCDD receptor is located in the cytosol is supported by the apparent cytosol-to-nucleus translocation (see Section III,A,1) of the TCDD–receptor complex (Okey *et al.*, 1979, 1980). Recent studies on the distribution of the *Ah* receptor between the cytoplasmic and nuclear compartments in mouse hepatoma cell lines suggest that in broken cells both unoccupied receptors and TCDD–receptor complexes redistribute between the nuclear and cytosolic fractions, presumably due to the dilution effect of the volume of buffer used for homogenization (Whitlock and Galeazzi, 1984). When the effects of dilution are minimized, 80–100% of the receptors are found in the nuclear fraction, suggesting that in the intact cell the *Ah* receptors are located within the nucleus. Enhanced affinity of the nuclear binding of the receptor–ligand complex versus unbound receptor would account for the apparent time-dependent cytosol-to-nucleus translocation (i.e., less receptor would redistribute to the cytosolic fraction during homogenization as a function of ligand binding).

### 4. Tissue Distribution of the Ah Receptor

**Species Differences.** The relative concentrations of the *Ah* receptor in hepatic and extrahepatic tissues of various species have been compared. The concentration of the hepatic receptor (fmoles/mg protein) in the Sprague–Dawley rat (61 ± 5 ) is similar to that detected in the livers from C57BL/6J mice (74 ± 10) (Gasiewicz and Neal, 1982). The receptor is not detectable in the livers from DBA/2J mice (see Section II,A). Hepatic receptor concentrations in the B6D2F1/J heterozygote (23 ± 2), a cross between C57BL/6J and DBA/2J inbred mice, is intermediate between that of the parent strains. Similar results have been reported by Mason and Okey (1982). In an extensive interspecies study, it was found that receptor concentrations (Table III) and binding affinity for [³H]TCDD in hepatic cytosol fractions of the guinea pig, rat, monkey, mouse, and hamster are similar (Gasiewicz, 1983).

Mason and Okey (1982) surveyed various tissues of the Sprague–Dawley rat and the C57BL/6J mouse for the *Ah* receptor. The *Ah* receptor was found in the liver, lung, thymus, intestine, kidney, and prostate (mouse only) (Table III) but was not detected in the testes, heart, pancreas, muscle, adrenal glands, or brain. In contrast, Carlstedt-Duke (1979) measured low levels of receptor in testes, brain, and skeletal muscle cytosol fractions from Sprague–Dawley rats. In the guinea pig and hamster, the *Ah* receptor is found in the liver, lung, thymus, intestine (guinea pig only), kidney, spleen, testes (guinea pig only), heart (guinea pig only), and brain (cerebellum) (Gasiewicz, 1983). In these animals, the *Ah* receptor is not detected in the pancreas, muscle, or adrenal glands. Thus, interspecies differences exist with regard to both the tissue distribution of the *Ah* receptor and, in those tissues possessing the receptor, the relative concentrations of this protein (Table III). The available data indicate that the *Ah* receptor is a

TABLE III

Relative Tissue Concentrations of *Ah* Receptor in Various Mammalian Species[a]

| Tissues | Guinea pig[b] (Hartley strain) | Hamster[b] (golden syrian) | Rat[c] (Sprague– Dawley) | Mouse[c] (C57BL/6J) |
|---|---|---|---|---|
| Liver | 59.0 ± 11.2 (8) | 60.6 ± 22.5 (7) | 39 ± 1.9 (8) | 32 ± 1.5 (8) |
| Lung | 85.9 ± 28.0 (4) | 34.5 ± 19.6 (4) | 47 ± 4.3 (3) | 23 ± 6.4 (3) |
| Thymus | 47 ± 7 (8) | 5.1 ± 6.3 (4) | 54 ± 3.9 (3) | 8 ± 2.2 (5) |
| Intestine | 17.7 (2) | ND | 15 ± 1.7 (4) | 8 ± 2.3 (3) |
| Kidney | 23.9 ± 19.3 (4) | 12.7 (2) | 1.2 ± 0.9 (3) | 10 ± 0.4 (2) |
| Spleen | 17.0 (2) | 5.7 ± 5.4 (3) | — | — |
| Testes | 50.1 ± 7.2 (4) | ND[d] | ND | ND |
| Heart | 16.1 (2) | ND | ND | ND |
| Pancreas | ND | ND | ND | ND |
| Muscle | ND | ND | ND | ND |
| Adrenal | ND | ND | ND | ND |
| Prostate | — | — | — | 1.3 ± 0.8 (4) |
| Brain | — | — | ND | ND |
| Midbrain | ND | ND | — | — |
| Cerebrum | 11.3 (2) | 13.6 (1) | — | — |
| Medulla | ND | ND | — | — |
| Cerebellum | 11.7 (2) | ND | — | — |
| Hypothalamus | ND | ND | — | — |

[a] Expressed as fmoles/mg of cytosolic protein and mean ± SD of (n) determinations.
[b] Gasiewicz (1983).
[c] Mason and Okey (1982).
[d] ND, not detectable.

necessary but not sufficient requirement for a tissue-specific toxic response to TCDD (Gasiewicz and Rucci, 1984a) (see Section III,C).

## II. BIOCHEMICAL CHARACTERIZATION OF THE *Ah* RECEPTOR

### A. Assay Methods

The rationale for measuring the specific binding of TCDD is analogous to the approach used for steroid hormone receptors. It is assumed that a small pool of high-affinity (specific) binding sites for TCDD exists within a much larger pool of low-affinity (nonspecific) sites. The first assay reported for measuring the *Ah* receptor used dextran-coated charcoal (Poland *et al.*, 1976a). A fixed concentration of [$^3$H]TCDD (not exceeding 1 n*M*) is incubated with the appropriate subcellular fraction (typically the cytosol fraction), and the unbound radioligand is removed by adsorption onto dextran-coated charcoal. The radioactivity not adsorbed represents total binding. A second portion of the cytosol fraction is incubated with the same concentration of [$^3$H]TCDD plus a 100- to 200-fold molar excess of unlabeled ligand (blocking the small pool of specific sites) and treated with dextran-coated charcoal. In these samples, nonadsorbed radioactivity gives a value for nonspecific binding. The difference, total minus nonspecific, represents specific binding.

Other assays used to measure the *Ah* receptor include sucrose density gradient analysis (Okey *et al.*, 1979), adsorption of the receptor on hydroxyapatite (Gasiewicz and Neal, 1982), isoelectric focusing (Carlstedt-Duke *et al.*, 1978), precipitation of the TCDD–receptor complex with protamine sulfate (Denison *et al.*, 1984), and gel permeation high-performance liquid chromatography (Gasiewicz and Rucci, 1984b). A sucrose density gradient profile for the specific binding of [$^3$H]TCDD to bone marrow cytosol fractions from C57BL/6J mice is shown in Fig. 2. The relative insolubility of TCDD in aqueous solutions (<0.2 ng/ml; International Agency for Research on Cancer, 1977) limits the ability of these assays to detect low-affinity ($K_D \gg 1$ n$M$) binding sites (e.g., cytosol fractions from DBA/2 mice). The solubility of TCDD is greater in buffers containing ampholytes or in the presence of protein (Poland *et al.*, 1976a). Binding studies in which the total concentration of [$^3$H]TCDD and unlabeled TCDD (or other halogenated analogues such as the tetra-, penta-, or hexachlorobiphenyls) exceeds 0.1 $\mu M$ should always provide data on the total radioactivity in solution before binding is assayed. The diminished solubility of [$^3$H]TCDD in the presence of high concentrations (>0.1 $\mu M$) of certain unlabeled ligands can result in apparent specific binding, as determined by the difference in radioactivity measured in samples incubated with radioligand alone versus those incubated with radioligand plus unlabeled ligand.

Sucrose density gradient analysis is one of the most widely used methods in studies on the *Ah* receptor. Typically, gradients run in swinging bucket rotors are centrifuged for 16 hr (48,000 rpm) at 2°C. Although the rate of dissociation of the TCDD–receptor complex is not known, it is possible that a prolonged centrifugation time can decrease the sensitivity of this assay. In this regard, comparison of the concentration of the progesterone receptor in human breast cancer biopsies measured using a swinging bucket rotor (16 hr centrifugation) versus a vertical tube rotor (3 hr centrifugation) indicates that prolonged centrifugation results in a greater than twofold underestimation of the receptor concentration, presumably due to dissociation of the receptor–ligand complex (Powell *et al.*, 1979). Similar studies carried out with the *Ah* receptor demonstrate that comparable levels of receptor are measured using either rotor (Tsui and Okey, 1981), suggesting that the *Ah* receptor–ligand complex does not rapidly dissociate in high-density sucrose.

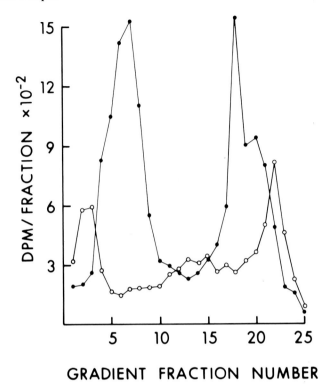

**Fig. 2.** Specific binding of [³H]TCDD to bone marrow cytosol fractions from C57BL/6 mice. Bone marrow was aspirated from the femurs of C57BL/6J mice (Greenlee *et al.*, 1981), and cytosol fractions were prepared and incubated with 1 n*M* [³H]TCDD (●) or with 1 n*M* [³H]TCDD plus 100 n*M* unlabeled TCDD (○). Labeled cytosolic fractions were treated with charcoal–dextran and specific binding of [³H]TCDD was measured by sucrose density gradient analysis, as described by Okey *et al.* (1979).

## B. Purification

There have been no significant advances in the purification of the *Ah* receptor since its initial identification by Poland *et al.* (1976a). A 5- to 10-fold purification can be achieved by the rather simple procedure of ammonium sulfate fractionation. The *Ah* receptor precipitates in the 30–55% fraction. A Scatchard plot of the partially purified receptor from C57BL/6 murine hepatic cytosol is shown in Fig. 3. The value obtained for the total number of binding sites (595 fmoles/ mg protein) represents a 10-fold purification.

Hannah *et al.* (1981) have resolved binding proteins for [³H]TCDD and [³H]MC in livers from C57BL/6J and B6D2 mice into three major components, separable by gel permeation chromatography. One of the binding components

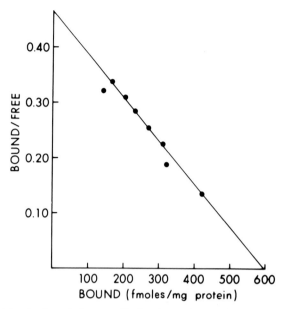

**Fig. 3.** Scatchard plot of the specific binding of [³H]TCDD to partially purified murine hepatic receptor. A 30–55% ammonium sulfate fraction was prepared from hepatic cytosol of C57BL/6J mice, and the specific binding of [³H]TCDD was assayed as described previously ($K_d$ = 0.59 nM and n = 595 fmoles/mg protein) (Greenlee and Poland, 1979).

(designated peak II) appeared to be the *Ah* receptor and was detected in C57BL/6 (responsive) but not DBA/2J (nonresponsive) mice. Peak II (245,000 molecular weight) eluted at concentrations of 0.16–0.20 *M* NaCl on anion exchange resins and had a Stokes radius of approximately 75 Å. The third component (peak III) eluted at lower NaCl concentrations and was found in hepatic cytosol fractions from both C57BL/6 and DBA/2 mice. Binding of [³H]TCDD to this component was not saturable. The first component (peak I) appeared to be an aggregate.

## C. Physical Properties of Crude and Partially Purified Receptors

Physical properties of the *Ah* receptor isolated from various tissues and species are summarized in Table IV. The *Ah* receptor is a high-molecular-weight, apparent oligomer, sedimenting at 9 S in low-salt buffers and 4–7 S in high-salt buffers. The more rapid sedimentation of the receptor under conditions of low-ionic strength is probably due to aggregation. Equilibrium dissociation constants range from 0.1–2 n*M* (Poland and Knutson, 1982), and the total number of hepatic binding sites ranges from 30–60 fmoles/mg protein (Gasiewicz, 1983).

**TABLE IV**

**Summary of the Physical Properties of the *Ah* Receptor in Various Species and Tissues[a]**

| Species | Tissue | MW | $K_d$ | n | s | Ref. |
|---|---|---|---|---|---|---|
| C57BL/6J Mouse | Liver (cytosol) | 245,000 | — | — | 9 (low-salt) 7.5 (high-salt) | b |
| Sprague–Dawley rat | Liver (cytosol) | 280,000 | 0.13 | 98 | — | c |
| | Thymus (cytosol) | — | 0.36 | 68 | 4–5 (high-salt) | d |
| | Thymus (nuclear) | — | — | — | 4–5 (high-salt) | d |
| Human | Lymphocyte (cytosol) | — | — | 0–42 | — | e |

[a] Abbreviations: MW, molecular weight; $K_d$, equilibrium dissociation constant (nM); n, concentration of *Ah* receptor (fmoles/mg protein); and s, sedimentation coefficient.

[b] Hannah *et al.* (1981).

[c] Gasiewicz and Rucci, 1984b.

[d] Lund *et al.* (1982).

[e] Carlstedt-Duke *et al.* (1982).

Species differences in the concentrations of the receptor in other tissues are given in Table III. Isoelectric focusing of the receptor from rat hepatic cytosol following limited trypsinization was carried out (Carlstedt-Duke *et al.*, 1978). A single peak focusing at p*I* 5.15–5.25 was observed. In Fig. 4 isoelectric focusing of the receptor from calf thymus is shown. The receptor was not trypsinized and focused as two peaks with p*I* values of 5.6 and 6.1, suggesting that the *Ah* receptor is composed of at least two protein subunits (see Section III,A,2).

## D. Stereospecificity

Halogenated aromatic compounds which bind to the *Ah* receptor with high affinity are (1) planar and (2) have ring systems which fit within a 3 × 10 Å rectangle with at least three of the four corners of that rectangle occupied by halogen atoms (Fig. 1) (Poland *et al.*, 1976a,b; Poland and Glover, 1977; Goldstein, 1980). While broadly applicable, this model does not explain the binding activity observed with halogenated naphthalenes such as 2,3,6,7-tetrabromona-

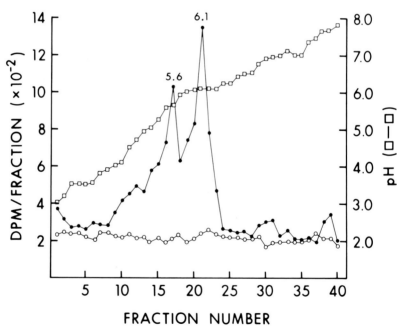

**Fig. 4.** Isoelectric focusing of the TCDD receptor. Cytosol fractions were prepared from calf thymus and incubated with 1 nM [³H]TCDD (●) or with 1 nM [³H]TCDD plus 100 nM unlabeled TCDD (○). After treatment with charcoal–dextran (Greenlee and Poland, 1979), the labeled cytosol fractions were analyzed by gel isoelectric focusing, as described by Brewer *et al.* (1974).

phthalene (3 × 8.06 Å; Poland and Knutson, 1982) and the PAH inducers. McKinney and McConnell (1982) proposed that molecular polarizability was the underlying electronic basis for binding to the *Ah* receptor and suggested that for certain compounds with high-binding activity (e.g., 2,3,7-tribromodibenzo-*p*-dioxin and 2,3,5,7,8-pentabromonaphthalene) a triangular rather than a rectangular fit to the *Ah* receptor was possible.

Recent investigations on a series of nonplanar halogenated biphenyl isomers (Bandiera *et al.*, 1982a,b; Parkinson *et al.*, 1980a,b, 1982; Sawyer and Safe, 1982) have reported that isomers with halogen atoms located on one and even two of the ortho carbons (positions 2,2′, 6, or 6′; Fig. 1) can induce cytochrome $P_1$-450-mediated activities and bind to the *Ah* receptor. Ortho substitution markedly increases the energy barriers to internal rotation, preventing planar conformations (McKinney *et al.*, 1983). Accordingly, the nonplanar ortho-substituted polyhalogenated biphenyls are weak inducers; for example, the $EC_{50}$ value for the induction of AHH activity in H-4-II-E cells by 3,3′,4,4′,5-pentachlorobiphenyl is $2.4 \times 10^{-10}\ M$ compared to a value of $7.11 \times 10^{-7}$ M for 2,3,3′,4,4′,5-hexachlorobiphenyl (Sawyer and Safe, 1982). The affinity of nonplanar polychlorinated biphenyls reported to be ligands for the *Ah* receptor (measured in rat hepatic cytosol fractions) is also significantly lower (one to three orders of magnitude) than the affinity of TCDD or the planar compounds, 3,3′,4,4′-tetrachlorobiphenyl or 3,3′,4,4′,5-pentachlorobiphenyl (Bandiera *et al.*, 1982a). Based on comparison of the structure–induction and structure–toxicity relationships for several polyhalogenated biphenyl isomers, McKinney *et al.* (1985) proposed that a second receptor, with stereospecificity distinct from that of the *Ah* induction receptor, mediates toxicity (see Section III,C).

## III. BIOLOGY OF THE *Ah* RECEPTOR

### A. Induction of Cytochrome(s) $P_1$-450

TCDD induces cytochrome $P_1$-450 and several genetically linked monooxygenase activities in both PAH-responsive and nonresponsive mice (Section I,A,1) (Poland *et al.*, 1974; Lang and Nebert, 1981). The spectral properties of the cytochrome P-450 species induced by TCDD or MC in either responsive or nonresponsive mice are essentially the same (Poland *et al.*, 1974). Cytochrome P-448 is a second form of cytochrome P-450 inducible by TCDD and PAH compounds (Negishi and Nebert, 1979). The induction of both cytochromes $P_1$-450 and P-448 is regulated by the *Ah* locus in the mouse (Negishi and Nebert, 1979), however, the expression of each protein (or group of proteins) is under different temporal control. Ontogenetically, cytochrome $P_1$-450 is inducible from several days (rat and mouse) to weeks (rabbit) earlier than cytochrome P-448 (Atlas *et al.*, 1977; Guenthner and Nebert, 1978).

## 1. Nuclear Binding of the TCDD–Ah Receptor Complex

Okey *et al.* (1979, 1980) have provided evidence for the apparent nuclear translocation of the TCDD–cytosol binding species in murine liver and continuous cell lines in culture and have shown that the nuclear TCDD–receptor complex is extractable by high-salt, similar to the behavior of the steroid hormone receptors. Carlstedt-Duke *et al.* (1981) reported that the rat hepatic TCDD–receptor complex binds to DNA-cellulose and that the receptor must first bind TCDD before it can interact with DNA. Limited digestion of crude receptor preparations with trypsin does not significantly alter ligand binding but destroys the ability of the receptor–ligand complex to bind to DNA-cellulose (Carlstedt-Duke *et al.*, 1981). These findings suggest that the *Ah* receptor is an allosteric protein with distinct ligand and chromatin binding domains.

Studies on the expression of cytochrome $P_1$-450 messenger RNA in inbred mice (Tukey *et al.*, 1982) and murine hepatoma cell lines (Israel and Whitlock, 1983) indicate that the amount of TCDD–receptor complex in the nucleus correlates with the quantity of messenger RNA transcribed. The TCDD–receptor complex can be recovered from hepatic and nonhepatic nuclei of DBA/2 (nonresponsive) mice, following treatment with TCDD at doses which induce AHH activity in both PAH-responsive and nonresponsive strains (Section I,A,1) (Mason and Okey, 1982). Reconstitution of hepatic cytosol and nuclear fractions from C57BL/6 and DBA/2 mice indicate that the genetically determined difference in sensitivity to TCDD and responsiveness to PAH inducers is confined to the *Ah* receptor (ligand binding site) and is not due to altered nuclear acceptor sites: equivalent nuclear binding is observed in either C57BL/6 or DBA/2 nuclei reconstituted with charged (preincubated with [$^3$H]TCDD) C57BL/6 cytosol whereas reconstitution of C57BL/6 nuclei with charged DBA/2 cytosol (with no detectable receptor binding) results in greatly diminished nuclear binding (Greenlee and Poland, 1979).

## 2. Somatic Cell Genetic Analysis of AHH Induction Variants

AHH activity is inducible by TCDD in wild-type murine hepatoma (Hepa 1c1c7) cells (Miller and Whitlock, 1981). Two classes of induction variants have been isolated and characterized (Miller and Whitlock, 1981; Miller *et al.*, 1983). One class (Class I), expressing low levels of basal and induced AHH activity, possesses *Ah* receptors with altered binding of TCDD. Nuclear accumulation of TCDD–receptor complexes is normal in these cells. A second class (Class II), with no detectable basal or inducible AHH activity, possesses *Ah* receptors with normal binding properties for TCDD but greatly diminished nuclear accumulation of the TCDD–receptor complexes (Miller *et al.*, 1983). Both classes are recessive with respect to the wild type, as determined by cell fusion, and the results of complementation analyses indicate that the defects are located on different genes (Miller *et al.*, 1983). It has been postulated that Class I variants contain a lesion in the ligand binding domain and Class II variants contain a lesion in the chromatin binding domain (Whitlock and Galeazzi, 1984). The involvement of different genes in the expression of the variant phenotypes suggests that the *Ah* receptor is composed of at least two proteins: one containing the ligand binding site and one containing a chromatin binding site (Whitlock and Galeazzi, 1984). This hypothesis is supported by the findings on the binding of the rat hepatic TCDD–receptor complex to DNA (Section III,A,1) (Carlstedt-Duke *et al.*, 1981).

A third class of murine hepatoma variants has been isolated, which over-transcribe the cytochrome $P_1$-450 gene in response to TCDD (Jones *et al.*, 1984). Biochemical and genetic analyses indicate that the variant phenotype is codominant with respect to the wild-type, and it was suggested that the variant cells contain an altered *cis*-acting genomic element regulating the expression of the cytochrome $P_1$-450 gene (Jones *et al.*, 1984). A DNA fragment containing sequences upstream of the cytochrome $P_1$-450 gene has been isolated and shown to contain at least three functional *cis*-acting control elements: a putative promoter, an inhibitory domain which blocks promoter function, and a TCDD-responsive domain (Jones *et al.*, 1985). Both the inhibitory and TCDD-responsive domains are upstream of the promoter and are presumably controlled by *trans*-acting regulatory factors. The TCDD-responsive control element was postulated to interact with the *Ah* receptor–TCDD complex, releasing the promoter from inhibition and thus increasing transcription of the cytochrome $P_1$-450 gene (Jones *et al.*, 1985).

## B. Coordinate Gene Expression

Binding of the TCDD–receptor complex to chromatin is the presumed initiating event in the sequence leading to the *de novo* synthesis of several inducible enzyme proteins (Poland *et al.*, 1979; Nebert and Jensen, 1979). The enzymes known to be induced by TCDD and PAH compounds are listed in Table V. The coordinate expression of at least four of these proteins, cytochrome $P_1$-450 (Nebert *et al.*, 1972; Thomas *et al.*, 1972), UDPglucuronyltransferase (Owens, 1977), DT-diaphorase [NAD(P)H dehydrogenase (quinone)] (Kumaki *et al.*, 1977), and ornithine decarboxylase (Nebert *et al.*, 1980) has been shown to be regulated by the *Ah* receptor. Uroporphyrinogen decarboxylase activity is decreased after chronic treatment of responsive (C57BL/6) but not nonresponsive (DBA/2J) mice with TCDD (Jones and Sweeney, 1980). Decreased decarboxylase activity is not associated with changes in the amount of immunoreactive protein (Elder and Sheppard, 1982), suggesting that the *Ah* receptor is regulating the activity but not the synthesis of this enzyme.

TABLE V

Enzymes Induced by TCDD and PAH Compounds

| Enzyme | Cell fraction | Reference |
|--------|---------------|-----------|
| Cytochrome $P_1$-450[a] | Microsomal | Nebert et al. (1972) |
| UDPglucuronyltransferase[a] | Microsomal | Owens (1977) |
| Ornithine decarboxylase[a] | Cytosol | Nebert et al. (1980) |
| Glutathione S-transferase B | Cytosol | Kirsch et al. (1975) |
| Aldehyde dehydrogenase | Cytosol | Dietrich et al. (1978) |
| Choline kinase | Cytosol | Ishidate et al. (1980) |
| DT-diaphorase[a] | Cytosol | Beatty and Neal (1976) |
| [NAD(P)H dehydrogenase (quinone)] | | Kumaki et al. (1977) |
| δ-Aminolevulinate synthase | Mitochondrial | Poland and Glover (1973) |

[a] Shown to be regulated by the Ah locus in inbred strains of mice.

Few data are available on the regulation of the synthesis of the Ah receptor. The hepatic Ah receptor from Sprague–Dawley rats is independent of endocrine modulation, as judged by receptor concentrations measured before and after ablation of the testes, ovaries, adrenals, or hypophysis (Carlstedt-Duke et al., 1979). Treatment with PB results in approximately a 30% increase in the concentration of the Ah receptor in hepatic cytosol from C57BL/6J mice and a doubling in the concentration of the receptor in hepatic cytosol from Sprague-Dawley rats (Okey and Vella, 1984). The elevated concentration in the hepatic Ah receptor elicited by PB does not appear to result in an increased response to PAH inducers (Okey and Vella, 1984). In an earlier study, the number of high-affinity TCDD binding sites was reported to increase 60–140% above control levels in cultured H-4-II-E and Hepa-1 cells treated with MC or benz[a]anthracene, two compounds which bind to the Ah receptor (Guenthner and Nebert, 1977), suggesting that the Ah receptor regulated its own synthesis. However, in subsequent investigations using a different receptor binding assay, no increase in the number of high-affinity TCDD binding sites was observed in either the transformed liver cell lines treated with benz[a]anthracene (Okey et al., 1979) or in C57BL/6J (responsive) mice given β-naphthoflavone (an Ah receptor ligand) (Okey et al., 1980).

## C. Receptor-Mediated Toxicity

### 1. Acute Toxicity

The single dose of TCDD which produces acute lethality in a number of animal species varies widely. The $LD_{50}$ values (dose resulting in lethality in 50%

of the treated animals) for TCDD are given in Table VI. Of the animal species so far examined, the guinea pig ($LD_{50}$ = 2 $\mu$g/kg) is the most sensitive (McConnell *et al.*, 1978b) and the hamster ($LD_{50}$ >3000 $\mu$g/kg) is the least sensitive (Olson *et al.*, 1980b; Henck *et al.*, 1981). The $LD_{50}$ values for TCDD in other species are between these two extremes (Table VI; Neal, 1985).

In different strains of mice the concentrations of *Ah* receptor or the affinity of the *Ah* receptor for TCDD correlates with the ability of TCDD to induce AHH activity (Poland *et al.*, 1976a) and to cause acute lethality (Neal *et al.*, 1982), thymic involution (Poland and Glover, 1977), cleft palate (Poland and Glover, 1978), and hepatic porphyria (Jones and Sweeney, 1980). However, this apparent correlation between the level of the *Ah* receptor and various adverse biological effects seen in murine strains is not always observed in other animal species (cf. Tables III and VI). In this regard, TCDD produces thymic atrophy in all animals after either chronic or acute (single exposure) dosing regimens (also see Section III,C,2,b). The $ED_{50}$ values (single dose) for TCDD-induced thymic atrophy in the guinea pig (1 $\mu$g/kg) are much less than for the rat (15 $\mu$g/kg) and hamster (>3000 $\mu$g/kg) (Gasiewicz, 1983). The concentrations of *Ah* receptor in the thymus of the rat and guinea pig are similar and are 10 times greater than in the hamster (Table III). Thus, differences in the sensitivity to TCDD-induced thymic atrophy in the guinea pig (the most sensitive species) versus the hamster (the least sensitive) appear to be determined by the relative concentrations of *Ah*

TABLE VI

Single Dose $LD_{50}$ Values for TCDD

| Species | Route | $LD_{50}$ ($\mu$g/kg) | References |
|---|---|---|---|
| Guinea pig | Oral | 2 | McConnell *et al.* (1978b) |
| Monkey | Oral | 50 | McConnell *et al.* (1978a) |
| Rat | | | |
|     Adult male | IP | 60 | Beatty *et al.* (1978) |
|     Weanling male | IP | 25 | Beatty *et al.* (1978) |
|     Weanling male | IP | 44 | Beatty *et al.* (1978) |
|     Adult female | IP | 25 | Beatty *et al.* (1978) |
| Rabbit | Oral | 115 | Schwetz *et al.* (1973) |
| Rabbit | Skin | 275 | Schwetz *et al.* (1973) |
| Mouse | | | |
|     C57BL/6J | IP | 132 | Gasiewicz *et al.* (1983) |
|     DBA/2J | IP | 620 | Gasiewicz *et al.* (1983) |
|     B6D2F1/J | IP | 300 | Gasiewicz *et al.* (1983) |
| Hamster | IP | >3000 | Olson *et al.* (1980b) |
| Hamster | Oral | 5051 | Henck *et al.* (1981) |

receptor, whereas in the rat, decreased sensitivity to TCDD is not associated with differences in the concentration of *Ah* receptor.

Rose *et al.* (1976), Poiger and Schlatter (1979), and Olson *et al.* (1980a) have shown that TCDD is slowly metabolized in a number of species. The cytochrome *P*-450 monooxygenase system is primarily responsible for the metabolism of TCDD, and the major metabolites are phenolic derivatives (Poiger and Schlatter, 1979; Olson *et al.*, 1983). To test whether the parent compound or metabolites were responsible for acute toxicity, the $LD_{50}$ values for TCDD in control weaning rats and in weanling rats pretreated with phenobarbital (PB) (50 mg/kg/3 days), MC (40 mg/kg), or TCDD (5 μg/kg), inducers of the cytochrome *P*-450 monooxygenase system, were determined (Beatty *et al.*, 1978). In the rats pretreated with PB, MC, or TCDD, the $LD_{50}$ values are increased relative to controls. These data suggest that metabolism leads to a decrease in the acute toxicity of TCDD and that the parent compound is probably responsible for the acute toxicity. This conclusion is consistent with the findings of Poiger and Buser (1983) who administered the metabolites of [³H]TCDD excreted in the bile of a dog to guinea pigs. On a molar basis (based on radioactivity), the metabolites were more than a 100 times less toxic than the parent compound (TCDD).

## 2. Chronic Toxicity

A comparison of some of the histopathological findings of the lesions produced by TCDD in humans and in mice is given in Table VII. Symptoms reported following human exposure to TCDD include weight loss, impaired liver function, hepatic porphyria, general malaise, peripheral neuropathies, and chloracne (Kimbrough, 1980). Chloracne is one of the most sensitive and widespread responses to TCDD in humans (Kimmig and Schultz, 1957; Crow, 1970; Taylor, 1974, 1979) but is expressed in only a limited number of animal species (e.g., rabbits, monkeys, and hairless mice) (Ingami *et al.*, 1969; McConnell *et al.*, 1978a; Kimbrough, 1980). The skin lesions which characterize chloracne include thikening of the epidermis (acanthosis), hyperkeratosis, and squamous metaplasia of the epithelial lining of the sebaceous glands (Kimbrough, 1980; McConnell *et al.*, 1978a). Acneform lesions develop as hair follicles dilate and become filled with keratin and sebaceous glands become cystic (Kimbrough, 1980).

In animal models, the available data on structure–activity relationships indicate that the chloracneogenic potential of halogenated aromatic compounds correlates with the relative affinity of these compounds for the *Ah* receptor (measured in hepatic cytosol from C57BL/6 mice) (Schwetz *et al.*, 1973; Poland *et al.*, 1979; McNulty *et al.*, 1980; Poland and Knutson, 1982). The *Ah* receptor has been detected in murine epidermis (Knutson and Poland, 1982), in cultured human epidermal cells (W. F. Greenlee, unpublished observations), and in human squamous cell carcinoma (SCC) lines of epidermal origin (Hudson *et al.*,

TABLE VII

**Comparison of the Histopathology of TCDD-Induced Lesions in Humans and Mice**

| Target | Mice | Humans |
|---|---|---|
| Immune system | Thymic atrophy[a] | Not known |
| | Suppressed cell immunity | Not known |
| | Induction of T-suppressor cells | Not known |
| Skin | Epidermal hyperplasia (hr/hr)[b] | Chloracne[c] |
| | | Interfollicular hyperplasia |
| | Promotion of skin papillomas in HRS/J mice | Hair follicle atrophy |
| | | Sebaceous metaplasia |
| Fetus | Cleft palate[a] | Not known |
| | Hydronephrosis | |

[a] Segregates with the *Ah* locus in C57BL/6 × DBA/2 (Poland and Glover, 1980; Clark *et al.*, 1983).

[b] Segregates with *Ah* locus in hr/hr hairless mice (Knutson and Poland, 1982).

[c] *Ah* receptor detected in cultured normal epidermal cells (W. F. Greenlee, unpublished observations) and SCC lines derived from the epidermis and tongue (Table I) (Hudson *et al.*, 1983).

1983). As discussed in the following section and in Section IV,A, the *Ah* receptor appears to play an important role in the expression of TCDD-induced skin toxicity.

**a. Epidermal Hyperplasia (Mice).** Recent investigations have shown that TCDD produces epidermal hyperplasia (Knutson and Poland, 1982) and promotes keratinizing skin papillomas (Poland *et al.*, 1982) in hairless mice (HRS/J) bearing the recessive mutation *hr/hr*. Both hairless (*hr/hr*) and haired (*hr/+*) HRS/J mice possess the *Ah* receptor. TCDD does not produce epidermal hyperplasia in either C57BL/6 (*Ah*-positive) or DBA/2 (*Ah*-negative) strains. However, in crosses between hairless (*hr/hr*) C57BL/6 and DBA/2 mice, TCDD-induced epidermal hyperplasia segregates with the *Ah* locus (Knutson and Poland, 1982). It was postulated that in the mouse this response to TCDD requires interaction between at least two regulatory genes, *Ah* and *hr* (Knutson and Poland, 1982).

**b. Thymus Atrophy: Evidence for Involvement of the *Ah* Receptor.** The adverse actions of TCDD on the immune system are selective for cells participating in the differentiation and expression of cell-mediated (T lymphocyte) immunity and T lymphocyte-dependent antibody responses (Table VII) (Dean *et al.*, 1982; Sharma *et al.*, 1978; Vecchi *et al.*, 1983; Vos, 1977). Studies comparing immunosuppression in adult and perinatally exposed animals indicate that the developing immune system is a particularly sensitive target for TCDD, with thymic atrophy being the most common pathologic finding (Gupta *et al.,* 1973; Luster *et al.,* 1982). The histopathology of TCDD-induced thymic atrophy is similar to that seen in cortisone-treated animals (Bach, 1975) and is characterized by a depletion of cortical thymocytes (Vos, 1977).

The *Ah* receptor is detected in thymic cytosol and nuclear fractions from rats (Lund *et al.,* 1982) and in the thymus from strains of mice most sensitive to TCDD (C57BL/6) but not in strains which are less sensitive to TCDD and nonresponsive to PAH inducers (DBA/2) (Poland and Glover, 1980). In crosses and backcrosses between these murine strains, it has been found that TCDD-induced thymic atrophy segregates with the *Ah* locus (Poland and Glover, 1980). Halogenated aromatic compounds which induce thymic atrophy compete with [$^3$H]TCDD for binding to the cytosolic receptor (measured in murine liver) (Poland and Glover, 1977, 1980). These data indicate that thymic atrophy induced by TCDD and halogenated analogues is mediated by the *Ah* receptor; however, unlike epidermal hyperplasia (Section III,C,2,a), the induction of thymic atrophy does not appear to require other regulatory loci (e.g., thymic atrophy occurs in both haired and hairless C57BL/6 mice) (Poland and Knutson, 1982).

**c. Cleft Palate.**    Teratogenicity of TCDD to mice was first reported by Courtney and Moore (1971). Cleft palate and hydronephrosis are the predominant lesions (Table VII) and it was noted that C57BL/6 mice are more sensitive than DBA/2 mice. Poland and Glover (1978) found that TCDD-induced cleft palate segregates with the *Ah* locus. In a recent study, Dencker and Pratt (1981) showed that midgestational embryos of murine strains sensitive to TCDD have high concentrations of the *Ah* receptor in the maxillary processes and secondary palatal shelves prior to palate closure. These findings strongly support a role for the *Ah* receptor in the expression of cleft palate.

## IV. MECHANISM(S) OF ACTION: CELL BIOLOGY AND BIOCHEMICAL APPROACHES

### A. Cultured Human Keratinocytes

Normal human epidermal cells, usually obtained from neonatal foreskin (Rheinwald and Green, 1975), and squamous cell carcinoma lines, derived from tumors of the epidermis and tongue (Rheinwald and Beckett, 1980, 1981), can be serially cultivated in the presence of lethally irradiated 3T3 murine fibroblast feeder cells. Stratified colonies consisting of a basal layer of proliferating cells and upper layers of terminally differentiating cells appear in culture. As cells migrate to the upper layers, they lose their ability to divide and begin the process of terminal differentiation. This is marked by an increase in cell size, changes in keratin expression, and the formation of cross-linked cornified envelopes (Green, 1980).

## 1. Evidence for the Ah Receptor

The responsiveness of cultured human SCC lines to TCDD was determined by measuring the induction of ECOD activity (Fig. 5) (Hudson *et al.*, 1983). In four of the SCC lines, the $EC_{50}$ value for the induction of ECOD activity by TCDD is approximately 1 n$M$, whereas in one line, the $EC_{50}$ value is 0.1 n$M$. In each of the less sensitive lines, a concentration of 0.1 n$M$ TCDD elicits less than 5% of the maximally induced enzyme activity. Using a concentration of [³H]TCDD (1 n$M$) estimated to produce 85–90% saturation of specific binding sites (as calculated from the reported $K_D$ value in murine liver; Poland *et al.*, 1976a), the *Ah* receptor is detected in the cytosolic fraction from all of the SCC lines (Table II). Under these conditions, the relative amount of receptor measured in each line

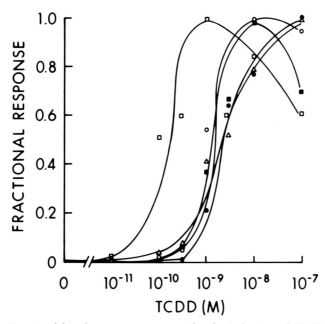

**Fig. 5.**   Fractional log dose–response curves for the induction of ECOD activity in human SCC cell lines by TCDD. Confluent cultures were treated with the indicated concentrations of TCDD for 48 hr and assayed for ECOD activity. Absolute values were converted to fractional responses by equating maximally induced activity to 1.0 and control activity to 0. Symbols: SCC-9 (○); SCC-13 (●); SCC-15 (□); SCC12F (■); and SCC-4 (△). [From Hudson *et al.* (1983).]

correlates with maximally induced ECOD activity (Hudson *et al.*, 1983). These data indicate that human cell lines derived from the epidermis (a target for TCDD toxicity) contain the *Ah* receptor and show differential sensitivity to TCDD as judged by the $EC_{50}$ values for the induction of ECOD activity analogous to the murine strain differences in sensitivity regulated by the *Ah* locus.

## 2. Hyperkeratinization in Vitro

TCDD enhances stratification and induces hyperkeratinization (as judged by the intensity of Rhodanile Blue staining) in murine XB teratoma cells (Knutson and Poland, 1980) and decreases cell proliferation at confluence in late-passage XB cells (Gierthy and Crane, 1984). TCDD stimulates the activity of trans-glutaminase, the enzyme involved in cross-linking of envelope precursor protein, in epidermal cells cultured from BALB/c mice (Puhvel *et al.*, 1984), but in a human malignant keratinocyte cell line, TCDD inhibits the induction of certain differentiation markers by hydrocortisone (Rice and Cline, 1984). In postcon-fluent cultures of normal human epidermal cells, TCDD stimulates [$^3$H]thymidine incorporation (Milstone and LaVigne, 1984), whereas treatment of early-passage human epidermal cells at confluence with TCDD results in enhanced differentiation (Fig. 6) analogous to hyperkeratinization observed *in vivo* (Greenlee *et al.*, 1985c; Osborne and Greenlee, 1985). Enhanced differ-entiation responses to TCDD in normal human epidermal cells and in a human SCC line (SCC-12F) with growth requirements similar to normal epidermal cells are associated with inhibition of EGF receptor binding and EGF-stimulated DNA synthesis (Greenlee *et al.*, 1985c; Hudson *et al.*, 1985a; Osborne and Greenlee, 1985).

## 3. Regulation of Biochemical Mediators of Proliferation and Differentiation by the Ah Receptor

Epidermal growth factor promotes proliferation and increases the lifetime of serially cultivated epidermal cells (Rheinwald and Green, 1977). In low-density cultures, EGF reduces the proportion of highly differentiated cells (Sun and Green, 1976), whereas in confluent cultures, EGF-induced proliferation is char-acterized by increased production of terminally differentiated squames (Green, 1977).

Treatment of normal human epidermal cells or SCC-12F cells with TCDD results in a concentration-dependent decrease in the specific binding of EGF to a value 40% of the control with an $EC_{50}$ of approximately 1 n$M$ (Hudson *et al.*, 1985a; Osborne and Greenlee, 1985). Scatchard analysis of EGF binding has been carried out in SCC-12F cells and indicates that TCDD exposure results in a loss of high-affinity ($K_d = 0.28$ n$M$) EGF binding sites (Hudson *et al.*, 1985a). Down-modulation of EGF binding by TCDD is accompanied by an inhibition of EGF-stimulated DNA synthesis.

Down-regulation of EGF binding in normal human epidermal cells or SCC-12F cells treated with 2,3,7,8-tetrachlorodibenzofuran, a stereoisomer of TCDD (see Fig. 1), is equal to that with TCDD, whereas no significant alteration in EGF binding occurs in cells treated with the nonstereoisomer 2,7-dichlorodibenzo-*p*-dioxin (Hudson *et al.*, 1985a; Osborne and Greenlee, 1985). The observed stereospecificity and potency ($EC_{50}$ = 1 n*M*) suggest that the down-modulation of EGF receptors in these cells by TCDD is mediated by the *Ah* receptor.

TCDD produces a prolonged decrease in EGF binding in SCC-12F cells (40% of control up to 10 days after removal of TCDD from the culture medium) (Hudson *et al.*, 1985a). In contrast, in these same cells benzo[*a*]pyrene (BP), an agent which binds to the *Ah* receptor (Poland *et al.*, 1976a) and down-modulates EGF binding in hepatoma cell lines (Karenlampi *et al.*, 1983) and C3H/10T1/2 cells (Ivanovic and Weinstein, 1982), produces a transient (24 hr) decrease in EGF binding (Hudson *et al.*, 1985a). The available data indicate that TCDD (Kimmig and Schultz, 1957; Taylor, 1974) but not BP evokes a hyperkeratinization response *in vivo*, and it has been postulated that sustained refractoriness of epidermal cells to growth factors such as EGF may be an important regulatory event in the genesis of hyperkeratinization (Greenlee *et al.*, 1985c).

Treatment of cultured human epidermal cells with agents which increase the intracellular concentration of cyclic AMP (cAMP) or with the cAMP analogues 8-bromo-cAMP and dibutyryl-cAMP stimulates proliferation (Green, 1978). The actions of TCDD on the regulation of adenylate cyclase activity in human SCC lines with different growth responses to TCDD have been examined. Under appropriate culture conditions, treatment of line SCC-9 with TCDD enhances colony expansion, whereas in line SCC-12F, TCDD inhibits colony expansion (Rice and Greenlee, 1982). TCDD does not markedly inhibit EGF binding in SCC-9 cells (cf. with the actions in SCC-12F cells). However, in these cells TCDD appears to act through a unique membrane–receptor mechanism to stimulate adenylate cyclase activity nearly twofold (Greenlee *et al.*, 1983). The mechanisms for the modulation of adenylate cyclase activity by TCDD are not known. Preliminary data suggest that TCDD may enhance the responsiveness of the adenylate cyclase system to hormone activators such as epinephrine (W. F. Greenlee, unpublished observations).

Thus, the initial findings in normal human epidermal cells and SCC lines suggest that the hyperkeratinization and hyperplastic responses of epidermal cells to TCDD result from actions of TCDD mediated by the *Ah* receptor to two (or more) membrane sites. Hyperkeratinization may result from the prolonged down-modulation of receptors for growth factors such as EGF and subsequent enhanced commitment of proliferating basal cells to terminal differentiation. Hyperplasia, a response commonly seen in the interfollicular epidermis (Kimbrough, 1980), may result from modulation of adenylate cyclase activity, in-

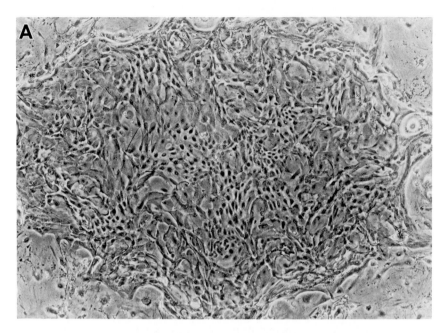

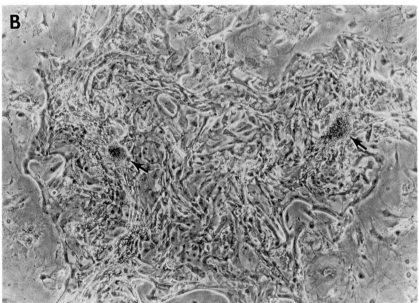

**Fig. 6.** Enhanced stratification in colonies of normal human epidermal cells treated with TCDD. Cells were plated and grown as described previously (Rheinwald and Green, 1975). Two days after plating, medium containing TCDD (10 nM) or solvent vehicle (0.1% p-dioxane) was added. The medium was replaced every 3–4 days with fresh growth medium containing the appropriate additions. After a 2-week treatment period, colonies

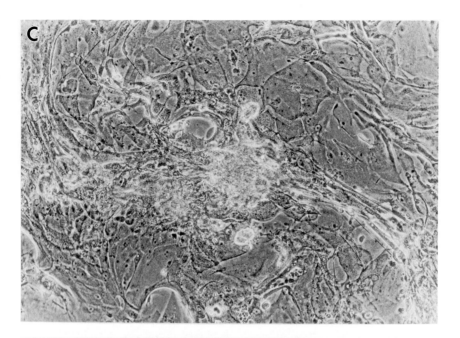

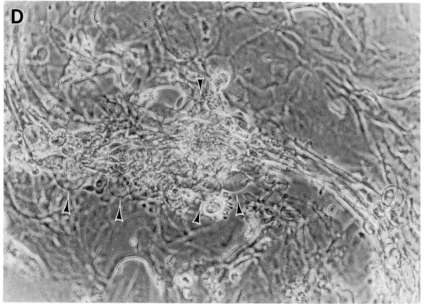

of living cells were photographed under phase-contrast. (A) Control (×40); (B) TCDD (×40), arrows indicate areas of enhanced stratification; (C) TCDD (×100), area shown is that indicated by the arrow to the left in (B); and (D) TCDD (×100), same field as in (C) but focused on a more superficial cell layer. Note the loosely adherent squames. [From Greenlee *et al.* (1985c).]

creasing the intracellular concentration of cAMP, a positive mediator of epidermal cell proliferation. The difference in the responses of SCC-9 cells versus SCC-12F and normal epidermal cells to TCDD suggests that the action of TCDD on a specific target cell depends on the program of regulatory mechanisms controlling cell division and differentiation in that cell.

## B. *In Vitro* Models for Thymic Atrophy

### 1. *Cultured Thymic Epithelial Cells: T-Lymphocyte Differentiation*

The differentiation of T lymphocytes requires trafficking of committed progenitor cells originating from the bone marrow to both cortical and medullary regions of the thymus (Ezine *et al.*, 1984). Intrathymic differentiation of T-lymphocyte precursors is a multistage process dependent on cell–cell contact between precursors and the thymic reticulum (reviewed in Bach and Papiernik, 1980) and on humoral factors produced by medullary thymic epithelial (TE) cells (Stutman, 1978; Haynes, 1984). Precursor maturation is marked by changes in the expression of membrane antigens, decreased sensitivity to corticosteroids, and enhanced responsiveness to mitogens (Stutman, 1978; Kruisbeck, 1979). The majority of precursor cells do not survive transit through the thymus (Weiss, 1983). The available data suggest that TCDD-induced thymic atrophy, characterized by a depletion of cortical thymocytes (T-lymphocyte precursors) (Section III,C,1,a), results from actions on progenitor cells in the bone marrow, impaired differentiation of intrathymic precursor cells (specifically by actions on the thymic epithelium), or both (Clark *et al.*, 1983; Poland and Knutson, 1982; Vos, 1977).

Coculture of thymocytes on TE monolayers provides an *in vitro* system for studying the cellular and molecular mechanisms of TCDD-induced thymic atrophy. Thymic epithelial cells are cultured under conditions which select against thymic fibroblasts and macrophages (Boniver *et al.*, 1981). In Fig. 7, cultured TE cells from C57BL/6 (*Ah*-positive, Section I,A,1) mice are shown after incubation with guinea pig prekeratin antibody followed by fluorescein-labeled anti-guinea pig antibody. The presence of prekeratin filaments confirms the epithelial origin of these cells.

Thymocytes from C57BL/6 mice cocultivated on syngeneic TE monolayers show a 5- to 10-fold enhanced response to the mitogens concanavalin A (Con A)

**Fig. 7.** Demonstration of prekeratin filaments in cultured thymic epithelial cells from C57BL/6 mice. Prekeratin filaments were visualized with anti-prekeratin antibody using an immunofluorescence sandwich technique. Formed prekeratin filaments are present only in epithelial cells (courtesty of R. D. Irons, CIIT).

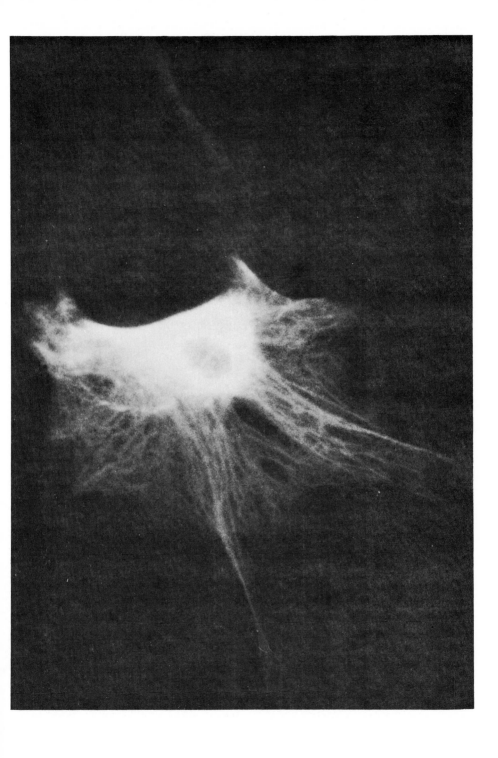

and phytohemagglutinin (PHA), respectively (Table VIII). This response is also observed after incubation in TE-conditioned medium, suggesting that the cultured TE cells produce humoral factors which promote the appearance of mitogen-responsive cells (Greenlee *et al.*, 1985a). The enhanced mitogen response is inhibited in thymocytes cocultivated with TCDD-treated TE monolayers (40% of control with 10 n*M* TCDD) (Table VIII). TCDD does not produce detectable cytotoxicity in TE cells, and mitogen-induced blastogenesis is not suppressed in thymocytes incubated in medium collected from TCDD treated cultures or in TE-conditioned medium to which TCDD (10 n*M*) has been added directly (Greenlee *et al.*, 1985a). These findings indicate that TE cells are a target site for TCDD: one consequence of this action appears to be the altered maturation of T-lymphocyte precursors dependent on direct cell–cell contact between thymocytes and TE cells. The concentration dependence (Table VIII) and stereospecificity (Greenlee *et al.*, 1985a) for the suppression of TE-dependent maturation of thymocytes by TCDD *in vitro* are the same as reported for TCDD-induced thymic atrophy *in vivo* (Poland and Glover, 1980) and suggest involvement of the *Ah* receptor. Measurement of the *Ah* receptor in cytosol

**TABLE VIII**

**Lymphoproliferative Responses of Thymocytes from C57BL/6 Mice Cocultivated on TCDD-Treated Syngeneic TE Monolayers[a]**

| Concentration of TCDD (nM) | TE monolayer | [³H]Thymidine incorporation | |
|:---:|:---:|:---:|:---:|
| | | Con A | PHA |
| 0 | − | 18,253 | 2,297 |
| 0 | + | 87,120 | 25,888 |
| | | (100)[b] | (100) |
| 0.1 | + | 77,258 | 19,848 |
| | | (89) | (77) |
| 1 | + | 58,215 | 10,709 |
| | | (67) | (47) |
| 10 | + | 34,306 | 9,402 |
| | | (39) | (36) |

[a] Confluent cultures of TE cells were treated with either solvent vehicle (DMSO, 0.1%) or the indicated concentrations of TCDD for 48 hr and then were washed two times with medium containing 5% fetal calf serum. Thymocytes were cocultivated with the treated TE monolayers for 48 hr and removed for determination of lymphoproliferative responses to the mitogens, Con A and PHA (Greenlee *et al.*, 1985a).

[b] Numbers in parentheses indicate the percentage of the control response of thymocytes cocultivated on untreated TE monolayers.

fractions prepared from whole thymus, thymocytes, and cultured TE cells indicates that there is a selective concentration of this receptor in the epithelial target cells (Greenlee *et al.*, 1985a). It has been proposed that impaired thymocyte maturation resulting from the actions of TCDD on TE target cells could lead to increased thymocyte death and the depletion of cortical thymocytes characteristic of TCDD-induced thymic atrophy (Greenlee *et al.*, 1985a).

The presence of the *Ah* receptor in bone marrow (Fig. 2) and human peripheral blood lymphocytes (Carlstedt-Duke *et al.*, 1982) suggests that TCDD may also act directly on lymphocyte targets, independent of its actions on thymic epithelium. Heterogeneity of target sites is supported further by the observed differences in sensitivity to toxic responses produced by TCDD, such as thymic atrophy ($ED_{50}$ = 10 nmoles/kg; Poland and Glover, 1980) and induction of T-suppressor cells ($ED_{50}$ = 0.01 nmoles/kg; Clark *et al.*, 1983).

## C. Altered Differentiation in Epidermal and Thymic Epithelial Target Cells

The proliferating cells of the epidermis are confined to the basal layer (Section IV,A). As cells migrate to upper strata and begin to terminally differentiate, they lose the capacity to divide and undergo several morphologic and biochemical changes, including the formation of cross-linked envelopes (Green, 1977). Recently it has been shown that keratinocyte maturation can also be monitored with monoclonal antibodies directed against human thymus (reviewed in Haynes, 1984). Among the several antibodies characterized are those designated TE-4, TE-8, and TE-15 (Haynes, 1984). Basal cells react with TE-4, whereas differentiating keratinocytes react with TE-8 and TE-15. Thymic epithelium, although not stratified, appears to be in a constant state of differentiation as indicated by a similar pattern of reactivity with these same monoclonal antibodies (Haynes, 1984).

It has been proposed that the toxic responses of the epidermis (hyperkeratinization) and thymus (atrophy) to TCDD result from altered patterns of differentiation in the respective epithelial target cells (Greenlee *et al.*, 1985b). The available data indicate that the altered regulation of epithelial cell differentiation is mediated at least in part by the *Ah* receptor and it was postulated that the initial regulatory changes occurred at the transition of the epidermal basal cell (or its thymic equivalent) to a cell irreversibly committed to terminal differentiation (Greenlee *et al.*, 1985b). In the epidermis enhanced differentiation would be an important component in the expression of hyperkeratinization, whereas in the thymus it is assumed that, during the process of TCDD-induced differentiation, TE cells lose the ability to support T-lymphocyte maturation (Section IV,B). In support of the hypothesis that basal cells are a primary target, a recent study has shown that the *Ah* receptor-mediated down-modulation of the EGF receptor in cultures of human SCC-12F cells occurs predominantly in the basal cell population (Hudson *et al.*, 1985b).

## D. *In Vitro* Models for Cleft Palate

As noted in Section III,C,2,c, TCDD-induced cleft palate in inbred strains of mice is mediated by the *Ah* receptor (Poland and Glover, 1978; Dencker and Pratt, 1981). During the embryonic development of the palate, the palatal shelves elevate to a horizontal position above the tongue and then establish contact (Pratt, 1983). The epithelial cells at the midline fuse and then undergo programmed cell death, at which time the mesenchyme of the two shelves become contiguous (Pratt, 1983). Chemically induced cleft palate can result from actions preventing the elevation of the palatal shelves, inhibition of epithelial fusion, or both. For TCDD, the production of hyperkeratinization in the midline palatal epithelial cells would prevent the convergence of the palatal shelves.

Recently, Grove and Pratt (1983) have described a system for culturing embryonic mouse palatal epithelial cells. High concentrations of the *Ah* receptor are found in the midline palate (Dencker and Pratt, 1981). Using cultured palatal epithelial cells, it should be possible to determine if these cells contain the *Ah* receptor and to dissect receptor-mediated events which may alter the normal epithelial fusion process.

## E. Transformation *in Vitro*

### *Promotion of Transformed Foci in C3H/10T1/2 Cells*

TCDD is a potent carcinogen (Kociba *et al.*, 1978), but it is not metabolized to species which covalently bind to DNA (less than 1 molecule of TCDD per $10^{11}$ nucleotides; Poland and Glover, 1979). TCDD acts as a tumor promoter in a two-stage model of liver carcinogenesis in the rat (Pitot *et al.*, 1980) and when painted on the skin of HRS/J hairless mice initiated with *N*-methyl-*N'*-nitro-*N*-nitrosoguanidine (MNNG) (Poland *et al.*, 1982). Treatment of MNNG-initiated C3H/10T1/2 cells with TCDD (0.4 p$M$ to 4 n$M$) promotes the formation of transformed foci (Abernethy *et al.*, 1984; 1985). A maximum ninefold increase in the number of foci is observed at TCDD concentrations $\geq$40 p$M$. Comparison with the potent tumor promoter 12-*O*-tetradecanoylphorbol 13-acetate (TPA) indicates that TCDD produces a maximal response at a 10,000-fold lower concentration than the optimal concentration of TPA. The *Ah* receptor has been detected in C3H/10T1/2 cells (Ho *et al.*, 1983) and in cloned sublines (Okey *et al.*, 1983). The putative role of the *Ah* receptor in mediating the expression of transformed foci in these cells has not been established.

## F. Model for the Regulation of the Expression of Multiple Gene Batteries

It has been postulated that in the mouse, the *Ah* locus, either singly or in concert with other regulatory genes, controls at least two distinct pleiotropic responses: a limited, but widely expressed gene battery which includes the structural genes for cytochrome(s) $P_1$-450 (see Table IV); and in a few organs such as skin and thymus, a second gene battery regulating cell proliferation and differentiation (Poland and Knutson, 1982). A model for the proposed regulation of multiple gene batteries by the *Ah* locus is shown in Fig. 8. In the model depicted,

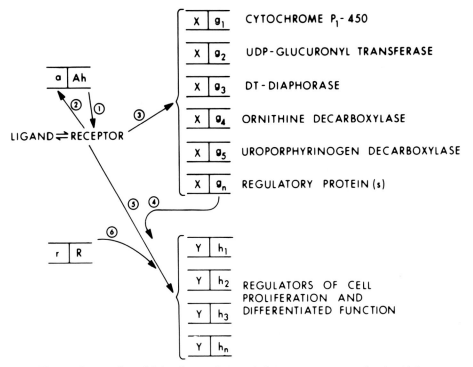

**Fig. 8.** Proposed model for the regulation of pleiotropic responses by the *Ah* locus. The *Ah* locus codes for the receptor (pathway 1) which is the stereospecific recognition site for TCDD and other ligands (Section II,D). The receptor–ligand complex interacts with a common nuclear receptor site (X) to regulate the coordinate expression of a gene battery (battery X) coding for several enzymes, some of which participate in xenobiotic metabolism (pathway 3), and may also regulate the expression of the *Ah* gene itself (pathway 2). In certain tissues, such as the skin and thymus, the *Ah* receptor regulates an additional gene battery (battery Y) coding for proteins essential in the control of cell growth and/or differentiation (pathway 5). Regulation of battery Y by the *Ah* receptor may be dependent on products of battery X, other regulatory loci (R), or both (pathways 4 and 6). See text for details. [Adapted from Poland and Knutson (1982).]

the *Ah* receptor–ligand complex interacts with a common nuclear receptor site (X) to regulate the coordinate expression of battery X, which includes the structural gene for cytochrome(s) $P_1$-450. Battery Y codes for proteins controlling cell growth and differentiation. Regulation of the expression of battery Y by the *Ah* receptor may be dependent on products of battery X, other regulatory loci (R), such as the *hr* locus in the hairless mouse, or both. At this time, specific regulatory proteins associated with battery Y and regulated by the *Ah* receptor have not been conclusively identified; however, the ability of TCDD (through the *Ah* receptor) to alter the response of epidermally derived cells to EGF through modulation of the membrane EGF receptor (Section IV,A,3) is providing some initial insights. The loss of high-affinity EGF binding sites could result from a TCDD-induced shift (through some unidentified mechanism) in the culture population to one with a greater proportion of highly differentiated cells which do not express EGF receptors. Alternatively, enhanced differentiation can be a direct response to the down-modulation of EGF receptors by TCDD (Hudson *et al.*, 1985a,b). In the latter case, it could be postulated that the EGF receptor (and possibly receptors for other growth factors) is a product of battery Y.

## V. CONCLUDING COMMENTS

Several *in vitro* systems have been described which should provide insight into the mechanism(s) by which the *Ah* receptor regulates proliferation and differentiation in target cells. Cultured human epidermal cells are of particular interest, since chloracne and associated proliferative changes in the interfollicular epidermis are among the most sensitive and widespread responses to TCDD in humans and appear to be mediated by the *Ah* receptor. These systems should also prove useful for screening chloracneogens and quantitating risk by measuring responses relevant to the spectrum of skin-associated toxicities, such as cell growth rates, DNA synthesis, keratin expression, cross-linked envelope formation, EGF binding, and stimulation of adenylate cyclase activity. The differences in sensitivity for induction of ECOD activity by TCDD (Fig. 5) and differential growth responses to TCDD (Section IV,A,2) observed in human SCC lines suggest that these cell culture systems maintain phenotypes *in vitro* potentially relevant to genetically determined differences in susceptibility to TCDD toxicity *in vivo*.

TCDD suppresses cellular immunity (Table VII), and in mice, the data indicate that TCDD-induced thymic atrophy and TCCD-enhanced T-suppressor cell activity are mediated by the *Ah* receptor (Section III,C,2,b). There have been no adverse human health effects attributed to the actions of TCDD on the immune system; however, the extraordinary potency of TCDD as a tumor promoter in animals (Pitot *et al.*, 1980; Poland *et al.*, 1982) and the evidence that T lymphocytes (cytotoxic effector cells) are one of the major cell types participating in immune surveillance against neoplasms (Herberman and Ortaldo, 1981) suggest that modulation of T-lymphocyte maturation (and differentiation) by the *Ah* receptor (Section IV,B) may be one of the mechanisms associated with the carcinogenic potential of TCDD.

There has been much speculation regarding potential endogenous ligands for the *Ah* receptor and on the putative physiologic role of this protein. This chapter has reviewed the studies indicating that the *Ah* receptor functions not only as the induction protein for cytochrome $P_1$-450 and other xenobiotic metabolizing enzymes, but also that it regulates (presumably at the level of gene expression) critical events in selected target cells to alter normal patterns of proliferation and differentiation. It is possible, as suggested in a review by Poland and Knutson (1982), that the endogenous ligand for the receptor no longer exists. Thus, cells exposed to TCDD are provided an inappropriate signal resulting in altered gene expression and subsequent disruption of the normal homeostatic control of proliferation and differentiation. An important component in the actions of TCDD is the ability of this compound to produce sustained responses, a characteristic which is presumably distinct from the actions of a physiologic ligand (Poland and Knutson, 1982). Another feature is latency; i.e., many of the toxic responses appear several days after exposure. Although much of the previous research has focused on control of inducible genes by the *Ah* receptor, repression of specific genes by the *Ah* receptor–TCDD complex may be of equal or greater importance in the expression of certain toxic responses. In this case one would predict latency determined by the half-life of regulatory proteins coded by the repressed genes. Answers to many of these questions await further research. The development of *in vitro* systems which permit modeling of specific responses to TCDD should provide fundamental insight at the cellular and molecular level of both toxic and potential biologic processes regulated by the *Ah* receptor.

## REFERENCES

Abernethy, D. J., Huband, J. C., Greenlee, W. F., and Boreiko, C. J. (1984). The effect of TCDD upon the transformation; initiation, and promotion of C3H/10T1/2 cells. *Environ. Mutagen.* **6,** 461.

Abernethy, D. J., Greenlee, W. F., Huband, J. C., and Boreiko, C. J. (1985). 2,3,7,8-Tetrachlorodibenzo-p-dioxin (TCDD) promotes the transformation of C3H10T1/2 cells. *Carcinogenesis* (in press).

Atlas, S. A., Vesell, E. S., and Nebert, D. W. (1976). Genetic control of interindividual variations in the inducibility of aryl hydrocarbon hydroxylase in cultured human lymphocytes. *Cancer Res.* **36,** 4619–4630.

Atlas, S. A., Boobis, A. R., Felton, J. S., Thorgeirsson, S. S., and Nebert, D. W. (1977). Ontogenic expression of polycyclic aromatic compound-inducible monooxygenase activities and forms of cytochrome *P*-450 in the rabbit. Evidence for temporal control and organ specificity of two genetic regulatory systems. *J. Biol. Chem.* **252,** 4712–4721.

Bach, J. F. (1975). Cortisteroids. *In* "The Mode of Action of Immunosuppressive Agents" (A. Neuberger and E. L. Tatum, eds.), pp. 21–91. North-Holland Publ., Amsterdam.

Bach, J. F., and Papiernik, M. (1980). Cellular and molecular signals in T cell differentiation. *Ciba Found. Symp.* **84**, 215–235.

Bandiera, S., Safe, S., and Okey, A. B. (1982a). Binding of polychlorinated biphenyls classified as either phenobarbitone-, 3-methylcholanthrene- or mixed-type inducers to cytosolic *Ah* receptor. *Chem.-Biol. Interact.* **39**, 259–277.

Bandiera, S., Sawyer, T., Campbell, M. A., Robertson, L., and Safe, S. (1982b). Halogenated biphenyls as AHH inducers: effects of different halogen substituents. *Life Sci.* **31**, 517–525.

Beatty, P., and Neal, R. A. (1976). Induction of DT-diaphorase by 2,3,7,8-tetrachlorodibenzo-*p*-dioxin (TCDD). *Biochem. Biophys. Res. Commun.* **68**, 197–204.

Beatty, P. W., Vaughn, W. K., and Neal, R. A. (1978). Effect of alteration of rat hepatic mixed-function oxidase (MFO) activity on the toxicity of 2,3,7,8-tetrachlorodibenzo-*p*-dioxin (TCDD). *Toxicol. Appl. Pharmacol.* **45**, 513–519.

Boniver, J., Declere, A., Dailey, M. O., Honsi, C., Liberman, M., and Kaplan, H. S. (1981). Macrophage and lymphocyte-depleted thymus reticulo-epithelial cell cultures: establishment and functional influence on T-lymphocyte maturation, C-type virus expression and lymphomatous transformation *in vitro. Thymus* **2**, 193–213.

Brewer, J. M., Pesce, A. J., and Ashworth, R. B. (1974). "Experimental Techniques in Biochemistry." Prentice-Hall, Englewood Cliffs, New Jersey.

Brown, S., Wiebel, F. J., Gelboin, H. V., and Minna, J. D. (1976). Assignment of a locus required for flavoprotein-linked monooxygenase expression to human chromosome 2. *Proc. Natl. Sci. U.S.A.* **73**, 4628–4632.

Carlstedt-Duke, J. M. B. (1979). Tissue distribution of the receptor for 2,3,7,8-tetrachlorodibenzo-*p*-dioxin in the rat. *Cancer Res.* **39**, 3172–3176.

Carlstedt-Duke, J., Elfstrom, G., Snochowski, M., Hogberg, B., and Gustafsson, J.-A. (1978). Detection of the 2,3,7,8-tetrachlorodibenzo-*p*-dioxin (TCDD) receptor in rat liver by isoelectric focusing in polyacrylamide gels. *Toxicol. Lett.* **2**, 365–373.

Carlstedt-Duke, J. M. B., Elfstrom, G., Hogberg, B., and Gustafsson, J.-A. (1979). Ontogeny of the rat hepatic receptor for 2,3,7,8-tetrachlorodibenzo-*p*-dioxin and its endocrine independence. *Cancer Res.* **39**, 4653–4656.

Carlstedt-Duke, J. M. B., Harnemo, U.-B., Hogberg, B., and Gustafsson, J.-A. (1981). Interaction of the hepatic receptor protein for 2,3,7,8-tetrachlorodibenzo-*p*-dioxin with DNA. *Biochim. Biophys. Acta* **672**, 131–141.

Carlstedt-Duke, J., Kurl, R., Poellinger, L., Gillner, M., Hansson, L.-A., Toftgard, R., Hogberg, B., and Gustafsson, J.-A. (1982). The detection and function of the cytosolic receptor for 2,3,7,8-tetrachlorodibenzo-*p*-dioxin (TCDD) and related cocarcinogens. *In* "Chlorinated Dioxins and Related Compounds: Impact on the Environment" (O. Hutzinger, R. W. Frei, E. Merian, and F. Pocchiari, eds.), pp. 355–365. Pergamon, New York.

Clark, D. A., Sweeney, G., Safe, S., Hancock, E., Kilbourn, D. G., and Gauldie, J. (1983). Cellular and genetic basis for suppression of cytotoxic T-cell generation by haloaromatic hydrocarbons. *Immunopharmacology* **6**, 143–153.

Courtney, K. D., and Moore, J. A. (1971). Teratology studies with 2,4,5-trichlorophenoxyacetic acid and 2,3,7,8-tetrachlorodibenzo-*p*-dioxin. *Toxicol. Appl. Pharmacol.* **20**, 396–403.

Crow, K. D. (1970). Chloracne. *Trans. St. John's Hosp. Dermatol. Soc.* **56**, 77–99.

Dean, J. H., Luster, M. I., and Boorman, G. A. (1982). Immunotoxicology. *In* "Immunopharmacology" (P. Sirois and M. Rola-Pleszczynski, eds.), pp. 349–397. Elsevier, New York.

Dencker, L., and Pratt, R. M. (1981). Association between the presence of the Ah receptor in embryonic murine tissues and sensitivity to TCDD-induced cleft palate. *Teratog., Carcinog., Mutagen.* **1**, 399–406.

Denison, M. S., Fine, J., and Wilkinson, C. F. (1984). Protamine sulfate precipitation: a new assay for the *Ah* receptor. *Anal. Biochem.* **142**, 28–36.

Dietrich, R. A., Bludeau, P., Roger, M., and Schmuck, J. (1978). Induction of aldehyde dehydrogenases. *Biochem. Pharmacol.* **27**, 2343–2347.

Elder, G. H., and Sheppard, D. M. (1982). Immunoreactive uroporphyrinogen decarboxylase is unchanged in porphyria caused by TCDD and hexachlorobenzene. *Biochem. Biophys. Res. Commun.* **109**, 113–120.

Ezine, S., Weissman, I. L., and Rouse, R. V. (1984). Bone marrow cells give rise to distinct cell clones within the thymus. *Nature* **309**, 629–631.

Gasiewicz, T. A. (1983). Receptors for 2,3,7,8-tetrachlorodibenzo-*p*-dioxin: their inter- and intra-species distribution and relationship to the toxicity of this compound. *Proc. Annu. Conf. Environ. Toxicol. 13th,* Dayton, Ohio, pp. 259–269.

Gasiewicz, T. A., and Neal, R. A. (1982). The examination and quantitation of tissue cytosolic receptors for 2,3,7,8-tetrachlorodibenzo-*p*-dioxin using hydroxylapatite. *Anal. Biochem.* **124**, 1–11.

Gasiewicz, T. A., and Rucci, G. (1984a). Cytosolic receptor for 2,3,7,8-tetrachlorodibenzo-*p*-dioxin. Evidence for a homologous nature among various mammalian species. *Mol. Pharmacol.* **26**, 90–98.

Gasiewicz, T. A., and Rucci, G. (1984b). Examination and rapid analysis of hepatic cytosolic receptors for 2,3,7,8-tetrachlorodibenzo-*p*-dioxin using gel-permeation high performance liquid chromatography. *Biochim. Biophys. Acta* **798**, 37–45.

Gasiewicz, T. A., Geiger, L. E., Rucci, G., and Neal, R. A. (1983). Distribution, excretion, and metabolism of 2,3,7,8-tetrachlorodibenzo-*p*-dioxin in C57BL/6J, DBA/2J, and B6D2F1/J mice. *Drug Metab. Dispos.* **11**, 397–403.

Gierthy, J. F., and Crane, D. (1984). Reversible inhibition of *in vitro* epithelial cell proliferation by 2,3,7,8-tetrachlorodibenzo-*p*-dioxin. *Toxicol. Appl. Pharmacol.* **74**, 91–98.

Goldstein, J. A. (1980). Structure–activity relationships for the biochemical effects and relationships to toxicity. *In* "Halogenated Biphenyls, Terphenyls, Naphthalenes, Dibenzodioxins, and Related Products" (R. D. Kimbrough, ed.) pp. 151–190. Elsevier/North-Holland, New York.

Green, H. (1977). Terminal differentiation of cultured human epidermal cells. *Cell* **11**, 405–419.

Green, H. (1978). Cyclic AMP in relation to proliferation of the epidermal cell: a new view. *Cell* **15**, 801–811.

Green, H. (1980). The keratinocyte as differentiated cell type. *Harvey Lect.* **74**, 101–139.

Greenlee, W. F. (1985). Molecular mechanisms of immunosuppression induced by 12-*O*-tetradecanoylphorbol-13-acetate and 2,3,7,8-tetrachlorodibenzo-*p*-dioxin. *In* "Toxicology of the Immune System" (J. H. Dean, A. Munson, and M. Luster, eds.). Raven Press, New York (in press).

Greenlee, W. F., and Poland, A. (1979). Nuclear uptake of 2,3,7,8-tetrachlorodibenzo-*p*-dioxin in C57BL/6J and DBA/2J mice. Role of the hepatic cytosol receptor protein. *J. Biol. Chem.* **254**, 9814–9821.

Greenlee, W. F., Sun, J. D., and Bus, J. S. (1981). A proposed mechanism of benzene toxicity: formation of reactive intermediates from polyphenol metabolites. *Toxicol. Appl. Pharmacol.* **59**, 187–195.

Greenlee, W. F., Young, M. J., Atkins, W. M., Hudson, L. G., Dorflinger, L., and Toscano, W. A. (1983). Regulation of adenylate cyclase activity in cultured human epithelial cells by 2,3,7,8-tetrachlorodibenzo-*p*-dioxin (TCDD). *Toxicol. Lett.* **18**, 5.

Greenlee, W. F., Dold, K. M., Irons, R. D., and Osborne, R. (1985a). Evidence for direct action of 2,3,7,8-tetrachlorodibenzo-p-dioxin (TCDD) on thymic epithelium. *Toxicol. Appl. Pharmacol.* (in press).

Greenlee, W. F., Dold, K. M., and Osborne, R. (1985b). A proposed model for the actions of

2,3,7,8-tetrachlorodibenzo-p-dioxin (TCDD) on epidermal and thymic epithelial target cells. *In* "Biologic Mechanisms of Dioxin Action" (A. Poland and R. Kimbrough, eds.), Banbury Report, Vol. 18, pp. 435–440. Cold Spring Harbor, New York.

Greenlee, W. F., Osborne, R., Dold, K. M., Hudson, L. G., and Toscano, W. A., Jr. (1985c). Toxicity of chlorinated aromatic compounds in animals and humans: *in vitro* approaches to toxic mechanisms and risk assessment. *Environ. Health Perspect.* (in press).

Grove, R. I., and Pratt, R. M. (1983). Growth and differentiation of embryonic mouse palatal epithelial cells in primary culture. *Exp. Cell Res.* **148**, 195–205.

Guenthner, T. M., and Nebert, D. W. (1977). Cytosolic receptor for aryl hydrocarbon hydroxylase induction by polycyclic aromatic compounds. Evidence for structural and regulatory variants among established cell culture lines. *J. Biol. Chem.* **252**, 8981–8989.

Guenthner, T. M., and Nebert, D. W. (1978). Evidence in rat and mouse liver for temporal control of two forms of cytochrome $p$-450 inducible by 2,3,7,8,-tetrachlorodibenzo-$p$-dioxin. *Eur. J. Biochem.* **91**, 449–456.

Gupta, B. N., Vos, J. G., Moore, J. A., Zinkl, J. G., and Bullock, B. C. (1973). Pathologic effects of 2,3,7,8-tetrachlorodibenzo-$p$-dioxin in laboratory animals. *Environ. Health Perspect.* **5**, 125–140.

Hannah, R. R., Nebert, D. W., and Eisen, H. J. (1981). Regulatory gene product of the *Ah* complex: Comparison of 2,3,7,8-tetrachlorodibenzo-$p$-dioxin and 3-methylcholanthrene binding to several moieties in mouse liver cytosol. *J. Biol. Chem.* **256**, 4584–4590.

Haynes, B. N. (1984). The human thymic microenvironment. *Adv. Immunol.* **36**, 87–142.

Henck, J. M., New, M. A., Kociba, R. J., and Rao, K. S. (1981). 2,3,7,8-Tetrachlorodibenzo-$p$-dioxin: acute oral toxicity in hamsters. *Toxicol. Appl. Pharmacol.* **59**, 405–407.

Herberman, R. B., and Ortaldo, J. R. (1981). Natural killer cells: their role in defenses against diseases. *Science (Washington, D.C.)* **214**, 24–30.

Ho, D., Gill, K., and Fahl, W. E. (1983). Benz[*a*]anthracene and 3-methylcholanthrene induction of cytochrome *P*-450 in C3H/10T1/2 mouse fibroblasts. *Mol. Pharmacol.* **23**, 198–205.

Hudson, L. G., Shaikh, R., Toscano, W. A., Jr., and Greenlee, W. F. (1983). Induction of 7-ethoxycoumarin *O*-deethylase activity in cultured human epithelial cells by 2,3,7,8-tetrachlorodibenzo-$p$-dioxin (TCDD): evidence for TCDD receptor. *Biochem. Biophys. Res. Commun.* **115**, 611–617.

Hudson, L. G., Toscano, W. A., Jr., and Greenlee, W. F. (1985a). Regulation of epidermal growth factor binding in a human keratinocyte cell line by 2,3,7,8-tetrachlorodibenzo-$p$-dioxin. *Toxicol. Appl. Pharmacol.* **77**, 251–259.

Hudson, L. G., Toscano, W. A., and Greenlee, W. F. (1985b). 2,3,7,8-Tetrachlorodibenzo-$p$-dioxin (TCDD) modulates epidermal growth factor (EGF) binding to basal cells from a human keratinocyte cell line. *Toxicologist* **4**, 37.

Ingami, K., Koga, T., Kikuchi, M., Hashimoto, M., Takahashi, H., and Wada, K. (1969). Experimental study of hairless mice following admistration of rice oil used by a "Yusho" patient. *Fukuoka Acta Med.* **60**, 549–553.

International Agency for Research on Cancer (1977). "Some Fumigants, the Herbicides 2,4-D and 2,4,5-T, Chlorinated Dibenzodioxins and Miscellaneous Industrial Chemicals," IARC Monographs on the Evaluation of Carcinogenic Risk of Chemicals to Man, Vol. 15, pp. 41–102. IARC, Geneva.

Ishidate, K., Tsuruoka, M., and Nakazawa, Y. (1980). Induction of choline kinase by polycyclic aromatic hydrocarbon carcinogens in rat liver. *Biochem. Biophys. Res. Commun.* **96**, 946–952.

Israel, D. I., and Whitlock, J. P., Jr. (1983). Induction of mRNA specific for cytochrome $P_1$-450 in wild type and variant mouse hepatoma cells. *J. Biol. Chem.* **258**, 10390–10394.

Ivanovic, V., and Weinstein, I. B. (1982). Benzo[*a*]pyrene and other inducers of cytochrome $P_1$-450 inhibit binding of epidermal growth factor to call surface receptors. *Carcinogenesis* **3**, 505–510.

Jones, K. G., and Sweeney, G. P. (1980). Dependence of the porphyrogenic effect of 2,3,7,8-tetrachlorodibenzo-*p*-dioxin upon inheritance of aryl hydrocarbon hydroxylase responsiveness. *Toxicol. Appl. Pharmacol.* **53**, 42–49.

Jones, P. B. C., Miller, A. G., Israel, D. I., Galeazzi, D. R., and Whitlock, J. P., Jr. (1984). Biochemical and genetic analysis of variant mouse hepatoma cells which overtranscribe the cytochrome $P_1$-450 gene in response to 2,3,7,8-tetrachlorodibenzo-*p*-dioxin. *J. Biol. Chem.* **254**, 12357–12363.

Jones, P. B. C., Galeazzi, D. R., Fisher, J. M., and Whitlock, J. P., Jr. (1985). Control of gene expression by dioxin: functional analysis of a TCDD-responsive regulatory element upstream of the cytochrome $P_1$-450 gene. *Science (Washington, D.C.)* **227**, 1499–1502.

Karenlampi, S. O., Eisen, H. J., Hankinson, O., and Nebert, D. W. (1983). Effects of cytochrome $P1$-450 inducers on the cell-surface receptors for epidermal growth factor, phorbol 12,13-dibutyrate, or insulin of cultured mouse hepatoma cells. *J. Biol. Chem.* **258**, 10378–10383.

Kellerman, G., Luyten-Kellerman, M., and Shaw, C. R. (1973). Genetic variation of aryl hydrocarbon hydroxylase in human lymphocytes. *Am. J. Hum. Genet.* **25**, 327–331.

Kimbrough, R. D. (1980). Occupational exposure. *In* "Halogenated Biphenyls, Terphenyls, Naphthalenes, Dibenzodioxins and Related Products" (R. D. Kimbrough, ed.), pp. 373–397. Elsevier/North-Holland, New York.

Kimmig, J., and Schultz, K. H. (1957). Chlorinated aromatic cyclic ethers as the cause of chloracne. *Naturwissenschaften* **44**, 337–338.

Kirsch, R., Fleischner, G., Kamisaka, K., and Arias, I. M. (1975). Structural and functional studies of ligandin a major renal organic anion-binding protein. *J. Clin. Invest.* **55**, 1009–1019.

Knutson, J. C., and Poland, A. (1980). Keratinization of mouse teratoma cell line XB produced by 2,3,7,8-tetrachlorodibenzo-*p*-dioxin: an *in vitro* model of toxicity. *Cell* **22**, 27–36.

Knutson, J. C., and Poland, A. (1982). Response of murine epidermis to 2,3,7,8-tetrachlorodibenzo-*p*-dioxin: interaction of the *Ah* and *hr* loci. *Cell* **30**, 225–234.

Kociba, R. J., Keyes, D. G., Beyer, J. E., Carreon, R. M., Wade, C. E., Dittenber, D. A., Kalnins, R. P., Frauson, L. E., Park, C. N., Barnard, S. D., Hummel, R. A., and Humiston, C. G. (1978). Results of a two-year chronic toxicity and oncogenicity study of 2,3,7,8-tetrachlorodibenzo-*p*-dioxin in rats. *Toxicol. Appl. Pharmacol.* **46**, 279–303.

Kruisbeck, A. M. (1979). Thymic factors and T cell maturation *in vitro:* a comparison of the effects of thymic epithelial cultures with thymic extracts and thymus dependent serum factors. *Thymus* **1**, 163–185.

Kumaki, K. Jensen, N. M., Shire, J. G. M., and Nebert, D. W. (1977). Genetic differences in induction of cytosol reduced NAD(P): menadione oxidoreductase and microsomal aryl hydrocarbon hydroxylase in the mouse. *J. Biol. Chem.* **251**, 157–165.

Lang, M. A., and Nebert, D. W. (1981). Structural gene products of the *Ah* locus. Evidence for many unique *P*-450-mediated monooxygenase activities reconstituted from 3-methylcholanthrene-treated C57BL/6N mouse liver microsomes. *J. Biol. Chem.* **256**, 12058–12067.

Lund, J., Kurl, R. N., Poellinger, L., and Gustafsson, J.-A. (1982). Cytosolic and nuclear binding proteins for 2,3,7,8-tetrachlorodibenzo-*p*-dioxin in the rat thymus. *Biochim. Biophys. Acta* **716**, 16–23.

Luster, M. I., Dean, J. H., and Boorman, G. A. (1982). Altered immune functions in rodents treated with 2,3,7,8-tetrachlorodibenzo-*p*-dioxin, phorbol-12-myristate-13-acetate, and benzo(*a*)pyrene. *In* "Environmental Factors in Human Growth and Development" (V.R. Hunt, M. K. Smith, and D. Worth, eds.), Banbury Report, Vol. 11, pp. 199–213. Cold Spring Harbor Lab., Cold Spring Harbor, New York.

McConnell, E. E., Moore, J. A., and Dalgard, D. W. (1978a). Toxicity of 2,3,7,8-tetrachlorodiben-zo-*p*-dioxin in Rhesus monkeys (*Macaca mulatta*) following a single oral dose. *Toxicol. Appl. Pharmacol.* **43,** 175–187.

McConnell, E. E., Moore, J. A., Haseman, J. K., and Harris, M. W. (1978b). The comparative toxicity of chlorinated dibenzo-*p*-dioxins in mice and guinea pigs. *Toxicol. Appl. Pharmacol.* **44,** 335–356.

McKinney, J. D., and McConnell, E. (1982). Structural specificity and the dioxin receptor. *In* "Chlorinated Dioxins and Related Compounds: Impact on the Environment" (O. Hutzinger, R. W. Frei, E. Merian, and F. Pocchiari, eds.), pp. 367–381. Pergamon, New York.

McKinney, J. D., Gottschalk, K. E., and Pedersen, L. (1983). A theoretical investigation of the conformation of polychlorinated biphenyls (PCB's). *J. Mol. Struct.* **104,** 445–450.

McKinney, J. D., and Chae, K., McConnell, E. E., Birnbaum, L. S. (1985). Structure-induction versus structure-toxicity relationships for polychlorinated biphenyls and related aromatic hydrocarbons. *Environ. Health Perspect.* (in press).

McNulty, W. P., Becker, G. M., and Cory, H. T. (1980). Chronic toxicity of 3,4,3′,4′- and 2,5,2′,5′-tetrachlorobiphenyls in Rhesus Macaques. *Toxicol. Appl. Pharmacol.* **56,** 182–190.

Mason, M. E., and Okey, A. B. (1982). Cytosolic and nuclear binding of 2,3,7,8-tetrachlorodiben-zo-*p*-dioxin to the *Ah* receptor in extra-hepatic tissues of rats and mice. *Eur. J. Biochem.* **123,** 209–215.

Miller, A. G., and Whitlock, J. P., Jr. (1981). Novel Variants in benzo(*a*)pyrene metabolism. Isolation by fluorescence-activating cell sorting. *J. Biol. Chem.* **256,** 2433–2437.

Miller, A. G., Israel, D., and Whitlock, J. P., Jr. (1983). Biochemical and genetic analysis of variant mouse hepatoma cells defective in the induction of benzo(*a*)pyrene-metabolizing enzyme activity. *J. Biol. Chem.* **258,** 3523–3527.

Milstone, L. M., and LaVigne, J. F. (1984). 2,3,7,8-Tetrachlorodibenzo-*p*-dioxin induces hyperplasia in confluent cultures of human Keratinocytes. *J. Invest. Dermatol.* **82,** 532–534.

Neal, R. A. (1985). Mechanisms of the biological effects of PCB's polychlorinated dibenzo-*p*-dioxins and dibenzofurans. *Environ. Health Perspect.* (in press).

Neal, R. A., Olson, J. R., Gasiewicz, T. A., and Geiger, L. E. (1982). The toxicokinetics of 2,3,7,8-tetrachlorodibenzo-*p*-dioxin in mammalian systems. *Drug Metab. Rev.* **13,** 355–385.

Nebert, D. W. (1979). Genetic differences in the induction of monooxygenase activities by polycyclic aromatic compounds. *Pharmacol. Ther.* **6,** 395–417.

Nebert, D. W., and Jensen, N. M. (1979). The *Ah* locus: genetic regulation of the metabolism of carcinogens, drugs, and other environmental chemicals by cytochrome *P*-450-mediated monooxygenases. *CRC Crit. Rev. Biol.* **6,** 401–437.

Nebert, D. W., Goujon, F. M., and Gielen, J. E. (1972). Aryl hydrocrabon hydroxylase induction by polycyclic hydrocarbons: simple autosomal dominant trait in the mouse. *Nature New Biol. (London),* **236,** 107–110.

Nebert, D. W., Jensen, N., Perry, J., and Oka, T. (1980). Association between ornithine decarboxylase induction and the *Ah* locus in mice treated with polycyclic aromatic compounds. *J. Biol. Chem.* **255,** 6836–6842.

Negishi, M., and Nebert, D. W. (1979). Structural gene products of the *Ah* locus. Genetic and immunochemical evidence for two forms of mouse liver cytochrome *P*-450 induced by 3-methylcholanthrene. *J. Biol. Chem.* **254,** 11015–11023.

Okey, A. B., and Vella, L. M. (1984). Elevated binding of 2,3,7,8-tetrachlorodibenzo-*p*-dioxin and 3-methylcholanthrene to the *Ah* receptor in hepatic cytosols from phenobarbital-treated rats and mice. *Biochem. Pharmacol.* **33,** 531–538.

Okey, A. B., Bondy, G. P., Mason, M. E., Kahl, G. F., Eisen, H. J., Guenthner, T. M., and Nebert, D. W. (1979). Regulatory gene product of the *Ah* locus. Characterization of the cytosolic inducer–receptor complex and evidence for its nuclear translocation. *J. Biol. Chem.* **254,** 11636–11648.

Okey, A. B., Bondy, G. P., Mason, M. E., Nebert, D. W., Forster-Gibson, C. J., Muncan, J., and Dufresne, M. J. (1980). Temperature-dependent cytosol-to-nucleus translocation of the *Ah* receptor for 2,3,7,8-tetrachlorodibenzo-*p*-dioxin in continuous cell culture lines. *J. Biol. Chem.* **255**, 11415–11422.

Okey, A. B., Mason, M. E., Gehly, B., Heidelberger, C., Muncan, J., and Dufresne, M. J. (1983). Defective binding of 3-methylcholanthrene to the *Ah* receptor within C3H10T1/2 Clone 8 mouse fibroblasts in culture. *Eur. J. Biochem.* **132**, 219–227.

Okuda, T., Vesell, E. S., Plotkin, E., Tarone, R., Bast, R. D., and Gelboin, H. V. (1977). Interindividual and intraindividual variations in aryl hydrocarbon hydroxylase in monocytes from monozygotic and dizygotic twins. *Cancer Res.* **37**, 3904–3911.

Olson, J. R., Gasiewicz, T. A., and Neal, R. A. (1980a). Tissue distribution, excretion, and metabolism of 2,3,7,8-tetrachlorodibenzo-*p*-dioxin (TCDD) in Golden Syrian Hamster. *Toxicol. Appl. Pharmacol.* **56**, 78–85.

Olson, J. R., Holscher, M. A., and Neal, R. A. (1980b). Toxicity of 2,3,7,8-tetrachlorodibenzo-*p*-dioxin in the Golden Syrian hamster. *Toxicol. Appl. Pharmacol.* **55**, 67–78.

Olson, J. R., Gasiewicz, T. A., Geiger, L. E., and Neal, R. A. (1983). The metabolism of 2,3,7,8-tetrachlorodibenzo-*p*-dioxin in mammalian systems. *In* "Accidental Exposure to Dioxins" (F. Coulston and F. Pocchiari, eds.), pp. 81–103. Academic Press, New York.

Osborne, R., and Greenlee, W. F. (1985). 2,3,7,8-Tetrachlorodibenzo-*p*-dioxin (TCDD) enhances terminal differentiation of cultured human epidermal cells. *Toxicol. Appl. Pharmacol.* **77**, 434–443.

Owens, I. S. (1977). Genetic regulation of UDP-glucuronosyltransferase induction by polycyclic aromatic compounds in mice. *J. Biol. Chem.* **252**, 2827–2833.

Parkinson, A., Cockerline, R., and Safe, S. (1980a). Polychlorinated biphenyl isomers and congeners as inducers of both 3-methylcholanthrene- and phenobarbitone-type microsomal enzyme activity. *Chem.-Biol. Interact.* **29**, 277–289.

Parkinson, A., Robertson, L., Safe, L., and Safe, S. (1980b). Polychlorinated biphenyls as inducers of hepatic microsomal enzymes: structure–activity rules. *Chem.-Biol. Interact.* **30**, 271–285.

Parkinson, A., Robertson, L., Uhlig, L., Campbell, M. A., and Safe, S. (1982). 2,3,4,4',5-pentachlorobiphenyl: differential effects on C57BL/6J and DBA/2J inbred mice. *Biochem. Pharmacol.* **31**, 2830–2833.

Pitot, H. C., Goldsworthy, T., Campbell, H. A., and Poland, A. (1980). Quantitative evaluation of the promotion by 2,3,7,8-tetrachlorodibenzo-*p*-dioxin of hepatocarcinogenesis from diethylnitrosamine. *Cancer Res.* **40**, 3616–3620.

Poiger, H., and Buser, H. R. (1983). Structure elucidation of mammalian TCDD metabolites. *In* "Human and Environmental Risks of Chlorinated Dioxins and Related Compounds" (R. Tucker, A. Young, and A. Gray, eds.), pp. 483–492. Plenum, New York.

Poiger, H., and Schlatter, C. (1979). Biological degradation of TCDD in rats. *Nature (London)* **281**, 706–707.

Poland, A., and Glover, E. (1973). Chlorinated dibenzo-*p*-dioxins: potent inducers of δ-aminolevulinic acid synthetase and aryl hydrocarbon hydroxylase. II. A study of the structure–activity relationship. *Mol. Pharmacol.* **9**, 736–747.

Poland, A., and Glover, E. (1974). Comparison of 2,3,7,8-tetrachlorodibenzo-*p*-dioxin, a potent inducer of aryl hydrocarbon hydroxylase, with 3-methylcholanthrene. *Mol. Pharmacol.* **10**, 349–359.

Poland, A., and Glover, E. (1975). Genetic expression of aryl hydrocarbon hydrolxylase by 2,3,7,8-tetrachlorodibenzo-*p*-dioxin: evidence for a receptor mutation in genetically non-responsive mice. *Mol. Pharmacol.* **11**, 389–398.

Poland, A., and Glover, E. (1977). Chlorinated biphenyl induction of aryl hydrocarbon hydroxylase activity: a study of the structure–activity relationship. *Mol. Pharmacol.* **13**, 924–938.

Poland, A., and Glover, E. (1979). An estimate of the maximum *in vivo* covalent binding of 2,3,7,8-

tetrachlorodibenzo-*p*-dioxin to rat liver protein, ribosomal RNA, and DNA. *Cancer Res.* **39**, 3341–3344.

Poland, A., and Glover, E. (1980). 2,3,7,8-Tetrachlorodibenzo-*p*-dioxin: segregation of toxicity with the *Ah* locus. *Mol. Pharmacol.* **17**, 86–94.

Poland, A., and Knutson, J. C. (1982). 2,3,7,8-Tetrachlorodibenzo-*p*-dioxin and related halogenated aromatic hydrocarbons: examination of the mechanism of toxicity. *Annu. Rev. Pharmacol. Toxicol.* **22**, 517–554.

Poland, A. P., Glover, E., Robinson, J R., and Nebert, D. W. (1974). Genetic expression of aryl hydrocarbon hydroxylase activity. Induction of monooxygenase activities and cytochrome $P_1$-450 formation by 2,3,7,8-tetrachlorodibenzo-*p*-dioxin in mice genetically ''non-responsive'' to other aromatic hydrocarbons. *J. Biol. Chem.* **249**, 5599–5606.

Poland, A., Glover, E., and Kende, A. S. (1976a). Stereospecific, high affinity binding of 2,3,7,8-tetrachlorodibenzo-*p*-dioxin by hepatic cytosol. Evidence that the binding species is receptor for induction of aryl hydrocarbon hydroxylase. *J. Biol. Chem.* **251**, 4936–4946.

Poland, A., Glover, E., Kende, A. S., DeCamp, M., and Giandomenico, C. M. (1976b). 3,4,3′,4′-Tetrachloroazoxybenzene and azobenzene: potent inducers of aryl hydrocarbon hydroxylase. *Science (Washington, D.C.)* **194**, 627–630.

Poland, A., Greenlee, W. F., and Kende, A. S. (1979). Studies on the mechanism of action of the chlorinated dibenzo-*p*-dioxins and related compounds. *Ann. N.Y. Acad. Sci.* **320**, 214–230.

Poland, A., Palen, D., and Glover, E. (1982). Tumor promotion by TCDD in skin of HRS/J hairless mice. *Nature (London)* **300**, 271–273.

Powell, B., Garola, R. E., Chamness, G. C., and McGuire, W. L. (1979). Measurement of progesterone receptor in human breast cancer biopsies. *Cancer Res.* **39**, 1678–1682.

Pratt, R. M. (1983). Mechanisms of chemically-induced cleft palate. *Trends Pharmacol. Sci.* **4**, 160–162.

Puhvel, S. M., Ertl, D. C., and Lynberg, C. A. (1984). Increased epidermal transglutaminase activity following 2,3,7,8-tetrachlorodibenzo-*p*-dioxin: *In vivo* and *in vitro* studies with mouse skin. *Toxicol. Appl. Pharmacol.* **73**, 42–47.

Rheinwald, J. G., and Beckett, M. A. (1980). Defective terminal differentiation in culture as a consistent and selectable character of malignant human keratinocytes. *Cell* **22**, 629–632.

Rheinwald, J. G., and Beckett, M. A. (1981). Tumorigenic keratinocyte lines requiring anchorage and fibroblast support from human squamous cell carcinomas. *Cancer Res.* **41**, 1657–1663.

Rheinwald, J. G., and Green, H. (1975). Serial cultivation of strains of human epidermal keratinocytes: the formation of keratinizing colonies from single cells. *Cell* **6**, 331–334.

Rheinwald, J. G., and Green, H. (1977). Epidermal growth factor and the multiplication of cultured human epidermal keratinocytes. *Nature (London)* **265**, 421–424.

Rice, R. H., and Cline, P. R. (1984). Opposing effects of 2,3,7,8-tetrachlorodibenzo-*p*-dioxin and hydrocortisone on growth and differentiation of cultured malignant human keratinocytes. *Carcinogenesis* **5**, 367–371.

Rice, R. H., and Greenlee, W. F. (1982). Effects of 2,3,7,8-tetrachlorodibenzo-*p*-dioxin in cultured human epithelial target cells: modulation by hydrocortisone. *Toxicologist* **2**, 463.

Robinson, J. R., Considine, N., and Nebert, D. W. (1974). Genetic expression of aryl hydrocarbon hydroxylase induction: evidence for the involvement of other genetic loci. *J. Biol. Chem.* **249**, 5851–5859.

Rose, J. Q., Ramsey, J. C., Mentzler, T. A., Hummel, R. A., and Gehring, P. J. (1976). The fate of 2,3,7,8-tetrachlorodibenzo-*p*-dioxin following single and repeated oral doses to the rat. *Toxicol. Appl. Pharmacol.* **36**, 209–226.

Sawyer, T., and Safe, S. (1982). PCB isomers and congeners: induction of aryl hydrocarbon hydroxylase and ethoxyresorufin *O*-deethylase enzyme activities in rat hepatoma cells. *Toxicol. Lett.* **13**, 87–94.

Schwetz, B. A., Norris, J. M., Sparschy, G. L., Rowe, V. K., Gehring, P. J., Emerson, J. L., and Gerbig, C. G. (1973). Toxicology of chlorinated dibenzo-*p*-dioxins. *Environ. Health Perspect.* **5,** 87–99.

Sharma, R. P., Kociba, R. J., and Gehring, P. J. (1978). Immunotoxicologic effects of 2,3,7,8-tetrachlorodibenzo-*p*-dioxin in laboratory animals. *Toxicol. Appl. Pharmacol.* **45,** 333.

Stutman, O. (1978). Intrathymic and extrathymic T-cell maturation. *Immunol. Rev.* **42,** 138–184.

Sun, T.-T., and Green, H. (1976). Differentiation of the epidermal keratinocyte in cell culture: formation of the cornified envelope. *Cell* **9,** 511–521.

Taylor, J. S. (1974). Chloracne—a continuing problem. *Cutis* **13,** 585–591.

Taylor, J. S. (1979). Environmental chloracne: update and overview. *Ann. N.Y. Acad. Sci.* **320,** 295–307.

Thomas, P. E., and Hutton, J. J. (1973). Genetics of aryl hydrocarbon hydroxylation induction in mice: additive inheritance in crosses between C3H/HeJ and DBA/2J. *Biochem. Genet.* **8,** 249–257.

Thomas, P. E., Kouri, R. E., and Hutton, J. J. (1972). The genetics of aryl hydrocarbon hydroxylase induction in mice: a single gene difference between C57CL/6J and DBA/2J. *Biochem. Genet.* **6,** 157–168.

Tsui, H. W., and Okey, A. B. (1981). Rapid vertical tube rotor gradient assay for binding of 2,3,7,8-tetrachlorodibenzo-*p*-dioxin to the *Ah* receptor. *Can. J. Physiol. Pharmacol.* **59,** 927–931.

Tukey, R. H., Hannah, R. R., Negishi, M., Nebert, D. W., and Eisen, H. J. (1982). The *Ah* locus: correlation of intranuclear appearance of inducer–receptor complex with induction of cytochrome $P_1$-450 mRNA. *Cell* **31,** 275–284.

Vecchi, A., Sironi, M., Canegrati, M. A., Recchia, M., and Garattini, S. (1983). Immunosuppressive effects of 2,3,7,8-tetrachlorodibenzo-*p*-dioxin in strains of mice with different susceptibility to induction of aryl hydrocarbon hydroxylase. *Toxicol. Appl. Pharmacol.* **68,** 434–441.

Vos, J. G. (1977). Immune suppression as related to toxicology. *CRC Crit. Rev. Toxicol.* **5,** 67–101.

Weiss, L. (1983). The life cycle of blood cells. *In* "Histology: Cell and Tissue Biology" (L. Weiss, ed.), 5th Ed., pp. 474–497. Elsevier, New York.

Whitlock, J. P., and Galeazzi, D. R. (1984). TCDD receptors in wild type and variant mouse hepatoma cells: nuclear location and strength of nuclear binding. *J. Biol. Chem.* **259,** 980–985.

Wiebel, F. J., Hlavica, P., and Grzeschik, K. H. (1981). Expression of aromatic polycyclic hydrocarbon-induced monoxygenase (aryl hydrocarbon hydroxylase) in man × mouse hybrids is associated with human chromosome 2. *Hum. Genet.* **59,** 277–280.

# 5

# Interactions of Animal Viruses with Cell Surface Receptors

JAMES C. PAULSON
Department of Biological Chemistry
UCLA School of Medicine
Los Angeles, California

## I. INTRODUCTION

Virus receptors are the cell surface components to which a virus will attach prior to penetration and infection. Although this definition is widely used by virologists and microbiologists (Lonberg-Holm and Philipson, 1974), it bears repeating in a book which is primarily devoted to receptors of physiological

**131**

THE RECEPTORS, VOL. II

ligands. For physiological ligand–receptor interactions, the cell surface receptor is a protein with a binding domain which interacts with a soluble ligand. For virus–receptor interactions, it is the viral attachment protein that has a binding site capable of interacting with a cell surface ligand, which is either protein, carbohydrate, or lipid. Thus, the viral attachment protein is the active agent which confers specificity in the virus–receptor interaction.

In recent years, there have been exciting advances pertaining to the structure of animal virus attachment proteins, the nature of the cell surface receptors which they recognize, and the roles of cellular receptors in viral penetration, host range, and tissue tropism. The purpose of this chapter is to attempt to cover each of these areas. Due to the broad scope, each topic will emphasize viruses which best illustrate new concepts, insights, or advances in descriptive biology. For this reason, it is recognized at the outset that the emphasis may at times appear uneven and that important contributions may have been missed. Throughout, the reader will be referred to many excellent reviews. Of particular value for a comprehensive overview of earlier work in the field are the monographs by Lonberg-Holm and Philipson (1974) and Dales (1973).

To integrate the work on the cell surface receptors of various groups of viruses, it has been useful to make a distinction between the receptor, the cell surface component to which the virus binds, and the receptor determinant, the limited structure recognized by the viral attachment protein. Accordingly, the review has been divided into five major sections; viral attachment proteins; analysis of cell surface receptors; viral receptor determinants; the role of receptors in viral penetration, and the role of receptors in host range and tissue tropism.

## II. VIRAL ATTACHMENT PROTEINS

### A. Membrane-Enveloped Viruses

Viral attachment proteins of membrane-enveloped viruses are, in general, glycoproteins which can in some cases be visualized by electron microscopy as spikes protruding from the lipid bilayer. An electron micrograph depicting the glycoprotein spikes of influenza virus is shown in Fig. 1. The properties of the attachment proteins for several groups of membrane-enveloped viruses are presented in this section. Although the main emphasis will be a description of the viral attachment protein itself, other viral glycoproteins which have functions related to viral attachment and penetration will also be discussed. The two viral glycoproteins of influenza viruses, the hemagglutinin and neuraminidase, will be presented in considerable detail since the three-dimensional crystal structures of both are now available. Many general aspects discussed for the influenza

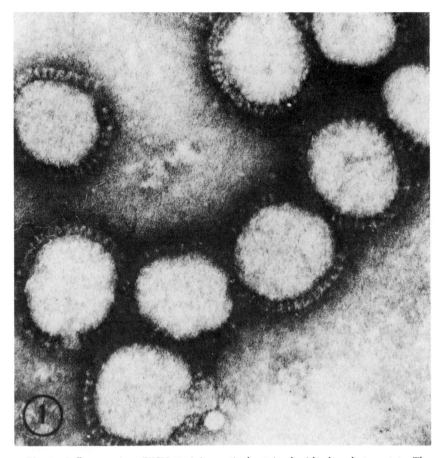

**Fig. 1.** Influenza virus (WSN strain) negatively stained with phosphotungstate. The virions are approximately 100 nm in diameter, and the glycoprotein spikes radiating from the viral envelope are 10–12 nm in length. (From Compans *et al.*, 1970.)

glycoproteins, such as overall biosynthesis, will not be repeated for other viral glycoproteins unless there are important differences.

## 1. Myxoviruses

The myxoviruses are comprised of the influenza A, B, and C viruses. The hemagglutinin of the influenza A and B viruses is perhaps the best characterized of all of the viral attachment proteins. The name hemagglutinin comes from its ability to mediate viral agglutination of erythrocytes (Hirst, 1941; McClelland and Hare, 1941). It is one of two viral glycoproteins that make up the spiked projections on the viral envelope, the other is the neuraminidase. These two glycoproteins are present at about 500 and 100 copies per virus particle, respectively.

Discovery of the viral neuraminidase was based on the observation that once bound to erythrocytes, influenza virus could elute (Hirst, 1942a,b), due to a viral receptor destroying enzyme (RDE). An analogous RDE was found in the culture filtrate of *Vibrio cholerae* by Burnet and Stone (1947). These enzymes were also active against mucoprotein inhibitors of hemagglutination and released a substance ultimately identified as *N*-acetylneuraminic acid (Gottschalk and Lind, 1949; Klenk *et al.*, 1955), a sialic acid. This important finding simultaneously identified the receptor-destroying enzymes as neuraminidases (or sialidases) and provided strong evidence that sialic acid was an essential receptor determinant of the hemagglutinin.

**a. Biosynthesis of the Hemagglutinin and Neuraminidase.** The hemagglutinin is a membrane glycoprotein and thus follows the typical biosynthetic route of other glycoproteins destined for the plasma membrane or for secretion. The hemagglutinin is initially synthesized with a hydrophobic signal peptide of about 15 amino acids which presumably directs synthesis to the endoplasmic reticulum and is soon removed by a peptidase (Waterfield *et al.*, 1979). In the endoplasmic reticulum, four to seven asparagine residues are glycosylated by en bloc transfer of an oligosaccharide containing glucose, mannose, and *N*-acetyl-glucosamine ($Glc_3Man_gGlcNAc_2$) from a lipid precursor (Schwartz *et al.*, 1977; Nakamura and Compans, 1979). After translocation to the Golgi, the oligosaccharides are processed to the typical high-mannose and complex types (for review see Hubbard and Ivatt, 1981). To some extent, the precise oligosaccharide structures present on the mature hemagglutinin (Matsumoto *et al.*, 1983) are dependent on the host in which the virus is grown (Nakamura and Compans, 1979b). It is presumably in the Golgi where attachment of fatty acids (Schmidt *et al.*, 1979; Schmidt and Schlesinger, 1979) and sulfation of oligosaccharides (Nakamura and Compans, 1977; Compans and Pinter, 1975) occurs. Finally, the hemagglutinin is routed to the plasma membrane where it participates in the budding of new virus particles.

One of the most important posttranslational modifications of the hemagglutinin is cleavage of the precursor polypeptide (HAO) by a host protease to yield two polypeptides HA1 and HA2 joined by disulfide bonds. This cleavage is essential for infectivity (Lazarowitz and Choppin, 1975; Klenk *et al.*, 1975) and has been reported to occur both in the Golgi or smooth endoplasmic reticulum (Lohmeyer and Klenk, 1979) and at the cell surface (Lazarowitz *et al.*, 1971; Hay, 1974). Some cell lines are able to effect cleavage of only a few influenza strains (Klenk *et al.*, 1975). The ability of a host to cleave the viral hemagglutinin may be an important factor in pathogeneity of avian influenza viruses (Bosch *et al.*, 1979). Some laboratory cell lines, such as MDCK cells, are able to produce infectious virus particles with cleaved hemagglutinin for most strains of influenza but only when trypsin is added to the culture media (Tobita *et al.,* 1975).

Although biosynthesis of the neuraminidase has not been studied as thoroughly, few significant differences are expected in glycosylation and routing through subcellular organelles. However, unlike the hemagglutinin, the $NH_2$-terminal hydrophoboc peptide which likely serves as a ''signal sequence'' is not cleaved off but remains embedded in the membrane (Block, 1981). Moreover, no host-dependent cleavage of the neuraminidase polypeptide has been observed.

**b. Functions of the Hemagglutinin.** The hemagglutinin and neuraminidase have been purified to homogeneity in biologically active form, demonstrating that their activities are separate and distinct (Compans and Choppin, 1975). The hemagglutinin has two main functions. In addition to mediating attachment to cell surface receptors, it appears to initiate fusion of the influenza virus membrane with that of endocytic vesicles or lysosomes through a low-pH induced fusion activity (see Section V,B,3). This then allows the influenza genome to be released into the cytoplasm where replication may ensue. While both uncleaved (HAO) and cleaved (HA1 and HA2) forms of the hemagglutinin mediate viral attachment, only the cleaved form exhibits fusion activity.

**c. Functions of the Neuraminidase.** The neuraminidase is generally thought to aid in the release of the budding virus from the cellular surface by destroying endogenous receptors and perhaps also to prevent irreversible adsorption to mucus glycoproteins (Compans and Choppin; 1975; Krizanova and Rathova, 1969). It also prevents attachment of sialic acid to the complex-type oligosaccharides of the hemagglutinin (Klenk, *et al,* 1970; Matsumoto *et al.,* 1983). Indeed, in the absence of viral neuraminidase, virions are sialylated and form large aggregates (Palase *et al.,* 1974). It has also been observed that some viral neuraminidases, like that of the WSN virus, are unique in promoting the production of infectious viruses in some hosts. In particular, the proper neuraminidase appears to allow efficient cleavage of the viral hemagglutinin by a host protease (Schulman and Palase, 1977; Ghendon *et al.,* 1979; Sugiura and Ueda, 1980; Nakajima and Sugiura, 1980). The property of the neuraminidase that accounts for this interesting observation has not yet been elucidated.

**d. Antigenic Properties of Influenza Virus Glycoproteins.** The hemagglutinin and neuraminidase are also the major surface antigens of the virus (for review see Webster and Laver, 1975; Webster *et al.,* 1982). There are 13 major antigenic subtypes for the hemagglutinin and 9 subtypes for the neuraminidase. Only three of the hemagglutinin subtypes (H1, H2 and H3) and two of the neuraminidase subtypes (N1 and N2) have appeared in viruses which circulated in human populations in this century. The other subtypes are found in viruses which infect animal species, birds, pigs, horses, and seals, with most subtypes found in viruses isolated from birds (Laver and Webster, 1979). Clinically, hemagglutinin antibodies are the most important, since they typically neutralize infection by the virus while neuraminidase antibodies alone do not. Emergence of a new hemagglutinin subtype in the human influenza viruses, called antigenic shift, occurs once every 10–30 years. In addition to the antigenic shift, the influenza virus gradually accumulates point mutations, allowing selection of antigenic variants sufficiently different from the parent strain that reinfection of a host previously infected with a virus of that subtype can occur; this is called antigenic drift. The combination of antigenic shift and antigenic drift is responsible for the uncontrolled recurring epidemics in man (Stuart-Harris and Schild, 1976).

**e. Structure of the Hemagglutinin.** The combination of interest in the biological functions and antigenic properties of the hemagglutinin and neuraminidase have resulted in extensive structural analysis by many laboratories. This has culminated in reports of the three-dimensional crystal structures for both the H3 hemagglutinin (Wilson *et al.*, 1981; Wiley *et al.*, 1981) and N2 neuraminidase (Varghese *et al.*, 1983; Colman *et al.*, 1983) as shown in Figs. 2 and 3, respectively.

Structural analysis of the hemagglutinin was performed by Wilson *et al.* (1981) on crystals of the bromelain fragment of a recombinant virus X-31 (H3N2). The native hemagglutinin is a trimer of 225,000 molecular weight. Each subunit has two polypeptides, HA1 (328 residues) and HA2 (221 residues). The H3 hemagglutinin contains about 19% carbohydrate and is glycosylated at seven asparagine residues. Bromelain treatment of the influenza virus cleaves the hemagglutinin at the membrane, releasing the trimer largely intact from a hydrophobic peptide of 5400 molecular weight which anchors the hemagglutinin to the membrane (Waterfield *et al.*, 1979).

The antigenic sites of the H3 hemagglutinin have been mapped (Wiley *et al.*, 1981) by taking advantage of extensive sequence information on antigenic-drift strains and antigenic variants obtained by selection with monoclonal antibodies (Fig. 2). Since the introduction of the H3 hemagglutinin in the human virus population in 1968, there have been several worldwide epidemics with the concomitant isolation of viruses with antigenically distinct hemagglutinins. Sequence data from representative drift strains from 1972, 1975, 1977, and 1979 have revealed that a relatively small number of amino acid substitutions were required to effect the changes in antigenicity. In addition, laboratory variants selected with monoclonal antibodies differed by a single amino acid, resulting in a total absence of antibody binding (Webster *et al.*, 1982). When localized on the three-dimensional structure, the amino acid changes in the drift strains and laboratory variants were found in four major areas or "sites." Localization of

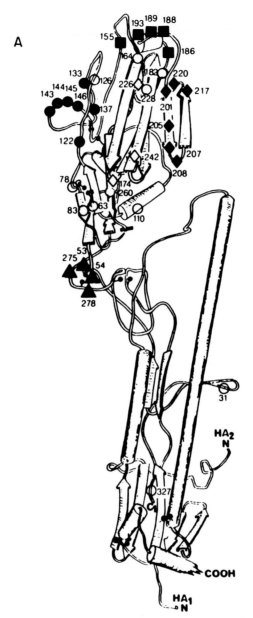

**Fig. 2.** Three-dimensional structure of the influenza H3 hemagglutinin. (A) A tracing of the polypeptide chain of hemagglutinin monomer with stretches of α-helix depicted as solid cylinders and β-pleated sheet as flat arrows. Amino acid changes responsible for antigenic variation in field isolates, from 1968 to 1977, and in selected laboratory variants are indicated in solid symbols, ●, Site A; ■, Site B; ▲, Site C, and ♦, Site D. Other amino acid changes, indicated by open symbols, are antigenically neutral. (From Wiley *et al.*, 1982. Reprinted by permission from *Nature* **289**, 373–378. Copyright © 1982 Macmillan Journals Limited.) (B) Selected amino acid residues surrounding the sialic acid-binding

(*continued*)

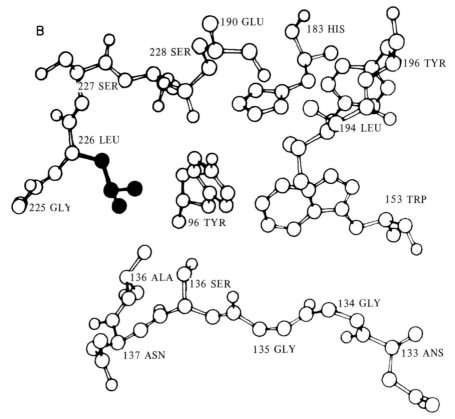

**Fig. 2.**   (cont.)
site. The change of amino acid 226 from Leu to Gln results in a change in the specificity of
H3 hemagglutinins from preferential binding of the SAα2,6Gal linkage to the SAα2,3Gal
linkage, respectively (see Section III,3). (From Rogers et al. (1983a). Reprinted by permission from *Nature* **304**, 76–77. Copyright © 1983 Macmillan Journals Limited.)

the antigenic sites of the H1 hemagglutinin of A/PR/8/34, when fitted to the
crystal structure of the H3 hemagglutinin, yielded similar conclusions (Caton *et
al.*, 1982).

Several structural features relevant to biological activities are of particular
interest. The globular head at the distal end of the molecule was proposed to
contain the receptor binding site based on a pocket of conserved amino acid
residues as shown in Fig. 2 and similarities to the sialic acid binding site of wheat
germ agglutinin (Wilson *et al.*, 1981). Further evidence for the identity of this
pocket as the binding site was the observation that a receptor variant of X-31 had
a single amino acid change at residue 226 (Leu to Gln), which is in the midst of
the conserved amino acid residues (Rogers *et al.*, 1983a; see Section IV,C,3).
The region thought to play a role in the membrane fusion activity is the $NH_2$

terminus of the HA2 peptide (Waterfield *et al.*, 1979; Skehel *et al.*, 1982) located near the viral membrane (Fig. 2). Its proposed role in membrane fusion is discussed in Section V,B,3.

**f. Structure of the Neuraminidase.** The recent crystal structure of the N2 neuraminidase by Varghese *et al.* (1983) and Colman *et al.* (1983) shown in Fig. 3 allows a detailed look at the structure of the other membrane glycoprotein of the influenza virus envelope. The neuraminidase is a tetramer of 240,000 molecular weight, consisting of a slender stalk embedded in the membrane with a

**Fig. 3.** Three-dimensional structure of the influenza neuraminidase. The neuraminidase tetramer is viewed from above. Each subunit highlights by bold symbols certain structural features of the molecule, disulfide bonds (top left), carbohydrate attachment sites (bottom left), metal ligands, (bottom right), and some conserved acidic and basic amino acids surrounding the sialic acid-binding site marked with a star (top right). (From Varghese *et al.*, 1983. Reprinted by permission from *Nature* **303**, 35–40. Copyright © 1983 Macmillan Journals Limited.)

"box-shaped" or "pinwheel-like" head. Four asparagine-linked oligosaccharides per subunit are located at the apical surface, most distal from the membrane. Amino acids involved in antibody binding are also concentrated on the apical surface and surround a cleft region comprising the sialic acid binding site. The sialic acid binding site was established based on the localization of conserved amino acid residues in the cleft and by direct localization of bound sialic acid through analysis of a Fourier difference map of X-ray diffraction data collected from crystals soaked in 0.5 m$M$ sialic acid (Colman $et\ al.$, 1983). Influenza virus neuraminidases also exhibit a $Ca^{2+}$ ion requirement for activity (Wilson and Rafelson, 1967; Dimmock, 1971; Carroll and Paulson, 1982). Metal binding sites localilzed for $Sm^{3+}$ near the fourfold axis are presumed to be the binding sites for $Ca^{2+}$ (Fig. 3).

**g. Influenza C.**　Unlike the influenza A and B viruses, the influenza C viruses do not utilize sialic acid-containing receptors (Kendal, 1975; O'Callaghan $et\ al.$, 1977). Yet fundamental properties of the structure and function of the viral glycoproteins are similar. The influenza C hemagglutinin requires proteolytic cleavage to activate a low-pH hemolysin activity and infectivity (Ohuchi $et\ al.$, 1982). While the virus has no neuraminidase, it is able to elute from erythrocytes and inactivates a glycoprotein inhibitor of hemagglutination (Styk, 1963; O'Callaghan and Labat, 1983). Thus, the influenza C viruses appear to have a receptor-destroying enzyme of as yet undefined specificity.

**h. Future Perspectives.**　It is truly remarkable that the three-dimensional structures of the influenza A hemagglutinin and neuraminidase have both been solved. These structures offer an unprecedented opportunity to study the functions of these proteins in viral attachment and penetration at the molecular level. It is expected that the availability of these structures coupled with the tremendous interest in the molecular biology of influenza viruses may evolve rational approaches to study important but subtle aspects of the biological functions previously ignored because of their complexity. Thus, major new insights may ultimately be gained into such difficult problems as the roles of these two proteins in host range and tissue tropism (Section VI).

## 2. Paramyxoviruses

Paramyxoviruses were originally grouped with the myxoviruses based on their common properties of agglutinating erythrocytes via sialic acid-containing receptors until it became clear that there were major physical and biological differences (Andrews $et\ al.$, 1955). Paramyxoviruses include many viruses which bind sialic acid-containing receptors like the Newcastle disease virus (NDV), Sendai virus, mumps virus, and simian virus 5 (SV5) viruses. In addition, there is a distinct subgroup, the moribilliviruses (notably measles and canine distemper), which do not bind sialic acid receptors (for review see Choppin and Compans, 1975; Choppin and Scheid, 1980).

The membrane glycoproteins of NDV, Sendai virus, and SV5 have been the most thoroughly studied. Like the myxoviruses, there are two membrane glycoproteins, HN and F1, which have receptor binding, neuraminidase, and fusion activities. In contrast, however, the receptor binding or hemagglutinin activity and the neuraminidase activity are both carried on the same glycoprotein, the HN protein. The fusion activity is carried on the other glycoprotein, the F protein.

Activation of the fusion activity of the F protein requires a proteolytic cleavage converting the intact polypeptide F0 to two peptides joined by a disulfide bond F1,2 (Homma and Ohuchi, 1973; Scheid and Choppin, 1974). The failure of some hosts to accomplish this proteolytic activation has been proposed to be a major factor in host restriction and tissue tropism of these viruses (Choppin and Scheid, 1980).

Because the F protein is active at neutral pH (Huang *et al.*, 1981), direct fusion of the virus with the host plasma membrane may occur (Dales, 1973; Fan and Sefton, 1978). Sequence analysis of the $NH_2$ terminus of the F2 peptide has revealed a stretch of hydrophobic and nonpolar amino acids thought to play a primary role in the fusion process (Gething *et al.*, 1978; Scheid *et al.*, 1978; Richardson *et al.*, 1980). The role of the F2 protein in penetration of the host cell is discussed in Section V,A,1.

### 3. Poxviruses

Poxviruses, the largest animal viruses, have a rather unique mode of replication (for review see Moss, 1974; Dales *et al.*, 1976; Choppin and Sheid, 1980). Initially virus particles are assembled intracellularly in cytoplasmic "factories" capable of de novo membrane synthesis. The intracellular particles, bound by a unit membrane (Ichihashi *et al.*, 1971), are readily recovered from lysed cells and are infectious. Membranes of intracellular virus particles contain one or two glycoproteins and several other nonglycosylated proteins (Holowczak, 1972; Sarov and Joklik, 1972; Payne, 1979). Poxviruses can also acquire a second membrane envelope by budding into Golgi vesicles or budding from the plasma membrane. This is the major form of the virus detected extracellularly (Appleyard *et al.*, 1971; Payne and Norrby, 1976).

Vaccine virus has long been known to produce hemagglutination activity (Nagler, 1942). The hemagglutinin is found in the membrane envelope of the extracellular forms of virus but not the intracellular forms (Payne and Norrby, 1976). The membrane envelope contains up to seven glycoproteins (Payne, 1978, 1979). A glycoprotein with a molecular weight of 85,000–89,000 has been identified as the hemagglutinin (HA), based on its presence and absence, respectively, in cells infected with $HA^+$ and $HA^-$ strains of vaccine virus

(Dales *et al.*, 1976; Payne, 1979; Shida and Dales, 1981). Moreover, only an 89,000 molecular weight glycoprotein remains bound to rooster erythrocytes after adsorption of virus and washing with dilute detergent (Payne, 1979). Since the HA is not present in the intracellular virus particles, it is presumably synthesized and inserted into the plasma membrane or other cellular membrane and is subsequently acquired by the virus during budding. The hemagglutinin contains oligosaccharides both N-linked to asparagine and O-linked to threonine and serine (Shida and Dales, 1981). Inhibition of N-linked glycosylation with tunicamycin yielded a 65,000 molecular weight protein with full HA activity. However, cleavage of O-linked oligosaccharides with mixed glycosidases abolished hemagglutination, suggesting their importance in the expression of biological activity.

At present, the relative roles of the intracellular virus particles and membrane-enveloped extracellular particles in natural infections is not clear. Both are infectious and appear to enter cells by fusion with the plasma membrane or with phagocytic vesicles (Dales, 1973; Payne and Norrby, 1978). However, the mechanism of penetration appears to differ in several respects (Payne and Norrby, 1978), perhaps due in part to different modes of attachment to the cell surface. Based on antibody protection experiments, it appears that the enveloped extra-cellular virus containing the hemagglutinin plays the major role in infections of animal hosts (Boulter, 1969; Appleyard *et al.*, 1971; Turner and Squires, 1971).

## 4. Rhabdoviruses

Among the best studied rhabdoviruses are vesicular stomatitis virus (VSV) and rabies virus (for review see Wagner, 1975). The attachment protein has been conclusively identified as the single envelope glycoprotein called G protein (Kelley *et al.*, 1972; Wiktor *et al.*, 1972).

Because VSV infects a variety of cell lines, shuts down host protein synthesis, and itself contains a single glycoprotein, it has been used by many laboratories in model systems to study the biosynthesis of membrane glycoproteins. Thus, the biosynthesis of the G protein has been extensively studied (see, e.g., reviews in Hubbard and Ivatt, 1981; Rothman, 1981). The native glycoprotein has two asparagine-linked oligosaccharides. The oligosaccharides of the mature virus grown in BHK cells are of the triantennary-complex type with terminal sialic acid residues (Reading *et al.*, 1978). In some strains of VSV, the presence of sialic acid on the virus has been found to be required for infectivity (Schloemer and Wagner, 1974, 1975a). However, in other strains, fully infectious virions could be obtained when grown in the presence of tunicamycin which totally blocks N-linked glycosylation. (Gibson *et al.*, 1979). In addition to glycosylation, evidence for covalent attachment of fatty acids has also been obtained (Dunphy *et al.*, 1981).

As described for the influenza hemagglutinin, the G protein is anchored to the membrane by a hydrophobic peptide (Schloemer and Wagner, 1975b) and is transported to the plasma membrane where viruses are formed by budding. As demonstrated by Compans *et al.* (1981) in confluent polarized epithelial cells, the VSV G protein is transported to the basal surface of the cell while the influenza hemagglutinin is transported to the apical surface. This is likely the determining factor for the budding of VSV and influenza at the basal and apical surfaces, respectively. At present, the mechanism of intracellular routing which underlies this fascinating observation is unknown.

## 5. *Togaviruses*

Togaviruses are a large family of viruses divided into two main groups; the alphaviruses or group A arboviruses, which include such widely studied viruses as Sindbis and Semliki forest virus, and the flaviviruses, which include the dengue and yellow fever viruses (Pfefferkorn and Shapiro, 1974; Strauss and Strauss, 1977). Semliki forest virus has three envelope glycoproteins (E1, E2, and E3) of which two, E2 and E3, appear to result from cleavage of a common precursor, NP62, during maturation. Sindbis virus has only two glycopeptides, E1 and E2. However, E2 arises from cleavage of a polypeptide precursor (PE2) as with Semliki forest virus. Thus, it has been suggested that E2 in the two viruses are analogous and that the other cleavage product remains associated with Semliki forest virus as E3 but is lost from Sindbis virus. In one report, E1 and E2 of Sindbis have been separated by isoelectric focusing, and the hemagglutination activity resided in E1 (Dalrymple *et al.*, 1976). Thus, it is possible that E1 is the viral attachment protein. The biological role of E2 is not yet clear, although antibodies to the E2 protein neutralize infection.

## 6. *Herpesviruses*

Herpesviruses are large DNA viruses, the most familiar being herpes simplex (type 1 and type 2) and Epstein–Barr virus (EBV). The herpesviruses are quite complex with 25–34 virally encoded structural proteins, 5–10 of which may be associated with the membrane envelope (for review see Roizman, 1977). Most of the attention concerning the functions of attachment and penetration has been focused on the viral glycoproteins. Herpes simplex type 1 has four major glycoproteins gA, gB, gC, and gD, and at least one of these are important in viral attachment to cells (Spear, 1980; Norrild, 1980). While the attachment protein has not been identified, gB appears to be involved in viral penetration since a mutant in gB was found to attach to cells but was unable to penetrate (Sarmiento *et al.*, 1979; Spear, 1980). Epstein–Barr virus has five envelope-associated proteins, VE1–VE5, and all but one, VE3, appears to be a glycoprotein (Epstein and Achong, 1979; Wells *et al.*, 1982). Following separation and partial purification of the Nonidet-P40 solubilized proteins by gel exclusion chromatography, Wells *et al.* (1982) incorporated each into liposomal vesicles and tested their ability to interact with cell surface receptors. None alone mediated attachment of vesicles to cells. However, vesicles containing both VE1 and VE2 bound to cells, suggesting that these two glycoproteins are involved in viral attachment. The precise roles of these glycoproteins in recognition of receptors and stabilization of binding remain to be established.

## 7. Retroviruses

The retroviruses are a large group of oncogenic RNA viruses in which replication occurs through a DNA intermediate produced by a virally encoded reverse transcriptase (Guilden, 1980). Of the two main groups, type C and type B viruses, the former have been the best characterized with respect to their viral attachment proteins. Type C viruses are observed to have knob-like or spiked projections embedded in the membrane envelope. The projections are formed by the major envelope glycoprotein of the virus (Rifkin and Compans, 1971) and have been shown to be responsible for viral attachment to cells. Indeed, viruses lacking the glycoprotein spikes do not adsorb to cells and are not infectious (DeGiuli et al., 1975), and antibody prepared from the purified glycoprotein causes neutralization of infection (Hunsmann et al., 1974; Ikeda et al., 1975). The major glycoproteins of type C avian viruses like Rous sarcoma virus have a molecular weight of 85,000 (gp85) while those of the mammalian viruses like Friend murine leukemia virus (F-MULV) and Rauscher murine leukemia virus (R-MULV) have a molecular weight of 70,000–71,000 (gp70/gp71). However, Schneider et al. (1980) have suggested that gp71 of F-MULV is actually one subunit of a two-subunit polypeptide gp85. The intact glycoprotein gp85 is composed of gp71 joined in disulfide linkage to a hydrophobic peptide gp15E which presumably anchors the complex to the viral envelope (Schneider et al., 1979, 1980; Schneider and Hunnsman, 1978). Minor glycoproteins gp35 and gp45 have also been demonstrated in avian and mammalian viruses, respectively (see Guilden, 1980, for review). The retrovirus glycoproteins (gp70 or gp85) can be readily purified in soluble form free of detergent with retention of their ability to bind to bell surface receptors (DeLarco and Todaro, 1976; Bishayee et al., 1978; Kalyanaraman et al., 1978; Robinson et al., 1980).

## B. Viruses with Protein Capsids

With the exception of adenoviruses, an unambiguous identification of the attachment proteins of nonenveloped viruses has been difficult. However, some progress has also been made in identification of the binding proteins of picornaviruses, papovaviruses, and reoviruses.

## 1. Adenoviruses

The adenoviruses are a large group of viruses classified by serotype. At least 33 human adenoviruses are known and over 80 which infect a variety of mam-

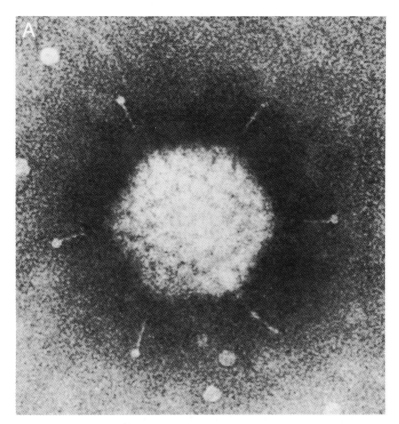

**Fig. 4.** Structure of the adenovirus capsid. (A) An electron micrograph readily shows the penton fiber projecting from the vertices of the capsid. The terminal knobs are thought to contain the receptor site. (From Valentine and Pereira, 1965.) (B) A diagram depicts the proposed organization of the adenovirus structural proteins and their relative abundance after separation by SDS–gel electrophoresis. The capsid proteins are the hexons, II; penton base, III; and the penton fiber, IV. (From Everitt *et al.*, 1975.)

malian and avian hosts. The capsid of adenoviruses is an icosahedron composed of 252 capsomers. Of these, 240 are hexons, each of which is bound by 6 other capsomers. The other 12 capsomers are found at the vertices and are bound by 5 hexons. As depicted in Fig. 4, the hexons are basically globular in shape while the pentons have a spiked fibrous projection clearly seen in electron micrographs (Fig. 4A). The subunit structures of the hexons and pentons have been extensively studied and thoroughly reviewed (see Boulanger and Lonberg-Holm, 1981).

Of particular interest are the pentons which consist of a base unit noncova-

B

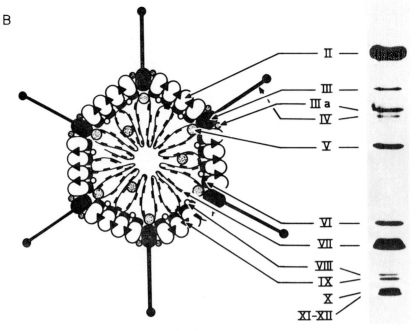

**Fig. 4.** (cont.)

lently joined to a thin shaft with a terminal knob called the fiber. Considerable evidence points to the terminal knob region of the fiber as having the binding site for cell surface receptors. Electron micrographs at early stages of virus cell interaction show attachment of the distal part of the fiber with the cell surfaces (Morgan *et al.,* 1969; Dales, 1973). Hemagglutination of erythrocytes by adenovirus can be blocked by type-specific antibodies (Pereira and DeFigueiredo, 1962; Valentine and Pereira, 1965; Norrby, 1966) shown by electron microscopy to bind to the terminal knob region of the fiber (Norrby, 1969). Type-specific antibodies also neutralize adenovirus infections (Norrby *et al.,* 1969; Della-Porta and Westaway, 1977). Finally, highly purified fiber binds tightly to cells and neutralizes infection at $10^5$ fibers per cell, suggesting that the purified fiber actively retains binding and prevents infection by blocking cell surface receptor sites (Philipson *et al.,* 1968).

Adenovirus infection of tissue culture cells results in an overproduction of fiber such that only 1–5% is incorporated into virions and the rest accummulates in the medium (White *et al.,* 1969; Everitt *et al.,* 1971). Thus, the fiber can be readily purified. As will be discussed in Section III,B,1, several laboratories have used purified fiber in attempts to isolate cell surface receptors of tissue culture cells (Meager *et al.,* 1976; Hennache and Boulanger, 1977; Svensson *et al.,* 1981).

## 2. Picornaviruses

Picornaviruses or small RNA viruses include enteroviruses (polioviruses, coxsackieviruses), cardioviruses [encephalomyocarditis (EMC)], rhinoviruses, and pathoviruses [foot and mouth disease virus (FMDV)]. As for adenoviruses, the picornaviruses have a protein capsid with icosohedral symmetry. It is composed of 60 units with three to four polypeptides each. Each capsid unit is initially synthesized as a single polypeptide ($M_r$=80,000–120,000) which is usually found cleaved into four peptides (VP1,VP2,VP3,VP4) in the mature virion. In some cases there is no cleavage between VP2 and VP4, giving a capsid unit with three polypeptides VP0, VP1, and VP3 (for review see Boulanger and Lonberg-Holm, 1981; Crowell and Landau, 1983).

There is still considerable uncertainty as to which subunit or combination of subunits are responsible for attachment to cell surface receptors. The fact that virions treated to remove VP4 were incapable of attaching to cells was initially taken as evidence that VP4 was the attachment protein (Breindl, 1971; Lonberg-Holm et al., 1976). However, it is now generally believed that VP4 is not involved since some picornaviruses lacking VP4 still bind to cells and several lines of evidence suggest that VP4 is buried in the capsid (reviewed in Boulanger and Lonberg-Holm, 1981).

Although it is attractive to think that the viral attachment protein would occur at the vertices of the capsid, there is some disagreement about this. Several models have been proposed for the arrangement of the subunits VP1–VP4 in the picornavirus capsid (Rueckert, 1976; Crowell and Landau, 1983). Neutralizing antibodies were found to bind to trypsin-sensitive sites on the vertices of FMDV (Brown and Small, 1970; Burroughs et al., 1971; Cavanaugh et al., 1977), and the trypsin-sensitive protein was reported to be VP1. However, VP3 was also reported to be trypsin sensitive (Bachrach et al., 1975; Kaaden et al., 1977). The VP2 subunit has also been suggested to be located at the vertices for several enteroviruses (Philipson et al., 1973; Martin and Johnston, 1972; Crowell and Landau, 1983). Boulanger and Lonberg-Holm (1981) have pointed out that some of the apparent discrepancies reported by different labs may be in the identification of VP1–VP4 on gels. Thus, for different viruses or even different gel systems it will be important to ascertain which polypeptides are functionally equivalent. To date, however, the picornavirus attachment proteins have not been identified with certainty.

## 3. Papovaviruses

This group of oncogenic viruses are the smallest viruses which contain a double-stranded DNA genome. The papovaviruses are divided into two groups, the papilloma group and the polyoma group. The most detailed information is available for the polyoma group including polyomavirus, SV40, and the human papovavirus BK (BKV) which will be considered here. Electron micrographs of negatively stained polyomavirus and BKV reveal a regular capsid studded with knob-like capsomers (Wildy *et al.*, 1960). Analysis of the micrographs by Klug and Finch (1965) and by Finch (1974) using three-dimensional image reconstruction suggested that there were 72 knobs arranged on the surface of a right-handed $T = 7$ icosohedral lattice. Of these 72 morphological units, 60 were hexavalent (bound by six other capsomers) and 12 were pentavalent (bound by five capsomers). This arrangement has been verified by X-ray diffraction analysis of crystals of polyomavirus capsids which yielded the structure at 22.5 Å shown in Fig. 5. Viruses of the papilloma group appear to have a similar structure (Finch and Crawford, 1975). Morphological considerations suggest that the capsid knobs would be likely candidates for the viral attachment proteins.

Polyomavirus has seven proteins with only three being coded for by the virus genome, VP1, VP2, and VP3. The others are mainly cellular histones associated with the viral DNA (Frearson and Crawford, 1972). Finch (1974) has proposed that 60 of the capsid knobs could be composed of hexamers of VP1 and the other

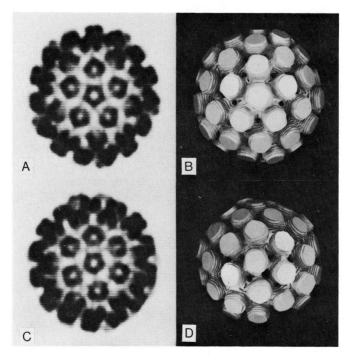

**Fig. 5.** Views of the polyomavirus capsid (495 Å) down the fivefold axis (A, B) and the sixfold axis (C, D). Computer graphics projections are shown on the left (A, C) and photographs of a model on the right (B, D). (From Rayment *et al.*, 1982. Reprinted by permission from *Nature* **295**, 110–115. Copyright © 1982 Macmillan Journals Limited.)

12 of pentamers of VP2 and/or VP3. However, several lines of evidence suggest that the capsid is composed of VP1 only. Papovaviruses produce empty capsids in addition to intact viruses. Purified empty capsids contain only VP1 yet have normal morphology (Rayment et al., 1982) and like the intact virions cause hemagglutination of erythrocytes (Walter and Etchison, 1977). Results of the X-ray crystallographic analysis of Rayment et al. (1982) suggest that all 72 knoblike capsomers are pentamers of VP1. Thus, while the 60 hexavalent and 12 pentavalent capsomers each have the same subunit composition, the subunits assume different configurations when bounded by six or by five other capsomers, respectively. This model requires 360 VP1 monomers, which agrees with the estimated number of VP1 copies per virion (Freidman and David, 1972) and is consistent with the molecular weight of $15 \times 10^6$ for the empty capsid (Rayment et al., 1982).

It is noteworthy that Bolen and Consigli (1979) have demonstrated differences in the receptor binding properties of intact virions and empty capsids; while both agglutinated guinea pig erythrocytes, striking differences in binding to murine kidney cells were found. In particular, competition of the binding of $^{125}$I-labeled virus was nearly 10,000 times more efficient with unlabeled virus than with capsids. Although the basis of this finding is not yet clear, further studies have suggested that VP1 in the virus is subject to proteolytic cleavage by virion-associated protease while VP1 of the empty capsid is not (Bolen and Consigli, 1980). Indeed, antibodies causing inhibition of hemagglutination bound to an affinity column containing empty capsids, while neutralizing antibodies directed to a peptide of VP1 did not bind. Thus, it is possible that a proteolytically modified form of VP1 is responsible for high-affinity binding to receptors on host cells, but this remains to be established.

### 4. Reoviruses

The Reoviridae are a large family of viruses of which the genus *Orthoreovirus* consisting of avian and mammalian reoviruses is the best studied. Reoviruses are unique in having a double-stranded RNA genome in 10–12 segments (for review see Ramig and Fields, 1977). The genome is surrounded by two protein capsids, an inner protease-resistant capsid and an outer capsid which is removed by lysosomal proteases after the virus has penetrated the host cell (see Section V,B,4). Presumably it is the outer capsid which normally interacts with host cell receptors. However, in some cases virus particles in which the outer capsid is removed by treatment with chymotrypsin have been found to be fully infectious (Spendlove et al., 1970).

The symmetry of the outer capsid of reoviruses has been studied from electron micrographs of negatively stained virus preparations. It is generally agreed that the capsid has icosahedral symmetry. However, the triangulation number, which describes the arrangement of the capsomers on the surface of the icosahedran

(Caspar and Klug, 1962), has been proposed to be $T = 9$ (Jordan and Mayor, 1962; Vasquez and Tournier, 1962) and $T = 12$ (Luftig *et al.*, 1972). More recently, Metcalf (1982) has examined images produced by a mica sandwich freeze-etching technique using a low-pass optical filtering method to eliminate detail below 4 nm and consequently highlight regular features. This analysis revealed 72 morphological units, 60 doughnut-shaped rings bounded by 6 other rings and 12 concave pentameric craters arranged in a $T = 13,1$ icosahedral lattice.

Little is known about the polypeptide composition of the structures which make up the outer capsid. Reoviruses have 7–10 structural polypeptides which fall into three classes; large ($\lambda$), middle ($\mu$), and small ($\sigma$) molecular weight (Ramig and Fields, 1977). Upon removal of the outer capsid by digestion with chymotrypsin, three polypeptide species ($\mu2$, $\sigma1$, and $\sigma3$) are lost, suggesting that they make up the capsid structure (Joklik, 1972; Shatkin and LaFiandra, 1972). Lee *et al.* (1981) have suggested that $\sigma1$ may mediate reovirus binding to the cell surface. However, the localization of $\sigma1$ (30 copies per virion) to specific structures on the capsid and its interactions with the other capsid proteins remain to be established.

## III. IDENTIFICATION OF CELL SURFACE RECEPTORS

It is a fair assumption that viruses evolved attachment proteins to recognize and bind to the normal components of a cell surface. Most of the research in viral receptors has been directed toward one of two goals, identification of the cell surface molecule to which the virus binds and elucidation of the functional group that interacts with the binding site of the attachment protein. For the sake of clarity, these two aspects of virus receptors will be considered separately. The following Section (IV) will be devoted to the functional groups or receptor determinants of viral receptors. In this section, progress toward and methods for identifying the specific molecules to which a virus binds will be reviewed. Research on cellular receptors of animal viruses has been conducted both in model systems and on intact host cells. Each has advantages and are discussed separately in this section.

## A. Model Systems

The two most widely used model systems are erythrocytes and artificial membranes. While it is not possible to study viral infection in either case, analysis of the interaction of viruses with these simple systems has been valuable in establishing the criteria to evaluate functional receptors of host cells.

## 1. Erythrocytes

Analysis of the cell surface components that mediate viral attachment to erythrocytes was a natural extension of the use of hemagglutination as a rapid assay for quantitation of viruses. For excellent reviews of the extensive early work on erythrocyte receptors see Howe and Lee (1972), Bachi *et al.* (1977), Lonberg-Holm and Philipson (1974), and Burness (1981). Several erythrocyte components have been investigated as viral receptor sites.

**a. Glycophorin A.** The major glycoprotein of the erythrocyte membrane is glycophorin. Since glycophorin A, and the related glycoproteins, glycophorin B and C, contain nearly 90% of the total cell-surface sialic acid (Bachi *et al.*, 1977), they are prime candidates for the primary attachment site of animal viruses that bind sialic acid-containing receptor determinants. A major advance in the purification of glycophorin A (reviewed in Furthmayr, 1981) was the use of the chaotropic salt lithium diiodosalicylate in its extraction from the erythrocyte membranes (Marchesi and Andrews, 1971). Human glycophorin A is a highly glycosylated polypeptide of 131 amino acids, containing 28% sialic acid and 66% total carbohydrate by weight (Furthmayr, 1981). It has 15 oligosaccharides O-linked to threonine or serine and one oligosaccharide N-linked to asparagine. The structure of the carbohydrate groups will be discussed in Section IV,C,2. Glycophorin A is a transmembrane protein with a short carboxyl-terminal domain inside the cell and an amino-terminal domain containing the carbohydrate extending from the surface of the cell (Marchesi *et al.*, 1973).

Consistent with its proposed role as the cell surface receptor of viruses, purified glycophorin is a potent inhibitor of hemagglutination for influenza virus (Marchesi and Andrews, 1971; Enegren and Burness, 1977), encephalomyocarditis (EMC) virus (Enegren and Burness, 1977), and small-plaque polyomavirus (Cahan *et al.*, 1983; see Section IV,C,3). Pardoe and Burness (1981) also demonstrated that EMC virus would bind to a column containing glycophorin A adsorbed to wheat germ agglutinin (WGA)–Sepharose but would not bind to WGA–Sepharose alone.

Glycophorins from erythrocytes of other species can be isolated in a manner similar to that described by Marchesi and Andrews (1971) for human glycophorin. Analysis by SDS–gel electrophoresis shows considerable dissimilarity from species to species (Furthmayr, 1981). Sequence studies of Homma *et al.* (1980) and Murayama *et al.* (1981, 1982, 1983b) show that the major sialoglycoproteins of humans, pigs, horses, and dogs are similar in their high content of glycosylated threonine and serine residues and in their hydrophobic domain but otherwise have no significant sequence homology. While they all have high-carbohydrate content (30–60%), it appears that their oligosaccharide structures also differ (Thomas and Winzler, 1969; Fukuda *et al.*, 1981; Murayama *et al.*, 1983a).

Knowledge of the structural diversity of animal glycophorins may yield valuable insights into the receptor specificities of animal viruses. Indeed, purified bovine glycophorin is not an inhibitor of hemagglutination of influenza virus but is a potent inhibitor of two paramyxoviruses, Sendai virus and Newcastle disease virus (Suzuki *et al.,* 1983). Yet all three of these viruses bind sialic acid-containing receptors. Another paramyxovirus, measles virus, does not exhibit sialic acid-dependent binding to cells (Choppin and Scheid, 1980). However, it agglutinates the erythrocytes of monkeys but not other species, and monkey glycophorin is an inhibitor of measles virus hemagglutination (Fenger and Howe, 1979). To date, the nature of the measles virus receptor determinant is unknown.

**b. Band 3.** Band 3 is the major erythrocyte membrane protein comprising nearly 25% of the total protein. It is a 95,000 dalton transmembrane protein that contains a single polyglycosyl asparagine-linked oligosaccharide with two sialic acid residues (Drickamer, 1978; Finne *et al.,* 1978; Jarnefelt *et al.,* 1978). Nigg *et al.* (1980) have examined the ability of Sendai virus and an influenza virus to interact with Band 3 *in situ* by inspection of intramembrane particles present on the protoplasmic surface of freeze-fractured membranes. Only Sendai virus caused aggregation and redistribution of the particles. Since Band 3 is a major component of these particles, these results were suggested to provide direct evidence for interaction of Sendai virus with Band 3.

**c. Other Erythrocyte Components.** Minor sialoglycoproteins of erythrocytes may also be attachment sites for viruses which exhibit sialic acid-dependent binding. This was illustrated by Maeda *et al.* (1979) using erythrocytes of the rare blood type En(a$^-$). These individuals have cells with no glycophorin A. Yet Sendai virus was able to bind equally well to these cells as to those containing glycophorin A. Thus, while glycophorin A may be the primary attachment site for Sendai virus, other glycoproteins (including glycophorin B and C) may also serve to bind virus.

Adenovirus Type 7 (Ad7) receptor on monkey erythrocytes was partially characterized by Neurath *et al.* (1969) after solubilization from erythrocyte ghosts with sodium deoxycholate. As judged by inhibition of hemagglutination, the receptor material could be partially resolved from influenza virus receptors by gel exclusion chromatography. The adenovirus inhibitor was inactivated by a variety of proteases and treatment with potassium periodate, but not with *V. cholerae* sialidase. Thus, the erythrocyte receptor of Ad7 appeared to be different than for influenza virus.

Glycolipids are a common constitutent of erythrocyte membranes but have not as yet been implicated as the primary receptors for these cells. As will be discussed subsequently, glycolipids containing sialic acid (gangliosides) are found to bind to Sendai virus and influenza virus. However, in human erythrocytes gangliosides account for less than 5% of the total sialic acid (Furthmayr,

1981). Thus, it is possible that binding to erythrocyte gangliosides by these viruses may be masked by binding to glycoproteins (Umeda *et al.*, 1984).

## 2. Artificial Membranes, Liposomes

Haywood (1974a,b) first demonstrated that Sendai virus could both bind and fuse to liposomes into which sialoglycolipids (gangliosides) had been incorporated. This suggested that gangliosides may serve as receptors for Sendai virus and demonstrated that artificial membrane systems may be a convenient means of studying virus–receptor interactions *in vitro* (for review see Tiffany, 1977; Burness, 1981). Influenza virus was shown to bind to liposomes containing either gangliosides or sialoglycoproteins, demonstrating that both cell surface components may potentially serve as cell surface receptors (Tiffany and Blough, 1971; Sharom *et al.*, 1976; Baron and Blough, 1983). While liposomes cannot replicate virus, they represent a well-defined system to study viral attachment and viral fusion for membrane-enveloped viruses, (Haywood and Boyer, 1982; Maeda *et al.*, 1981; Oku *et al.*, 1981, 1982; Umeda *et al.*, 1984).

## B. Receptors of Host Cells

Analysis of cell surface receptors of host cells is somewhat more complex, hampered in many cases by a lack of suitable assays for receptor activity. However, in recent years several approaches have shown considerable promise.

## 1. High-Affinity Receptors of Viral Attachment Proteins

Using the assumption that receptors will bind the viral attachment protein with high affinity, potential cell surface receptors for several viruses have been identified. In most cases, putative receptor proteins have been isolated from detergent-solubilized host cell membranes, usually tissue culture cells, by affinity chromatography on an insolublized matrix containing covalently bound viral attachment protein or by immunoprecipitation of preformed receptor–attachment protein complexes with antiattachment protein antibody. Several examples are given later.

**a. Adenoviruses.** As described in Section II,B,1, the attachment proteins of adenoviruses are the penton fibers which project from the vertices of the protein capsid (see Fig. 4). Hennache and Boulanger (1977) examined solubilized KB cell membrane proteins for their ability to form immunoprecipitable complexes with adenovirus type 2 (Ad2) fiber and antifiber antibody. In addition, they isolated receptor–fiber complexes joined by a cleavable diimidoester cross-linking reagent. From analysis of the fiber-associated proteins by SDS–gel electrophoresis, they concluded that three polypeptides with molecular weights of 78,000, 42,000, and 34,000 were likely candidates for the receptor. Similarly,

Meager *et al.* (1976) found that heterogeneous KB cell glycoproteins of approximately 100,000 daltons bound with high affinity to a column of Ad5 fiber–Sepharose. Svensson *et al.* (1981) analyzed Triton-X-100-solubilized HeLa cell membrane proteins which bound to an affinity matrix consisting of Ad2 virions coupled to AH–Sepharose. A single major protein of 40,000–42,000 daltons was found to bind with high affinity. The isolated material appeared to be a glycoprotein based on adsorption to wheat germ lectin–Sepharose. Consistent with its proposed role as a receptor, it inhibited binding of Ad2 virions to cells.

**b. Retroviruses.** The glycoprotein attachment proteins of retroviruses can be purified with retention of full binding activity toward cell surface receptors (DeLarco and Todaro, 1976; Bishayee *et al.*, 1978; Kalyanaraman *et al.*, 1978). Purified viral attachment protein gp85 from Friend murine leukemia virus (F-MULV) was found to form a specific immunoprecipitable complex with a 14,000 dalton protein from detergent-solubilized membranes of mouse spleen leukocytes (Robinson *et al.*, 1980). It was concluded that the protein was not related to H2 histocompatibility antigens previously suggested to be potential receptors for F-MULV (Bubbers *et al.*, 1978; Schrader *et al.*, 1975). Immunoprecipitation was also employed to identify a putative 190,000 dalton receptor of Maloney leukemia virus gp71 from murine thymus cells (Choppin *et al.*, 1981; Schaffer-Deshayes *et al.*, 1981). A peptide of 10,000 daltons that specifically bound to gp70 of Rauscher murine leukemia virus was identified in the culture medium of murine fibroblasts by Landen and Fox (1980). It inhibited binding of gp70 to cells and was suggested to be the receptor of the virus which was shed by fibroblasts into the culture medium.

**c. Togaviruses.** Helenius *et al.* (1978) have proposed that human (HLA) and murine (H2) histocompatibility antigens may serve as cell surface receptors for Semliki forest virus (SFV). Histocompatibility antigens were the major membrane proteins isolated on an SFV spike protein affinity column and were the predominant proteins isolated by immunoprecipitation of soluble spike protein–membrane protein complexes with spike protein antibodies. Moreover, histocompatibility antigens incorporated into liposomes inhibited adsorption of SFV to cells. The presence of histocompatibility antigens on most mammalian cell lines was proposed to account for the wide host range of SFV. However, it was argued by Oldstone *et al.* (1980) that other membrane components in addition to histocompatibility antigens must also be able to serve as cell surface receptors for SFV. Indeed certain cell lines, murine F9 and PCC4 teratocarcinoma cells, do not express H2 histocompatibility antigens but were infected by SFV.

### 2. *Epstein–Barr Virus Receptor: Antireceptor Antibody*

Epstein–Barr virus (EBV), a herpesvirus, has one of the most restricted host ranges of all viruses, infecting only B lymphocytes of humans and other primates. Moreover, the host range appears to be determined almost entirely by the presence of specific receptors on the B cells and the lack of these receptors on the other cells (see Section VI). This prompted Jønsson et al. (1982) to investigate the relationship between EBV receptors and the B cell-specific complement receptors for C3b and C3d. Using a quantitative assay for EBV receptors based on the binding of [125]I-labeled EBV to cells (Koide et al., 1981), a strong correlation between the presence of Cd3 receptors and EBV receptors was shown for a variety of lymphocyte cell lines with variable expression of both components. No correlation was found for EBV and Cb3 receptors. Based on this correlation, the C3d receptor was postulated to be the cell surface receptor for EBV.

Subsequent reports by Simmons et al. (1983) and Hutt-Fletcher et al. (1983) suggest that the EBV receptor is not the complement receptor for C3d. The EBV receptor could be largely solubilized from Raji cells under conditions (10% glycerol) that did not deplete the receptor for C3d. While solubilized EBV receptor inhibited EBV binding to cells, it did not inhibit the interaction of C3d with its receptor. Examination of clonal somatic cell hybrids from human lymphoblast and mouse myeloma cell parents revealed that EBV receptors and C3d receptors are not coexpressed and are likely coded for on different chromosomes (Hutt-Fletcher et al., 1983).

Polyclonal antibody was prepared to the receptor extract from Raji cells and extensively adsorbed by EBV receptor minus cells (V-698 and K652). The antibody, when preadsorbed to receptor-positive cells, blocked binding of EBV virus. Immuno-precipitation of Triton-X-100-solubilized, [125]I-labeled, membrane proteins from Raji cells with EBV receptor antibody resulted in the selection of a single protein species of 150,000 daltons (Simmons et al., 1983). The molecular weight of the presumed receptor did not correspond to that reported for the C3d receptor ($M_r$ 72,000; Lambris et al., 1981).

Because the restricted host range of EBV is receptor mediated and limited to B lymphocytes, it is likely that the EBV receptor is a B lymphocyte-specific membrane protein (Jønsson et al., 1982). The weight of the evidence now suggests that the complement receptors for C3b and C3d are not the receptors for EBV. However, in view of the wide range of tools now available to investigate the EBV receptor, its identity may be known in the foreseeable future.

### 3. Receptors of Sendai and Influenza Viruses: Glycoproteins or Glycolipids?

Myxoviruses and paramyxoviruses bind sialic acid-containing receptors on cells. As described in the preceding section (Section III,A,1), the predominant erythrocyte receptor of these viruses is probably glycophorin, the major erythrocyte glycoprotein. However, in vitro, influenza and Sendai virus also bind gangliosides (Bogoch, 1957; Haywood, 1974a,b; Tiffany and Blough, 1971; Holmgren et al., 1980). These viruses also exhibit specificity for the sialyloligosaccharide sequence which will serve as the receptor determinant (see Section IV). But which cell surface components, glycoproteins or glycolipids, function as host cell receptors? There are conflicting views on this point.

Wu *et al.* (1980) found that Sendai virus attachment to HeLa cells could be abolished with either neuraminidase or trypsin, suggesting that the receptor was a sialoglycoprotein. The HeLa cell membranes were also fractionated into glycolipid and glycoprotein fractions which were then incorporated into liposomes and tested as inhibitors of hemagglutination. The glycoproteins were potent inhibitors while gangliosides exhibited little or no inhibition.

In contrast, Markwell *et al.* (1981; 1984a,b) examined gangliosides as receptors by applying them to tissue culture cells rendered insusceptible to infection by treatment with sialidase. Exogenous gangliosides restored full susceptibility to infection for sialidase-treated MDBK, HeLa, and MDCK cells. Moreover, the efficacy of receptor restoration by various gangliosides was directly correlated with their ability to bind Sendai virus *in vitro* (see Section IV,C,3). Thus, gangliosides apparently were able to mediate binding and penetration of cells. It was presumed that the gangliosides added to the cells were functionally inserted into the lipid bilayer of the plasma membrane. Although viral attachment mediated by gangliosides adsorbed to cell surface proteins could not be totally excluded, trypsin digestion of ganglioside-treated MDBK cells did not prevent infection (Markwell *et al.,* 1981). The amount of gangliosides required to restore susceptibility to infection was calculated to be equivalent to the amount naturally present in MDBK cells (Markwell *et al.,* 1981). Moreover, gangliosides accounted for 15–60% of the total cell surface sialic acid in the three cell lines studied (Markwell *et al.,* 1982).

At present the findings of Wu *et al.* (1980) and Markwell *et al.* (1981, 1984a) regarding the role of gangliosides as Sendai virus receptors cannot be easily reconciled. However, it is possible that both glycoproteins and glycolipids serve as receptors and that the relative contribution of the two components as physiological receptors will differ from cell to cell. Reports by Bergelson *et al.* (1982) and Bukrinskaya *et al.* (1982) suggest that gangliosides applied to sialidase-treated tissue culture cells can stimulate influenza binding and penetration. Thus, there is also a question of whether glycoproteins or gangliosides are receptors for influenza viruses. A consideration of how glycoprotein versus glycolipid receptors may influence viral penetration is given in Section V.

## IV. CELL SURFACE RECEPTOR DETERMINANTS OF ANIMAL VIRUSES

It is well documented that host range and tissue tropism of animal viruses are in many instances determined by recognition of receptors on host cells (see Section VI). It is also apparent that very closely related viruses within the picornavirus (Crowell and Landau, 1983), reovirus (Tardieu *et al.*, 1982), retro-virus (Weiss, 1981), and myxovirus (Stone, 1951; Carroll *et al.*, 1981a; Rogers and Paulson, 1983; Rogers *et al.*, 1983a) families exhibit differences in receptor-binding properties. Yet in most cases, the basis for these differences is unknown.

One exception is for the myxoviruses, for which it has been shown that receptor-binding variants of influenza A viruses of the H3 serotype are due to substitution of a single amino acid in the receptor-binding pocket (Rogers *et al.*, 1983a). The result of a change from Leu to Gln at amino acid 226 results in a change in preferential binding of the oligosaccharide receptor sequence from SA$\alpha$2,6Gal to SA$\alpha$2,3Gal. In this case, the difference in receptor-binding properties of two variants has been explained in terms of the limited portion of structure recognized by the receptor-binding pocket, i.e., the receptor determinant, not by identification of discrete cell surface molecules to which the viruses bind. Indeed, since sialyloligosaccharides are a common component of glycoproteins and glycolipids, it is likely that the cell surface receptors of influenza viruses would be heterogeneous and would differ from cell to cell.

It is reasonable to expect that receptor variants in other groups of closely related viruses are also due to subtle changes in the receptor-binding pockets. If so, the receptors will likely be related in structure but have small differences which can be distinguished by the viral binding proteins. Thus, to fully under-stand the roles of receptor recognition of animal viruses in host range and tissue tropism, it will be necessary to elucidate the details of their receptor determi-nants.

For the purposes of this chapter, a viral receptor determinant is defined as the limited portion of the cell surface receptor recognized by the binding site of the viral attachment protein. In principle, this could be any structural component found on the cell surface. Evidence for utilization of proteins, lipids, and carbo-hydrates as cell surface receptor determinants of animal viruses will be presented later. It will be evident that most is currently known about receptor determinants of viruses that recognize cell surface carbohydrate groups, particularly those that require sialic acid for receptor recognition.

## A. Polypeptide Receptor Determinants

Many viruses utilize cell surface proteins or glycoproteins as receptors. This is generally taken to be the case if viral adsorption is abolished by protease treat-ment of the cells. For example, in this way proteins have been identified to be receptors for Rauscher murine leukemia virus (Kalyanaraman *et al.*, 1978; Bishayee *et al.*, 1978), for enterovirus (Zajac and Crowell, 1965), for adenovirus types 2 and 5 (Philipson *et al.*, 1968), and for herpes simplex virus (Blomberg, 1979). Direct interactions of viral attachment proteins with specific cell surface glycoproteins have also been demonstrated and have been reviewed in Section

III. However, in no case has a polypeptide sequence been identified which serves as a specific receptor determinant of a virus.

The need to ascertain the specific structural features of a protein recognized by a virus can be seen from widely cited reports on Semliki forest virus (SFV) receptors. As described in Section III,B,1, Helenius *et al.* (1978) have provided strong evidence that the spike proteins of SFV, a togavirus, specifically bind to histocompatibility antigens on a variety of human (HLA-A and HLA-B) and murine (H2K and H2D) cell lines and concluded that these proteins are the cell surface receptors for the virus. In contrast, Oldstone *et al.* (1980) demonstrated that murine F9 and PCC4 teratoma cells which lacked H2 antigens were readily infected by SFV. They concluded, therefore, that the major transplantation antigens were not "specific" receptors of SFV. While these results appear contradictory, it is possible that the receptor determinants recognized by SFV are preferentially found on histocompatibility antigens in some cell lines and the same or similar receptor determinants are found on unrelated cell surface proteins in other cell lines.

An immediate problem in defining a polypeptide receptor determinant is the fact that most cell surface proteins are glycoproteins. Thus, identification of a virus receptor as a "protein" does not usually distinguish between the receptor determinant being the carbohydrate or polypeptide portion of the molecule.

Even if binding to carbohydrates can be excluded, identification of the limited aspect of protein structure recognized by the virus may be technically difficult. To illustrate the problem, a single example can be cited from a bacteriophage receptor which has been extensively characterized. Certain T-even like bacteriophages are known to utilize the outer membrane proteins of *Escherichia coli* OmpA as a receptor binding site (Datta *et al.*, 1977; Manning *et al.*, 1976; Van Alphen *et al.*, 1977). Several mutant alleles of OmpA produce a protein which does not serve as a bacteriophage receptor. One of these has been characterized by Cole *et al.* (1983) to have an amino acid substitution of Gly to Arg at position 70. Thus, a single amino acid in the receptor determinant of bacteriophages can alter receptor binding. It is entirely feasible that similar examples of specificity will ultimately be found for polypeptide receptor determinants of animal viruses.

## B. Lipid Receptor Determinants

Lipids have been implicated as components of the receptors of animal viruses if viral binding is lost upon treatment of cells with lipases (for review see Lonberg-Holm and Philipson, 1974; Bishayee *et al.*, 1978; Kalyanaraman *et al.*, 1978). Interaction of Sindbis virus with phospholipid vesicles has also been reported (Mooney *et al.*, 1975). Yet in these cases there is little direct evidence that a lipid moiety serves as the receptor determinant on host cells.

Recently, Schlegel *et al.* (1983) have reported that phosphatidylserine is a potential cell surface binding site for vesicular stomatitis virus (VSV). Phosphatidylserine, but not other lipids and gangliosides tested, could inhibit binding and plaque formation on Vero monkey cells at 1–10 $\mu M$. Moreover, lipid vesicles containing phosphatidylserine specifically bound to purified virions. It was concluded from these results that phosphatidylserine may be a component of the cell surface receptor of the virus.

Glycolipids have also been implicated as receptor determinants of animal viruses (Haywood, 1974a,b; Holmgren *et al.*, 1980; Markwell *et al.*, 1981; Bergelson *et al.*, 1982). However, in these instances binding is mediated by the carbohydrate group.

## C. Carbohydrate Receptor Determinants

Oligosaccharides present on glycoproteins (S. Kornfeld and Kornfeld, 1976; R. Kornfeld and Kornfeld, 1980; Berger *et al.*, 1982) and glycolipids (Ledeen and Yu, 1982) are ubiquitous and abundant cell surface constituents of higher organisms. It is therefore not surprising that some animal viruses have evolved to utilize these structures as attachment sites on host cells. At the present time, most is known about viruses for which binding is dependent on terminal sialic acid residues (for reviews see Hirst, 1959; Cohen, 1963; Gottschalk, 1966; Hoyle, 1968; Howe and Lee, 1972; Dales, 1973; Lonberg-Holm and Philipson, 1974; Bachi *et al.*, 1977; Burness, 1981; Dimmock, 1982; Tardieu *et al.*, 1982).

### 1. Evidence for Sialic Acids as Receptor Determinants of Animal Viruses

As described in Section II,A,1, sialic acids were established as receptor determinants of influenza virus when it was discovered that receptor-destroying enzymes present in the virus (Hirst, 1942a,b) and in the culture filtrates of *V. cholerae* (Burnett and Stone, 1947) were actually neuraminidases (Gottschalk and Lind, 1949; Klenk *et al.*, 1955; Blix *et al.*, 1957). Although the ability of neuraminidase to destroy influenza virus receptors was initially demonstrated with erythrocytes, it was subsequently found to prevent adsorption to and susceptibility to infection of laboratory hosts, including the chorioallantoic membrane of chicken embryos (Stone, 1948a) and the epithelium of mouse lung (Fazekas de St. Groth, 1948a,b; Stone, 1948b). This established that the sialic acid-containing receptors of erythrocytes were analogous to the receptors on host cells.

**a. Sialidase Sensitivity.** Following the elegant early work with influenza viruses, bacterial neuraminidases (or sialidases) have been used to test the role of sialic acids as receptor determinants for other groups of animal viruses. As summarized in Table I, many viruses have been found to exhibit a dependence on sialic acids for attachment to cell surface receptors. These include membrane-enveloped viruses of the myxovirus and paramyxovirus families and nonenveloped viruses of the papovavirus, adenovirus, picornavirus, and reovirus families. It is of interest that in no case do all of the viruses of a family exhibit dependence on sialic acid for receptor recognition. Indeed, while influenza A and B viruses do, influenza C viruses clearly do not (Hirst, 1950; Kendal, 1975; O'Callaghan *et al.*, 1977). Similarly, while Newcastle disease, Sendai, and Simian virus 5 viruses exhibit sialic acid dependence, other members of the paramyxovirus family, notably the measles viruses, have no dependence (for review see Choppin and Compans, 1975). For the other families, viruses that adsorb to sialic acid receptors appear to be the exception rather than the rule.

**b. Glycoprotein Inhibitors.** Additional evidence for sialic acids as receptor determinants was obtained in studies with glycoprotein inhibitors of hemagglutination (for reviews see Gottschalk, 1966; Krizanova and Rathova, 1969; Howe and Lee, 1972; Burness, 1981). In principle, sialyloligosaccharides on soluble glycoproteins would interfere with viral binding to the cell surface by acting as a receptor analogue to block the receptor site on the viral attachment protein. The role of sialic acid was typically demonstrated by the loss of inhibitory potency upon treatment of the glycoprotein with sialidase. Some glycoproteins combined so avidly with influenza virus that they were potent inhibitors of infection (Burnett, 1948; Tamm and Horsfall, 1951; Choppin and Tamm, 1960a; Cohen and Biddle, 1960). Although glycoprotein inhibitors of influenza viruses have been the most extensively studied, inhibitors for polyomavirus (Hartley *et al.*, 1959; Mori *et al.*, 1962) and cardiovirus (Mandel and Racker, 1953; Enegren and Burness, 1977) have also been reported.

**c. Sialyloligosaccharide Binding Specificity.** Suggestions that myxoviruses and paramyxoviruses differed in their detailed specificity toward sialyloligosaccharide receptors came from several observations. Virus strains were found to differ in their ability to agglutinate erythrocytes of different species (Burnet and Bull, 1943). A receptor gradient was described by Stone (1951) whereby erythrocytes exposed to the neuraminidase of one virus could no longer be agglutinated by that virus but were still agglutinable by other viruses. Influenza viruses were also found to differ in their sensitivity to glycoprotein inhibitors of hemagglutination and infection (Stone, 1949; Tamm and Horsfall, 1951; Choppin and Tamm, 1959, 1960a; Cohen and Belyavin, 1959; Gottschalk, 1966).

## 2. Diversity of Sialyloligosaccharide Structure in Glycoproteins and Glycolipids

The fact that myxo- and paramyxoviruses can exhibit dramatic differences in receptor-binding properties suggests that receptor recognition extends beyond the terminal sialic acid. Sialic acids occur at terminal positions on a diverse group of

**TABLE I**

**Cellular Receptors of Animal Viruses Exhibiting Sensitivity to *V. cholerae* Sialidase**

| Virus | Cell type or host | Property inhibited by sialidase treatment | Reference |
|---|---|---|---|
| Myxoviruses | | | |
| Influenza A | Chicken erythrocytes | Agglutination | Burnet and Stone (1947) |
| Influenza B | Mouse lung | Viral adsorption | Fazekas de St. Groth (1948a,b) |
| | Chick embryo/mice | Infection | Stone (1948a,b) |
| | Mouse embryo cells | Infection | Mori et al. (1962) |
| Paramyxoviruses | | | |
| Newcastle disease | Chicken erythrocytes | Agglutination | Burnett and Stone (1947) |
| | Chicken embryo | Infection | Stone (1948a) |
| Sendai (HJV) | HeLa cells | Infection | Markus (1959) |
| | MDBK cells | Infection | Markwell and Paulson (1980) |
| Papovaviruses | | | |
| Polyoma | Erythrocytes | Agglutination | Eddy et al. (1958) |
| | Murine fibroblasts (3T6) | Infection | Fried et al. (1981) |
| BK | Human erythrocytes | Agglutination | Mäntyjärvi et al. (1972) |
| | | | Seganti et al. (1981) |
| Adenoviruses | | | |
| Adenovirus type 7 | Rhesus monkey erythrocytes | Agglutination | Wadell (1969) |
| Picornaviruses | | | |
| Cardiovirus | Erythrocytes | Agglutination | Verlinde and DeBann (1949) |
| | | | Angel and Burness (1977) |
| Enterovirus 70 | HeLa cells | Infection | Kozda and Jungeblut (1958) |
| Mengovirus | Human erythrocytes | Agglutination | Utagawa et al. (1982) |
| Bovine enterovirus | Human fibroblasts | Infection | Stoner et al. (1973) |
| Equine rhinovirus | Human erythrocytes | Agglutination | Stott and Killington (1972) |
| Reoviruses | | | |
| Reovirus 3 | Bovine erythrocytes | Agglutination | Gomatos and Tamm (1962) |

oligosaccharides of glycoproteins, glycolipids (S. Kornfeld and Kornfeld, 1976; R. Kornfeld and Kornfeld, 1980; Berger *et al.*, 1982; Ledeen and Yu, 1982), and also on proteoglycans (Hascall, 1981). In recent years, some details of the sialyloligosaccharide-binding specificities of animal viruses have emerged, as will be discussed in Section IV,C,3. In the following discussion, a brief overview of the sialyloligosaccharide structures found in glycoproteins and glycolipids is presented. For simplicity, proteoglycans will be considered as glycoproteins, since the oligosaccharides containing sialic acid are similar. Glycoprotein oligosaccharides are divided into two main groups based on their attachment in N-glycosidic linkage to asparagine or O-glycosidic linkage to threonine and serine. Representative N-linked and O-linked oligosaccharides are shown in Tables II and III.

All asparagine-linked oligosaccharides (Table II) have a common core of three mannose and two *N*-acetylglucosamine residues ($Man_3GlcNAc_2$). This stems from the fact that synthesis of these structures is initiated in the endoplasmic reticulum by en bloc transfer of a common lipid linked intermediate ($Glc_3,Man_9GlcNAc_2$) which is processed to the final structure during transit of the glycoprotein through the Golgi apparatus (for review see Hubbard and Ivatt, 1981). Oligosaccharides which contain sialic acid are of the "complex type" and usually contain from two to four branches of the disaccharide Galβ1,4GlcNAc or Galβ1,3GlcNAc attached to the terminal-mannose residues of the $Man_3GlcNAc_2$ core structure. Attachment of sialic acid and fucose at different positions of these chains gives rise to the variety of structures shown (S. Kornfeld and Kornfeld, 1976; R. Kornfeld and Kornfeld, 1980; Berger *et al.*, 1982; Finne, 1982).

Oligosaccharides O-linked to threonine and serine are synthesized primarily in the Golgi apparatus by sequential addition of sugars, starting with the attachment

**TABLE II**

**Asparagine-Linked Oligosaccharides Containing Sialic Acid**

| Terminal branches[a] | | Common core |
|---|---|---|
| SAα2,6Galβ1,4GlcNAc- | | |
| SAα2,3Galβ1,4GlcNAc- | Manα1,3 | |
| SAα2,8SAα2,3Galβ1,4GlcNAc- | Manα1,6 | Manβ1,4GlcNAcβ1,4GlcNAcβAsn |
| SAα2,3Galβ1,4 — GlcNAc<br>Fucα1,3 | | |
| SAα2,3Galβ1,3GlcNAc- | | |
| SAα2,4Galβ1,3GlcNAc- | | |
| SAα2,3Galβ1,3 — GlcNAc-<br>SAα2,6 | $n = 2-4$ | |

[a] Two to four terminal branches are attached to the Manα1,3Man and Manα1,6Man branches of the core (R. Kornfeld and Kornfeld, 1980; Berger *et al.*, 1982; Finne, 1982).

**TABLE III**

Sialyloligosaccharides O-Linked to Threonine or Serine

| Sialyloligosaccharide | Reference |
|---|---|
| SAα2,6GalNAcαThr/Ser | a |
| Galβ1,3 \\ <br>             >GalNAcαThr/Ser <br> SAα2,8SAα2,6 ⁄ | b |
| SAα2,3Galβ1,3 \\ <br>            >GalNAcαThr/Ser <br> SAα2,6 ⁄ | a,c |
| GalNAcβ1,4 \\ <br>           >Galβ1,3 \\ <br> SAα2,3 ⁄             >GalNAcαThr/Ser <br>                SAα2,6 ⁄ | d |
| SAα2,3Galβ1,3 \\ <br>                   >GalNAcαThr/Ser <br> SAα2,3Galβ1,4GlcNAcβ1,6 ⁄ | e |
| SAα1,4GlcNAcβ1,3 \\ <br>           >Galβ1,4GlcNAcβ1,3 \\ <br> SAα2,6 ⁄          SAα2,6 ⁄ Galβ1,4GlcNAcβ1,3 \\ <br>                     SAα2,6 ⁄ >GalNAcαThr/Ser | f |

a S. Kornfeld and Kornfeld (1976) and R. Kornfeld and Kornfeld (1980).
b Kiang et al. (1982).
c Thomas and Winzler (1969).
d Blanchard et al. (1983).
e Akiyama et al. (1983).
f Slomiany and Slomiany (1978).

of *N*-acetylgalactosamine (Berger *et al.*, 1982). These oligosaccharides vary greatly in size and structure. However, most of the sialyloligosaccharide sequences found to date are represented in the structures shown in Table III (Kornfeld and Kornfeld, 1976; Berger *et al.*, 1982; Slomiany and Slomiany, 1978; Akiyama *et al.*, 1983; Blanchard *et al.*, 1983).

Gangliosides, glycolipids containing sialic acid, are also common substituents of the cell surface. A ganglioside consists of a carbohydrate group attached to the ceramide moiety which anchors it to the cell membrane. The most common gangliosides fall into several groups based on the neutral tetrasaccharide-core structure attached to ceramide. Representative structures from each group are shown in Table IV (see also Ledeen and Yu, 1982). Gangliosides of the gangliotetrasylceramide series have the neutral core Galβ1,3GalNAcβ1, 4Galβ1,4Glc-. This is the only core structure which can have sialic acid attached to either the terminal galactose or internal galactose residues. Gangliosides with up to five sialic acids have been reported in this series. They are widely distributed in many tissues and cell types but are most abundant in brain. The nomenclature of Svennerholm (1963) is more often used than that adopted by the IUPAC–IUB (1977).

The other major group is based on the neutral core lacto-*N*-neotetrasylceramide

**TABLE IV**

**Examples of Ganglioside Structure**[a]

| | | Name | |
|---|---|---|---|
| Core | | Common[b] | IUPAC–IUB[c] |
| Gangliotetraose core | | | |
| | SAα2,3Galβ1,4Glc-Cer | GM3 | III³NeuAc-Lac |
| Galβ1,3GalNAcβ1,4<br>SAα2,3 | Galβ1,4Glc-Cer | GM1 | II³NeuAc-GgOse₄ |
| SAα2,3Galβ1,3GalNAcβ1,4<br>SAα2,3 | Galβ1,4Glc-Cer | GD1a | IV³NeuAc,II³NeuAc-GgOse₄ |
| Galβ1,3GalNAcβ1,4<br>SAα2,8SAα2,3 | Galβ1,4Glc-Cer | GD1b | II³NeuAc₂-GgOse4 |
| SAα2,3Galβ1,3GalNAcβ1,4<br>SAα2,8SAα2,3 | Galβ1,4Glc-Cer | GT1b | IV³NeuAc,II³NeuAc₂-GgOse₄ |
| SAα2,8SAα2,3Galβ1,3GalNAcβ1,4<br>SAα2,8SAα2,3 | Galβ1,4Glc-Cer | GQ1b | IV³NeuAc₂,II³NeuAc₂-GgOse₄ |
| Lacto-N-neotetraose core | | | |
| SAα2,6Galβ1,4GlcNAcβ1,3Galβ1,4Glc-Cer | | 6'LM1 | IV⁶NeuAC-nLcOse₄ |
| SAα2,3Galβ1,4GlcNAcβ1,3Galβ1,4Glc-Cer | | 3'LM1 | IV³NeuAc-nLcOse₄ |
| SAα2,8SAα2,3Galβ1,4GlcNAcβ1,3Galβ1,4Glc-Cer | | | IV³NeuAc₂-nLcOse₄ |
| Lacto-N-tetraose core | | | |
| SAα2,3Galβ1,3<br>Fucα1,4 | GlcNAcβ1,3Galβ1,4Glc-Cer | | IV³NeuAc,III⁴FucLcOse₄ |

[a] For additional structures, see Ledeen and Yu (1982).
[b] Svennerholm (1963).
[c] IUPAC–IUB (1977).

(Galβ1,4GlcNAcβ1,3Galβ1,4Glc-). These typically have only one or two sialic acids attached in the NeuAcα2,3Gal or NeuAcα2,8NeuAcα2,3Gal sequence to the terminal galactose residue. They are found primarily in tissues other than brain and in some cell types like erythrocytes they are the predominant ganglioside species (Watanabe *et al.*, 1979). Gangliosides of this series with longer core structure have also been reported (Ledeen and Yu, 1982), but the terminal sequences containing sialic acid are the same. The single example with the lacto-N-tetrasylceramide core (Galβ1,3GlcNAcβ1,3Galβ1,4Glc) was only recently discovered as a tumor-associated antigen of colonocarcinoma (Magnani *et al.*, 1982).

A final source of structural diversity is the sialic acids themselves. As shown in Fig. 6, sialic acid is a charged nine carbon ketosugar formed by condensation of pyruvic acid with N-acetylmannosamine. At least 17 distinct sialic acids have been identified by Schauer and colleagues (Schauer *et al.*, 1974; Schauer, 1982). The amino group at C-5 always contains either an N-acetyl or N-glycolyl group giving the two major sialic acid species. The other derivatives arise from sub-

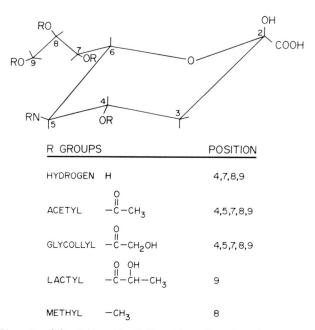

| R GROUPS | | POSITION |
|---|---|---|
| HYDROGEN | H | 4,7,8,9 |
| ACETYL | $-\overset{\overset{O}{\|\|}}{C}-CH_3$ | 4,5,7,8,9 |
| GLYCOLLYL | $-\overset{\overset{O}{\|\|}}{C}-CH_2OH$ | 4,5,7,8,9 |
| LACTYL | $-\overset{\overset{O}{\|\|}}{C}-\overset{\overset{OH}{\|}}{CH}-CH_3$ | 9 |
| METHYL | $-CH_3$ | 8 |

**Fig. 6.** Diversity of the sialic acids. Sialic acids are based on the 9-carbon keto sugar neuraminic acid. In nature, the amino group at position 5 is always acylated as N-acetylneuraminic acid or N-glycolyl neuraminic acid. Other derivatives occur through O-acylation or O-methylation at positions 4, 7, 8, and 9 (see Schauer, 1982).

stitution of the hydroxyl groups at positions 4,7,8, and 9 with acetyl, glycolyl, lactyl, or methyl groups (Schauer, 1982). Most of the sialic acid derivatives occur as minor components or are not widely distributed. However, major differences between species are evident. For example, man produces *N*-acetylneuraminic acid almost exclusively. In contrast, the major sialic acid in porcine species is *N*-glycolylneuraminic acid. Sialic acids of bovine species in addition contain one or two O-acetyl groups at positions 7,8, and 9 and equine species have sialic acids with an O-acetyl group at position 4 (Schauer, 1982). Unless otherwise specified, the general term sialic acid (SA) will refer to *N*-acetylneuraminic acid or will be used when the type of sialic acid has not been established.

## 3. Analysis of Sialyloligosaccharide Receptor Specificity

From the foregoing discussion, it is clear that a complete analysis of the interaction of a virus with all possible sialyloligosaccharide structures that occur in nature would be a formidable task. However, two approaches have made progress in this direction by attempting to define major binding specificities of viruses toward some of the more common sialyloligosaccharide structures.

One approach is based on the fact that gangliosides can be purified to homoge-

**TABLE V**

**Specificity of Purified Mammalian Sialyltransferases**

| Sialyltransferase (ST) | Acceptor sequence | Product[a] | Product abbreviation | Reference |
|---|---|---|---|---|
| I Galβ1,4GlcNAc α2,6 ST | Galβ1,4GlcNAc- | SAα2,6Galβ1,4GlcNAc- | SAα2,6Gal(I) | b,c |
| II Galβ1,3(4)GlcNAc α2,3 ST | Galβ1,3GlcNAc- | SAα2,3Galβ1,3GlcNAc- | SAα2,3Gal(II) | b |
|  | Galβ1,4GlcNAc- | SAα2,3Galβ1,4GlcNAc |  |  |
| III Galβ1,3GalNAc α2,3 ST | Galβ1,3GalNAc- | SAα2,3Galβ1,3GalNAc- | SAα2,3Gal(III) | d |
| IV GalNAc α2,6 ST | GalNAcαThr/Ser | SAα2,6GalNAcαThr/Ser | SAα2,6GalNAc(IV) | e |
|  | Galβ1,3GalNAcαThr/Ser | Galβ1,3  GalNAcαThr/Ser |  |  |
|  |  | SAα2,6 |  |  |

[a] Products are found as terminal sequences in glycoproteins and glycolipids as shown in Tables II, III, and IV. To date, sialyltransferases I and II have not been shown to utilize glycolipid acceptor substrates.

[b] Weinstein et al. (1982a,b).

[c] Paulson et al. (1977a,b).

[d] Sadler et al. (1979b) and Rearick et al. (1979).

[e] Sadler et al. (1979c).

neity (Ledeen and Yu, 1982). This enables them to be used individually in model systems to test the ability of viruses to interact with defined sialyloligosaccharide structures such as those depicted in Table IV. Gangliosides were first proposed to be receptors for myxoviruses by Bogoch (1957) and for paramyxoviruses by Haywood (1974a,b). More recently, highly purified gangliosides have been used to investigate the detailed binding specificities of Sendai virus (Holmgren *et al.*, 1980; Markwell *et al.*, 1981) and influenza virus (Bergelson *et al.*, 1982), as will be discussed in more detail later.

Another approach takes advantage of the recent availability of highly purified mammalian sialyltransferases (Beyer *et al.*, 1981; Weinstein *et al.*, 1982a; 1982b). These enzymes catalyze the general reaction

$$CMP\text{-}SA + HO\text{-}Acceptor + SA\text{-}O\text{-}Acceptor + CMP \tag{1}$$

in which the acceptor is an oligosaccharide attached to glycoprotein or glycolipid. In general, these enzymes exhibit strict specificity for the terminal-oligosaccharide sequence recognized as an acceptor, a summary of the specificities of four sialyltransferases purified to date is given in Table V. The specificity of each enzyme ensures that it will form sialyloligosaccharides of a defined sequence even in complex systems containing acceptor sequences of other sialyltransferases.

The use of these enzymes to analyze receptor specificity of myxoviruses and paramyxoviruses was demonstrated initially by enzymatic modification of intact human erythrocytes (Sadler *et al.* 1979a; Paulson *et al.*, 1979) according to the scheme shown in Fig. 7. The cells are first treated with *V. cholerae* sialidase to remove sialic acid and abolish viral adsorption. Aliquots of the asialo cells can then be reacted with CMP-SA and one of the purified sialyltransferases to generate sialyloligosaccharides of defined sequence. In this way several cellular preparations with different sequences may be prepared. Use of such derivatized cells in hemagglutination assays has revealed strict and varied specificities of the viral attachment proteins for sialyloligosaccharide receptor determinants (Paulson *et al.*, 1979; Carroll *et al.*, 1981a; Cahan and Paulson, 1980; Rogers and Paulson, 1983). The same methodology has been applied to tissue culture cells to investigate receptor determinants that lead to infection of host cells (Markwell and Paulson, 1980; Fried *et al.*, 1981).

Although the sialyltransferases elaborate a defined terminal sequence with a high degree of fidelity, reactions with sialidase-treated cells may be expected to involve several or many cell surface components, with oligosaccharides containing the acceptor sequence of the enzyme. For the sialyltransferases listed in Table V, the sequences shown as products for enzymes I and II are typically found on N-linked oligosaccharides and those for enzymes III and IV are found on O-linked oligosaccharides. However, this is not an absolute rule. For example, inspection of Tables II, III, and IV reveals that the sequence SAα2,

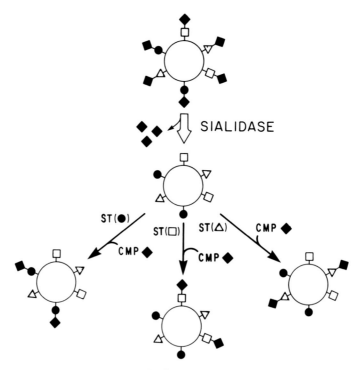

**Fig. 7.** Enzymatic modification of cell surface sialyloligosaccharides. Cells are modified to carry sialyloligosaccharides of defined sequence by removal of sialic acid with sialidase, and then reaction with CMP-SA and a specific sialyltransferase (ST). Symbols: sialic acid (◆), and sialyloligosaccharides (●——◆), (□——◆), and (△——◆).

3Galβ1,4GlcNAc occurs as a terminal sequence on N-linked oligosaccharides, O-linked oligosaccharides, and on glycolipids. Sialyltransferase II, therefore, may transfer to all three classes of oligosaccharides, although this has not been formally demonstrated (Weinstein *et al.*, 1982b). Similarly, the sequence SAα2,3Galβ1,3GalNAc, a product of sialyltransferase III, is found on O-linked oligosaccharides and on gangliosides. In this case, transfer to both glycoprotein and glycolipid acceptor substrates has been demonstrated (Rearick *et al.*, 1979). Due to these considerations, viral-receptor determinants elucidated by the use of sialylated erythrocytes or tissue culture cells are most valuable for defining the terminal oligosaccharide structure recognized by the viral binding site.

Of the animal viruses which exhibit a dependence on sialic acid for binding to cells, the receptor determinants of myxoviruses, paramyxoviruses, and polyomavirus of the papovavirus family have been the most extensively studied. Early evidence for recognition of specific receptor determinants and more recent attempts to define receptor specificity using newer approaches described previously are presented later for each of these and other virus families.

**a. Myxoviruses.** Attempts to explain the different reactivities of influenza virus strains with erythrocytes and with glycoprotein inhibitors of hemagglutination and infection stimulated numerous investigations into the interactions of influenza viruses with sialyloligosaccharides. Treatment of glycoprotein inhibitors with periodate ($IO_4^-$) was found to abolish the binding to some viruses but not others (see, e.g., Burnet, 1948; Takátsy et al., 1959; Takátsy and Barb, 1959; Levinson et al., 1969). The loss of inhibitory activity on periodate treatment was shown by Suttajit and Winzler (1971) to correspond to periodate cleavage of the eighth and ninth carbons, leaving the pyranose ring intact. Levinson et al. (1969) showed that the potent horse serum inhibitor of avid H2 strains was totally inactive against an influenza B virus. Its insensitivity to inactivation by sialidase was suggested to be due to the presence of 4-O-acetyl-N-acetylneuraminic acid. The elegant studies of Drzeniek (1972) demonstrated that influenza virus and paramyxovirus neuraminidases exhibited remarkable specificities for sialyloligosaccharides as substrates, preferentially hydrolyzing the SAα2,3Gal linkage. Zakstelskaya et al. (1972) synthesized a variety of sialic acid-containing disaccharides and tested them for interaction with influenza hemagglutinins and neuraminidases. Of the compounds tested, NeuAcα2,3Glc was found to be a potent inhibitor of hemagglutination and infection while NeuAcα2,3GlcNAc, NeuAcα2,6Glc, and NeuAcα2,6Gal were devoid of antihemagglutinin activity but were inhibitors of some of the viral neuraminidases.

The early studies of Bogoch demonstrated that bovine brain ganglioside preparations mixed with influenza (A/PR/8/34) virus could inhibit hemagglutination and neurotoxicity in mice (Bogoch, 1957; Bogoch et al., 1959) and thus provided evidence for the potential role of gangliosides as cellular receptors. More recently, Bergelson et al. (1982) and Bukrinskaya et al. (1982) have examined the effect of gangliosides on the binding of influenza virus (fowl plague) to Ehrlich ascites cells and chick embryo fibroblasts, respectively. Sialidase-treated cells with low capacity to adsorb viruses were incubated with various gangliosides and then tested for their ability to bind viruses. In order of increasing effectiveness, gangliosides GM1, GD1a, and GT1b increased the capacity of viral adsorption while GM2 was without effect. Ganglioside GT1b, the most effective, was also found to stimulate penetration of the viral genome to the nucleus and partially restore virus-induced hemolysis to sialidase-treated erythrocytes. Huang (1982, 1983) has demonstrated that neutral glycolipids which terminate with galactose-inhibited influenza virus induce hemolysis. However, in this case the mechanism of inhibition is presumably not mediated through the sialic acid binding site of the hemagglutinin.

Using enzymatically modified erythrocytes, Paulson et al. (1979) showed that two influenza virus hemagglutinins could exhibit markedly different specificities for receptor determinants on erythrocyte glycoproteins. To date, over 20 influenza A viruses of human and animal origin have been examined for their

ability to agglutinate erythrocytes modified to contain sialyloligosaccharide sequences elaborated by each of the sialyltransferases listed in Table V (Carroll *et
al.*, 1981b; Rogers and Paulson, 1983). In general, three main binding types
were identified, based on the binding of viruses to erythrocytes containing the
NeuAcα2,6Galβ1,4GlcNAc [SAα1,6Gal(I)] or NeuAcα2,3Galβ1,3GalNAc
[SAα2,3Gal(III)] sequences. Viruses exhibited either strong preferential binding to one of the two sialic acid linkages or bound both. Viruses which exhibited preferential binding to the SAα2,3Gal(III) linkage also bound cells containing the SAα2,3Galβ1,4GlcNAc [SAα2,3Gal(II)] sequence, while those
with preferential binding to the SAα2,6Gal(I) linkage exhibited variable adsorption to these cells. Most viruses showed little or no agglutination of cells
containing the Galβ1,3(NeuAcα2,6)GalNAc [SAα2,6GalNAc(IV)] sequence
(Rogers and Paulson, 1983).

Analysis of the erythrocyte-receptor specificity of the A/RI/5$^+$/57 (RI/5$^+$)
and A/RI/5$^-$/57 (RI/5$^-$) viruses isolated by Choppin and Tamm (1960a,b)
revealed additional information about their biological properties (Carroll *et al.*,
1981a). These viruses were originally isolated based on sensitivity (RI/5$^+$) and
insensitivity (RI/5$^-$) to neutralization by horse serum. In addition, once bound
to erythrocytes, the RI/5$^-$ virus rapidly eluted while the RI/5$^+$ virus remained

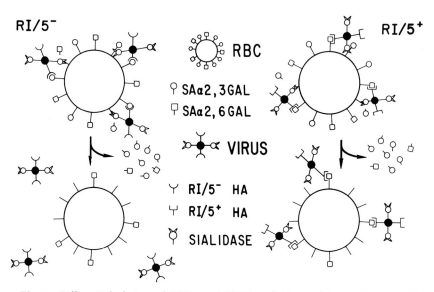

**Fig. 8.** Differential elution of RI/5$^-$ and RI/5$^+$ and viruses from erythrocytes. Interaction of the two viruses with a simplified red blood cell (RBC) containing two sialic
acid linkages is shown. Elution of RI/5$^-$ and not RI/5$^+$ is due to hemagglutinins of
different receptor specificity and a common sialidase which preferentially cleaves the
SAα2,3Gal linkage (●), the receptor determinant of RI/5$^-$. (From Carroll *et al.*, 1981a.)

TABLE VI

Correlation of Receptor Specificity and Hemagglutination Inhibition (HAI) by Horse-Serum Glycoproteins for Influenza Viruses of the H3 Serotype

| Influenza virus | Erythrocyte-receptor specificity (HA titer)[a] | | | | Hemagglutination inhibition (HAI titer)[b] |
|---|---|---|---|---|---|
| | Native | Asialo | SAα2,3Gal(III) | SAα2,6Gal(I) | |
| Human isolates | | | | | |
| A/Hong Kong/8/68 | 256 | 0 | 0 | 256 | 2048 |
| A/Aichi/2/68 | 1024 | 0 | 0 | 1024 | 2048 |
| A/Memphis/102/72 | 256 | 0 | 0 | 256 | 2048 |
| A/ Victoria/3/75 | 512 | 0 | 0 | 512 | 2048 |
| A/Texas/1/77 | 512 | 0 | 0 | 512 | 4096 |
| Avian isolates | | | | | |
| A/duck/Ukraine/1/63 | 256 | 0 | 128 | 8 | 64 |
| A/duck/Memphis/928/74 | 2048 | 0 | 2048 | 256 | 32 |
| A/duck/New York/6874/78 | 512 | 0 | 256 | 128 | 32 |
| Equine isolates | | | | | |
| A/Equine/Miami/1/63 | 1024 | 0 | 1024 | 0 | 64 |

[a] Receptor-specificity experiments were performed in microtiter wells using either native erythrocytes, erythrocytes treated with sialidase (asialo), or asialo erythrocytes resialylated with purified sialyltransferases (Table V) to contain the SAα2,3Gal(III) or SAα2,6Gal(I) linkages. Hemagglutination (HA) is expressed as the reciprocal of the maximum dilution still giving complete agglutination ($0 \leq 2$).

[b] HAI titer is expressed as the reciprocal of the highest dilution of horse serum causing inhibition of native cell agglutination by four hemagglutination units of virus. Adapted from Rogers and Paulson (1983) and Rogers et al. (1983b).

firmly bound (Choppin and Tamm, 1960a). These two viruses were found to exhibit strict specificity for their receptor determinants, the RI/5$^-$ virus binding the SAα2,3Gal(III) linkage and the RI/5$^+$ virus binding the SAα2,6Gal(I) linkage (Carroll et al., 1981a). Furthermore, the neuraminidase of both viruses exhibited a marked specificity for hydrolysis of the SAα2,3Gal linkage. As diagrammed in Fig. 8, these results suggested that the differential elution of the RI/5$^-$ and RI/5$^+$ viruses from erythrocytes was because the receptor determinant of the RI/5$^-$ virus was susceptible to hydrolysis while that of the RI/5$^+$ virus was resistant.

One of the most interesting findings in a survey of viral receptor specificity (Rogers and Paulson, 1983) was that influenza A isolates containing the H3 hemagglutinin revealed different receptor specificities which correlated with species of origin. As shown in Table VI, human isolates from 1968 to 1977 agglutinated erythrocytes containing the SAα2,6Gal(I) linkage but not those containing the SAα2,3Gal(III) linkage. In contrast, avian and equine isolates preferentially agglutinated cells containing the SAα2,3Gal(III) linkage. Human isolates were also found to be sensitive to inhibition of hemagglutination by horse serum while avian and equine isolates were only weakly inhibited (Table VI). The active principle in horse serum was found to be $\alpha_2$-macroglobulin (Rogers *et al.*, 1983b; T. Pritchet, unpublished observations), as found earlier for influenza viruses with the H2 hemagglutinin (Levinson *et al.*, 1969).

Taken together, these results suggested that receptor specificity and inhibitor sensitivity were related. This was confirmed by growth of A/Memphis/102/72 in MDCK cells or the recombinant virus X-31 (H3N2) *in ovo* in the presence and absence of horse serum or equine $\alpha_2$-macroglobulin (Rogers *et al.*, 1983a; G. Rogers, unpublished observations). Progeny virus grown in the absence of horse serum exhibited binding properties of the parent virus (Table VI; Rogers *et al.*, 1983b) while virus propagated in the presence of horse serum exhibited binding properties characteristic of the avian and equine isolates. Thus, it appeared that horse serum or equine $\alpha_2$-macroglobulin supressed the growth of the SA$\alpha$2, 6Gal(I) inhibitor-sensitive binding type allowing the growth to predominance of a variant exhibiting the SA$\alpha$2,3Gal(III)-specific, inhibitor-insensitive phenotype.

Sequence analysis of the hemagglutinins of the egg-grown X-31 variants and cloned variants representing the parent and inhibitor-insensitive binding types from A/Memphis/102/72 revealed a single nucleotide substitution which resulted in a change at amino acid 226 from leucine, for the parent virus, to glutamine, for the SA$\alpha$2,3Gal(III)-specific inhibitor-sensitive variants (Rogers *et al.*, 1983a). This residue is located at the center of the proposed receptor binding site in the crystal structure of the X-31 hemagglutinin reported by Wilson *et al.* (1981), as depicted in Fig. 2, Section II. Hemagglutinin sequences have also been reported for several of the other H3 strains, for which binding properties are listed in Table V. Three of the human isolates, A/Aichi/2/68, A/Victoria/3/74, and A/Texas/1/77, have leucine and one avian isolate, A/duck/Ukraine/1/63, has glutamine (reviewed in Ward and Copheide, 1981b). In each case, the same correspondence between binding properties and the amino acid at position 226 is seen. These observations demonstrate that dramatic changes in receptor-binding properties of the influenza hemagglutinin can be attributed to a single amino acid in the receptor-binding pocket.

Because receptor specificity correlated with species of origin for viruses bearing the H3 hemagglutinin, it has been proposed that sialyloligosaccharides elaborated by each host species may exert different selective pressures, resulting in the growth to predominance of the receptor-specific variant with binding properties optimal for growth in that host (Rogers and Paulson, 1983; Rogers *et al.*, 1983b). The significance of such selection in natural infections of influenza is discussed in Section VI.

**b. Paramyxoviruses.**    Binding to cellular receptors by paramyxoviruses is mediated by the viral HN protein (see Section II). Evidence for receptor specificity in Sendai virus was obtained by Haywood (1974a,b) in studies examining inhibition of hemagglutination by liposomes containing gangliosides. Liposomes alone caused no inhibition of hemagglutination nor did liposomes containing GM1, but inhibition was observed with liposomes containing di- and tri-sialogangliosides. Since gangliosides also mediated Sendai virus binding and fusion with liposomes, they were proposed to be receptors on the cell surface of host cells (Haywood, 1974b). Fidgen (1975) provided further evidence for Sendai virus-receptor specificity from analysis of hemagglutination titers of erythrocytes subjected to graded hydrolysis by *V. cholerae* sialidase. It was proposed that Sendai virus bound sialic acid in an α2,3 linkage to the penultimate sugar, since Drzeniek (1972) had earlier shown that *V. cholerae* neuraminidase preferentially hydrolyzed this linkage.

Both Sendai virus and Newcastle disease virus were found to exhibit strict specificity for binding of the SAα2,3Gal(III) sequence on enzymatically derivatized erythrocytes (Paulson *et al.*, 1979). Agglutination of sialidase-treated human erythrocytes was restored by resialylation with purified sialyltransferases to contain the SAα2,3Gal(III) linkage but not the SAα2,6Gal(I) or SAα2,6Gal-NAc(IV) sequences (see Table V for full sequences). Similarly, susceptibility of MDBK cells to infection was abolished by sialidase treatment and, specifically, restored by resialylation to contain the SAα2,3Gal(III) but not the SAα2,6Gal(I) sequence (Markwell and Paulson, 1980). This demonstrated that Sendai virus exhibited similar specificity for both erythrocyte receptors and host cell receptors that mediate infection.

Holmgren *et al.* (1980) also found strict specificity for the binding of Sendai virus to gangliosides adsorbed to plastic dishes. Sendai virus bound to gangliosides of the ganglio series containing the terminal sequence SAα2,3Galβ1,3GalNAc (e.g., GD1a, GT1b) but not analogous structures without the terminal sialic acid (e.g., GM1, GD1b). The importance of the SAα2,3Gal linkage as a receptor determinant of Sendai virus was in agreement with the results obtained with resialylated erythrocytes and MDBK cells. In addition, however, gangliosides containing the terminal sequence SAα2,8SAα2,3Gal (e.g., GQ1b, GP1c) bound Sendai virus with 100-fold higher affinity. Direct evidence for gangliosides bearing the SAα2,3Gal or SAα2,8SAα2,3Gal sequences to serve as receptor determinants on host cells was obtained by Markwell *et al.* (1981; 1984a,b), as discussed earlier in Section III,B,3. As shown in Table VII, the ability of gangliosides to mediate adsorption of Sendai virus to plastic plates, as reported by Holmgren *et al.* (1980), correlated with their ability to restore susceptibility of sialidase-treated MDBK cells to infection.

In addition to Sendai virus adsorption to glycolipids of the ganglio (Galβ1, 3GalNAc) series, Hansson *et al.* (1984) and Umeda *et al.* (1984) have recently demonstrated the interaction of Sendai virus with gangliosides of the lacto (Galβ1,4GlcNAc) series. Indeed, Sendai virus bound more avidly to sia-lyosylparagloboside (SAα2,3Galβ1,4GlcNAcβ1,3Galβ1,4Glc-ceramide) incor-

TABLE VII

Gangliosides Restore Sendai Virus Receptors to Asialo-MDBK Cells[a]

| Cell preparation | Ganglioside tested | Structure | Amount added (ng) | Virus produced (HA titer) |
|---|---|---|---|---|
| Native | — | | | 256 |
| Asialo | — | | | 0 |
| Asialo | GM1 | ●—□—●—■ (▲ below ●) | 2500 | 0 |
| Asialo | | | 250 | 0 |
| Asialo | GD1a | ▲—●—□—●—■ (▲ below ●) | 2500 | 128 |
| Asialo | | | 250 | 4 |
| Asialo | | | 25 | 0 |
| Asialo | GD1b | ●—□—●—■ (▲▲ below ●) | 2500 | 0 |
| Asialo | GT1b | ▲—●—□—●—■ (▲▲ below ●) | 2500 | 64 |
| Asialo | | | 250 | 2 |
| Asialo | | | 25 | 0 |
| Asialo | GQ1b | ▲—▲—●—□—●—■ (▲▲ below ●) | 250 | 128 |
| Asialo | | | 25 | 64 |
| Asialo | | | 2.5 | 4 |

[a] Asialo-(sialidase treated) MDBK cells were exposed (20 min) to various gangliosides and washed before inoculation with Sendai virus (10 min). Progeny virus in the culture fluid were assessed by hemagglutination (HA) after further incubation (48 hr) at 37°C. HA titers are expressed as the maximum dilution which still gives agglutination, and 0 indicates no agglutination with two-fold dilution of culture fluid. Ganglioside structures are shown in symbol form for which ▲ is SA, ● is Gal, □ is GalNAc, and ■ is Glc (see Table IV). Adapted from Markwell et al. (1981).

porated in liposomes (Umeda et al., 1984) or supported on a thin-layer chromatography plate (Hansson et al., 1984) than the gangliosides of the ganglio series. In fact, Hansson et al. (1984) saw no binding to ganglio series gangliosides. Binding of Sendai virus to sialyloligosaccharides of the lacto series has also been proposed to account for the avid interaction of Sendai virus with bovine erythrocyte glycophorin (Suzuki et al., 1984).

There are conflicting reports on the ability of Sendai virus to bind sialyloligosaccharides containing sialic acids other than NeuAc. Several reports have suggested that the Sendai virus fails to agglutinate equine erythrocytes (Kuroya et al., 1953; Jensen et al., 1955; Yamamoto et al., 1974), presumably due to the presence of N-glycolylneuraminic acid (NeuGc) instead of N-acetylneuraminic acid (Yamakawa, 1966; Suzuki et al., 1983). In contrast, Suzuki et al. (1983) reported Sendai-virus agglutination of equine, bovine, and porcine erythrocytes which all contained predominately (90%) NeuGc. Furthermore, purified bovine glycophorin, which contains primarily NeuGc in O-linked oligosaccharides of the proposed structure NeuGc- (Galβ1,4GlcNAc)$_{1-2}$ -Galβ1,3GalNAc-, was a potent inhibitor of hemagglutination (Fukuda *et al.*, 1981; Suzuki *et al.*, 1983). Differences in receptor-binding properties noted between laboratories could be due to variations in Sendai virus strains, as seen with influenza viruses, but there is currently no direct evidence to support this possibility (Hansson *et al.*, 1984).

In summary, the consensus is that the receptor determinants of Sendai virus are the sequences SAα2,3Gal and SAα2,8SAα2,3Gal (Paulson *et al.*, 1979; Holmgren *et al.*, 1980; Markwell and Paulson, 1980; Markwell *et al.*, 1981, 1982). Either sequence on gangliosides is normally present in sufficient quantities in MDBK cells, HeLa cells, and MDCK cells to fully support susceptibility to infection (Markwell *et al.*, 1981; 1984a,b). In gangliosides, the neutral disaccharide following sialic acid is Galβ1,3GalNAc, yielding the terminal sequence SAα2,3Galβ1,3GalNAc also found in O-linked oligosaccharides of glycoproteins. This sequence on the O-linked oligosaccharides of glycophorin appears to mediate binding of Sendai virus to human erythrocytes (Bachi *et al.*, 1977; Paulson *et al.*, 1979; Umeda *et al.*, 1984). The SAα2,3Gal linkage is also found in the sequence SAα2,3Galβ1,4GlcNAc in both glycoproteins and glycolipids (Table II, III, and IV), as is the potential ''high-affinity'' analogue SAα2,8SAα2,3Galβ1,4GlcNAc, and the ability of these sequences to serve as receptor determinants of the Sendai virus has been demonstrated (Umeda *et al.*, 1984; Hansson *et al.*, 1984; Suzuki *et al.*, 1984). It is clear, however, that certain sialyloligosaccharides do not serve as receptor determinants of Sendai virus, notably the SAα2,6Galβ1,4GlcNAc and SAα2,6GalNAc sequences common to N- and O-linked oligosaccharides of glycoprotein, respectively (Paulson *et al.*, 1979; Markwell and Paulson, 1980), and gangliosides GM1 and GD1b (Holmgren *et al.*, 1980; Markwell *et al.*, 1981, 1984 a,b; Umeda *et al.*, 1984). It will be of interest in the future to understand the structural basis for this specificity in the receptor site of the HN protein.

**c. Papovaviruses.** For papovaviruses, the receptor determinants of polyomavirus have been the most extensively studied. Mori *et al.* (1962) concluded that polyomavirus receptor sites on guinea pig erythrocytes were similar to those of an influenza virus (PR8) but that the former were more sensitive to hydrolysis by *V. cholerae* sialidase and treatment with periodate. In contrast, the receptors of polyomavirus on host cells (primary mouse embryo cells) were less sensitive to sialidase. Polyomavirus-receptor determinants were investigated by Cahan and Paulson (1980) using native, asialo, and resialylated human erythrocytes. A remarkable specificity for binding the SAα2,3Gal(III) linkage was found. As determined by hemagglutination assays, no binding to cells containing the SAα2,6Gal(I) and SAα2,6GalNAc(IV) linkages was observed. Treatment of erythrocytes with Newcastle disease virus, which is specific for hydrolysis of the

SAα2,3Gal linkage (Drzeniek, 1972; Paulson *et al.*, 1982), prevented hemag-glutination by polyomavirus but not a variety of influenza viruses including PR8. Unlike polyomavirus, PR8 also bound receptor determinants containing the Neu-Acα2,6Gal(I) linkage (Cahan and Paulson, 1980; Rogers and Paulson, 1983). Thus, it appears that sialidase more readily removes erythrocyte receptors of polyomavirus compared to PR8, as first observed by Mori *et al.* (1962), since the sialyloligosaccharide sequence SAα2,3Gal, which serves as its receptor determi-nant, is preferentially cleaved by the *V. cholerae* and viral sialidases (Drzeniek, 1972).

Based on a comparison between binding of polyoma virions and empty capsids to guinea pig erythrocytes and mouse kidney cells, Bolen and Consigli (1979) suggested that erythrocyte receptors differed from host cell receptors that medi-ate infection. To examine this, Fried *et al.* (1981) tested the ability of poly-omavirus to infect native, asialo, and resialylated 3T6 cells. Sialidase treatment abolished susceptibility to infection which was subsequently restored by re-sialylation to contain the SAα2,3Gal(III) linkage but not the SAα2,6Gal(I) link-age. Thus, the sialyloligosaccharide-receptor determinant found to mediate ag-glutination of erythrocytes was also found to support polyomavirus infection of 3T6 cells.

In addition to the wild-type large-plaque polyomavirus, two binding variants have been described; the small-plaque variant (Diamond and Crawford, 1964; Eddy, 1969) and the Py235 variant (Basilico and DiMayorca, 1974). The small-plaque variant appears to be the most "avid" binding of the three, since ag-glutination to guinea pig erythocytes is least affected by temperature (4–37°C) and pH (6.5–8.0). In contrast, the binding of the Py235 variant is markedly decreased at pH above 7.3 and binds weakly at temperatures >4°C (Basilico and DiMayorca, 1974). The large-plaque polyomavirus exhibits binding properties intermediate between the small-plaque and Py235 variants. Similar binding dif-ferences were observed with host cells. The Py235 virus exhibited impaired adsorption to BHK cells but not to murine 3T3 cells (Basilico and DiMayorca, 1974).

Comparison of all three variants for binding to guinea pig erythrocytes and native, asialo, and resialylated human erythrocytes revealed some striking dif-ferences (Cahan *et al.*, 1983). All three viruses agglutinated guinea pig cells. However, while the small-plaque variant agglutinated human erythrocytes with equal titer, agglutination of human cells by the large-plaque virus was markedly reduced, and agglutination by the Py235 variant was nonexistent. Yet, as shown in Table VIII, all three binding variants agglutinated with high-titer human erythrocytes resialylated to contain the NeuAcα2, 3Galβ1,3GalNAc sequence [SAα2,3Gal(III)].

To account for this unexpected observation, it was reasoned that while the major sialylated species on the resialylated cells was NeuAcα2,3Galβ1,3GalNAc on O-

TABLE VIII

**Hemagglutination of Native and Derivatized Human Erythrocytes by Binding Variants of Polyomavirus**

| Erythrocyte preparation | Sialic acid linkage[a] | Polyomavirus variant (HA titers) | | |
|---|---|---|---|---|
| | | Large-plaque | Small-plaque | Py235 |
| Native | Multiple | 64 | ≤4096 | <16 |
| Sialidase-treated | — | <2 | <2 | <16 |
| Resialylated | SAα2,6Gal(I) | 2 | <2 | <16 |
| Resialylated | SAα2,3Gal(III) | ≥4096 | 2048 | ≥4096 |
| Resialylated | SAα2,6GalNAc(IV) | <2 | <2 | <16 |

[a] Human erythrocyte derivatives were prepared and hemagglutination assays performed as in Table VI except at 4°C. Adapted from Cahan et al. (1983).

linked oligosaccharides of glycophorin (Paulson et al., 1979), the major O-linked oligosaccharide on native glycophorin was the disialylated structure Neu-Acα2,3Galβ1,3(NeuAcα2,6)GalNAc (Thomas and Winzler, 1969). To test this possibility, glycophorin was purified (Marchesi and Andrews, 1971), the sialidase treated, and resialylated derivatives analogous to the enzymatically modified erythrocytes were prepared. These were then examined as inhibitors of hemagglutination for all three binding variants as shown in Table IX. While the resialylated derivative was a potent inhibitor for all three variants, native glycophorin was a potent inhibitor only for the small-plaque variant. Thus, it appeared that while all three recognized the sequence NeuAcα2,3Gal-β1,3GalNAc as a receptor determinant, the presence of sialic acid in the Neu-Acα2,6GalNAc sequence blocked binding by the large-plaque and Py235 variants.

These results in large part accounted for the major differences in binding properties of the three variants with human erythrocytes. The primary distinction between the large-plaque and Py235 variants appeared to be mainly avidity for receptors rather than specificity. Additional experiments with other glycoprotein inhibitors of hemagglutination (e.g., $\alpha_1$-acid glycoprotein) suggested that all three viruses could also bind the NeuAcα2,3Galβ1,4GlcNAc sequence on N-linked oligosaccharides. Given these facts, observed variation in the adsorption of polyomaviruses to erythrocytes and host cells may well be due to differential expression of the various sialyloligosaccharide structures that serve as cell surface-receptor determinants.

**d. Picornaviruses.**    Of the picornaoviruses which exhibit a dependence on sialic acid for binding to cells, encephalomyocarditis (EMC) has been the subject of most investigations. Yet, little is currently known about the details of its

TABLE IX

Inhibition of Polyomavirus Hemagglutination by Native and Enzymatically Modified Derivatives of Glycophorin[a]

| Inhibitor | Oligosaccharide sequence | Polyomavirus variant (HAI titers) | | |
|---|---|---|---|---|
| | | Large-plaque | Small-plaque | Py235 |
| Glycophorin | SA$\alpha$2,3Gal$\beta$1,3 ⟍ <br>           ⟋ GalNAc- <br> SA$\alpha$2,6 | 0 | 4096 | 0 |
| Sialidase-treated glycophorin | Gal$\beta$1,3GalNAc- | 0 | 0 | 0 |
| Resialylated glycophorin | SA$\alpha$2,3Gal$\beta$1,3GalNAc- | 4096 | 1024 | 4096 |

[a] Oligosaccharide sequences shown are the major O-linked oligosaccharides present in the glycophorin preparations. Hemagglutination inhibition (HAI) titers are expressed as the maximum dilution which gave complete inhibition of hemagglutination of guinea pig erythrocytes. Each inhibitor was compared at approximately 0.1 mg. Adapted from Cahan et al. (1983).

binding specificity (Burness, 1981). Angel and Burness (1977) showed that dramatic differences in the ability of EMC to agglutinate erythrocytes of various species was not related to the sialic acid content of the cells. Human glycophorin was a potent inhibitor of hemagglutination (Enegren and Burness, 1977), suggesting that it was the erythrocyte receptor on human erythrocytes. Graded digestion of erythrocytes or purified glycophorin with V. cholerae sialidase abolished EMC adsorption when only 30–40% of the sialic acid was removed (Burness and Pardoe, 1981). Based on the specificity of the V. cholerae sialidase (Drzeniek, 1972), this may indicate preferential adsorption to SA$\alpha$2,3Gal linkages. It is of interest that while most available information is consistent with the binding of EMC to sialyloligosaccharide receptors, there are several reports with apparently contradictory results. Inhibition of hemagglutination by an intestinal mucopolysaccharide was not sensitive to sialidase treatment (Mandel and Racker, 1953), and susceptibility of HeLa cells to infection by EMC virus was abolished by treatment of the cells with sialidase, but similar treatment of mouse L cells had no effect (Kozda and Jungeblut, 1958).

## 4. Other Carbohydrate-Receptor Determinants

The potential for other oligosaccharide structures to serve as receptor determinants has not been fully explored. However, several reports suggest that mannose-containing oligosaccharides may be receptors for Japanese encephalitis virus (Nozima et al., 1968; Homma, 1968; Yasui et al., 1969). It is likely that some animal viruses recognize other classes of carbohydrate receptors. In view of the instrumental role of bacterial sialidases in elucidating the importance of sialic acids as receptor determinants, it may be fruitful to consider the use of glycosidases with other specificities to identify additional carbohydrate-receptor determinants.

## V. ROLE OF RECEPTORS IN VIRAL PENETRATION

Following attachment of a virus to the host cell surface, it must penetrate the membrane barrier in order to initiate viral replication. Several general mechanisms accomplish this purpose. In the following discussion, the major entry mechanisms will be briefly discussed with emphasis on examples that illustrate the roles of cell surface receptors. For a more detailed survey of early and current work on viral penetration, the reader is referred to several excellent reviews (Dales, 1973; Lonberg-Holm and Philipson, 1974, 1980; Howe *et al.*, 1980; Choppin and Scheid, 1980; Marsh *et al.*, 1982a; Dimmock, 1982).

## A. Direct Penetration of the Plasma Membrane

### 1. Membrane Fusion by Enveloped Viruses

Several paramyxoviruses, Sendai, SV5, NDV, and measles viruses are generally accepted to be capable of penetrating cells by direct fusion of the viral envelope with the plasma membrane (Poste and Pasternak, 1978; Choppin and Scheid, 1980). However, membrane fusion may not be the only mode of entry of these viruses in all cells (for review see Dales, 1973). While the mechanism of fusion is not completely understood, it is clear that the fusion activity resides in the F glycoprotein of the viral envelope, as mentioned earlier in Section II,A,2. Compelling evidence for the role of the F protein in fusion comes from analysis of abortive infections. In productive infection, the F protein is initially synthesized as a single polypeptide ($F_0$) and is then cleaved by a host protease into two disulfide-linked polypeptides ($F_{1,2}$). However, in abortive infections, noninfectious virus particles are produced due to a lack of cleavage of the precursor protein $F_0$ (Homma and Ohuchi, 1973; Scheid and Choppin, 1974). The resulting viral particles are able to attach to cells but do not fuse. Subsequent *in vitro* treatment of the inactive virions by mild digestion with trypsin cleaves $F_0$ into $F_{1,2}$ and simultaneously activates the fusion activity and the ability of the virus to cause infection. Sequence analysis of the $NH_2$ terminus formed in $F_1$ has revealed a hydrophobic stretch of at least 5–18 amino acids for Sendai, SV5, and NDV viruses (Gething *et al.*, 1978; Scheid *et al.*, 1978; Richardson *et al.*, 1980). As judged from circular-dichroism spectra and detergent-binding studies, the proteolytic cleavage apparently causes a conformational change resulting in exposure of this hydrophobic sequence (Hsu *et al.*, 1981). Such observations have prompted the suggestion that fusion is initiated by insertion of the hydrophobic $NH_2$ terminus of $F_1$ into the host plasma membrane (Gething *et al.*, 1978; Choppin and Scheid, 1978; Hsu *et al.*, 1981).

Of course the $F_{1,2}$ protein cannot initiate fusion before attachment of the virus which is mediated by the HN protein (Section II,B,2). The separate roles of the HN protein for viral attachment and the $F_{1,2}$ protein for fusion have been readily demonstrated in several systems. Yamamoto *et al.* (1974) observed that Sendai virus was unable to bind to horse erythrocytes, but when attachment was achieved by bridging of viral and cell surface glycoproteins with the plant lectin concanavalin A, hemolysis and fusion were observed. In other studies, liposomes containing both the HN and $F_{1,2}$ viral glycoproteins were found to cause membrane fusion (Hosaka and Shimizu, 1972a,b; Volsky and Loyter, 1978; Hsu *et al.*, 1979) while those containing only the $F_{1,2}$ protein did not unless attachment was induced using wheat germ agglutinin (Hsu *et al.*, 1979). An interesting temperature-sensitive mutant of Sendai virus (ts 271), when grown at a nonpermissive temperature, produces virus particles which contain no HN protein (Portner *et al.*, 1975). While these virions were unable to infect most cell lines which serve as hosts for the wild-type virus, a hepatoma cell line, HepG2 (Schwartz *et al.*, 1981), was readily infected (Markwell *et al.*, 1983). In this case, binding and subsequent infection by the virus was apparently mediated by adsorption to a cell surface-binding protein specific for terminal-galactose residues which are found on the N-linked oligosaccharides of the fusion protein (Yoshima *et al.*, 1981b).

As discussed earlier, the HN protein of Sendai virus binds to sialyloligosaccharide-receptor determinants that are found on both glycoproteins and gangliosides (Sections III,B,3 and IV,C,3). Thus, both glycoproteins and gangliosides potentially may serve as cell surface receptors for Sendai virus. However, while gangliosides alone appear capable of mediating viral attachment and infection of host cells (Markwell *et al.*, 1981, 1984a,b), the ability of glycoproteins alone to serve as natural receptors of influenza virus has not been unequivocally established. In this regard, it has been suggested that both glycoproteins and ganglioside receptors may play important roles in viral attachment and penetration (Haywood, 1974b; Holmgren *et al.*, 1980). Thus, it is envisioned that binding of the virus to glycoprotein receptors would be the most likely site of initial contact and attachment, while binding to ganglioside receptors would subsequently bring the viral and cell membranes into close proximity. While this is an attractive hypothesis, experiments cited above suggest that glycoprotein receptors may allow binding in sufficient proximity to the membrane for fusion to occur. Indeed, membrane fusion is observed when viral attachment is mediated by plant lectins (Yamamoto *et al.*, 1974) or by a host cell surface carbohydrate-binding protein instead of the viral HN protein (Markwell *et al.*, 1983). Definitive evidence for the relative roles of ganglioside and glycoprotein receptors in the membrane fusion of host cells by Sendai virus may yet be forthcoming.

## 2. Direct Entry by Nonenveloped Viruses

A few reports have indicated that adenoviruses (Morgan et al., 1969) and picornaviruses (Dunnebacke et al., 1969) can cross the plasma membrane directly. However, there is general agreement that viruses in these groups are usually taken up by endocytosis and that the viral genome gains entry to the cytoplasm by penetration of endocytic vesicles or lysosomes (Dales, 1973; Howe et al., 1980; Choppin and Scheid, 1980).

## B. Adsorptive Endocytosis

## 1. Morphological Evidence for Adsorptive Endocytosis

Examination of the early events in viral interactions with cells by electron microscopy has provided convincing evidence that many animal viruses once bound to the cell surface are rapidly endocytosed in the form of membrane-bound vesicles (Dales, 1973; Howe et al., 1980; Marsh et al., 1982a). In various reports, this process has been called viropexis, phagocytosis, pinocytosis, and adsorptive endocytosis. While viropexis is a general term (Dales, 1973), phagocytosis generally implies engulfment of large particles and portions of membrane, pinocytosis describes nonspecific pinching off of fluid-filled vesicles, and adsorptive endocytosis or receptor-mediated endocytosis refers to active uptake of ligands bound to receptors (Steinman et al., 1983). In some reports, these terms have been used without regard to a specific mechanism. Since most viruses can be presumed to bind to cell surface receptors, adsorptive endocytosis will be used throughout this discussion to describe the process of viral uptake by vesicle formation. It is recognized, however, that viruses may differ in their detailed mechanisms of endocytosis.

The process of adsorptive endocytosis for ligand–membrane receptor complexes has been extensively studied for such physiological ligands as low-density lipoprotein (Brown and Goldstein, 1979; Goldstein et al., 1979), galactose-terminated glycoproteins (Ciechanover et al., 1983; Geuze et al., 1983), and transferrin (Klausner et al., 1983). In general, uptake is very rapid and occurs through formation of clathrin-coated vesicles which bud from specialized coated pits at localized regions of the cell surface. Adsorptive endocytosis of some viruses resembles the receptor-mediated endocytosis of simpler ligands. Electron microscopic examination of vesicles forming arond adsorbed viruses often shows the electron-dense thickening on the cytoplasmic side of the plasma membrane characteristic of coated pits (Dales, 1973; Howe et al., 1980). Examples are shown in Fig. 9 for several membrane-enveloped viruses (Marsh et al., 1982a).

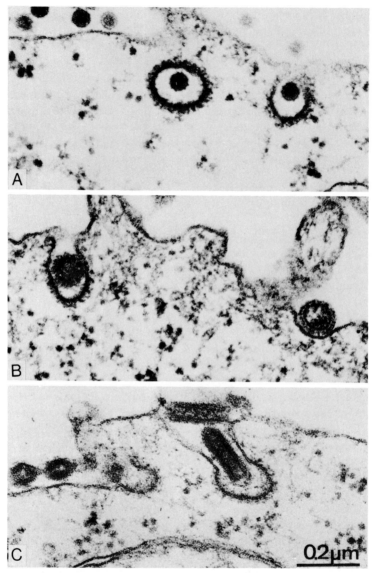

**Fig. 9.** Endocytosis of membrane-enveloped animal viruses. (A) Binding and internalization at coated pits is shown for Semliki forest virus, (B) influenza virus, and (C) vesicular stomatitus virus. (From Marsh *et al.*, 1982a.)

Subsequent to endocytosis, vesicles containing viruses appear to exhibit similar intracellular pathways to those formed during classical receptor-mediated endocytosis. This will be discussed in greater detail in Section V,B,3.

## 2. Role of Receptors in Adsorptive Endocytosis

The molecular properties of receptors which can mediate rapid endocytosis via coated pits are not yet well-defined. For example, it is not known what features are required to promote migration of the receptor–ligand complex to the coated pits. Is a transmembrane protein required or can any cell surface molecule, including glycolipids, mediate this activity? It is clear that not all ligands adsorbed to cell surface receptors are internalized at the high-rate characteristic of receptor-mediated endocytosis (Ahmed et al., 1981; Youngdahl-Turner et al., 1979). In one interesting case, a human fibroblast mutant cell line deficient in LDL (low density lipoprotein) uptake has been characterized to bind LDL, but the receptor–ligand complex fails to be internalized (Brown and Goldstein 1976; Goldstein et al., 1977). Thus by analogy, the potential exists for a virus to bind to receptors which cannot participate in receptor-mediated endocytosis. An example of this may be the inability of Epstein–Barr virus to infect the lymphoblastoid T-cell line Molt 4 (Menezes et al., 1977). These cells bind the EB virus to specific receptors, but the virus is not subsequently endocytosed.

It is likely that viral receptors involved in adsorptive endocytosis need only provide attachment sites and participate in an active-uptake process. In one report, dengue virus, a togavirus, was demonstrated to infect human monocytes by two separate receptor systems (Daughaday et al., 1981). The natural receptors could be removed by trypsin, making the cells insusceptible to infection. Susceptibility to infection could be restored by the addition of nonneutralizing antibody to the virus. Results suggested that the antibody formed a bridge between the virus and the Fc receptor on monocytes. Thus, the Fc receptor was able to replace the natural receptor of dengue virus in promoting infection.

A major difference between adsorptive endocytosis of most physiological ligands and viruses is that viruses bind with very high valency. Indeed, the potential exists for literally hundreds of receptors to be bound simultaneously. As shown for Sendai virus by Haywood (1974a,b; Haywood and Boyer, 1981) using liposomes containing ganglioside receptors, the cooperative binding of hemagglutinins with receptors can cause deformation of a membrane in the direction of engulfment of a virus particle. Obviously, such interactions could influence the process of endocytosis. This phenomenon has been discussed by Matlin et al. (1981) to account for the fact that influenza viruses appear to be endocytosed in smooth vesicles in addition to the coated vesicles presumed to have originated from coated pits. Despite major insights gained into the mechanisms of viral endocytosis in recent years, much is yet to be learned to fully understand the roles of receptors in mediating this process.

Following adsorptive endocytosis, the membrane-enveloped viruses appear to differ fundamentally from the nonenveloped viruses in the manner in which they are released from the endocytic vesicle. Thus, it is useful to discuss the details of adsorptive endocytosis of the two groups separately.

### 3. Fusion of Membrane-Enveloped Viruses with Endocytic Vesicles and Lysosomes

**a. Low-pH-Activated Fusion Activities.** In recent years, it has become clear that a number of membrane-enveloped viruses exhibit fusion activities similar to Sendai virus and others, paramyxoviruses. In contrast to the Sendai virus which fuses with membranes at neutral pH, many viruses exhibit fusion activities only at acidic pH. This has been demonstrated for myxoviruses, including influenza A, B, and C viruses (Maeda and Ohnishi, 1980; Maeda et al., 1981; Huang et al., 1981; Matlin et al., 1981; Ohuchi et al., 1982; Lenard and Miller, 1981), togaviruses such as Semliki forest virus, Sindbis virus, and rubella virus (Väänänen and Kääriäinen, 1979, 1980; White and Helenius, 1980; White et al., 1980; Helenius et al., 1980a), and the rhabdoviruses, vesicular stomatitis virus and rabies virus (White et al., 1981; Mifume et al., 1982; Matlin et al., 1982). In many of these studies, the fusion activity was detected by the ability of the viruses to cause hemolysis of erythrocytes at acidic pH (pH 4.5–5.5). More importantly, the fusion of these viruses with the plasma membrane of host cells could be induced by brief treatment of adsorbed virus for 0.5–1 min at acidic pH (White and Helenius, 1980; White et al., 1980, 1981; Matlin et al., 1981, 1982; Mifume et al., 1982; Yoshimura et al., 1982). Such "nonphysiological" fusion has resulted in infection of the cells in several instances, suggesting that the introduction of the genome into the cytoplasm is sufficient for viral replication to proceed (White et al., 1980; Yoshimura et al., 1982).

Such observations with Semliki forest virus prompted Helenius and co-workers to suggest that the normal route of infection was endocytosis, followed by transport via phagocytic vesicles to lysosomes where the acidic pH would activate fusion, and the resultant extrusion of the viral genome into the cytoplasm (Helenius et al., 1980a,b; Marsh and Helenius, 1980; White et al., 1980). The proposed pathway was subsequently suggested to be similar for influenza virus (Maeda and Ohnishi, 1980; Maeda et al., 1981; Matlin et al., 1981; Huang et al., 1981) VSV and rabies virus (White et al., 1981; Mifume et al., 1982), and Sindbis virus (Talbot and Vance, 1980, 1982). The original concept involving lysosomes as the site of viral fusion has been expanded to accommodate findings obtained from analysis of the pathway of receptor-mediated endocytosis of physiological ligands. Endocytic vesicles appear to fuse first with larger vacuoles or endosomes where the ligand and receptor dissociate and the ligand may be further routed to lysosomes (Dunn et al., 1980; Pastan and Willingham, 1981;

Baenziger and Fiete, 1982; Geuze *et al.*, 1983). Endosomes, like lysosomes, are acidic, pH 5.0 (Tycko and Maxfield, 1982), and thus represent alternative sites for the low-pH activation of viral fusion. Indeed, evidence is accummulating that viruses may follow similar endocytic pathways to those of physiological ligands taken up by receptor-mediated endocytosis. In particular, direct evidence has been obtained for fusion of Semliki forest virus at a prelysosomal site during the normal course of infection (Marsh *et al.*, 1982b, 1983).

Despite the fact that the fusion activity of influenza virus is activated at low pH, Huang *et al.* (1980a) have observed fusion of reconstituted viral envelopes with cells at neutral pH. Under these conditions, the fusion reaction was shown to require both hemagglutinin and neuraminidase (Huang *et al.*, 1980b). At present, the relationship of fusion observed in this reconstituted system with that of the intact virus is not clear. However, such model systems may yield new insights into the nature of the fusion reaction.

The mechanism of influenza virus fusion with membranes of acidic vesicles is largely analogous to that described for paramyxovirus fusion with the plasma membrane (Section V,A,1). For influenza virus the fusion activity resides in hemagglutinin. As for the Sendai virus $F_{1,2}$ protein, the influenza virus hemagglutinin must be cleaved by a host protease to yield two polypeptides (HA1 and HA2) joined by disulfide bonds in order for the fusion activity to be expressed. Moreover, as with $F_2$ there is a stretch of hydrophobic residues at the amino terminus of HA2 (see Section II,A,1.). However, as pointed out by Maeda and Ohnishi (1980), unlike the $F_2$ protein in which the first 18 residues are uncharged, the amino terminus of the HA2 has two glutamic acid residues. Thus, it is possible that low pH aids in the protonation of these charged residues to activate the fusion activity.

As seen in the crystal structure of the hemagglutinin reported by Wilson *et al.* (1981), the amino terminus of HA2 is located near the base of the hemagglutinin spike. Low pH (4.9–5.2) has been shown by Skehel *et al.* (1982) to induce a conformational change in the hemagglutinin, exposing hydrophobic regions, as judged from binding of lipid vesicles and nonionic detergent. Furthermore, newly exposed trypsin-sensitive sites in the acid-treated HA indicate that there is a conformational change in both HA1 and HA2. At present, the precise meaning of these conformational changes with respect to the fusion process is not clear. For example, it has not been established if fusion is purely a physical-chemical event of if an enzymatic activity such as a phospholipase is activated (Skehel *et al.*, 1982).

**b. Fusion of the Plasma Membrane versus Endocytosis.** The fact that many membrane-enveloped viruses exhibit fusion activities which differ mainly in their pH optima provides a unifying concept for the problem of penetration of the host-membrane barrier. In every case, the viral genome crosses either the plasma membrane or a membrane bound organelle of the endocytic pathway.

It is unlikely that all viruses will fuse strictly with the plasma membrane or strictly with membranes of acidic intracellular organelles. Indeed, many membrane-enveloped viruses have been observed to variably penetrate by both fusion and endocytosis (Dales, 1973). For example, paramyxoviruses in different studies appeared to penetrate cells solely by fusion with the plasma membrane, solely by endocytosis, or by some of each. In particular, Newcastle disease virus was found to penetrate by endocytosis (Silverstein and Marcus, 1964) or by both endocytosis and fusion (Meiselman *et al.*, 1967). Such differences may be due in part to variable fusigenic activities of different NDV strains. Newcastle disease virus strains are known to exhibit a large variation in overall hemolytic activity (Clavell and Bratt, 1972) and exhibit broad pH optima with peak hemolytic activity ranging from pH 5.5–7.0. It is easy to see how such variation could influence the relative amounts of endocytosis versus fusion with the plasma membrane. Another factor, however, may be the relative sensitivity of different cell lines to fusion with the plasma membrane. In this regard, Polos and Gallaher (1979, 1982) have shown that a ricin-resistant clone of Chinese hamster ovary (CHO) cells (clone 15-B) is resistant to fusion by NDV. Yet this cell line exhibited as much attachment of NDV and was as susceptible to infection by NDV as wild-type CHO cells. Although the reason for the decreased sensitivity to fusion has not been elucidated, the resistance would likely favor endocytosis of the virus. All of these considerations taken together simply emphasize the fact that fusion with the plasma membrane or fusion after endocytosis are not necessarily mutually exclusive pathways. For a given virus or group of viruses, strain differences and host differences may influence the predominant route of viral penetration.

## 4. Uncoating of Nonenveloped Viruses

The penetration via endocytic vesicles and uncoating of viruses with protein capsids is less well understood than for the membrane-enveloped viruses. Adsorptive endocytosis is clearly observed in electron micrographs, and in some cases, endocytic vesicles appear to form at coated pits (Dales, 1973; Howe *et al.*, 1980). Multivalent binding of enteroviruses and rhinoviruses to cell surface receptors has been postulated to aid in the deformation of the plasma membrane during vesicle formation as shown in Fig. 10 (Boulanger and Lonberg-Holm, 1981). However, it is not at present clear to what degree the mechanism of adsorptive endocytosis for viruses with protein capsids differs, if any, from that of membrane-enveloped viruses. Once inside the cell, the virus must escape either the endocytic vesicle, the endosome, or the lysosome. In addition, some minimal "uncoating" of the protein capsid must occur for the genome to become accessible to host replication machinery. These processes are poorly understood

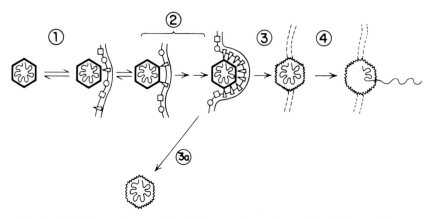

**Fig. 10.** Binding and penetration of picornaviruses. Interaction of virus with receptors (1) is proposed to aid in invagination of the plasma membrane (2) and conformational alteration of the virion to the A particle (3), which may elute as the noninfectious virus (3a). Uncoating (4) usually occurs inside the cell. (From Boulanger and Lonberg–Holm, 1981.)

for most nonmembrane-enveloped viruses. However, some progress has been made in analysis of the uncoating of picornaviruses and reoviruses.

**a. Picornaviruses.** Much of the literature on picornavirus penetration of host cells has been recently reviewed by Boulanger and Lonberg-Holm (1981) and by Crowell and Landau (1983). Penetration by these viruses is difficult to study since as little as 5% of the adsorbed virus will actually penetrate the cell (Mandel, 1965). Most of the remainder elutes from the cell as "A particles" which have altered properties from the native virus. The "A particles" are noninfectious, have distinct antigenic determinants, and usually have lost one of the capsid polypeptides, VP4. Conversion of virus to the "A particle" appears to be initiated upon interaction with cell surface receptors (DeSena and Mandel, 1976, 1977; Guttman and Baltimore, 1977). The "A particles" have also been shown to exhibit increased hydrophobicity, which may aid their crossing a membrane barrier (Lonberg-Holm *et al.*, 1976). Further processing consistent with the loss of an additional capsid polypeptide, VP2, has been observed (DeSena and Mandel, 1976, 1977; Guttman and Baltimore, 1977; McGeady and Crowell, 1981).

Due to these and other observations, Crowell and Siak (1978) have proposed that interaction of picornaviruses with their receptors may be sufficient to initiate the uncoating steps illustrated in Fig. 10. While this could be initiated at the cell surface, it is generally agreed that the final uncoating steps culminating in the release of RNA occur after endocytosis. To date, alternative mechanisms of uncoating such as involvement of proteases have not been excluded (Boulanger and Lonberg-Holm, 1981; Crowell and Landau, 1983).

**b. Reoviruses.** Uncoating of reoviruses occurs by a distinctly different mechanism (for review see Silverstein *et al.*, 1976). After endocytosis, reovirus appears in phagosomes (endosomes?) and subsequently in lysosomes. Here the outer capsid is digested by proteases, leaving a subviral particle consisting of a protein shell and the double-stranded RNA genome. The inner protein shell is resistant to further proteolysis and, in addition, protects the genome from cleavage by nucleases (Chang and Zweerink, 1971). This proteolytic step can be mimicked *in vitro* by digestion with chymotrypsin, yielding subviral particles which are fully infectious (Joklik, 1972; Shatkin and LaFiandra, 1972). Proteolysis is absolutely essential to initiate the replicative cycle, since five of the six viral replicative enzymes are activated in the process (Shatkin and Sipe, 1968).

Since viral transcriptase is not active at the acidic pH of lysosomes, it is thought that the subviral particle must exit from the lysosome for replication to proceed (Silverstein *et al.*, 1976). Apparently the subviral particles are able to penetrate membranes directly (Borsa *et al.*, 1979), suggesting that once formed in the lysosome they may exit spontaneously. Further work is needed to substantiate this attractive possibility.

**c. Summary.** From these two examples it may be seen that the uncoating events of viruses with protein capsids may differ from each other in significant respects. It will be of interest to determine if they also exhibit distinct or common mechanisms of adsorptive endocytosis and to what degree the uptake of these viruses resembles that of enveloped viruses.

## VI. ROLE OF VIRAL RECEPTORS IN HOST RANGE AND TISSUE TROPISM

The role of viral receptors in determining the host range and tissue preference of a virus has been well documented in several instances. One of the main goals in virus receptor research has been to elucidate the basis for such selectivity. Of course, there are many factors which determine the host range of a virus in addition to receptor interactions. As mentioned previously, viruses may bind to cells but fail to penetrate the membrane barrier (Choppin and Scheid, 1980; Bosch *et al.*, 1979). Failure to penetrate the cell could in some cases be a receptor defect. However, abortive infections may also result from any stage of viral replication if the host cell machinery is not compatible with the synthesis and assembly of viral components. These other factors influencing viral virulence will be largely excluded from consideration in the following discussion.

Several lines of evidence have been used to document the role of receptors in determining the host range of a virus. These will be briefly reviewed later with examples.

## A. Correlation between Cellular Receptors and Susceptibility to Infection

Some of the earliest evidence for specific receptors mediating the host range of a virus came from analysis of picornavirus infection of tissue cultured cell lines by McLaren et al. (1959, 1960) and Holland et al. (1959a,b). A good correlation was found for the ability of cells to adsorb virus and to support viral replication. Moreover, cell lines resistant to infection by intact virus became "infected" and produced progeny virus upon exposure to the viral ribonucleic acids which were presumably taken up nonspecifically by the cells (Holland et al., 1959a,b; Wilson et al., 1977, 1979). Poliovirus was also shown to bind to minced or homogenized human or monkey brain and intestine tissues, which are naturally infected, and did not bind to most other tissues examined (Holland, 1961; Kunin and Jordan, 1961; Harter and Choppin, 1965). It was proposed that receptor specificity of enteroviruses was a major factor in determining tissue tropism (Holland, 1961).

Experiments demonstrating a correlation between the presence of receptors on host cells and susceptibility to infection have now been reported for many picornaviruses, and the results have been extensively reviewed (Lonberg-Holm and Philipson, 1974; Crowell et al., 1981; Crowell and Landau, 1983). Binding of radiolabeled virus to cells can be demonstrated to approach saturation, and the number of receptor sites for various virus cell combinations are estimated to be $10^3$–$10^6$ sites per cell (Crowell, 1966; Lonberg-Holm and Philipson, 1974). By testing the ability of one virus to block the binding of another, viruses can be grouped into receptor families. Although there are some cases in which viral adsorption to cells does not result in infection (Crowell and Landau, 1983), it is generally agreed that infection of a cell requires specific cell surface receptors and that the presence and absence of receptors on a cell is a major factor in its host range.

Another striking example of tissue tropism based on the presence of cell surface receptors is found with mammalian reoviruses. While reovirus type 1 infects ependymal cells of the brain causing ependymitis in humans and hydrocephalas in mice, type 3 infects neurons causing fatal encephalitis in mice (Margolis and Kilham, 1969; Weiner et al., 1977; Spriggs et al., 1983). The property of tissue tropism has been mapped to the viral hemagglutinin (Weiner et al.,

1977, 1980). Binding of reovirus to isolated human and murine ependymal cells has been examined using an immunofluorescence technique. Consistent with their known tissue tropisms, reovirus type 1 binds the ependymal cells, producing intense staining where negligible binding of reovirus type 3 is observed (Tardieu and Weiner, 1982).

Binding of other types of viruses to host cells has also been examined, but direct correlations between the presence and absence of receptors on cells susceptible and resistant to infection either have not been as extensively documented (e.g., adenoviruses; see Boulanger and Lonberg-Holm, 1981) or are simply not apparent. The host range of retroviruses is presumed to be largely due to the presence of cell surface receptors, as will be discussed later. Yet little difference could be detected in the binding of intact virus to susceptible and resistant cells (Piraino, 1967; Steck and Rubin, 1965). Nonspecific binding of virus may mask specific binding, since susceptible cells in some cases do appear to bind more virus than resistant cells (Notter *et al.*, 1982). However, for retroviruses and other virus groups, other approaches have also been successfully applied to demonstrate the roles of cellular receptors in host range.

## B. Infection of Resistant Cells through Alternate Receptors

### 1. *Retroviruses*

Avian retroviruses have been classified into major subgroups (A–E) based on their ability to infect genetically defined chicken embryo fibroblasts maintained in tissue culture. Cells are designated by the viral subgroups to which they are resistant. It appears that the resistance and susceptibility of the cells is determined primarily by the presence or absence of specific receptors on the cell surface (for review see Weiss, 1981). Compelling evidence that this is the case comes from experiments demonstrating infection of resistant cells by attachment of a virus through receptors other than those recognized by their own binding proteins. For example, inactivated Sendai virus has been found to mediate Rous sarcoma virus (RSV) infection of cells which are normally resistant (Weiss, 1969; Hanafusa *et al.*, 1970). Presumably, this is due to the binding of the Sendai virus HN protein to sialyloligosaccharides on the cell and on the glycoproteins of RSV thereby bridging RSV to the cell. The fusion protein (F) of Sendai virus may also aid in fusion of the RSV envelope with the plasma membrane, but this has not been established.

Overcoming resistance to infection has also been accomplished with pseudotype viruses produced by co-infection of cells with an envelope-defective RSV and chicken leukemia viruses. The pseudotype virions produced contain the genome of RSV and the viral attachment proteins of the leukemia virus. Cells normally resistant to RSV are infected by the pseudotype virions if they contain receptors for the leukemia virus attachment proteins (Rubin, 1965; Vogt and Ishizaki, 1965; Weiss, 1969). Pseudotype virions produced with viral attachment proteins of retroviruses and the genome of vesicular stomatitis virus (VSV), a rhabdovirus, have also been used to demonstrate the role of receptors in determining resistance. Although intact VSV replicates in most cells, the pseudotype virus will only infect cells which carry receptors recognized by the retrovirus attachment proteins (Zavada, 1972; Love and Weiss, 1974; Boettiger et al., 1975).

## 2. Epstein–Barr Virus

As discussed in Section III,B,2, the herpesvirus Epstein–Barr virus (EBV) has one of the most restricted host ranges of all of the animal viruses. It infects only B lymphocytes of humans and other primates. Although its receptor has not yet been identified (Section III,B,2), it is thought to be a B-cell specific protein or glycoprotein (Jønsson et al., 1982). The lack of EBV receptors on other cells appears to play a role in its limited host range.

Several reports have demonstrated the ability of EBV to infect other cell types when penetration of the membrane barrier is facilitated with Sendai virus glycoproteins. Co-reconstitution of EBV with detergent-solubilized HN and F proteins of Sendai virus yields a virus preparation that infects not only B-cell lines but a variety of other cell lines, including human and mouse T lymphocytes and mouse spleen cells (Shapiro et al., 1981). Similarly, pretreatment of receptor-negative cells with inactivated Sendai virus or reconstituted Sendai virus envelopes rendered the cells susceptible to infection by BV virus (Khelifa and Menezes, 1983). Pretreatment of EBV with sialidase inhibited its binding to the Sendai virus-treated cells, suggesting that binding was mediated through attachment of the Sendai virus HN protein to sialic acids on the EBV glycoproteins.

Implantation of EBV receptors into receptor-negative cells has also been reported (Volsky et al., 1980; Shapiro and Volsky, 1983). Membrane fragments from receptor-positive B-cell lines were reconstituted into liposomes containing Sendai virus glycoproteins and fused with receptor-negative cells. Subsequent infection by EBV virus was proposed to be due to interaction with B cell receptors. However, since Sendai virus glycoproteins alone can mediate EBV attachment to and penetration of receptor-negative cells (Khelifa and Menezes, 1983), an unequivocal assignment of the role of the B cell receptor in these experiments cannot be made. Indeed, EBV virus attachment to cells alone apparently is not sufficient for viral penetration. Attachment of EBV to receptor-negative cells with the plant lectin concanavalin A fails to result in infection (Khelifa and Menezes, 1982). As previously mentioned in Section V, the T-cell line Molt 4 has EBV receptors, but the virus does not penetrate (Menezes et al., 1977). In contrast, Molt 4 exposed to Sendai virus glycoproteins is infected

(Khelifa and Menezes, 1983; Volsky *et al.*, 1980). Thus, it appears that the fusion protein ($F_{1,2}$) of Sendai virus may be important to these EBV infections of receptor-negative cells. It will be important in the future to establish the relative roles of viral attachment and viral penetration as factors in restricting the host range of EBV.

An apparent paradox of EBV biology is the presence of EBV DNA and EBV nuclear antigen in nasopharyngeal carcinoma (Zur Hausen *et al.*, 1970; Nonoyama *et al.*, 1973; Klein *et al.*, 1974), since normal epithelium of the nasopharynx is not infected by EBV (Glazer *et al.*, 1980; Shapiro and Volsky, 1983). This is of particular interest in view of the potential role of EBV as the causative agent of this neoplasm. Based on the types of experiments described earlier, several proposals for the apparent contradiction have been put forth. Khelifa and Menezes (1983) have suggested that natural infections of the nasopharynx with other viruses may result in the appearance of viral attachment proteins in the membrane that could mediate binding of EBV analogous to that seen with the Sendai virus glycoproteins implanted in receptor-negative cells. Alternatively, Shapiro and Volsky (1983) have suggested that EBV-infected B lymphocytes may fuse (Bayliss and Wolf, 1980) with the nasopharyngeal epithelium, yielding either direct infection or implantation of the B cell receptors. Regardless of the mechanism, Shapiro and Volsky (1983) have shown that nasopharyngeal epithelium can be infected by EBV provided that the means for EBV attachment and penetration are provided.

### 3. SV40

In an interesting report, Loyter *et al.* (1983) have demonstrated SV40 infection of a resistant rat fibroblast cell line F1′ 1–4. Two approaches were used: in one, SV40 DNA was delivered to cells in liposomes containing Sendai virus glycoproteins. As many as 20% of the cells could be shown to be infected by this method, as judged by production of SV40 T antigen. In separate experiments, resistant cells were also implanted with receptors by fusion with liposomes prepared from Sendai virus glycoproteins and membrane fragments from receptor-positive cells. The resulting implanted cells could be infected by native SV40 virus. Infection presumably occurred through natural receptors of SV40, since no infection was detected in untreated cells or in cells which had been "mock implanted" with liposomes containing Sendai virus glycoproteins alone.

## C. Receptor Variants Influencing Host Adaptation and Host Range

In addition to the variation in receptor specificity observed between families of animal viruses or even different viruses within one family, there are a number of cases in which even very closely related viruses have been found to exhibit demonstrable differences in receptor-binding properties. To some extent, this has been discussed in Section IV. In some instances, such receptor variants have been fortuitously isolated in the laboratory. There are other examples, however, for which receptor variants have arisen by adaptation of a virus from one host to another. The following examples illustrate the ways in which receptor variants have been detected and to what extent the altered binding properties affect their interactions with host cells.

## 1. Picornaviruses

Passage of picornaviruses in various cell lines has been documented to result in the selection of atypical variants suited for growth in those cells (for review see Crowell and Landau, 1983). In most cases, the basis for selection has not been established. However, serial passage of nonhemagglutinating (Ha$^-$) coxsackie group B viruses in rhabdosarcoma cells results in the selection of a hemagglutinating (Ha$^+$) variant (Crowell and Landau, 1978, 1979, 1983). While the Ha$^+$ virus infects both HeLa cells and rhabdosarcoma cells, the Ha$^-$ strain infects only the former. It has been suggested that selection of the Ha$^+$ variant may be a receptor-mediated event (Crowell and Landau, 1983).

Coxsackie virus variants have been implicated in the etiology of virally induced juvenile-onset diabetes (Craighead, 1977, 1979; Notkins, 1977; Yoon et al., 1979). Juvenile-onset diabetes is characterized by a rapid onset following destruction of the pancreatic β islet cells, initiated by an environmental factor such as a viral infection and/or anti-β cell antibodies. In addition, a genetic factor predisposes individuals to the disease (Foster, 1983). In a murine model of the disease, serial passage of coxsackie virus in cultured pancreatic β cells results in increased β-cell tropism and the resultant onset of diabetes in the intact animal (Yoon et al., 1978). Although the factors which determine virally induced β-cell destruction are complex, it has been suggested that β-cell tropism may be due to receptor variants which bind to β- cell receptors with increased affinity (Chairez et al., 1978). In the human disease a well-documented case of isolation of coxsackie virus from the pancreas of a diabetic child has been reported (Yoon et al., 1979). The isolated virus induced diabetes in mice, suggesting a relationship between the causes of the human and murine diseases. At present, the possible role of viral receptors on pancreatic β- cells as a predisposing factor for juvenile-onset diabetes is not clear. Additional information concerning the nature of the host cell receptors of coxsackie viruses and their β-cell tropic variants will be required to fully evaluate these ideas.

## 2. Retroviruses

Genetically defined avian fibroblasts are designated by the retrovirus subgroups to which they are resistant. Thus, chicken (C) fibroblasts resistant to subgroup B and E retroviruses are designated C/BE (see Weiss, 1981). Tsichlis *et al.* (1980) have reported isolation of variants from subgroup B or D chicken avian viruses that have altered host range. While turkey fibroblasts T/BD are normally resistant to chicken subgroup B and D viruses, several B virus recombinants were found to infect T/BD fibroblasts. These unusual receptor variants were found to utilize the subgroup-B receptor on chicken (C/E) cells and, in addition, the subgroup-E receptor on turkey (T/BD) cells. Fingerprint analysis of the viral genome revealed a small alteration of the *env* gene coding for the envelope glycoprotein. Thus, it was proposed that the subgroup-B receptor on chicken cells is related in structure to the subgroup-E receptor on turkey cells, such that a small change in the viral attachment protein is sufficient to allow recognition of both receptors.

### 3. *Pancreatic Necrosis Virus*

Infectious pancreatic necrosis virus (IPNV) is considered a prototype of a new virus group. It has two double-stranded RNA segments and was originally isolated from trout (Dobos, 1976; Macdonald and Yamamoto, 1977). Two main host range variants are known; JV, which infects cultured fathead minnow (FHM) cells and chinook salmon embryo (CHSE) cells; and OV, which replicates only in CHSE cells. Darragh and Macdonald (1982) have demonstrated that the host range of these two variants is based on recognition of cell surface receptors. While both the JV and OV viruses rapidly adsorbed to FHM cells, only JV adsorbed to CHSE cells. Transcapsidation of the OV RNA into JV capsids resulted in "infection" of CSHE cells and synthesis of large amounts of OV capsid antigen, thus confirming that the inability of the OV variant to infect CHSE cells was its failure to adsorb to cell surface receptors.

### 4. *Myxoviruses*

As mentioned previously, influenza viruses cause disease in a large number of animal species. In general, viruses do not readily cross species barriers. However, transmission of influenza between mammalian species (i.e., human and swine) and between birds and lower mammals has been documented. Furthermore, reassortment of genes between viruses from different host species is believed to result in "new" influenza strains (Laver and Webster, 1979; Hinshaw and Webster, 1982). Because influenza viruses bind sialic acid-containing oligosaccharides and sialic acids are found as cell surface components of all host species, the presence or absence of receptors has not been considered a major factor in host range. Yet as described in Section IV,C, even closely related influenza virus strains can vary in their detailed receptor specificity. There are some indications that such variation in receptor may be important in the process adaptation of a virus from a natural host to a laboratory host or from one host species to another. Several examples of changes in receptor-binding properties observed during host adaptation are reviewed later.

Early studies by Burnet and Bull (1943) and by Stone (1951) documented a receptor shift upon adaptation of human influenza viruses (H1) to growth in chicken embryos. The original isolate (O virus) could only be grown in the amnion but after several passages could be propagated in the allantoic cavity as the derived or D virus. Concommitant with adaptation, the binding properties of the virus changed. Notably, the D virus agglutinated chicken and human erythrocytes with high titer while the O virus preferentially agglutinated human erythrocytes. The selection of the D variant in eggs was proposed to occur due to its more avid binding to receptors on fowl cells (Stone, 1951).

More recently, human influenza B viruses isolated and propagated by growth in MDCK cells have been found to undergo dramatic selection when adapted to growth in chicken embryos (Schild et al., 1983). Selection was shown to occur by the growth to predominance of a minor variant (about 2–3%) present in the MDCK-cell-grown virus. The hemagglutinin of the progeny virus from eggs had very different antigenic properties, as judged from a comparison of the MDCK-cell- and egg-grown viruses and their interactions with monoclonal antibodies. This was of considerable interest in view of the fact that influenza virus vaccine is produced from virus grown in eggs and that human-patient antisera reacted preferentially with the MDCK-cell-grown virus. Although the mechanism of selection is not known at present, one difference in receptor-binding properties of the egg- and MDCK-cell-grown viruses has been found. When examined for agglutination of erythrocytes of seven different species, the egg-grown viruses agglutinated the cells of all seven while the MDCK-cell-grown viruses failed to agglutinate those from sheep, horse, and rat. To account for the observation of nonimmune selection of antigenically distinct influenza B viruses in eggs and the differential agglutination of erythrocytes, it has been suggested that the receptor binding site and one of the antigenic sites of the viral hemagglutinin are structurally related (Schild et al., 1983). This would allow the potential for selection of antigenic variants based on selective pressures exerted at the level of viral adsorption to cell surface receptors.

Kilbourne (1978) described a case of genetic dimorphism in swine viruses whereby cloned viruses exhibited either high yield in eggs (H virus) and insensitivity to neutralization by antisera to A/sw/Cam/39 or low yield in eggs (L virus) and high sensitivity to neutralization by antibody. Genetic reassortment analysis revealed that the phenotypic properties of the two viruses were carried by the RNA segment coding for hemagglutinin. Although both the H and L forms of the virus are found in influenza virus isolates from swine, experimental infections of swine suggest that the H form is more virulent (Kilbourne et al., 1979, 1981). The primary structural difference between the H and L hemagglutinins appears to be a single amino acid change of glutamic acid (H) to glycine (L) at residue 155, which is located in a region near the receptor binding site (Kilbourne et al., 1983; Both et al., 1983). How this change is related to the difference in replication rates in various hosts and whether it is associated with quantifiable differences in receptor-binding properties remains to be determined.

Introduction of the human Hong Kong virus in the human population in 1968 was accompanied by a shift in antigenic specificity from the H2 to the H3 hemagglutinin, resulting in world-wide pandemic (Laver and Webster, 1979; Webster *et al.*, 1982). Comparison of the antigenic properties of the Hong Kong viruses with other human and animal influenza viruses revealed a close similarity between the H3 hemagglutinin and the hemagglutinins of avian (Hav7) and equine viruses (Heq2) isolated 5 years earlier in 1963 (Coleman *et al.*, 1968; Zakstelskaya *et al.*, 1969; Laver and Webster, 1973). The Hav7 and Heq2 hemagglutinins are now also classified as H3 hemagglutinins (World Health Organization, 1980). Analysis of the viral genome by cross-hybridization showed that seven of the eight RNA segments of the Hong Kong viruses were homologous to those of the previously circulating human strains (H2N2) and that gene segment 4 exhibited little homology to that of the human strain but 92% homology to an avian strain A/duck/Ukraine/1/63 (Scholtissek *et al.*, 1978). Confirmation of the close similarity of the human and avian virus hemagglutinins was provided by the complete amino acid sequences of A/Aichi/2/68 and A/duck/Ukraine/1/63 which showed 96% homology (Ward and Dolpheide, 1981a,b; Fang *et al.*, 1981). These observations strongly support the proposal (Scholtissek *et al.*, 1978; Laver and Webster, 1979) that the Hong Kong virus arose by genetic reassortment of the previously circulating human flu virus and an avian virus similar to A/duck/Ukraine/1/63 which contributed the hemagglutinin.

Despite the similarity in primary structure between the H3 hemagglutinins of human, avian, and equine viruses, fundamental differences in binding specificity were observed which correlated with species of origin (Rogers and Paulson, 1983; Rogers *et al.*, 1983a,b). As detailed in Section IV,C,3, human viruses exhibited preferential binding to the SAα2,6Gal linkage and were sensitive to inhibition of infection by the glycoprotein equine $\alpha_2$-macroglobulin, while avian and equine viruses preferentially bound the SAα2,3Gal linkage and were insensitive to inhibition by equine $\alpha_2$-macroglobulin. Sequence analysis of human H3 hemagglutinins and cloned variants with binding properties analogous to the avian and equine viruses suggested that the contrasting specificities are due in large part to a single amino acid in the receptor-binding pocket (Rogers *et al.*, 1983a). The presence of leucine at residue 226 was associated with binding specificity characteristic of the human viruses and glutamine at 226 with the specificity characteristic of avian viruses (see Fig. 2 in Section II,A,1).

If the H3 hemagglutinin of the human Hong Kong virus originated from an avian virus, why do the receptor specificities of viruses isolated from the two species differ so dramatically? Similar questions can be raised for the changes in receptor-binding properties exhibited in the other examples of host adaptation described earlier. One possibility is that sialyloligosaccharides present on extra-cellular and cell surface glycoproteins exert selective pressures that result in the growth to predominance of a receptor variant best suited for propagation in that host. It is well documented that sialyloligosaccharides of extracellular glycoproteins (Kornfeld and Kornfeld, 1980; Schauer, 1982; Yoshima *et al.*, 1981a) and membrane glycoproteins (Thomas and Winzler, 1969; Yoshima *et al.*, 1980; Murayama, 1983a) exhibit species differences (see also IV,C,2). Thus, selection could occur in one of two ways; a glycoprotein inhibitor of infection such as a mucin or other glycoprotein could selectively neutralize one of the receptor-binding types. The potential for this type of selection was demonstrated by Choppin and Tamm (1959, 1960a) in the selection of inhibition-sensitive variants from human viruses bearing the H2 hemagglutinin. Indeed, analogous selection using horse serum or equine $\alpha_2$-macroglobulin was used to select "avian type" receptor variants from human Hong Kong viruses (Rogers *et al.*, 1983a,b, 1984).

A second type of selection could be differential adsorption of receptor-specific variants to sialyloligosaccharides present on the host cell. This was also found to be a formal possibility when specific influenza viruses were tested for replication on MDCK cells enzymatically modified to contain either the $SA\alpha2,6Gal$ or $SA\alpha2,3Gal$ linkages (Carroll and Paulson, 1982, and unpublished observations). In general, the ability of a virus to infect derivatized tissue culture cells was largely determined by its receptor specificity, established using derivatized erythrocytes. Moreover, cells derivatized to contain the $SA\alpha2,3Gal$ linkage could preferentially select for a virus of the complimentary binding specificity when innoculated with a mixture of viruses exhibiting preferential binding to $SA\alpha2,3Gal$ or $SA\alpha2,6Gal$ linkages as receptor determinants.

As the foregoing discussion suggests, it is unlikely that the range of receptor specificities exhibited by influenza viruses are without consequence in the ability of a virus to replicate in various hosts. It will be of interest to determine to what degree such factors play a role in natural infections of influenza.


## VII. SUMMARY AND FUTURE PERSPECTIVES

It is evident from a basic consideration of the physical problem that viral infection is initiated by attachment of the virus to the host cell. It is now generally agreed, with few exceptions, that viral adsorption to cells is mediated by a viral attachment protein which binds to a specific cell surface receptor. The viral attachment protein has been identified for many animal viruses (Section II). Less is known about the cell surface receptors to which they bind (Section III). Nevertheless, it has been clearly established that receptor recognition is specific (Section IV) and can influence the host range of a virus (Section VI).

Details of the virus–cell interaction at the molecular level are beginning to emerge for myxoviruses and paramyxoviruses. This is due to major advances in the elucidation of the structure of the viral hemagglutinin, especially for influenza virus (Section II), to improved methods for analysis of receptor specificity (Section IV), and insights into the mechanisms of viral penetration of cellular membranes (Section V). Continued progress may be expected. For example, a more detailed look at how small changes in the binding pocket of influenza hemagglutinin influences receptor specificity is likely to be forthcoming. It may also be possible to determine the host-derived factors which appear to select receptor-specific variants upon host adaptation.

For other groups of viruses, increasing emphasis on elucidating the nature of the viral-receptor determinant may be warranted. Once this is known for one virus in a family, the basis for variation in receptor specificity can be explored in a systematic way. Adenoviruses, retroviruses, and togaviruses may be particularly well suited for further investigation since their binding proteins can be isolated in a form still able to bind cellular receptors (Sections II and III).

The value of elucidating the details of virus–host recognition probably goes far beyond understanding the host–parasite relationship. Indeed, the factors relevant to viral recognition, attachment, and penetration may be similar to those of the ligand–receptor and cell–cell interactions of higher organisms. To some extent this can already be seen. Thus, it can be anticipated that in the future concepts evolving from the field of virus receptors will merge with those from other types of receptor interactions represented in this treatise.

## REFERENCES

Ahmed, C. E., Sawyer, H. R., and Niswender, G. D. (1981). Internalization and degradation of human chorionic gonadotropin in ovine luteal cells: kinetic studies. *Endocrinology (Baltimore)* **109,** 1380–1387.

Akiyama, K., Schmidt, K., Haupt, H., Schwick, H. G., van Halbeek, H., and Vliegenthart, J. F. G. (1983). The structure of the *O*-glycosidic carbohydrate units of human plasma galactoprotein. *In* "Glycoconjugates" (M. A. Chester, D. Heinegärd, A. Lundblad, and S. Svensson, eds.), pp. 196. Rahms i Lund, Sweden.

Andrews, C. H., Bang, F. B., and Burnet, F. M. (1955). A short description of the myxovirus group (influenza and related viruses). *Virology* **1,** 176–181.

Angel, M. A., and Burness, A. T. H. (1977). The attachment of encephalomyocarditis virus to erythrocytes from several animal species. *Virology* **83,** 428–432.

Appleyard, G., Hapel, A., and Boulter, E. A. (1971). An antigenic difference between intracellular and extracellular rabbitpox virus. *J. Gen. Virol.* **13,** 9–17.

Bachi, T., Deas, J. E., and Howe, C. (1977). Virus–erythrocyte membrane interactions. *In* "Virus Infection and the Cell Surface" (G. Poste and G. L. Nicolson, eds.), pp. 83–128. Elsevier/North-Holland, New York.

Bachrach, H. L., Moore, D. M., McKercher, P. D., and Polatnick, J. (1975). Immune and antibody responses to an isolated capsid protein of foot-and-mouth disease virus. *J. Immunol.* **115,** 1636–1641.

Baenziger, J. O., and Fiete, D. (1982). Recycling of the hepatocyte asialoglycoprotein receptor does not require delivery of ligand to lysosomes. *J. Biol. Chem.* **257,** 6007–6009.

Baron, C. B., and Blough, H. A. (1983). Binding of influenza virus to a reconstituted receptor complex containing glycophorin. *Intervirology* **19,** 33–43.

Basilico, C., and DiMayorca, G. (1974). Mutant of polyoma virus with impaired adsorption to BHK cells. *J. Virol.* **13,** 931–934.

Bayliss, G. J., and Wolf, H. (1980). Epstein–Barr virus-induced cell fusion. *Nature (London)* **287,** 164–165.

Bergelson, L. D., Bukrinskaya, A. G., Provkazova, N. V., Shaposhnikova, G. I., Kocharov, S. L., Shevchenko, V. P., Kornilaeva, V., and Fomina-Ageeva, E. V. (1982). Role of gangliosides in reception of influenza virus. *Eur. J. Biochem.* **128,** 467–474.

Berger, E. G., Buddecke, E., Kamerling, J. P., Kobata, A., Paulson, J. C., and Vliegenthart, J. F. G. (1982). Structure, biosynthesis and functions of blycoprotein glycans. *Experientia* **38,** 1129–1258.

Beyer, T. A., Sadler, J. E., Rearick, J. I., Paulson, J. C., and Hill, R. L. (1981). Glycosyltransferases and their use in assessing oligosaccharide structure and structure–function relationships. *Adv. Enzymol.* **52,** 23–176. *Relat. Areas Mol. Biol.*

Bishayee, S., Strand, M., and August, J. T. (1978). Cellular membrane receptors for oncovirus envelope glycoprotein: Properties of the binding reaction and influence of different reagents on the substrate and the receptors. *Arch. Biochem. Biophys.* **189,** 161–171.

Blanchard, D., Carton, J.-P., Fournet, B., Montreuil, J., van Halbeek, H., and Vliegenthart, J. F. G. (1983). Primary structure of the oligosaccharide determinant of blood group Cad specificity. *J. Biol. Chem.* **258,** 7691–7695.

Blix, G., Gottschalk, A., and Klenk, E. (1957). Proposed nomenclature in the field of neuraminic and sialic acids. *Nature (London)* **179,** 1088–1088.

Block, J. (1981). Sequence variation at the 3′ ends of neuraminidase gene segements within and among the different NA subtypes. *In* "Genetic Variation Among Influenza Viruses" (D. P. Nayak, ed.), pp. 45–54. Academic Press, New York.

Blomberg, J. (1979). Studies on the attachment of Herpes simplex virus. Effect of trypsin. *Arch. Virol.* **61,** 201–206.

Boettiger, D., Love, D. N., and Weiss, R. A. (1975). Virus envelope markers in mammalian tropism of Avian RNA Tumor Viruses. *J. Virol.* **15,** 108–114.

Bogoch, S. (1957). Inhibition of viral hemagglutination by brain ganglioside. *Virology* **4,** 458–466.

Bogoch, S., Lynch, P., and Levine, A. S. (1959). Influence of brain ganglioside upon the neurotoxic effect of influenza virus in mouse brain. *Virology* **7,** 161–169.

Bolen, J. B., and Consigli, R. A. (1979). Differential adsorption of polyoma virions and capsids to mouse kidney cells and guinea pig erythrocytes. *J. Virol.* **32,** 679–683.

Bolen, J. B., and Consigli, R. A. (1980). Separation of neutralizing and hemagglutination-inhibiting antibody activities and specificity of antisera to sodium dodecyl sulfate derived polypeptides of polyoma virons. *J. Virol.* **34,** 119–129.

Borsa, J., Morash, B. D., Sargent, M. D., Copps, T. P., Lievaart, P. A., and Szekely, J. G. (1979). Two modes of reovirus entry into L cells. *J. Virol.* **34,** 119–129.

Bosch, F. X., Orlich, M., Klenk, H.-D., and Rott, R. (1979). The structure of the hemagglutinin, a determinant for the pathogenicity of influenza viruses. *Virology* **95,** 197–208.

Both, G. M., Shi, C. H., and Kilbourne, E. D. (1983). Hemagglutinin of swine influenza virus: A single amino acid change pleiotropically affects viral antigenicity and replication. *Proc. Natl. Acad. Sci. U.S.A.* **80,** 6996–7000.

Boulanger, P., and Lonberg-Holm, K. (1981). Componants of non-eveloped viruses which recognize receptors. *In* "Virus Receptors, Part 2: Animal Viruses" (K. Lonberg-Holm and L. Philipson, eds.), pp. 23–46. Chapman & Hall, London.

Boulter, E. A. (1969). Protection against poxviruses. *Proc. R. Soc. Med.* **62**, 295–297.

Breindl, M. (1971). VP4, the D reactive part of poliovirus. *Virology* **46**, 962–964.

Brown, F., and Small, C. J. (1970). Demonstration of three specific sites on the surface of foot-and-mouth-disease virus by antibody complexing. *J. Gen. Virol.* **7**, 115–127.

Brown, M. S., and Goldstein, J. L. (1976). Analysis of a mutant strain of human fibroblasts with a defect in the internalization of receptor-bound low density lipoprotein. *Cell* **9**, 663–674.

Brown, M. S., and Goldstein, J. L. (1979). Receptor-mediated endocytosis: insights from the lipoprotein system. *Proc. Natl. Acad. Sci. U.S.A.* **76**, 3330–3337.

Bubber, J. E., Chen, S., and Lilly, F. (1978). Nonrandom inclusion of H-2K and H-2D antigens in Friend virus particles from mice of various strains. *J. Exp. Med.* **147**, 340–351.

Bukrinskaya, A. G., Kornilaeva, G. V., Vorkunova, N. K., Timofeeva, N. G., Shaposhnikova, G. I., and Bergelson, L. G. (1982). Gangliosides, specific receptors for influenza virus. *Vopr. Virusol.* **27**, 661–666.

Burness, A. T. H. (1981). Glycophorin and sialylated components as receptors for viruses. *In* "Virus Receptors, Part 2: Animal Viruses" (K. Lonberg-Holm and L. Philipson, eds.), pp. 65–84, Chapman & Hall, London.

Burness, A. T. H., and Pardoe, I. U. (1981). Effect of enzymes on the attachment of influenza and encephalomyocarditis viruses to erythrocytes. *J. Gen. Virol.* **55**, 275–288.

Burnet, F. M. (1948). Mucins and mucoids in relation to influenza virus action IV. Inhibition by purified mucoid of infection and heamaggutination with the virus strain WSW. *Aust. J. Exp. Biol. Med. Sci.* **26**, 381–411.

Burnet, F. M., and Bull, D. R. (1943). Changes in influenza virus associated with adaptation to passage in chick embryos. *Aust. J. Exp. Biol. Med. Sci.* **21**, 55–69.

Burnet, F. M., and Stone, J. D. (1947). The receptor destroying enzyme of *V. cholerae*. *Aust. J. Exp. Biol. Med. Sci.* **25**, 227–233.

Burroughs, J. N., Rowlands, D. J., Sanger, D. V., Talbot, P., and Brown, F. (1971). Further evidence for multiple proteins in the foot-and-mouth disease virus particle. *J. Gen. Virol.* **13**, 73–84.

Cahan, L. D., and Paulson, J. C. (1980). Polyoma virus adsorbs to specific sialyloligosaccharides on erythrocytes. *Virology* **103**, 505–509.

Cahan, L. D., Singh, R., and Paulson, J. C. (1983). Sialyloligosaccharide receptors of binding variants of polyoma virus. *Virology* **130**, 281–289.

Carroll, S. M., and Paulson, J. C. (1982). Infection of host cells by influenza virus requires specific cell surface receptor determinants. *Fed. Proc., Fed. Am. Soc. Exp. Biol.* **41**, 3686. (Abstr.)

Carroll, S. M., Higa, H. H., and Paulson, J. C. (1981a). Different cell surface receptor determinants of antigenically similar influenza virus hemagglutinins. *J. Biol. Chem.* **256**, 8357–8363.

Carroll, S. M., Higa, H. H., Cahan, L. D., and Paulson, J. C. (1981b). Different sialyloligosaccharide receptor determinants of antigenically related influenza virus hemagglutinins. *In* "Genetic Variation Among Influenza Viruses" (D. Nayak and C. F. Fox, eds.), pp. 415–421. Academic Press, New York.

Caspar, D. L. D., and Klug, A. (1962). Physical principles in the construction of viruses. *Cold Spring Harbor Symp. Quant. Biol.* **27**, 1–24.

Caton, A. J., Brownlee, G. G., Yewdell, J. W., and Gerhard, W. (1982). The antigenic structure of influenza virus A/PR/8/34 hemagglutinin (H1 subtype). *Cell* **31**, 417–427.

Cavanagh, D., Sangar, D. V., Rowlands, D. J., and Brown, F. (1977). Immunogenic and cell attachment sites of FMDV: further evidence for their localization in a single capsid polypeptide. *J. Gen. Virol.* **35**, 149–158.

Chairez, R., Yoon, J. W., and Notkins, A. L. (1978). Virus-induced diabetes mellitus. X. Attachment of encephalomyocarditis virus and and permissiveness of cultured pancreatic beta cells to infection. *Virology* **85**, 606–611.

Chang, C.-T., and Zweerink, H. J. (1971). Fate of parental reovirus in infected cell. *Virology* **46**, 544–555.

Choppin, P. W., and Compans, R. W. (1975). Reproduction of paramyxoviruses. *In* "Comprehensive Virology" (H. Fraenkel-Conrat and R. R. Wagner, eds.), Vol. 4, pp. 95–178. Plenum, New York.

Choppin, P. W., and Scheid, A. (1980). The role of glycoproteins in adsorption penetration and pathogenicity of viruses. *Rev. Infect. Dis.* **2**, 40–61.

Choppin, P. W., and Tamm, I. (1959). Two kinds of particles with contrasting properties in influenza A virus strains from the 1957 pandemic. *Virology* **8**, 539–542.

Choppin, P. W., and Tamm, I. (1960a). Studies of two kinds of virus particles which comprise influenza A2 virus strains. I. Characterization of stable homogeneous substrains in reactions with specific antibody, mucoprotein inhibitors and erythrocytes. *J. Exp. Med.* **112**, 895–920.

Choppin, P. W., and Tamm, I. (1960b). Studies of two kinds of virus particles which comprise influenza A2 virus strains. II. Reactivity with virus inhibitors in normal sera. *J. Exp. Med.* **112**, 921–944.

Choppin, J., Schaffar-Deshayes, L., Debré, P., and Lévy, J.-P. (1981). Lymphoid cell surface receptor for Moloney Leukemia virus envelope glycoprotein gp-71. I. Binding characteristics. *J. Immunol.* **126**, 2347–2351.

Ciechanover, A., Schwartz, A. L., and Lodish, H. F. (1983). The asialoglycoprotein receptor internalizes and recycles independently of the transferin and insulin receptors. *Cell* **32**, 267–275.

Clavell, L. A., and Bratt, M. A. (1972). Hemolytic interaction of Newcastle disease virus and chicken erythrocytes. *Appl. Microbiol.* **23**, 461–470.

Cohen, A. (1963). Mechanisms of cell infection I. Virus attachment and penetration. *In* "Mechanisms of Virus Infection" (W. Smith, ed.), pp. 152–190. Academic Press, New York.

Cohen, A., and Belyavin, G. (1959). Hemagglutination inhibition of Asian influenza viruses: A new pattern of response. *Virology* **7**, 59–79.

Cohen, A., and Biddle, F. (1960). The effect of passage in different hosts on the inhibitor sensitivity of an Asian influenza virus strain. *Virology* **11**, 458–472.

Cole, S. T., Chen-Schmeisser, U., Hindennach, I., and Henning, V. (1983). Apparent bacteriophage-binding region of an *Escherichia coli* K-12 outer membrane protein. *J. Bacteriol.* **153**, 581–587.

Coleman, M. T., Dowdle, W. R., Pereira, H. G., Schild, G. C., and Chang, W. K. (1968). The Hong Kong/68 influenza A2 variant. *Lancet* **ii**, 1384–1386.

Colman, P. M., Varghese, J. N., and Laver, W. G. (1983). Structure of the catalytic and antigenic sites in influenza virus neuraminidase. *Nature (London)* **303**, 41–44.

Compans, R. W., and Choppin, P. W. (1975). Reproduction of myxoviruses. *In* "Comprehensive Virology" (H. Frankel-Conrat and R. R. Wagner, eds.), Vol 5, pp. 179–253. Plenum, New York.

Compans, R. W., and Pinter, A. (1975). Incorporation of Sulfate into Influenza Virus Glycoproteins. *Virology* **66**, 151–160.

Compans, R. W., Klenk, H.-D., Caliguisi, L. A., and Choppin, P. W. (1970). Influenza virus proteins. I. Analysis of polypeptides of the viron and identification of spike glycoproteins. *Virology* **42**, 880–889.

Compans, R. W., Roth, M. G., and Alonso, F. V. (1981). Directional transport of viral glycoproteins in polarized epithelial cells. *In* "Genetic Variation Among Influenza Viruses" (D. P. Nayak, ed.), pp. 213–231. Academic Press, New York.

Craighead, J. E. (1977). Viral diabetes. *In* "The Diabetic Pancreas" (B. W. Volk and K. F. Weltmann, eds.), pp. 467–488. Plenum, New York.

Craighead, J. E. (1979). Does insulin dependent diabetes have a viral etiology? *Hum. Pathol.* **10**, 267–278.

Crowell, R. L. (1966). Specific cell-surface alteration by enteroviruses as reflected by viral-attachment interference. *J. Bacteriol.* **91,** 198–204.

Crowell, R. L., and Landau, B. J. (1978). Picornaviridae: Enterovirus-coxsackievirus. *In* "CRC Handbook in Clinical Laboratory Science" (G. D. Hsiung and R. Green, eds.), Vol. 1, pp. 131–155. CRC Press, Boca Raton, Florida.

Crowell, R. L., and Landau, B. J. (1979). Receptors as determinants of cellular tropism in picornavirus infections. *In* "Receptors and Human Diseases" (A. G. Bearn and P. W. Choppin, eds.), pp. 1–33. Josiah Macy, Jr. Found., New York.

Crowell, R. L., and Landau, B. J. (1983). Receptors in the initiation of picornavirus infections. *In* "Comprehensive Virology" (H. Fraenkel-Conrat and R. R. Wagner, eds.), Vol. 18, pp. 1–42. Plenum, New York.

Crowell, R. L., and Siak, J.-S. (1978). Receptor for group B coxsackieviruses: Characterization and extraction from HeLa cell membranes. *Perspect. Virol.* **10,** 39–53.

Crowell, R. L., Landau, B. J., and Siak, J.-S. (1981). Picornavirus receptors in pathogenesis. *In* "Virus Receptors, Part 2: Animal Viruses" (K. Lonberg-Holm and L. Philipson, eds.), pp. 169–184. Chapman & Hall, London.

Dales, S. (1973). Early events in cell. Animal virus interaction. *Bacteriol. Rev.* **37,** 103–135.

Dales, S., Stern, W., Weintraub, S. B., and Huima, T. (1976). Genetically controlled surface modifications by poxviruses influencing cell–cell and cell–virus interactions. *In* "Cell Membrane Receptors for Viruses Antigens and Antibodies, Polypeptide Hormones and Small Molecules" (R. F. Beers, Jr. and E. G. Bassett, eds.), pp. 253–270. Raven, New York.

Dalrymple, J. M., Schlesinger, S., and Russel, P. K. (1976). Antigenic characterization of two Sindbis envelope glycoproteins separately by isoelectric focusing. *Virology* **69,** 93–103.

Darragh, A. E., and Macdonald, R. D. (1982). A host range restriction in infections pancreatic necrosis virus maps to the large RNA segment and involves virus attachment to the cell surface. *Virology* **123,** 264–272.

Datta, D. B., Arden, B., and Henning, U. (1977). Major proteins of the *Escherichia coli* outer envelope membrane as bacteriophage receptors. *J. Bacteriol.* **131,** 821–829.

Daughaday, C. C., Brandt, W. E., McCown, J. M., and Russell, P. K. (1981). Evidence for two mechanisms of dengue virus infection of adherent human monocytes: Trypsin sensitive virus receptors and trypsin resistant immune complex receptors. *Infect. Immunol.* **32,** 469–473.

DeGiuli, C., Kawai, S., Dales, S., and Hanafusa, H. (1975). Absence of surface projections on some non-infectious forms of RSV. *Virology* **66,** 253–260.

DeLarco, J., and Todaro, G. J. (1976). Membrane receptor for murine leukemia viruses: characterization using purified envelope glycoprotein gp71. *Cell* **8,** 365–371.

Della-Porta, A. J., and Westaway, E. G. (1977). A multi-hit model for the neutralization of animal viruses. *J. Gen. Virol.* **38,** 1–19.

DeSena, J., and Mandel, B. (1976). Studies on the *in vitro* uncoating of poliovirus. I. Characterization of the modifying factor and the modifying reaction. *Virology* **70,** 470–483.

DeSena, J., and Mandel, B. (1977). Studies on the *in vitro* uncoating of poliovirus. II. Characterization of the membrane modified particle. *Virology* **70,** 544–566.

Diamond, L., and Crawford, L. V. (1964). Some characteristics of large plaque and small plaque lines of polyoma virus. *Virology* **22,** 235–244.

Dimmock, N. J. (1971). Dependence of the activity of an influenza virus neuraminidase upon $Ca^{++}$. *J. Gen. Virol.* **13,** 481–483.

Dimmock, N. J. (1982). Initial stages in infection with animal viruses. *J. Gen. Virol.* **59,** 1–22.

Dobos, P. (1976). Size and structure of the genome of infections pancreatic necrosis virus. *Nucleic Acids Res.* **3,** 1903–1924.

Drickamer, K. L. (1978). Orientation of the Band 3 polypeptide from human erythrocyte membranes: Identification of the $NH_2$-terminal sequence and site of carbohydrate attachment. *J. Biol. Chem.* **253**, 7242–7248.

Drzeniek, R. (1972). Substrate specificities of neuraminidases. *Histochem. J.* **5**, 271–290.

Dunn, W. A., Hubbard, A. L., and Aronson, N. N. (1980). Low temperature selectively inhibits fusion between pinocytic vesicles and lysosomes during heterophagy of $^{125}I$-asialofetuin by the perfused rat liver. *J. Biol. Chem.* **255**, 5971–5978.

Dunnebacke, T. H., Levinthal, J. D., and Williams, R. C. (1969). Entry and release of poliovirus as observed by electron microscopy of cultured cells. *J. Virol.* **4**, 505–513.

Dunphy, W., Fries, E., Urbani, L., and Rothman, J. (1981). Early and late functions associated with the Golgi apparatus reside in distinct compartments. *Proc. Natl. Acad. Sci. U.S.A.* **78**, 7453–7457.

Eddy, B. E. (1969). "Polyoma Virus," Virology Monographs, Vol. 7. Springer-Verlag, Berlin and New York.

Eddy, B. E., Rowe, W. P., Hartley, J. W., Stewart, S. E., and Huebner, R. J. (1958). Hemagglutination with the SE Polyoma Virus. *Virology* **6**, 290–291.

Enegren, B. J., and Burness, A. T. H. (1977). Chemical structure of attachment sites for viruses on human erythrocytes. *Nature (London)* **268**, 536–537.

Epstein, M., and Achong, B. (1979). "The Epstein–Barr Virus." Springer-Verlag, Berlin and New York.

Everitt, E., Sundquist, B., and Philipson, L. (1971). Mechanism of the arginine requirement for adenovirus synthesis. *J. Virol.* **8**, 742–753.

Everitt, E., Lutter, L., and Philipson, L. (1975). Structural proteins of adenovirus. *Virology* **67**, 197–208.

Fan, D. P., and Sefton, B. M. (1978). The entry into host cells of Sindbic virus, vesicular stomatitis virus and Sendai virus. *Cell* **15**, 985–992.

Fang, R., Min Jou, W., Huylebroeck, D., Devos, R., and Fiers, W. (1981). Complete structure of A/duck/Ukraine/1/63 influenza hemagglutinin gene: Animal virus as progenitor of human H3 hong Kong 1968 influenza hemagglutinin. *Cell* **25**, 315–323.

Fazekas de St. Groth, S. (1948a). Destruction of influenza virus receptors in the mouse lung by an enzyme from *V. cholerae. Aust. J. Exp. Biol. Med.* **26**, 29–36.

Fazekas de St. Groth, S. (1948b). Regeneration of virus receptors in mouse lungs after artificial destruction. *Aust. J. Exp. Biol. Med.* **26**, 271–285.

Fenger, T. W., and Howe, C. (1979). Isolation and characterization of erythrocyte receptors for measles virus (4066a). *Proc. Soc. Exp. Biol. Med.* **162**, 299–303.

Fidgen, K. (1975). The action of *Vibrio cholerae* and *Corynebacterium diphtheriae* neuraminidases on the Sendai virus receptor of human erythrocytes. *J. Gen. Microbiol.* **89**, 48–56.

Finch, J. T. (1974). The surface structure of polyoma virus. *J. Gen. Virol.* **24**, 359–364.

Finch, J. T., and Crawford, L. (1975). Structure of small DNA-containing animal viruses. *In* "Comprehensive Virology" (H. Fraenkel-Conrat and R. R. Wagner, eds.), Vol. 5, pp. 119–154. Plenum, New York.

Finne, J. (1982). Occurrence of unique polysialosyl carbohydrate units in glycoproteins of developing brain. *J. Biol. Chem.* **257**, 11966–11970.

Finne, J., Krusius, T., Rauvala, H., Kekomäki, R., and Myllylä, G. (1978). Alkalai-stable blood grown A- and B-active polyglycosyl-peptides from human erythrocyte membrane. *FEBS LETT.* **89**, 111–115.

Foster, D. W. (1983). Diabetes mellitus. *In* "Metabolic Basis of Inherited Disease" (J. B. Stanbury, J. B., Wyngaarden, D. S. Frederickson, J. L. Goldstein, and M. S. Brown, eds.), pp. 99–117. McGraw-Hill, New York.

Frearson, D. M., and Crawford, L. V. (1972). Polyoma virus basic proteins. *J. Gen. Virol.* **14**, 141–155.

Fried, H., Cahan, L. D., and Paulson, J. C. (1981). Polyoma virus recognizes specific sialyloligosaccharide receptors on host cells. *Virology* **109**, 188–192.

Freidman, T., and David, D. (1972). Structural roles of polyoma virus proteins. *J. Virol.* **10**, 776–782.

Fukuda, K., Tomita, M., and Hamada, A. (1981). Isolation and structural studies of the neutral oligosaccharide units from bovine glycophorin. *Biochim. Biophys. Acta* **677**, 462–470.

Furthmayr, H. (1981). Glycophorin A: a model membrane glycoprotein. *Biol. Carbohydr.* **1**, 123–198.

Gething, M. J., White, J. M., and Waterfield, M. D. (1978). Purification of the fusion protein of Sendai virus: analysis of the $NH_2$-terminal sequence generated during precursor activation. *Proc. Natl. Acad. Sci. U.S.A.* **75**, 2737–2740.

Geuze, H. J., Slot, J. W., Strous, G. J. A. M., Lodish, H. F., and Schwartz, A. L. (1983). Intracellular site of asialo-glycoprotein receptor-ligand uncoupling: Double-label immunoelectron microscopy during receptor mediated endocytosis. *Cell* **32**, 277–287.

Ghendon, Y., Tuckova, E., Vonka, V., Klimov, A., Ginzberg, U., and Markushin, S. (1979). Replication of two influenza virus strains and a recombinant in HEF and LEP cells. *J. Gen. Virol.* **44**, 179–186.

Gibson, R., Schlesinger, S., and Kornfeld, S. (1979). The non-glycosylated glycoprotein of VSV is temperature sensitive and undergoes intracellular aggregation at elevated temperatures. *J. Biol. Chem.* **254**, 3600–3607.

Glazer, R., Lang, C. M., Lee, K. J., Schuller, D. E., Yacobs, D., and McQuettie, C. (1980). Attempt to infect nonmalignant nasopharyngeal cells from humans and squirrel monkeys with EBV. *J. Natl. Cancer Inst.* **64**, 1085–1090.

Goldstein, E. A., and Pons, M. W. (1970). The effect of polyvinylsulfate on the ribonucleoprotein of influenza virus. *Virology* **41**, 382–384.

Goldstein, J. L., Brown, M. S., and Stone, N. J. (1977). Genetics of the LDL receptor, evidence that the mutations affecting binding and internalization are allelic. *Cell* **12**, 629–641.

Goldstein, J. L., Anderson, R. G. W., and Brown, M. S. (1979). Coated pits, coated vesicles, and receptor mediated endocytosis. *Nature (London)* **279**, 679–685.

Gomatos, P. J., and Tamm, I. (1962). Reactive sites of reovirus type 3 and their interaction with receptor substances. *Virology* **17**, 455–461.

Gottschalk, A. (1966). Interaction between glycoproteins and viruses. *In* "The Amino Sugars" (R. W. Jeanloz and E. A. Balazs, eds.), Vol. 2B, pp. 337–359. Academic Press, New York.

Gottschalk, A., and Lind, P. E. (1949). Product of interaction between influenza virus enzyme and ovomucin. *Nature (London)* **164**, 232–233.

Guilden, R. V. (1980). Biology of RNA tumor viruses. *In* "The Molecular Biology of Animal Viruses" (D. P. Nayak, ed.), pp. 475–542. Decker, New York.

Guttman, N., and Baltimore, D. (1977). A plasma membrane componant able to bind and alter virons of poliovirus type 1: Studies on cell free alteration using a simplified assay. *Virology* **82**, 25–36.

Hanafusa, H., Miyamoto, T., and Hanafusa, T. (1970). A cell-associated factor essential for formation of an infectious form of Rous sarcoma virus. *Proc. Natl. Acad. Sci. U.S.A.* **66**, 314–321.

Hansson, G. C., Karlsson, K.-A., Larson, G., Strömberg, N., Thurin, J., Örvell, C., and Norrby, E. (1984). A novel approach to the study of glycolipid receptors for viruses, binding of Sendai virus to thin-layer chromatograms. *FEBS Lett.* **170**, 15–18.

Harter, D. H., and Choppin, P. W. (1965). Adsorption of attenuated and neuovirulent poliovirus strains to central nervous system tissues of primates. *J. Immunol.* **95**, 730–736.

Hartley, J. W., Rowe, W. P., Chanock, R. M., and Andrews, B. E. (1959). Studies of mouse polyoma virus infection. IV. Evidence for mucoprotein erythrocyte receptors in polyoma virus hemagglutination. *J. Exp. Med.* **110**, 81–91.

Hascall, V. (1981). Proteoglycans: structure and function. *Biol. Carbohydr.* **1**, 1–50.

Hay, A. J. (1974). Studies on the formation of the influenza virus envelope. *Virology* **60**, 398–418.

Haywood, A. M. (1974a). Characteristics of Sendai virus receptors in a model membrane. *J. Mol. Biol.* **83**, 427–436.

Haywood, A. M. (1974b). Fusion of Sendai viruses with model membranes. *J. Mol. Biol.* **87**, 625–628.

Haywood, A. M., and Boyer, B. P. (1981). Initiation of fusion and disassembly of Sendai virus membranes into liposomes. *Biochim. Biophys. Acta* **646**, 31–35.

Haywood, A. M., and Boyer, B. P. (1982). Sendai virus membrane fusion: Time course and effect of temperature pH, calcium and receptor concentration. *Biochemistry* **21**, 6041–6045.

Helenius, A., Morein, B., Fries, E., Simons, K., Robinson, P., Schirrmacher, V., Terhorst, C., and Strominger, J. L. (1978). Human (HLA-A and HLA-B) and murine (H-2K and H-2D) histocompatibility antigens are cell surface receptors for Semliki Forest virus. *Proc. Natl. Acad. Sci. U.S.A.* **75**, 3846–3850.

Helenius, A., Kartenbeck, J., Simons, K., and Fries, E. (1980a). On the entry of Semliki Forest virus into BHK cells. *J. Cell Biol.* **84**, 404–420.

Helenius, A., Marsh, M., and White, J. (1980b). The entry of viruses into animal cells. *Trends Biochem. Sci.* **5**, 104–106.

Hennache, B., and Boulanger, P. (1977). Biochemical study of KB-cell receptor for adenovirus. *Biochem. J.* **166**, 237–247.

Hinshaw, V. S., and Webster, R. G. (1982). The natural history of influenza A viruses. *In* "Basic and Applied Influenza Research" (A. S. Bear, ed.), pp. 79–104. CRC Press, Boca Raton, Florida.

Hirst, G. K. (1941). Agglutination of red cells by allantoic fluid of chick embryos infected with influenza virus. *Science (Washington, D.C.)* **94**, 22–23.

Hirst, G. K. (1942a). The quantitative determination of influenza virus and antibodies by means of red cell agglutinations. *J. Exp. Med.* **75**, 47–64.

Hirst, G. K. (1942b). Adsorption of influenza virus hemagglutinins and virus by red blood cells. *J. Exp. Med.* **76**, 195–209.

Hirst, G. K. (1950). The relationship of the receptors of a new strain of virus to those of the mumps-NDV-influenza groups. *J. Exp. Med.* **91**, 177–184.

Hirst, G. K. (1959). Virus–host cell relation. *In* "Viral and Rickettsial Infections of Man" (T. M. Rivers and F. L. Horsfall, eds.), pp. 96–144. Lippincott, Philadelphia, Pennsylvania.

Holland, J. J. (1961). Receptor affinities as major determinants of enterovirus tissue tropism in humans. *Virology* **15**, 312–326.

Holland, J. J., McLaren, L. C., and Syverton, J. T. (1959a). Mammalian cell–virus relationship. III. Poliovirus production by non-primate cells exposed to poliovirus ribonucleic acid. *Proc. Soc. Exp. Biol. Med.* **100**, 843–855.

Holland, J. J., McLaren, L. C., and Syverton, J. T. (1959b). The mammalian cell–virus relationship. IV. Infection of naturally insusceptible cells with enterovirus ribonucleic acid. *J. Exp. Med.* **110**, 65–80.

Holmgren, J., Svennerholm, L., Elwing, H., Fredman, P., and Strannegard, Ö. (1980). Sendai virus receptor: Proposed recognition structure based on binding to plastic-adsorbed gangliosides. *Proc. Natl. Acad. Sci. U.S.A.* **77**, 5693–5697.

Holowczak, J. A. (1972). Glycopeptides of vaccina virus I. preliminary characterization and hexosamine content. *Virology* **42**, 87–99.

Homma, K., Tomita, M., and Hamada, A. (1980). Amino acid sequence and attachment sites of oligosaccharide units of porcine erythrocyte glycophorin. *J. Biochem. (Tokyo)* **88**, 1679–1691.

Homma, M., and Ohuchi, M. (1973). Trypsin action on the growth of Sendai virus in tissue culture cells. III Structural difference of Sendai viruses grown in eggs and tissue culture cells. *J. Virol.* **12**, 1457–1465.

Homma, R. (1968). Reaction of Japanese encephalitis virus with mannan. *Acta Virol. (Engl. Ed.)* **12**, 385–396.

Hosaka, Y., and Shimizu, Y. K. (1972a). Artificial assembly of envelope particles of HVJ (Sendai virus). I. Assembly of hemolytic and fusion factors from envelopes solubilized by nonidel P40. *Virology* **49**, 627–639.

Hosaka, Y., and Shimizu, Y. K. (1972b). Artificial assembly of envelope particles of HVJ (Sendai virus). II. Lipid components for formation of this active hemolysin. *Virology* **49**, 640–646.

Howe, C., and Lee, L. T. (1972). Virus–erythrocyte interactions. *Adv. Virus Res.* **17**, 1–50.

Howe, C., Coward, J. E., and Fenger, T. W. (1980). Viral invasion: morphological, biochemical and biophysical aspects. *In* "Comprehensive Virology" (H. Fraenkel-Conrat and R. R. Wagner, eds.), Vol. 16, pp. 1–71. Plenum, New York.

Hoyle, L. (1968). "Influenza Viruses," Virology Monographs, Vol. 4. Springer-Verlag, Berlin and New York.

Hsu, M.-C., Scheid, A., and Choppin, P. W. (1979). Reconstitution of membranes with individual paramyxovirus glycoproteins and phospholipid in cholate solution. *Virology* **95**, 476–491.

Hsu, M., Scheid, A., and Choppin, P. W. (1981). Activation of the Sendai virus fusion protein (F) involves a conformational change with exposure of a new hydrophobic region. *J. Biol. Chem.* **256**, 3557–3563.

Huang, R. T. C. (1982). Myxovirus-induced membrane fusion mediated by phospholipids and neutral glycolipids. *Adv. Exp. Med. Biol.* **152**, 393–400.

Huang, R. T. C. (1983). Involvement of glycolipids in myxovirus-induced membrane fusion (hemolysis). *J. Gen. Virol.* **64**, 221–224.

Huang, R. T. C., Wahn, K., Klenk, H.-D., and Rott, R. (1980a). Fusion between cell membrane and liposomes containing the glycoproteins of influenza virus. *Virology* **104**, 294–302.

Huang, R. T. C., Rott, R., Wahn, K., Klenk, H.-D., and Kohana, T. (1980b). The function of the neuraminidase in membrane fusion induced by myxoviruses. *Virology* **107**, 313–319.

Huang, R. T., Rott, R., and Klenk, H.-D. (1981). Influenza viruses cause hemolysis and fusion of cells. *Virology* **110**, 243–247.

Hubbard, S. C., and Ivatt, R. J. (1981). Synthesis and processing of asparagine linked oligosaccharides. *Annu. Rev. Biochem.* **50**, 555–583.

Hunsmann, G., Moennig, V., Pister, L., Seifert, E., and Schafer, W. (1974). Properties of mouse leukemia viruses. VIII. The major viral glycoprotein of friend leukemia virus. Seroimmunological interfering and hemagglutinating capacities. *Virology* **62**, 307–318.

Hutt-Fletcher, L. M., Fowler, E., Lambris, J. D., Faighny, R. J., Simmons, J. G., and Ross, G. D. (1983). Studies of the Epstein Barr virus receptor found on Rajii cells: II. A comparison of lymphoyte binding sites for Epstein Barr virus and C3d. *J. Immunol.* **130**, 1309–1312.

Ichihashi, Y., Matsumoto, S., and Dales, S. (1971). Biogenesis of poxviruses: role of A-type inclusions and host cell membranes in virus dissemination. *Virology* **46**, 507–532.

Ikeda, H., Hardy, W., Tress, E., and Fleissner, E. (1975). Chromatographic separation and antigenic analysis of proteins of the oncornaviruses. V. Identification of a new murine viral protein. *J. Virol.* **16**, 53–61.

IUPAC–IUB Commission on Biochemical Nomenclature (1977). The nomenclature of lipids. *Lipids* **12**, 455–463.

Jarnefelt, J., Rush, J., Li, Y.-T., and Laine, R. A. (1978). Erythroglycan, a high molecular weight glycopeptide with a repeating structure Galβ1,4GlcNAc comprising more than one third of the protein bound carbohydrate of the erythrocyte stroma. *J. Biol. Chem.* **253**, 8006–8009.

Jensen, K. E., Minuse, E., and Ackerman, W. W. (1955). Serologic evidence of American experience with newborn pneumonitis virus (type Sendai). *J. Immunol.* **75**, 71–77.

Jønsson, V., Wells, A., and Klein, G. (1982). Receptors for the complement C3d componant and the Epstein–Barr virus are quantitatively coexpressed on a series of B-Cell lines and their somatic cell hybrids. *Cell. Immunol.* **72**, 263–276.

Joklik, W. (1972). Studies on the effect of chymotrypsin on reovirions. *Virology* **49**, 700–715.

Jordon, E. J., and Mayor, H. D. (1962). The fine structure of reovirus, a new member of the icosahedral series. *Virology* **17**, 597–599.

Kaaden, O. R., Adam, K.-H., and Strohmaier, K. (1977). Induction of neutralizing antibodies and immunity in vaccinated guinea pigs by cyanogen bromide-peptides of $VP_3$ of foot-and-mouth disease virus. *J. Gen. Virol.* **34**, 397–400.

Kalyanaraman, V. S., Sarngadharan, M. G., and Gallo, R. C. (1978). Characterization of Rauscher murine leukemia virus envelope glycoprotein receptor in membranes from murine fibroblasts. *J. Virol.* **28**, 686–696.

Kelley, J. M., Emerson, S. U., and Wagner, R. R. (1972). The glycoprotein of vesicular stomatitis virus is the antigen that gives rise to and racts with neutralizing antibody. *J. Virol.* **10**, 1231–1235.

Kendal, A. P. (1975). A comparison of "influenza C" with prototype myxoviruses: Receptor-destroying activity (neuraminidase) and structural polypeptides. *Virology* **65**, 87–99.

Khelifa, R., and Menezes, J. (1982). Epstein–Barr virus (EBV) lymphoid cell interactions. III. Effect of concanavalin A and saccharides on EBV penetration. *J. Virol.* **42**, 402–410.

Khelifa, R., and Menezes, J. (1983). Sendai virus envelopes can mediate Epstein–Barr virus binding to and penetration into Epstein–Barr virus receptor negative cells. *J. Virol.* **46**, 325–332.

Kiang, W.-L., Krusius, T., Finne, J., Margolis, R. U., and Margolis, R. K. (1982). Glycoproteins and proteoglycans of the chromaffin granule matrix. *J. Biol. Chem.* **257**, 1651–1659.

Kilbourne, E. D. (1978). Genetic dimorphism in influenza viruses: Characterization of stably associated hemagglutinin mutants differing in antigenicity and biological properties. *Proc. Natl. Acad. Sci. U.S.A.* **75**, 6258–6262.

Kilbourne, E. D., McGregor, S., and Easterday, B. C. (1979). Hemagglutinin mutants of swine influenza virus differing in replication characteristics in their natural hosts. *Infect. Immun.* **26**, 197–201.

Kilbourne, E. D., McGregor, S., and Easterday, B. C. (1981). Transmission in swine of hemagglutinin mutants of swine influenza virus. *In* "The Replication of Negative Strand Viruses" (D. H. L. Bishop and R. W. Compans, eds.), pp. 449–453. Elsevier/North-Holland, New York.

Kilbourne, E. D., Gerhard, W., and Whitaker, C. W. (1983). Monoclonal antibodies to the hemagglutinin Sa antigenic site of A/PR/8/34 influenza virus distinguish biologic mutants of swine influenza virus. *Proc. Natl. Acad. Sci. U.S.A.* **80**, 6399–6402.

Klausner, R. D., Renswoude, J. V., Ashwell, G., Kempf, C., Schechter, A. D., and Bridges, K. R. (1983). Receptor mediated endocytosis of transferin in K562 cells. *J. Biol. Chem.* **258**, 4715–4724.

Klein, G., Giovanella, B. C., Lindahl, T., Fralkow, P. J., Singh, S., and Stehlin, J. (1974). Direct evidence of the presence of Epstein–Barr virus DNA and nuclear antigen in malignant epithelial cells from patients with anaplastic carcinoma of the nasopharynx. *Proc. Natl. Acad. Sci. U.S.A.* **71**, 4737–4741.

Klenk, H. O., Compans, R. W., and Choppin, P. W. (1970). An electron microscopic study of the presence or absence of neuramino acid in enveloped viruses, *Virology* **42**, 1158–1162.

Klenk, E., Faillard, H., and Lempfrid, H. (1955). The enzymatic activity of influenza virus. *Hoppe-Seyler's Z. Physiol. Chem.* **301**, 235–246.

Klenk, H.-D., Rott, R., Orlich, M., and Blödorn, J. (1975). Activation of influenza A viruses by trypsin treatment. *Virology* **68**, 426–439.

Klug, A., and Finch, J. T. (1965). Structure of viruses of the papilloma-polyoma type. I. Human wart virus. *J. Mol. Biol.* **11**, 403–423.

Koide, N., Wells, A., Volsky, D. J., Shapiro, I. M., and Klein, G. (1981). The detection of Epstein–Barr virus receptors utilizing radiolabeled virus. *J. Gen. Virol.* **54**, 191–195.

Kornfeld, R., and Kornfeld, S. (1980). Structure of glycoproteins and their oligosaccharide units. *In*

"The Biochemistry of Glycoproteins and Proteoglycans," (W. Lennarz, ed.) pp. 1–34. Plenum, New York.

Kornfeld, S., and Kornfeld, R. (1976). Comparative aspects of glycoprotein structure. *Annu. Rev. Biochem.* **45**, 217–237.

Kozda, H., and Jungeblut, C. W. (1958). Effect of receptor-destroying enzyme on the growth of EMC virus in tissue culture. *J. Immunol.* **81**, 76–81.

Krizanova, O., and Rathova, V. (1969). Serum inhibitors of myxoviruses. *Curr. Top. Microbiol.* **42**, 125–151.

Kunin, C. M., and Jordon, W. S. (1961). *In vitro* adsorption of poliovirus by non-cultured tissues. Effect of species, age and malignancy. *Am. J. Hyg.* **73**, 245–257.

Kuroya, M., Ishida, N., and Shiratori, T. (1953). Pneumonia in newborn infants caused by virus. *Virus* **3**, 323–332.

Lambris, J. D., Dobson, N. J., and Ross, G. D. (1981). Isolation of lymphocyte membrane complement receptor type two (the C3d receptor) and preparation of receptor-specific antibody. *Proc. Natl. Acad. Sci. U.S.A.* **78**, 1828–1832.

Landen, B., and Fox, C. F. (1980). Isolation of BP gp70, a fibroblast receptor for the envelope antigen of Rauscher murine leukemia virus. *Proc. Natl. Acad. Sci. U.S.A.* **77**, 4988–4992.

Laver, W. G., and Webster, R. G. (1973). Studies on the origin of pandemic influenza. III. Evidence implicating duck and equine influenza viruses as possible progenitors of the Hong Kong strain of human influenza. *Virology* **51**, 383–391.

Laver, W. G., and Webster, R. G. (1979). Ecology of influenza viruses in lower mammals and birds. *Br. Med. Bull.* **35**, 29–33.

Lazarowitz, S. G., and Choppin, P. W. (1975). Enhancement of the infectivity of influenza A and B viruses by proteolytic cleavage of the hemagglutinin polypeptide. *Virology* **68**, 440–454.

Lazarowitz, S. G., Compans, R. W., and Choppin, P. W. (1971). Influenza virus structural and nonstructural proteins in infected cells and their plasma membranes. *Virology* **46**, 830–843.

Ledeen, R. W., and Yu, R. K. (1982). Gangliosides: Structure, isolation and analysis. *In* "Complex Carbohydrates," Part D (V. Ginsburg, ed.), Methods in Enzymology, Vol. 83, pp. 134–195. Academic Press, New York.

Lee, P. W. K., Hayes, E. C., and Joklik, W. K. (1981). Protein J1 is the reovirus cell attachment protein. *Virology* **108**, 156–163.

Lenard, J., and Miller, D. K. (1981). pH-dependent hemolysis by influenza virus, Semliki forest virus and Sendai virus. *Virology* **110**, 479–482.

Levinson, B., Pepper, D., and Belyevin, G. (1969). Substituted sialic acid prosthetic groups as determinants of viral hemagglutination. *J. Virol.* **3**, 477–483.

Lohmeyer, J., and Klenk, H.-D. (1979). A mutant of influenza virus with a temperature-sensitive defect in the posttranslational processing of the hemagglutinin. *Virology* **93**, 134–145.

Lonberg-Holm, K., and Philipson, L. (1974). Early interaction between animal viruses and cells. *In* "Monographs in Virology," Vol. 9, (J. L. Melnick, ed.). Karger, Basel.

Lonberg-Holm, K., and Philipson, L. (1980). Molecular aspects of virus receptors and cell surfaces. *In* "Cell Membranes and Viral Envelopes" (H. A. Blough and J. M. Tiffany, eds.), Vol. 2, pp. 789–848. Academic Press, New York.

Lonberg-Holm, K., Gosser, L. B., and Shimshick, E. J. (1976). Interaction of liposomes with subviral particles of poliovirus type 2 and rhinovirus type 2. *J. Virol.* **19**, 746–749.

Love, D. N., and Weiss, R. A. (1974). Pseudotypes of vesicular stomatitis virus determined by exogenous and endogenous Avian RNA Tumor Viruses. *Virology* **57**, 271–278.

Loyter, A., Vainstein, A., Graessmann, M., and Graessmann, A. (1983). Fusion-mediated infection of SV40-DNA: Introduction of SV40-DNA into tissue culture cells by the use of DNA-loaded reconstituted Sendai virus envelopes. *Exp. Cell Res.* **143**, 415–425.

Luftig, R. B., Kilham, S. S., Hay, A. J., Zweerink, H. J., and Joklik, W. K. (1972). An ultrastructural study of virions and cores of reovirus type 3. *Virology* **48**, 170–181.

McClelland, L., and Hare, R. (1941). The adsorption of influenza virus by red cells and a new *in vitro* method of measuring antibodies for influenza virus in the embryonated egg. *Can. J. Public Health* **32**, 530–538.

Macdonald, R. D., and Yamamoto, T. (1977). The structure of infections pancreatic necrosis virus RNA. *J. Gen. Virol.* **34**, 235–247.

McGeady, M. L., and Crowell, R. L. (1981). Proteolytic cleavage of VP1 in "A" particles of coxaskievirus B3 does not appear to mediate virus uncoating by HeLa cells. *J. Gen. Virol.* **55**, 439–450.

McLaren, L. C., Holland, J. J., and Syverton, J. T. (1959). The mammalian cell–virus relationship. I. Attachment of poliovirus to cultivated cells of primate and non-primate origin. *J. Exp. Med.* **109**, 475–485.

McLaren, L. C., Holland, J. J., and Syverton, J. T. (1960). The mammalian cell virus relationship. V. Susceptibility and resistance of cells *in vitro* to infection by coxsackie A9 virus. *J. Exp. Med.* **112**, 581–594.

Maeda, T., and Ohnishi, S. (1980). Activation of influenza virus by acidic media causes hemolysis and fusion of erythrocytes. *FEBS Lett.* **122**, 283–287.

Maeda, T., Kuroda, K., Ohnishi, S., Austee, D. J., and Tanner, M. J. A. (1979). Interaction between haemagglutinating virus of Japan and human En(19-) erythrocytes lacking major sialoglycoprotein. *FEBS Lett.* **98**, 157–160.

Maeda, T., Kawasaki, K., and Ohnishi, S. (1981). Interaction of influenza virus hemagglutinin with target membrane lipids is a key step in virus-induced hemolysis and fusion at pH 5.2. *Proc. Natl. Acad. Sci. U.S.A.* **78**, 4133–4137.

Mäntyjärvi, R. A., Arstila, P., and Meurman, O. H. (1972). Hemagglutination by BK virus, a tentative new member of the papovavirus group. *Infect. Immun.* **6**, 824–828.

Magnani, J. L., Nilsson, B., Brockhaus, M., Zopf, D., Steplewski, Z., Koprowski, H., and Ginsburg, V. (1982). A monoclonal antibody-defined antigen associated with gastrointestinal carrier is a ganglioside containing sialylated Lacto-*N*-fucopentaose II. *J. Biol. Chem.* **257**, 14365–14369.

Mandel, B. (1965). The fate of the inoculum in HeLa cells infected with poliovirus. *Virology* **25**, 152–154.

Mandel, B., and Racker, E. (1953). Inhibition of Theiler's encephalomyelitis virus (GDVII strain) of mice by an intestinal mucopolysaccharide. *J. Exp. Med.* **98**, 399–416.

Manning, P. A., Puspurs, A., and Reeves (1976). Outer membrane of *Escherichia coli* K-12: isolation of mutants with altered protein 3A by using host range mutants of bacteriophage K3. *J. Bacteriol.* **127**, 1080–1084.

Marchesi, V. T., and Andrews, E. P. (1971). Glycoproteins: Isolation from cell membranes with lithium diiodosalicylate. *Science (Washington, D.C.)* **174**, 1247–1248.

Marchesi, V. T., Jackson, R. L., Segrest, J. P., and Kahane, I. (1973). Molecular features of the major glycoproteins of the human erythrocyte membrane. *(Washington, D.C.) Fed. Proc.* **32**, 1833–1937.

Margolis, G., and Kilham, L. (1969). Hydrocephalus in Hamsters, Ferrets, Rats and Mice following inoculations with reovirus type I. II. Pathologic studies. *Lab. Invest.* **21**, 189–198.

Markus, P. I. (1959). Symposium on the biology of cells modified by viruses or antigens. IV. Single-cell techniques in tracing virus–host interactions. *Bacteriol. Rev.* **23**, 232–249.

Markwell, M. A. K., and Paulson, J. C. (1980). Sendai virus utilizes specific sialyloligosaccharides as host cell receptor determinants. *Proc. Natl. Acad. Sci. U.S.A.* **77**, 5693–5697.

Markwell, M. A. K., Svennerholm, L., and Paulson, J. C. (1981). Specific gangliosides function as host cell receptors for Sendai virus. *Proc. Natl. Acad. Sci. U.S.A.* **78**, 5406–5410.

Markwell, M. A. K., Portner, A., and Schwartz, A. L. (1983). Sendai virus host–cell interaction: Defining the roles of the viral glycoproteins in adsorption and fusion by means of the asialoglycoprotein receptor. *In* "Glycoconjugates" (M. A. Chester, D. Heinegärd, A. Lundblad, and S. Svensson, eds.), pp. 656–657. Rahms i Lund, Sweden.

Markwell, M. A. K., Fredman, P., and Svennerholm, L. (1984a). Specific gangliosides are receptors for Sendai virus. *In* ''Ganglioside, Structure, Function and Biomedical Potential'' (R. W. Ledeen, R. K. Yu, M. M. Rapport, K. Suzuki, and G. Tettamanti, eds.), pp. 369–379. Plenum, New York.

Markwell, M. A. K., Fredman, P., and Svennerholm, L. (1984b). Receptor ganglioside content of three hosts for Sendai virus MDBK, HeLa and MDCK cells. *Biochim. Biophys. Acta* **775,** 1–16.

Marsh, M., and Helenius, A. (1980). Adsorptive endocytosis of Semliki forest virus. *J. Mol. Biol.* **142,** 439–454.

Marsh, M., Matlin, K., Reggio, H., White, J., Kartenbeck, J., and Helenius, A. (1982a). Are lysosomes a site of enveloped-virus penetration? *Cold Spring Harbor Symp. Quant. Biol.* **46,** 835–843.

Marsh, M., Wellsteed, J., Kern, H., Harms, E., and Helenius, A. (1982b). Monensin inhibits Semliki Forest virus penetration into culture cells. *Proc. Natl. Acad. Sci. U.S.A.* **79,** 5297–5301.

Marsh, M., Bolzau, E., and Helenius, A. (1983). Penetration of Semliki Forest Virus from Acidic Prelysosome Vacuoles. *Cell* **32,** 931–940.

Martin, S. J., and Johnston, M. D. (1972). The selective release of proteins from a bovine enterovirus. *J. Gen. Virol.* **16,** 115–125.

Matlin, K. S., Reggio, H., Helenius, A., and Simons, K. (1981). Infectious entry pathway of influenza virus in a canine kidney cell line. *J. Cell Biol.* **91,** 601–603.

Matlin, K. S., Reggio, H., Helenius, A., and Simons, K. (1982). Pathway of vesicular stomatitis entry leading to infection. *J. Mol. Biol.* **156,** 609–631.

Matsumoto, A., Yoshima, H., and Kobata, A. (1983). Carbohydrates of influenza virus hemagglutinin: Structure of the whole neutral sugar chains. *Biochemistry* **22,** 188–196.

Meager, A., Butters, T. D., Mautner, V., and Hughes, R. C. (1976). Interactions of KB-cell glycoproteins with an adenovirus capsid protein. *Eur. J. Biochem.* **61,** 345–353.

Meiselman, N., Kohn, A., and Danon, D. (1967). Electron microscopic study of penetration of Newcastle disease virus cells leading to formation of polykaryocytes. *J. Cell Sci.* **2,** 71–76.

Menezes, J., Sergneurin, J. M., Patel, P., Bourkas, A., and Lenoir, G. (1977). Presence of Epstein–Barr virus receptors but absence of virus penetration, in cells of an EBV genome negative human lymphoblastoid T line (Molt 4). *J. Virol.* **22,** 816–821.

Metcalf, P. (1982). The semmetry of reovirus outer shell. *J. Ultrastruct, Res.* **78,** 292–301.

Mifume, K., Ohuchi, M., and Mannen, K. (1982). Hemolysis and cell fusion by Rhabdoviruses. *FEBS Lett.* **137,** 293–297.

Mooney, J. J., Dalrymple, S. M., Alving, C. R., and Russell, P. K. (1975). Interaction of Sindbis virus with liposomal model membranes. *J. Virol.* **15,** 225–231.

Morgan, C., Rosenkranz, H. S., and Mednis, B. (1969). Structure and development of viruses as observed in the electron microscope. X. Entry and uncoating of adenovirus. *J. Virol.* **4,** 777–796.

Mori, R. J., Schieble, H., and Ackermann, W. W. (1962). Reaction of polyoma and influenza viruses with receptors of erythrocytes and host cells. *Proc. Soc. Exp. Biol. Med.* **109,** 689–690.

Moss, B. (1974). Reproduction of poxviruses. In ''Comprehensive Virology'' (H. Frankel-Conrat and R. R. Wagner, eds.), Vol. 3, pp. 405–475. Plenum, New York.

Murayama, J.-I., Takeshita, K., Tomita, M., and Hamada, A. (1981). Isolation and characterization of two glycophorins from horse erythrocyte membranes. *J. Biochem. (Tokyo)* **89,** 1593–1598.

Murayama, J.-I., Tomita, M., and Hamada, A. (1982). Primary structure of horse erythrocyte glycophorin HA. Its amino acid sequence has a unique homology with those of human and porcine erythrocyte glycophorins. *J. Membr. Biol.* **64,** 205–215.

Murayama, J.-I., Fukuda, K., Yamashita, T., and Hamada, A. (1983a). Structural studies on sugar chains of glycophorins. *In* "Glycoconjugates" (A. Chester, D. Heinegärd, A. Lundblad, and S. Svensson, eds.), pp. 184–185. Rahms, Lund, Sweden.

Murayama, J.-I., Yamashita, T., Tomita, M., and Hamada, A. (1983b). Amino acid sequence and oligosaccharide attachment sites of the glycosylated domain of dog erythrocyte glycophorin. *Biochim. Biophys. Acta* **742**, 477–483.

Nagler, F. P. O. (1942). Application of Hirst's phenomenon to titration of vaccinia virus and vaccinia immune serum. *Med. J. Aust.* **1**, 281–283.

Nakajima, S., and Sugiura, A. (1980). Neurovirulence of influenza virus in mice. II. Mechanism of virulence as studied in a neuroblastoma cell line. *Virology* **101**, 450–457.

Nakamura, K., and Compans, R. W. (1977). The cellular site of sulfation of influenza viral glycoproteins. *Virology* **79**, 381–392.

Nakamura, K., and Compans, R. W. (1979). Biosynthesis of the oligosaccharides of influenza viral glycoproteins. *Virology* **93** 31–47.

Neurath, A. R., Hartzell, R. W., and Rubin, B. A. (1969). Solubilization and some properties of erythrocyte receptors for adenovirus type 7 hemagglutinin. *Nature* **221**, 1069–1071.

Nigg, E. A., Cherry, R. J., and Bachi, T. (1980). Influence of influenza and Sendai virus on the rotational mobility of Band 3 proteins in human erythrocyte membranes. *Virology* **107**, 552–556.

Nonoyama, M., Huang, C. A., Pagano, S. S., Klein, G., and Singh, S. (1973). DNA of Epstein–Barr virus detected in tissue of Barkitt's lymphoma and nasopharyngeal carcinoma. *Proc. Natl. Acad. Sci. U.S.A.* **70**, 3265–3268.

Norrby, E. (1966). The relationship between the soluble antigens and the virion of adenovirus Type 3. II. Identification and characterization of an incomplete hemaglutinin. *Virology* **30**, 608–617.

Norrby, E. (1969). The relationship between the soluble antigens and the virion of adenovirus Type 3. IV. Immunological complexity of soluble components. *Virology* **37**, 565–576.

Norrby, E., Wadell, G., and Marusyk, H. (1969). Fiber-associated incomplete and complete hemagglutinins of adenovirus Type 6. *Arch. Gesamte Virusforsch.* **28**, 239–244.

Norrild, B. (1980). Immunochemistry of herpes simplex glycoproteins. *Curr. Top. Microbiol. Immunol.* **90**, 378–387.

Notkins, A. L. (1977). Virus-induced diabetes mellitus. *Arch. Virol.* **54**, 1–17.

Notter, M. F. D., Leary, J. F., and Balduzzi, P. C. (1982). Adsorption of Rous Sarcoma virus to genetically susceptible and resistant chicken cells studied by laser flow cytometry. *J. Virol.* **41**, 958–964.

Nozima, T., Yasui, K., and Homma, R. (1968). Reaction of several saccharides with Japanese encephalitis virus. *Acta Virol. (Engl. Ed.)* **12**, 296–300.

O'Callaghan, R. J., and Labat, D. D. (1983). Evidence of a soluble substrate for the receptor destroying enzyme of influenza C virus. *Infect. Immun.* **39**, 305–310.

O'Callaghan, R. J., Loughlin, S. M., Labat, D. D., and Howe, C. (1977). Properties of influenza C grown in cell culture. *J. Virol.* **24**, 875–882.

Ohuchi, M., Ohuchi, R., and Mifume, K. (1982). Demonstration of hemolytic and fusion activities of influenza C virus. *J. Virol.* **42**, 1076–1079.

Oku, N., Nojima, S., and Inoue, K. (1981). Studies on the interaction of Sendai virus with liposomal membranes. Sendai virus-induced agglutination of liposomes containing glycophorin. *Biochim. Biophys. Acta* **646**, 36–42.

Oku, N., Nojima, S., and Inoue, K. (1982). Studies on the interaction of HVJ (Sendai virus) with liposomal membranes, HVJ-induced permeability increase of liposomes containing glycophorin. *Virology* **116**, 419–472.

Oldstone, M. B., Tishon, A., Dutko, R. J., Kennedy, S. I. T., Holland, J. J., and Lampert, P. W.

(1980). Does the major histocompatibility complex serve as a specific receptor for Semliki Forest virus? *J. Virol.* **34**, 256–265.

Palase, P., Tobita, K., Ueda, M., and Compans, R. W. (1974). Characterization of temperature sensitive influenza virus mutants defective in neuraminidase. *Virology* **61**, 397–410.

Pardoe, I. U., and Burness, A. T. H. (1981). The interaction of encephalomyocarditis virus with its erythrocyte receptor on affinity chromatography columns. *J. Gen. Virol. 57*, 239–243.

Pastan, I. H., and Willingham, M. C. (1981). Journey to the center of the cell: role of the receptosome. *Science (Washington, D.C.)* **214**, 504–509.

Paulson, J. C., Beranek, W. E., and Hill, R. L. (1977a). Purification of a sialyltransferase from bovine colostrum by affinity chromatography on CDP-agarose. *J. Biol. Chem.* **252**, 2356–2362.

Paulson, J. C., Rearick, J. I., and Hill, R. L. (1977b). Enzymatic properties of β-D-galactoside α2,6 sialyltransferases from bovine colostrum. *J. Biol. Chem.* **252**, 2363–2371.

Paulson, J. C., Sadler, J. E., and Hill, R. L. (1979). Restoration of specific myxovirus receptors to asialoerythrocytes by incorporation of sialic acid with pure sialyltransferases. *J. Biol. Chem.* **254**, 2120–2124.

Paulson, J. C., Weinstein, J., Dorland, L., van Halbeek, H., and Vliegenthart, J. F. G. (1982). Newcastle disease virus contains a linkage-specific glycoprotein sialidase: application to the localization of sialic acid residues in N-linked oligosaccharides of α1-acid glycoprotein. *J. Biol. Chem.* **257**, 12734–12738.

Payne, L. G. (1978). Polypeptide composition of extracellular enveloped vaccina virus. *J. Virol.* **27**, 28–37.

Payne, L. G. (1979). Identification of the vaccina hemagglutinin polypeptide from a cell system yielding large amounts of extracellular enveloped virus. *J. Virol.* **31**, 147–155.

Payne, L. G., and Norrby, E. (1976). Presence of hemagglutinin in the envelope of extracellular vaccina virus particles. *J. Gen. Virol.* **32**, 63–72.

Payne, L. G., and Norrby, E. (1978). Adsorption and penetration of enveloped and naked vaccina virus particles. *J. Virol.* **27**, 19–27.

Pereira, H. G., and DeFigueiredo, M. V. T. (1962). Mechanism of hemagglutination by adenovirus Types 1,2,4,5 and 6. *Virology* **18**, 1–8.

Pfefferkorn, E. R., and Shapiro, D. (1974). Reproduction of togaviruses. *In* "Comprehensive Virology" (H. Fraenkel-Conrat and R. R. Wagner, eds.), Vol. 2, pp. 171–230. Plenum, New York.

Philipson, L., Lonberg-Holm, K., and Petterson, V. (1968). Virus–receptor interaction in an adenovirus system. *J. Virol.* **2**, 1064–1075.

Philipson, L., Beatrice, S. T., and Crowell, R. L. (1973). A structural model of picornaviruses as suggested from an analysis of urea-degraded virions and procapsids of coxsackievirus B3. *Virology* **54**, 69–79.

Piraino, F. (1967). The mechanism of genetic resistance of chick embryo cells to infection by Rous sarcoma virus Bryan strain (BS-RSV). *Virology* **32**, 700–707.

Polos, P. G., and Gallaher, W. R. (1979). Insensitivity of a ricin-resistant mutant of Chinese hamster ovary cells to fusion by Newcastle disease virus. *J. Virol.* **30**, 69–75.

Polos, P. G., and Gallaher, W. R. (1982). Independent effects of phytohemagglutinin on cell fusion and attachment by Newcastle disease virus. *Virology* **119**, 268–275.

Pons, M. W. (1971). Isolation of influenza virus ribonucleoprotein from infected cells. Demonstration of the presence of negative stranded RNA in viral RNP. *Virology* **46**, 149–160.

Portner, A., Scrogs, R. A., Marx, P. A., and Kingsbury, D. W. (1975). A temperature sensitive mutant of Sendai virus with an altered hemagglutinin-neuraminidase polypeptide: Consequences for virus assembly and cytopathology. *Virology* **67**, 179–187.

Poste, G., and Pasternak, C. A. (1978). Virus-induced cell fusion. *Cell Surf. Rev.* **5**, 305–367.

Ramig, R. F., and Fields, B. N. (1977). Reoviruses. *In* "The Molecular Biology of Animal Viruses" (D. P. Nayak, ed.), pp. 383–434. Decker, New York.

Rayment, I., Baker, T. S., Casper, D. L. D., and Murakami, W. T. (1982). Polyoma virus capsid structure at 22.5 A resolution. *Nature (London)* **295**, 110–115.

Reading, C. L., Penhoet, E. E., and Ballou, C. E. (1978). Carbohydrate structure of vesicular stomatitis virus glycoprotein. *J. Biol. Chem.* **253**, 5600–5612.

Rearick, J. I., Sadler, J. E., Paulson, J. C., and Hill, R. L. (1979). Enzymatic characterization of β-D-galactoside α2,3 sialyltransferase from porcine submaxillary gland. *J. Biol. Chem.* **254**, 4444–4451.

Richardson, C. R., Scheid, A., and Choppin, P. W. (1980). Specific inhibition of paramyxovirus and myxovirus replication by oligopeptides with amino acid sequences similar to those at the N-termini of the $F_1$ or $HA_2$ viral polypeptides. *Virology* **105**, 205–222.

Rifkin, D. B., and Compans, R. W. (1971). Identification of spike proteins of Rous sarcoma virus. *Virology* **46**, 485–489.

Robinson, P. J., Hunsmann, G., Schneider, J., and Schirrmacher, V. (1980). Possible cell surface receptor for Friend murine leukemia virus isolated with viral envelope glycoprotein complexes. *J. Virol.* **36**, 291–294.

Rogers, G. N., and Paulson, J. C. (1983). Receptor determinants of human and animal influenza virus isolates: Differences in receptor specificity of the H3 hemagglutinin based on species of origin. *Virology* **127**, 361–373.

Rogers, G. N., Paulson, J. C., Daniels, R. S., Skehel, J. J., Wilson, I. A., and Wiley, D. C. (1983a). Single amino acid substitutions in the influenza hemagglutinin change the specificity of receptor binding. *Nature (London)* **304**, 76–78.

Rogers, G. N., Pritchett, T. J., Lane, J. L., and Paulson, J. C. (1983b). Differential sensitivity of human, avian and equine influenza A viruses to a glycoprotein inhibitor of infection: Selection of receptor specific variants. *Virology* **131**, 394–408.

Rogers, G. N., Wang, X., Pritchett, T. J., Haber, L. F., and Paulson, J. C. (1984). Selection of receptor variants from human and avian influenza isolates with the H3 hemagglutinin. *In* "Segmented Negative Stranded Viruses" (R. W. Compans and D. H. L. Bishop, eds.), pp. 239–246. Academic Press, San Diego, California. In press.

Roizman, B. (1977). The herpesviruses. *In* "The Molecular Biology of Animal Viruses" (D. P. Nayak, ed.). pp. 769–848. Decker, New York.

Rothman, J. E. (1981). The Golgi apparatus: Two organelles in tandem. *Science (Washington, D.C.)* **213**, 1212–1219.

Rubin, H. (1965). Genetic control of cellular susceptibility to pseudotypes of Rous sarcoma virus. *Virology* **26**, 270–276.

Rueckert, R. R. (1976). On the structure and morphogenesis of picornaviruses. *In* "Comprehensive Virology" (H. Fraenkel-Conrat and R. R. Wagner, eds.), Vol. 6, pp. 131–213. Plenum, New York.

Sadler, J. E., Paulson, J. C., and Hill, R. L. (1979a). The role of sialic acid in the expression of human MN blood group antigens. *J. Biol. Chem.* **254**, 2112–2119.

Sadler, J. E., Rearick, J. I., Paulson, J. C., and Hill, R. L. (1979b). Purification to homogeneity of a β-galactoside α2,3 sialyltransferase and partial purification of an α-N-acetylgalactosaminide α2,6 sialyltransferase from porcine submaxillary glands. *J. Biol. Chem.* **254**, 4434–4443.

Sadler, J. E., Rearick, J. I., and Hill, R. L. (1979c). Purification to homogeneity and enzymatic characterization of an α-N-acetylgalactosaminide α2,6 sialyltransferase from porcine submaxillary glands. *J. Biol. Chem.* **254**, 5934–5941.

Sarmiento, M., Haffey, M., and Spear, P. G. (1979). Membrane proteins specified by Herpes simplex viruses III. role of glycoprotein VP7 (B2) in viron infectivity. *J. Virol.* **29**, 1149–1158.

Sarov, I., and Joklik, W. K. (1972). Studies on the nature and localization of the capsid polypeptides of vaccina virions. *Virology* **50**, 579–592.

Schaffer-Deshayes, L., Choppin, J., and Levy, J.-P. (1981). Lymphoid cell surface receptor for Moloney Leukemia virus envelope glycoprotein gp71. II. Isolation of the receptor. *J. Immunol.* **126**, 2352–2354.

Schauer, R. (1982). Chemistry, metabolism and biological functions of sialic acids. *Adv. Carbohydr. Chem.* **40**, 132–235.

Schauer, R., Buscher, H.-P., and Casals-Stenzel, J. (1974). Sialic acids: Their analysis and enzymatic modification in relation to the synthesis of submandibular-gland glycoproteins. *Biochem. Soc. Symp.* No. 40, 87–116.

Scheid, A., and Choppin, P. W. (1974). Identification of biological activities of paramyxovirus glycoproteins. Activation of cell fusion, hemolysis and infectivity by proteolytic cleavage of an inactive precorsor protein of Sendai virus. *Virology* **56**, 475–490.

Scheid, A., Graves, M. C., Silver, S. M., and Choppin, P. W. (1978). Studies on the structure and function of paramyxovirus glycoproteins. *In* "Negative Strand Viruses and the Host Cell" (B. W. J. Mahy and R. D. Barry, eds.), pp. 181–193. Academic Press, New York.

Schild, G. C., Oxford, J. S., deJong, J. C., and Webster, R. G. (1983). Evidence for host–cell selection of influenza virus antigenic variants. *Nature (London)* **303**, 706–709.

Schlegel, R., Tralka, T. S., Willingham, M. C., and Pastan, I. (1983). Inhibition of VSV binding and infectivity by phosphotidyl-serine: Is phosphotidyl-serine a VSV binding site? *Cell* **32**, 639–646.

Schloemer, R. H., and Wagner, R. R. (1974). Sialoglycoprotein of vesicular stomatitis virus: role of the neuraminic acid in infection. *J. Virol.* **14**, 270–281.

Schloemer, R. H., and Wagner, R. R. (1975a). Cellular adsorption function of the sialoglycoprotein of vesicular stomatitis virus and its neuraminic acid. *J. Virol.* **15**, 882–893.

Schloemer, R. H., and Wagner, R. R. (1975b). Association of vesicular stomatitis virus glycoprotein with virion membrane: Characterization of the lipophilic tail fragment. *J. Virol.* **16**, 237–249.

Schmidt, M. F. G., and Schlesinger, M. J. (1979). Fatty acid binding to VSV glycoprotein: a new type of post-translational modification of the viral glycoprotein. *Cell* **17**, 813–819.

Schmidt, M. F. G., Bracha, M., and Schlesinger, M. J. (1979). Evidence for covalent attachment of fatty acids to Sindbis virus glycoproteins. *Proc. Natl. Acad. Sci. U.S.A.* **76**, 1687–1691.

Schneider, J., and Hunsmann, G. (1978). Surface expression of murine leukemia virus structural polypeptides on host cells and the virion. *Int. J. Cancer* **22**, 204–213.

Schneider, J., Schwarz, H., and Hunsmann, G. (1979). Rosettes from Friend leukemia virus envelope: preparation and physiochemical and partial characterization. *J. Virol.* **29**, 624–632.

Schneider, J., Falk, H., and Hunsmann, G. (1980). Envelope polypeptides of Friend leukemia virus: purification and structural analysis. *J. Virol.* **33**, 597–605.

Scholtissek, C., Rohde, W., von Hoyningen, V., and Rott, R. (1978). On the origin of the human influenza virus subtypes H2N2 and H3N2. *Virology* **87**, 13–20.

Schrader, J. W., Cunningham, B. A., and Edelman, G. M. (1975). Functional interactions of viral and histocompatibility antigens at tumor cell surfaces. *Proc. Natl. Acad. Sci. U.S.A.* **72**, 5066–5070.

Schulman, J. L., and Palase, P. (1977). Virulence factors of influenza A viruses neuraminidase required for plaque production in MDBK cells. *J. Virol.* **24**, 170–176.

Schwartz, A. L., Fridovich, S. E., Knowles, B. B., and Lodish, H. F. (1981). Characterization of the asialoglycoprotein Receptor in a continuous hepatoma line. *J. Biol. Chem.* **256**, 8878–8881.

Schwartz, R. T., Schmidt, M. F. G., Anwer, V., and Klenk, H.-D. (1977). Carbohydrates of influenza virus. I. Glycopeptides derived from viral glycoproteins after labeling with radioactive sugars. *J. Virol.* **23**, 217–226.

Seganti, L., Mastromarino, P., Superti, F., Simibaldi, L., and Orsi, N. (1981). Receptors for BK virus on human erythrocytes. *Acta Virol. (Engl. Ed.)* **25**, 177–181.

Shapiro, I. M., and Volsky, D. V. (1983). Infection of normal human epithelial cells by Epstein–Barr virus. *Science (Washington, D.C.)* **219**, 1225–1228.

Shapiro, I. M., Klein, G., and Volsky, D. J. (1981). Epstein–Barr virus co-reconstituted with Sendai virus envelopes infects Epstein–Barr virus receptor negative cells. *Biochim. Biophys. Acta* **676**, 19–24.

Sharom, F. J., Baratt, D. G., Thede, A. E., and Grant, C. W. M. (1976). Glycolipids in model membranes spin label and freeze-etch studies. *Biochim. Biophys. Acta* **455**, 485–492.

Shatkin, A. J., and LaFiandra, A. J. (1972). Transcription by infectious subviral particles of reovirus. *J. Virol.* **10**, 698–706.

Shatkin, A. J., and Sipe, J. D. (1968). RNA polymerase activity in purified reoviruses. *Proc. Natl. Acad. Sci. U.S.A.* **61** 1462–1469.

Shida, H., and Dales, S. (1981). Biogenesis of vaccina: Carbohydrate of the hemagglutinin molecule. *Virology* **111**, 56–72.

Silverstein, S. C., and Marcus, P. I. (1964). Early stages of Newcastle disease virus–HeLa cell interaction: an electron microscopic study. *Virology* **23**, 370–380.

Silverstein, S. C., Christman, J. K., and Acs, G. (1976). The reovirus replicative cycle. *Annu. Rev. Biochem.* **45**, 375–408.

Simmons, J. G., Lindsey, M., Hutt-Fletcher, M., Fowler, E., and Feighny, R. J. (1983). Studies of the Epstein–Barr receptor found on Raji cells. I. Extraction of receptor and preparation of anti-receptor antibody. *J. Immunol.* **130**, 1303–1308.

Skehel, J. J., Bayley, P. M., Brown, E. B., Martin, S. R., Waterfield, M. D., White, J. M., Wilson, I. A., and Wiley, D. C. (1982). Changes in the conformation of influenza virus hemagglutinin at the pH optimum of virus-mediated membrane fusion. *Proc. Natl. Acad. Sci. U.S.A.* **79**, 972–986.

Slomiany, S., and Slomiany, B. L. (1978). Structures of the acidic oligosaccharides isolated from rat sublingual glycoprotein. *J. Biol. Chem.* **253**, 7301–7306.

Spear, P. G. (1980). Herpesviruses. *In* "Cell membranes and Viral Envelopes" (H. A. Blough and J. M. Tiffany, eds.). Vol. 2, pp. 709–750. Academic Press, New York.

Spendlove, R. S., McClain, M. E., and Lennette, E. H. (1970). Enhancement of reovirus infectivity by extracellular removal or alteration of the virus capsid by proteolytic enzymes. *J. Gen. Virol.* **8**, 83–94.

Spriggs, D. R., Bronson, R. T., and Fields, B. N. (1983). Hemagglutinin variants of reovirus type 3 have altered central nervous system tropism. *Science (Washington, D.C.)* **220**, 505–507.

Steck, F. T., and Rubin, H. (1965). The mechanism of interference between an avian leukosis virus and Rous sarcoma virus. II. Early steps of infection by RSV of cells under conditions of interference. *Virology* **29**, 642–653.

Steinman, R. M., Mellman, I. S., Muller, W. A., and Cohen, Z. A. (1983). Endocytosis and the recycling of plasma membrane. *Cell* **96**, 1–27.

Stone, J. D. (1948a). Prevention of virus infection with enzyme of *V. cholerae*. I. Studies with viruses of mumps influenza group in chick embryos. *Aust. J. Exp. Biol. Med. Sci.* **26**, 49–64.

Stone, J. D. (1948b). Prevention of virus infection with enzyme of *V. cholerae*. II. Studies with influenza virus in mice. *Aust. J. Exp. Biol. Med. Sci.* **26**, 287–298.

Stone, J. D. (1949). Inhibition of influenza virus hemagglutination by mucoids. II. Differential behavior of mucoid inhibitors with indicator viruses. *Aust. J. Exp. Biol. Med. Sci.* **27**, 557–569.

Stone, J. D. (1951). Adsorptive and enzymatic behavior of influenza virus during the O-D change. *Br. J. Exp. Pathol.* **32**, 367–376.

Stoner, G. D., Williams, B., Kniazeff, A., and Shimkin, M. B. (1973). Effect of neuraminidase pretreatment of normal and transformed mammalian cells to bovine enterovirus 261. *Nature (London)* **245**, 319–320.

Stott, E. T., and Killington, R. A. (1972). Hemagglutination by rhinoviruses. *Lancet* **i**, 1369–1370.

Strauss, J. H., and Strauss, E. G. (1977). Togaviruses. *In* "The Molecular Biology of Animal Viruses" (D. P. Nayak, ed.), pp. 111–166. Decker, New York.

Stuart-Harris, C. H., ans Schild, G. C. (1976). The epidemiology of influenza, Parts I and II. *In* "Influenza: The Virus and the Disease," (C. H. Stuart-Harris, ed.), pp. 112–143. PSG Publ, Littleton, Massachusetts.

Styk, B. (1963). Effect of some inhibitor-destroying substances on the nonspecific inhibitor of influenza C virus present in normal rat serum. *Acta Virol. (Engl. Ed.)* **7**, 88–89.

Sugiura, A., and Ueda, M. (1980). Neurovirulence of influenza virus in mice. I. Neurovirulence of recombinants between virulent and avirulent virus strains. *Virology* **101**, 440–449.

Suttajit, M., and Winzler, R. J. (1971). Effect of Modification of $N$-acetylneuraminic acid on the binding of glycoproteins to influenza virus and on susceptibility to cleavage by neuraminidase. *J. Biol. Chem.* **246**, 3398–3404.

Suzuki, Y., Suzuki, T., and Matsumoto, M. (1983). Isolation and characterization of receptor sialoglycoprotein for hemagglutinating virus of Japan (Sendai Virus) from bovine erythrocyte membrane. *J. Biochem. (Tokyo)* **93**, 1621–1633.

Suzuki, T., Harada, M., Suzuki, Y., and Matsumoto, M. (1984). Incorporation of sialoglycoprotein containing lacto-series oligosaccharides into chicken asialoerythrocyte membranes and restoration of receptor activity toward hemagglutination virus of Japan (Sendai Virus). *J. Biochem.* **95**, 1193–1200.

Svennerholm, L. (1963). Chromatographic separation of human brain gangliosides. *J. Neurochem.* **10**, 613–623.

Svensson, U., Persson, R., and Everitt, E. (1981). Virus–receptor interaction in the adenovirus system: Identification of virion attachment proteins of the HeLa cell plasma membrane. *J. Virol.* **38**, 70–81.

Takátsy, G. Y., and Barb, K. (1959). On the normal serum inhibitors for the avid asian strains of influenza virus. *Acta Virol. (Engl. Ed.)* **3**, 71–77.

Takátsy, G. Y., Barb, K., and Farkas, E. (1959). A new erythrocyte receptor with exclusive affinity to avid strains of influenza A virus. *Acta Virol. (Engl. Ed.)* **3**, 79–84.

Talbot, P. J., and Vance, C. E. (1980). Sindbis virus infects BHK virus via a lysosomal route. *Can. J. Biochem.* **58**, 1131–1137.

Talbot, P. J., and Vance, D. E. (1982). Biochemical studies on the entry of Sindbis virus into BHK-21 cells and the effect of $NH_4Cl$. *Virology* **118**, 451–455.

Tamm, I., and Horsfall, F. I. (1951). A mucoprotein derived from human urine which reacts with influenza, mumps and Newcastle disease viruses. *J. Exp. Med.* **95**, 71–97.

Tardieu, M., and Weiner, H. L. (1982). Viral receptors on isolated murine and human ependymal cells. *Science (Washington, D.C.)* **215**, 419–421.

Tardieu, M., Epstein, R. L., and Weiner, H. L. (1982). Interaction of viruses with cel surface receptors. *Int. Rev. Cytol.* **80**, 27–61.

Thomas, D. B., and Winzler, R. J. (1969). Structural studies on human erythrocyte glycoproteins: Alkali labile oligosaccharides. *J. Biol. Chem.* **244**, 5943–5946.

Tiffany, J. M. (1977). The interaction of viruses with model membranes. *Cell Surf. Rev.* **2**, 157–194.

Tiffany, J. M., and Blough, H. A. (1971). Attachment of myxoviruses to artificial membranes: Electron microscopic studies. *Virology* **44**, 18–28.

Tobita, K., Sugiura, A., Enomoto, C., and Furoyama, M. (1975). Plaque assay and primary isolation of influenza A viruses in an established line of canine kidney cells (MDCK) in the presence of trypsin. *Med. Microbiol. Immunol.* **162**, 1–14.

Tsichlis, P. N., Conklin, K. F., and Coffin, J. M. (1980). Mutant and recombinant avian retroviruses with extended host range. *Proc. Natl. Acad. Sci. U.S.A.* **77,** 536–540.

Turner, G. W., and Squires, E. J. (1971). Inactivated smallpox vaccine: Immunigenicity of inactivated intracellular and extracellular vaccina virus. *J. Gen. Virol.* **13,** 19–25.

Tycko, B., and Maxfield, F. R. (1982). Rapid acidification of endocytic vesicles containing $\alpha_2$ macroglobulin. *Cell* **28,** 643–651.

Umeda, M., Nojima, S., and Inoue, K. (1984). Activity of human erythrocyte gangliosides as a receptor to HVJ. *Virology* **133,** 172–182.

Utagawa, E. T., Miyamura, K., Mukoyama, A., and Kono, R. (1982). Neuraminidase-sensitive erythrocyte receptor for enterovirus Type 70. *J. Gen. Virol.* **63,** 141–148.

Väänänen, P., and Kääriäinen, L. (1979). Hemolysis by two alphaviruses: Semliki Forest virus and Sindbis virus. *J. Gen. Virol.* **43,** 593–601.

Väänänen, P., and Kääriäinen, L. (1980). Fusion and hemolysis of erythrocytes caused by three togaviruses, Semliki forest, Sindbis and rubella. *J. Gen. Virol.* **46,** 467–475.

Valentine, R. C., and Pereira, H. G. (1965). Antigens and structure of the adenovirus. *J. Mol. Biol.* **13,** 13–20.

Van Alphen, L., Hawekes, L., and Lugtenberg, B. (1977). Major outer membrane protein d of *Escherichia coli* K-12. Purification and *in vitro* activity of bacteriophage K3 and F-pilus mediated conjugation. *FEBS Lett.* **75,** 285–290.

Varghese, J. N., Laver, W. G., and Coleman, P. M. (1983). Structure of the influenza virus glycoprotein antigen neuraminidase at 2.9 A resolution. *Nature (London)* **303,** 35–40.

Vasquez, C., and Tournier, P. (1962). The morphology of reovirus. *Virology* **17,** 503–510.

Verlinde, J. D., and DeBaan, P. (1949). Sur l'hemagglutination par des virus poliomyelitiques murins et la destruction enzymatique des recepteurs de virus poliomyelitique de la cellule receptive. *Ann. Inst. Pasteur, Paris* **77,** 632–641.

Vogt, P. K., and Ishizaki, R. (1965). Reciprocal patterns of genetic resistance to avian tumor virus in two lines of chickens. *Virology* **26,** 664–672.

Volsky, D. J., and Loyter, A. (1978). An efficient method for reassembly of fusogenic Sendai virus envelopes after solubilizatin of intact virions with Triton X-100. *FEBS Lett.* **92,** 190–194.

Volsky, D. J., Shapiro, I. M., and Klein, G. (1980). Transfer of Epstein–Barr virus receptors to receptor negative cells permits virus penetration and antigen expression. *Proc. Natl. Acad. Sci. U.S.A.* **77,** 5453–5457.

Wadell, G. (1969). Hemagglutination with adenovirus serotypes belonging to Rosen's sub-groups II and III. *Proc. Soc. Exp. Biol. Med.* **132,** 413–421.

Wagner, R. R. (1975). Reproduction of rhabdoviruses. *In* "Comprehensive Virology" (H. Fraenkel-Conrat and R. R. Wagner, eds.), Vol. 4, pp. 1–93. Plenum, New York.

Walter, G., and Etchison, D. (1977). Subunit interactions in polyoma virus structure. *Virology* **77,** 783–797.

Ward, C. W., and Dopheide, T. A. (1981a). Evolution of the Hong Kong influenza A subtype: Structural relationships between the hemagglutinin from A/duck/Ukraine/1/63 (Hav7) and the Hong Kong (H3) hemagglutinin. *Biochem. J.* **195,** 337–340.

Ward, C., and Dopheide, T. (1981b). The Hong Kong hemagglutinin. Structural relationships between the human (H3) hemagglutinins and the hemagglutinin from the putative progenitor strain, A/duck/Ukraine/1/63 (Hav7). *In* "Genetic Variation Among Influenza Viruses" (D. Nayak and C. F. Fox, eds.), pp. 323–340. Academic Press, New York.

Watanabe, K., Powell, M. E., and Hakomori, S. (1979). Isolation and characterization of gangliosides with a new sialosyl linkage and core structures. *J. Biol. Chem.* **254,** 8223–8229.

Waterfield, M. D., Espelie, K., Elder, K., and Skehel, J. J. (1979). Structure of the haemagglutinin of influenza virus. *Br. Med. Bull.* **35,** 57–63.

Webster, R. G., and Laver, W. G. (1975). Antigenic variation of influenza viruses. *In* "The

Influenza Viruses and Influenza'' (E. D. Kilbourne, ed.), pp. 269–314. Academic ¨ress, New York.

Webster, R. G., Laver, W. G., Air, G. M., and Schild, G. C. (1982). Molecular mechanisms of variation in influenza virus. *Nature (London)* **296**, 115–121.

Weiner, H. L., Drayna, D., Averill, D. R., and Fields, B. N. (1977). Molecular basis of reovirus virulence: Role of the S1 gene. *Proc. Natl. Acad. Sci. U.S.A.* **74**, 5744–5748.

Weiner, H. L., Ault, K. A., and Fields, B. N. (1980). Interaction of reovirus with cell surface receptors. I. Murine and human lymphocytes have a receptor for the hemagglutinin of reovirus type 3. *J. Immunol.* **124**, 2143–2148.

Weinstein, J., de Souza-e-Silva, U., and Paulson, J. C. (1982a). Purification of a Galβ1,4GlcNAc α2,6 sialyltransferase and a Galβ1,3(4)GlcNAc α2,3 sialyltransferase to homogeneity from rat liver. *J. Biol. Chem.* **257**, 13835–13844.

Weinstein, J., de Souza-e-Silva, U., and Paulson, J. C. (1982b). Sialylation of glycoprotein oligosaccharides N-linked to asparagine: Enzymatic characterization of Galβ1,3(4)GlcNAc α2,3 silayltransferase and a Galβ1,4GlcNAc α2,6 sialyltransferase from rat liver. *J. Biol. Chem.* **257**, 13845–13853.

Weiss, R. A. (1969). The host range of Bryan strain Rous sarcoma virus synthesized in the absence of helper virus. *J. Gen. Virol.* **5**, 511–528.

Weiss, R. A. (1981). Retrovirus receptors and their genetics. *In* ''Virus Receptors, Part 2: Animal Viruses'' (K. Lonberg-Holm and L. Philipson, eds.), pp. 185–202. Chapman & Hall, London.

Wells, A., Koide, N., and Klein, G. (1982). Two class viron envelope glycoproteins mediate Epstein–Barr virus binding to receptor positive cells. *J. Virol.* **41**, 286–297.

White, D. O., Scharff, M. D., and Maizel, J. V., Jr. (1969). The polypeptides of adenovirus III. Synthesis in infected cells. *Virology* **38**, 395–406.

White, J., and Helenius, A. (1980). pH-dependent fusion between the Semliki Forest virus membrane and liposomes. *Proc. Natl. Acad. Sci. U.S.A.* **77**, 3273–3277.

White, J., Kartenbeck, J., and Helenius, A. (1980). Fusion of Semliki forest virus with the plasma membrane can be indused by low pH. *J. Cell Biol.* **87**, 264–272.

White, J., Matlin, K., and Helenius, A. (1981). Cell fusion by Semliki Forest, influenza and vesicular stomatitis viruses. *J. Cell Biol.* **89**, 674–679.

Wildy, P., Stocker, M. G. P., Macpherson, I. A., and Horne, R. W. (1960). The fine structure of polyoma virus. *Virology* **11**, 444–457.

Wiley, D. C., Wilson, I. A., and Skehel, J. J. (1982). Structural identification of the antibody binding sites of Hong-Kong influenza hemagglutinin and their involvement in antigenic variation. *Nature (London)* **289**, 373–378.

Wilson, I. A., Skehel, J. J., and Wiley, D. C. (1981). Structure of the hemagglutinin membrane glycoprotein of influenza virus at 3A resolution. *Nature (London)* **289**, 366–373.

Wilson, T., Papahadjopoulos, D., and Taber, R. (1977). Biological properties of poliovirus encapsulated in lipid vesicles: Antibody resistance and infectivity in virus-resistant cells. *Proc. Natl. Acad. Sci. U.S.A.* **74**, 3471–3475.

Wilson, T., Papahadjopoulos, D., and Taber, R. (1979). The introduction of poliovirus RNA into cells via lipid vesicles (liposomes). *Cell* **17**, 77–84.

Wilson, V. W., and Rafelson, M. E. (1967). Studies on the neuraminidases of influenza virus III. Stimulation of activity by bivalent cations. *Biochim. Biophys. Acta* **146**, 160–166.

Witkor, T. J., Gyorgy, E., Schlumberger, H. D., Sokol, F., and Koprowski, H. (1972). Antigenic properties of rabies virus components. *J. Immunol.* **110**, 269–276.

World Health Organization (1980). A revision of the system of nomenclature for influenza viruses— a WHO memorandum. *Bull. W.H.O.* **58**, 294–295.

Wu, P.-S., Ledeen, R. W., Udem, S., and Isaacson, Y. A. (1980). Nature of the Sendai virus receptor: Glycoprotein versus ganglioside. *J. Virol.* **33**, 304–310.

Yamakawa, T. (1966). *In* "Lipoide" (C. E. Schutte, ed.). pp. 87–111. Springer-Verlag, Berlin and New York.

Yamamoto, K., Inoe, K., and Suzuki, K. (1974). Interaction of paramyxoviruses with erythrocyte membranes modified by concanavalin A. *Nature (London)* **250**, 511–513.

Yasui, K., Nozima, T., Homma, R., and Ueda, S. (1969). Effects of α-mannosidase on the active site of Japanese encephalitis viral receptor. *Acta Virol. (Prague)* **13**, 158–159.

Yoon, J.-W., Onodera, T., and Notkins, A. L. (1978). Virus-induced diabetes mellitus. XV. Beta cell damage and insulin-dependent hyperglycemia in mice infected with coxsackie virus B4. *J. Exp. Med.* **148**, 1068–1080.

Yoon, J.-W., Austin, M., Onodera, T., and Notkins, A. L. (1979). Virus induced diabetes mellitus. Isolation of a virus from the pancreas of a child with diabetic ketoacidosis. *N. Engl. J. Med.* **300**, 1173–1179.

Yoshima, H., Furthmayr, H., and Kobata, A. (1980). Structures of the asparagine-linked sugar chains of glycophorin A. *J. Biol. Chem.* **255**, 9713–9718.

Yoshima, H., Matsumoto, A., Mizuochi, T., Kawasaki, T., and Kobata, A. (1981a). Comparative study of the carbohydrate moieties of rat and human plasma $\alpha_1$-acid glycoprotein. *J. Biol. Chem.* **256**, 8476–8484.

Yoshima, H., Nakanishi, M., Okada, Y., and Kobata, A. (1981b). Carbohydrate structures of HVJ (Sendai virus) glycoproteins. *J. Biol. Chem.* **256**, 5355–5361.

Yoshimura, A., Kuroda, K., Kawasaki, K., Yamashina, S., Maeda, T., and Ohnishi, S. (1982). Infectious cell entry mechanism of influenza virus. *J. Virol.* **43**, 284–293.

Youngdahl-Turner, P., Mellman, I. S., Allen, R. H., and Rosenberg, L. E. (1979). Protein mediated vitamin uptake. Adsorptive endocytosis of the transcobalamin II-cobalamin complex by cultured human fibroblasts. *Exp. Cell Res.* **118**, 127–134.

Zajac, I., and Crowell, R. L. (1965). Effects of enzymes on the interaction of enteroviruses with living HeLa cells. *J. Bacteriol.* **89**, 574–582.

Zakstelskaya, L. Y., Eustigreeva, N. A., Isachenko, V. A., Shenderovitch, S. P., and Efimova, V. A. (1969). Influenza in the U.S.S.R.: New antigenic variant A2/Hong Kong/1/68 and its possible precursors. *Am. J. Epidemiol.* **90**, 400–405.

Zakstelskaya, L. Y., Molibog, E. V., Yakhno, M. A., Eustigreeva, N. A., Isachenko, V. A., Privalova, I. M., and Khorlin, A. Y. (1972). Use of synthetic inhibitors of neuraminidase and hemagglutinin for the study of the functional role of active subunits of membranes of myxo- and paramyxo-viruses. *Vopr. Virusol.* **17**, 223–228.

Zavada, J. (1972). Pseudotypes of vesicular stomatitis virus with the coat of murine leukaemia and of avian myeloblastosis viruses. *J. Gen. Virol.* **15**, 183–191.

Zur Hausen, H., Schulte-Holthausen, H., Klein, G., Henle, W., Henle, G., Clifford, P., and Santesson, L. (1970). EB-virus DNA in biopsies of Burkitt tumors and anaplastic carcinomas of the nasopharynx. *Nature (London)* **228**, 1056–1058.

# 6

# Studies on Insulin Receptors: Implications for Insulin Action

**YORAM SHECHTER**
Department of Hormone Research
The Weizmann Institute of Science
Rehovot, Israel

## I. INTRODUCTION

Although the mechanism of action of insulin is largely unknown, impressive achievements were made during recent years with respect to the structure, basic properties, and behavior of the insulin receptor. Rapid progress was made due to both important discoveries and the availability of new techniques. Some of the cornerstones and the main achievements of recent years, in somewhat chronological manner, are the following: quantitation of hormone–receptor interaction by reliable radiotracer-binding techniques (1,2); detergent extraction of the insulin

**221**

receptor with the retention of insulin binding (3); purification of the insulin receptor to homogeneity by conventional and affinity chromatography methods (4–8); identification and characterization of the basic subunit structures using photoaffinity labeling and cross-linking techniques (9–15); mobility and aggregation of the insulin receptor in the plane of the plasma membrane of intact cells, as determined by fluorescent photobleaching recovery and fluorescent microscopy techniques (19–21); elucidation of the biochemical basis for downregulation by isopycnic centrifugation (22–24); internalization of hormone–receptor complexes; degradation of insulin and recycling of the insulin receptors (20,25–30); protein kinase properties of the insulin receptor, the autophosphorylation of the β subunit (31–35); common features to insulin and insulin-like growth factor I (IGF-I) receptors (36–40); These topics and others will be the subject of this chapter.


## II. RECEPTOR PURIFICATION

The discovery that nonionic detergents could solubilize insulin receptors while preserving binding properties (3) made it possible to purify them by general methods and procedures used to isolate protein. Purification is still hampered, however, both by the relatively low concentration of receptor in the starting material and by its physical similarity to other plasma-membrane proteins. Conventional methods based on size and charge (i.e., gel filtration and ion-exchange chromatography), thus, usually resulted in only a 2- to 5-fold purification. Because of its glycoprotein nature (41), concanavalin A– or wheat germ agglutinin (WGA)–Sepharose affinity chromatography was possible (4,6,8) and an additional 7- to 10-fold purification of the receptor was obtained. The major purification step, however, is dependent upon the insulin receptor's biospecific properties and is accomplished by affinity chromatography on an insulin-agarose column (4,8). An additional 250- to 300-fold purification is obtained, out of the 2500- to 3000-fold purification needed to obtain essentially pure receptor protein (4,8).

Efficient affinity purification requires that the insulin molecule be separated from the matrix backbone by a long spacer or "arm" (42). Although it has not been studied extensively, lysine B-29 is at present the most suitable side chain for attaching insulin to the spacer (Y. Shechter, unpublished observations). Such attachment can best be achieved by blocking both $\alpha$-amino groups by citraconic anhydride (19), linking the hormone to the Sepharose–arm beads at pH 8.5, and then deblocking the amino groups at pH 3.5 (Y. Shechter, unpublished observations).

When Jacobs et al. (4) first purified insulin receptor, they found that 2.4 μg insulin were bound per milligram of purified liver protein, about 8% of the expected amount, assuming an insulin receptor molecular weight (MW) of

300,000. The receptor was purified to near homogeneity, as judged by a few criteria (4,9,43), and the lower theoretical binding capacity was due to partial denaturation of the receptor during elution with buffer containing 4.5 $M$ urea (44). This was improved by Fujita-Yamaguchi *et al.*, who were able to obtain pure insulin receptors from human placentas with a specific activity of 28.5 µg of insulin bound per milligram of protein. The product obtained was very close to the theoretical value expected from pure receptor (8).

Two contaminants which may be copurified with insulin receptors are insulinase and the receptors for the insulin-like growth factor I (IGF-I). Since it is not a glycoprotein, insulinase can be removed by passage through lectin–Sepharose columns (6). Although IGF-I receptor has only 1/200th of the affinity of the insulin receptor to insulin (39), it may remain as a low-level contaminant. Moreover, because of the similarity in their subunit structure, both insulin and IGF-I receptors look alike on SDS–polyacrylamide gels (45). Therefore, this factor should be considered if insulin receptor is purified from target cells enriched in IGF-I receptor protein.

## III. RECEPTOR STRUCTURE

The methods most frequently utilized to determine receptor structure are photoaffinity labeling and cross-linking techniques, and a fairly clear picture of both the structure and the properties of the insulin receptor has emerged from studies reported by a number of laboratories.

A molecular weight of about 350,000 for the native receptor has been obtained using various techniques, under nondenaturing conditions (7,47–49). Studies both by Jacobs *et al.* (9,43) and by Massague *et al.* (10,18) suggest that the receptor is a heterotetramer, of the form $(\alpha\beta)_2$, with four subunits covalently linked by interchain disulfide bonds (Fig. 1). The approximate molecular weights of the $\alpha$ and $\beta$ subunits are, respectively, 135,000 and 90,000. A smaller subunit ($M_r$ 45,000) appears to be a proteolytic fragment of $\beta$ (46).

Both $\alpha$ and $\beta$ subunits are glycoproteins (47–49), and the glycosidic moieties are located on the external surface of the cell (50,51). While the $\beta$ subunit binds directly to insulin, the $\alpha$ subunit is also likely to be involved in insulin binding. In addition to disulfide bonding, the subunits are linked to each other by strong noncovalent interactions and are dissociated into the corresponding subunits only after being both reduced and denatured by boiling in SDS and mercaptoethanol (9,17).

## Alternative Structure

The studies by Yip and Moule using two photoreactive derivatives of insulin (52) obtained evidence supporting a different type of insulin-receptor structure in

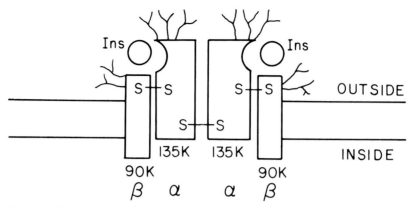

**Fig. 1.** Schematic model of the proposed subunit structure of the insulin (Ins) receptor, according to Jacobs et al. (9,43) and Massague et al. (10,18). Symbols: Branched structures, complex carbohydrate chains; S–S, disulfide bonds.

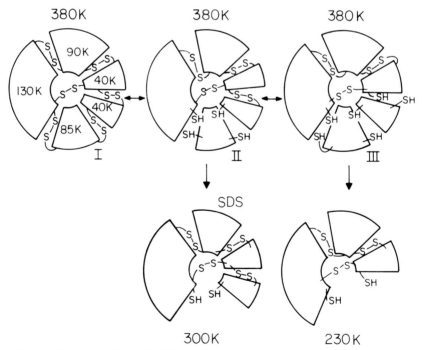

**Fig. 2.** The proposed model of the insulin receptor of rat adipocytes according to Yip et al. (11–14,52). The 380 kilodalton form consists of five subunits (see text). Upon reduction and exposure to SDS there is a dissociation of one or two subunits and the concomitant formation of the 300- and 230-kilodalton species, respectively.

fat cells. According to these authors, the receptor exists in three interconvertible redox forms of 380, 300, and 230 kilodaltons (Fig. 2). The interconversion of these three receptor species is the result of oxidation and reduction of the inter-subunit sulfhydryl groups. The reduction–reoxidation processes may be coupled to the metabolic redox state of the cell, through mechanisms involving the participation of glutathione (52). The 380,000 form consists of five subunits having molecular weights of 130,000, 90,000, 85,000, and two of 40,000 (52; Fig. 2). This receptor structure differs substantially from that obtained by Jacobs *et al.* (9,43) in liver membranes and Massague *et al.* (10,18) in fat cells, indicat-ing that further studies are required in this direction.

## IV. GLYCOPROTEIN NATURE OF THE INSULIN RECEPTOR

All of the evidence available indicates that the insulin receptor is a glycopro-tein. For example: (i) the soluble insulin receptor is adsorbed to wheat germ agglutinin– or concanavalin A–agarose columns and can be eluted with simple sugars (41,53); (ii) digestion of fat cells with a combination of neuraminidase and β-galactosidase destroys insulin binding (50,51); (iii) tunicamycin, an anti-biotic which inhibits protein glycosylation, prevents insulin receptor synthesis (54,55); (iv) both subunits biosynthetically incorporate galactose, glucosamine, fucose, and mannose (56); and (v) treatment with neuraminidase alters the mobil-ity of both α and β subunits (43,57,58). Thus, both subunits are glycoproteins, and the externally located glycoprotein portion of the receptor participates in insulin binding. Glycosylation may be necessary for delivery of newly synthe-sized receptor to the cell membrane and might also equip the receptor with features needed for insulin binding. In tunicamycin-treated cells, receptors accu-mulate at some intracellular site in a nonglycosylated form which is not able to bind insulin (55).

Based on the available evidence, it is reasonable to assume that the insulin binding site is located in close proximity to the saccharide moieties, however, the lectins, which can mimic insulin, do not compete with insulin binding to intact fat cells (59,60) or may partially compete provided that the fat cells have been preincubated with the lectins for up to 1 hr (60). The detergent-solubilized insulin receptor can be fractionated into hydrophilic and hydrophobic popula-tions (61). Binding of insulin to the hydrophobic population is blocked by con-canavalin A (62). These studies further support the contention that the insulin receptor is heterogenous.

## V. TYROSYL PHOSPHORYLATION

A recent finding of importance is the demonstration that the insulin receptor is an insulin-activated tyrosine-specific protein kinase (31,32). Formerly, it was assumed that the receptor was only an insulin-binding protein. An interesting feature is that insulin stimulates the autophosphorylation of the subunit of its own receptor in both intact cells (31,32) and in a cell-free system (33–35,63,65). Early studies were not conclusive as to whether the insulin receptor is a protein kinase or only a substrate for an intrinsic protein kinase. This issue was solved by the demonstration that highly purified insulin receptor from human placenta (8) retained protein kinase activity (66). In addition to the autophosphorylation event, a search should be made for other immediate cellular substrates of the insulin-receptor kinase. Insulin-receptor kinase also shares common substrate specificity properties and autophosphorylation with other tyrosyl-specific protein kinases, such as those coded by viral oncogenes (68,69) and the cellular growth factors epidermal growth factor (EGF) (69), platelet-derived growth factor (PDGF) (70), and IGF-I (71). All of the above have been implicated in initiating cellular proliferative responses. Insulin does not modulate proliferative response in its typical target tissues but is able to do so in certain cell types (see later).

It is possible that receptor autophosphorylation could participate in altering affinity to insulin, play a role in receptor internalization, or initiate some of insulin's biological responses. The latter possibility is supported by the demonstration that vanadate ions which mimic the biological effects of insulin (72–74) are also potent inhibitors of tyrosine-specific phosphoprotein phosphatase (75).

## VI. INTERRELATIONS OF THE INSULIN RECEPTOR TO OTHER RECEPTOR STRUCTURES

Thanks to cross-linking and photoaffinity-labeling techniques, the minimum subunit structures for receptor systems of several hormones are currently available. Of particular interest was the comparison of the insulin receptor to other receptors of polypeptide hormones which share amino-acid-sequence homology to insulin (i.e., IGF-I, IGF-II) (76,77), or have overlapping biological activities with insulin (i.e., IGF-I, IFG-II, and other growth factors) (85,88), or were demonstrated, like insulin, to have protein kinase activity (i.e., EGF, IGF-I, and PDGF; see previous paragraph). Figure 3 demonstrates that, with the exception of IGF-I, all of the other receptors consist of a single polypeptide chain of molecular weight ranging from 60,000 to 250,000 (78–81) and are, therefore, different in their basic structure from the insulin receptor. In contrast, the structural similarities between the insulin receptor and the IGF-I receptor are striking. Both migrate as high-molecular weight glycoproteins ($M_r$ 350,000) on SDS gels, in the absence of reductant, and dissociate under reducing and denaturing conditions into $\alpha$ and $\beta$ subunits which resemble each other in both types of receptors (36–40,45). The two receptors also resemble each other in their degree of susceptibility to reduction of disulfide bonds or receptor proteolysis (39,82). Two-dimensional peptide mapping of the subunits of insulin and IGF-I indicated that

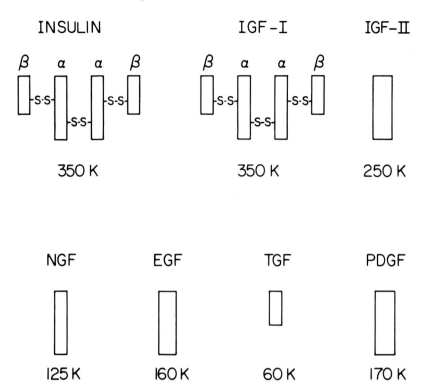

**Fig. 3.** The minimum subunit structures of several receptors for growth factors, as obtained by cross-linking and photoaffinity-labeling techniques (36–40,45,76–81). The receptors of IGF-I (insulin-like growth factor I); IGF-II (insulin-like growth factor II); NGF (nerve growth factor); EGF (epidermal growth factor); TGF (transforming growth factor) and PDGF (platelet-derived growth factor) are shown.

virtually all of the major peptides in the audioradiograms were identical (83), a criterion for a high degree of homology in the primary structure of both receptors. The striking homology strongly suggests either one of two possibilities: (i) the receptors are products of a common single gene, and the variance affinity toward the respective hormones is the consequence of different posttranslational modifications of the same basic polypeptide structure. (ii) The receptors are the products of two distinct genes which diverged from a common ancestral gene. In support of the latter possibility, treatment of IM-9 cells with monensin, an agent that inhibits post translational protein maturation, results in the accumulation of the polypeptide precursor for the insulin receptor (84). The latter is recognizable by insulin and antibodies to the insulin receptor but not by the antibody to the IGF-I receptor, thus indicating that the putative precursors for insulin and IGF-I are distinct polypeptides (84).

As expected from the similarities in receptor structure, insulin and IGF-I have low levels ($\sim 1\%$) of intrinsic cross-reactivity (40,85–88). Thus, high concentrations of insulin can produce mitogenic effects by interacting with the IGF-I receptors of chick embryo and human fibroblasts (40). Similarly, adipocytes (which lack receptors for IGF-I) can respond to IGF-I via their insulin receptors (87). In addition, in some tissues the insulin receptors can modulate mitogenic effects, while that of IGF-I can modulate metabolic effects. Thus, in H35 rat hepatoma, mouse embryonal carcinoma, and human endothelial cells, insulin exerts mitogenic effects at low concentrations via the insulin receptor (86–88), while in muscle and fibroblasts, IGF-I receptors mediate metabolic effects (88).

It can be concluded that each of the two receptors contains a minimum of one binding site with a common recognizable feature toward insulin and IGF-I and either two or one common functional entity which enables each receptor to promote growth and metabolic effects. The elucidation of the primary structures of these receptors and the overlapped-sequence homologies will be important to identify the functional entities of each receptor.

## VII. RECEPTOR-MEDIATED INTERNALIZATION AND DEGRADATION OF INSULIN

Intact cells but not broken cell preparations, degrade [125]I-insulin in direct proportion to the amount of insulin bound at steady state, by a process which is receptor-mediated (89–91). Following incubation of intact cells with insulin, the bound hormone gradually becomes resistant to acid dissociation and trypsin degradation (95). After a lag period, internalized insulin is extruded into the medium both as intact insulin and as its degradation products, the proportion of which vary depending on the cell type (28,92). Degradation can be blocked by inhibitors of lysosomal hydrolases, by uncouplers of oxidative phosphorylation, and by drugs which disassemble microtubules (26,89,93,94). Based on biochemical, electron, and fluorescent microscopy studies in several cell types, the following hypothetical route has been proposed for internalization and degradation of insulin. Insulin initially binds to receptors which are arranged in a diffuse pattern on the cell surface (21,95–97). Once occupied, the receptors are rapidly redistributed to form clusters or patches (21,27,98,99). In fibroblasts, IM-9 lymphocytes, and 3T3-L1 cells, the clusters form in coated pits, which later pinch off to form endocytic vesicles containing internalized insulin–receptor complexes (27,99). In liver and adipocytes, it has been suggested that insulin may not be internalized via coated pits (26,100). The endocytic vesicles develop an acidic pH (101) which causes insulin to dissociate from its receptor. It is postulated that the receptors are sequestered in one region of the vesicle. These eventually bud off, and, with the aid of the cytoskeleton, the derived vesicle, enriched in receptor, appears to be recycled back to the plasma membranes (29,102). The rate of receptor internalization is about the same as that of receptor recycling, since the number of surface receptors remains unaltered (29,103,104). The remaining vesicles, depleted of receptors but containing free insulin, fuse with lysosomes or lysosome-like vesicles and are degraded (105–107). If the process of separation is not complete, insulin may be extruded intact into the medium, while receptors may reach the lysosomes along with free insulin and be degraded.

The efficient processes of internalization and dissociation are, most probably, mechanisms to terminate the biological response after the free-hormone disappears from the circulation. In accordance with this, removal of extracellular insulin terminates bioprocesses with $t_{1/2} \simeq 10$ min (108), and insulin, which was covalently bound to fat cells, produces persistent activation after removal of the unbound hormone (109).

## VIII. DOWN-REGULATION

Down-regulation refers to a loss of surface-receptor sites when cells at 37°C are exposed to moderate or high concentrations of insulin. In addition to temperature and concentration, the process is time dependent and occurs in many cell types *in vivo* and *in vitro* (110–116).

In a series of studies, Lane and co-workers directly measured the rate of receptor synthesis and degradation which contributed a great deal to the understanding of down-regulation (22–24). In these studies, the receptor was biosynthetically labeled by incubating cells with amino acids composed of heavy isotopes. Labeled heavy receptor was separated from naturally occurring receptor by isopycnic centrifugation and quantitated by [125]I-insulin binding (22). Using this technique, Lane and co-workers found that in a variety of cell types total-insulin receptor turns over with a half-life time of about 7 hr (22,24). Out of several possibilities considered, insulin was found to decrease the number of surface receptors in most cell types by accelerating the rate of receptor degradation. An exception is chick hepatocytes in which insulin caused a decrease in cell surface sites but no change in total cellular receptors (23). Krupp and Lane concluded that in this cell line insulin induces translocation of receptors from the cell surface to an intracellular compartment (23).

For most other cell types, 3T3-L1 adipocytes are a more representative example of insulin-dependent receptor regulation. When insulin is removed from the medium of fully differentiated 3T3-L1 adipocytes, insulin receptors are doubled in number within 24 hr (up-regulation). When up-regulated 3T3-L1 cells are reexposed to insulin, down-regulation occurs with a $t_{1/2}$ of 2–3 hr (117). A shift in the $t_{1/2}$ value of the receptor from 8.1 to 14.8 hr occurred in the up-regulated cells, and this value returned to 6.9 hr in the down-regulated cells (117). Since under all conditions the rate of receptor synthesis is not modified, up-regulation was due to a decreased rate of receptor degradation whereas down-regulation was the consequence of an increased degradation rate.

Kosmacos and Roth who examined this process in IM-9 lymphocytes also concluded that down-regulation is due to an increased rate of receptor degradation (118). In this cell line, however, cycloheximide prevented insulin-induced down-regulation. Therefore, it is possible that in IM-9 lymphocytes down-regulation requires biosynthesis of protein(s) (118).

Several investigators have implicated down-regulation with receptor-mediated internalization of insulin. This, however, does not seem to be the case in most or all cell types, since, as discussed previously, the internalized receptor escapes degradation and is recycled to the plasma membrane. As for EGF, both internalized hormone molecules and receptors are ultimately degraded by lysosomes, and for this growth factor a good correlation exists between down-regulation and ligand degradation (119,120).

The biochemical basis for the down-regulation is not yet known. The increased susceptibility of the occupied receptors to the cellular-degrading system may originate from one or a combination of secondary events at the level of the receptors, such as partial reduction of inter- or intradisulfide bonds, change in the state of aggregation, or phosphorylation of receptors.

## IX. POSSIBLE INVOLVEMENT OF CYTOSKELETAL ELEMENTS IN INSULIN ACTION

As mentioned previously, insulin induces clustering and internalization of its own receptors (26,27,100). It also stimulates enzymatic systems located on internal organelles such as mitochondrial pyruvate dehydrogenase and the nuclear RNA polymerase II [reviewed in Czech (121)]. Because of such observations, the possible involvement of the cytoskeletal system in clustering, activation, and signal-transmission events has been frequently considered. Microtubule- and microfilament-disrupting agents, however, were unable to resolve these issues, because the agents used proved to be either ineffective or nonspecific. Intermediate (10 n$M$) filaments may still be involved, since the latter are not influenced by the routinely used disrupting agents (122). Development of the fluorescent photobleaching recovery technique allowed study of the lateral mobility of insulin receptor on the cellular membrane. In the case of fibroblasts, the rate of lateral mobility was found to be 3–5 $\times$ 10$^{-10}$ cm$^2$/sec (19), a value more than an order of magnitude slower than that of a lipid probe (123). Theoretically, a membrane protein is expected to have a $D$ value only slightly smaller than that of a lipid probe unless restricted by factors other than plasma membrane viscosity, such as possible association with the cytoskeleton (124). Indeed, proteins which are not restricted by the cytoskeletal meshwork or proteins which have been experimentally disconnected from the cytoskeleton have lateral mobilities close to those of lipid probes (125–127). Hence, possible involvement of the cytoskeletal system of the cell requires further studies. Techniques of permeating target cells under conditions in which the action of insulin is preserved together with the use of specific antibodies to cytoskeletal components may resolve this issue.

## X. RECEPTOR VALENCE

The proposed symmetrical structure of the insulin receptor having a molecular weight of 350,000–360,000 (7,47–49) (Fig. 1) suggests that two molecules of insulin bind to each receptor molecule. A study by Harrison et al. (128) supports this notion. The soluble $^{125}$I-insulin–receptor complex, in which only part of the sites are occupied, could be quantitatively immunoprecipitated by autoantibodies to the insulin receptor (12), even though these antibodies and insulin mutually competed for receptor binding. By saturating the sites with $^{125}$I-insulin, the fraction of insulin–receptor complex that could be immunoprecipitated declined (128). This was consistent with bivalency or multivalency of the insulin receptor, since immunoprecipitation of receptor along with insulin presumably occurred by interaction of the antibody with empty sites (128).

Assuming that the receptor for insulin is bivalent, a specific activity of 33 μg insulin should bind per milligram of receptor protein. Indeed, in the purified preparation of human-placenta-insulin receptor (8) which was eluted under mild conditions to prevent receptor denaturation a maximum insulin binding of 28.5 μg/mg of protein was obtained (8). It, therefore, seems that the present data are consistent with receptor bivalency. There is insufficient evidence at present for receptor multivalency. The temperature-dependent redistribution of the insulin receptors, followed by energy and cytoskeletal-dependent internalization (previous section) seems to be common to numerous hormones and other ligands as well. In addition to multivalency, a common signal to many occupied receptors (such as a common recognition site of receptors to components of the coated pit or the microtubular system) should be considered as well.

## XI. NONLINEAR SCATCHARD PLOTS

In most instances, Scatchard plots of insulin-receptor binding in whole cells and plasma membranes are concave (129–131). This may be ascribed to negative cooperativity in insulin-receptor binding (129,132) or binding-site heterogeneity, involving two major classes of binding sites with widely differing affinities and capacities. Varying affinities toward insulin may be expected in view of the complexity of the insulin receptors described in the previous sections. Heterogeneity may be originated from different contents of saccharide moieties, several redox forms of the insulin receptor, secondary alterations such as proteolysis, receptor phosphorylation, and dissociation into smaller subunits. In whole cells or plasma membranes, factors which may contribute to "negatively cooperative" behavior may exist such as multiple-interacting binding sites or secondary interactions of occupied receptors with noninsulin-binding regulatory proteins (133,134).

Such interactions with nonreceptor protein are not expected, however, to occur in the Triton X-100, highly purified, solubilized receptor. There is no evidence also for receptor–receptor interactions under the above circumstances (8). In spite of this, the highly purified soluble-insulin receptor showed a curvilinear Scatchard plot (8). This seems to limit the possibilities to two: negative cooperativity exists within the two binding sites of the heterotetrameric receptor molecule, or (more likely) the pure soluble-receptor preparation is heterogenous with respect to varying affinities toward insulin binding.

## XII. POSSIBLE ROLE FOR RECEPTOR CROSS-LINKING AND FOR NON-INSULIN-BINDING REGULATORY GLYCOPROTEIN

The demonstrations that epidermal growth factor (EGF) and insulin are mobile in the plane of the plasma membrane and that the homogeneously distributed receptors are clustered after being occupied by the respective hormones (19,20) raised the question whether cross-linking or aggregation phenomena are relevant for eliciting responses. As for immunoglobulins, there is convincing evidence that dimerization of two receptor molecules of immunoglobulin E is a sufficient signal for mediating histamine release from basophils and mast cells (135). Among the polypeptide hormones, the addition of anti-EGF antibody to fibroblasts that were incubated with low concentrations of EGF, which by itself were too low to elicit significant response, resulted in the stimulation of DNA synthesis (136). In addition, cyanogen bromide-cleaved EGF, a partial antagonist of EGF, regained its ability to elicit a biological response and to form clusters on the cell surface of fibroblasts when cross-linked with anti-EGF antibodies (136). Strong evidence for the relevance of cross-linking to activation also comes from another hormonal system. An antagonistic derivative of gonadotropin releasing hormone (GnRH) becomes an agonist when cross-linked by anti-GnRH antibody (137). In this study, it was also demonstrated that the linked receptors must be separated by 15–150 Å. No response was elicited by chemically cross-linked GnRH antagonist that was separated by 15 Å (137).

Several multivalent ligands, such as plant lectins and antibodies which presumably can cross-link insulin receptors, are able to mimic the biological actions of insulin (138–140). In two specific cases it was demonstrated that the monovalent Fab fragments were unable to produce responses but regained this ability when cross-linked by bivalent anti-Fab antibodies (140,141). In one study, antibodies to insulin increased the affinity of labeled hormone toward fibroblasts and liver membrane cells. The study suggested that the low-affinity sites for insulin in these target tissues are defective in their ability to self-aggregate (142).

While these studies suggest that receptor cross-linking may have a biological role in insulin action, other interpretations of the available experimental data should be considered. It is not clear whether other membrane components participate in insulin action. Cross-linking events may increase the dissociation or association of the insulin receptor with other membrane components. In fact, several studies have supplied experimental evidence to support the existence of a non-insulin-binding regulatory protein (effector E) as an integral part of the overall insulin machinery system (133,134,143).

Several mathematical models postulate a secondary interaction of hormone–receptor complex (HR) with effector to form the ternary complex hormone–receptor–effector (HRE) (144,145). The physiological response elicited by the hormone would be proportional to the amount of the ternary complex formed. Assuming such a model to be correct, some of the multivalent proteins that mimic insulin may do so by cross-linking receptor to effector or by perturbing the effector in some way that resembled its perturbation by the hormone–receptor complex.

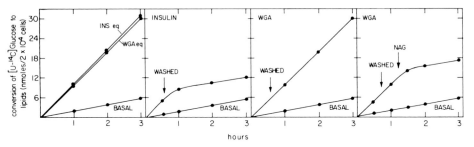

**Fig. 4.** Persistent activation of lipogenesis in wheat germ agglutinin-adsorbed adipocytes. From left to right; lipogenesis produced by insulin or WGA at equilibrium; decreased rate of lipogenesis after removal of insulin; persistent activation of lipogenesis after removal of unbound lectin; and termination of persistent lipogenesis by the addition of 50 mM N-acetyl-D-glucosamine (NAG) [according to Shechter (143)].

A recent study (143) has demonstrated that wheat germ agglutinin and other lectins, which mimic insulin, are not internalized and processed by adipocytes, as occurred with insulin itself (previous paragraph). As a consequence of this, the insulin–effector system is "permanently" activated in WGA-adsorbed adipocytes, in a lectin-free medium (143; Fig. 4). Termination is achieved at each state by the addition of $N$-acetylglucosamine (Fig. 4). The study supports the contention that other cell surface determinants (effectors) are involved in insulin action. The latter are not linked to a receptor-mediated terminating system.

## XIII. TRANSMEMBRANE SIGNALING

Shortly after the binding of insulin to its surface receptors, a wide variety of metabolic pathways and substrate fluxes are altered [reviewed in Czech (121)]. Any explanation of the mechanism of insulin action must take into account a system which transmits the signal obtained by the initial hormone–receptor interaction to plasma membrane transport system, as well as to cytoplasmic metabolic pathways and enzymatic systems which are located on internal organelles. With the currently available data, three such possibilities for trans-membrane signaling may be considered, although each one alone cannot account for the wide spectrum of insulin-dependent events. Therefore, two or more mechanisms may be operative, and they may work independently or in concert. The finding that the insulin receptor is an insulin-activated tyrosine-specific protein kinase (Section V) may be considered in the signal-transfer mechanism, since several insulin-responsive cellular enzyme systems have been shown to be regulated by covalent modification via phosphorylation reactions [reviewed in Czech (121)]. It is possible to imagine a cascade of phosphorylation–de-phosphorylation events initiated by the binding of insulin to the insulin-receptor kinase. A series of protein kinases and phosphate-transfer proteins may supply the system with the amplification and direction needed. The cytoskeleton (pre-vious paragraph) may also participate in transmembrane signaling. Supportive evidence for this may be derived from other hormonal systems. Cytochalasin B and, more specifically, the incorporation of anti-actin into adrenal gland cells by fusion techniques inhibited the action of adrenocorticotropic hormone and di-butyryladenosine 3′,5′-monophosphate on cholesterol transport from the cytoplasm to the mitochondrion and its processing (146,147). Similar studies with Leydig cells gave essentially the same results. Anti-actin-inhibited choles-terol transport to mitochondria, in response to lutropin (LH) and cyclic AMP in these cells (148). These studies suggest that microfilaments can play a role in regulating vectorial transport of signals through the cytoplasm and, thus, may also participate in insulin action in activating at least those enzymatic systems which are located on internal organelles.

A third possibility is the production of intracellular mediators. Studies from several laboratories have demonstrated that addition of insulin to a broken-cell preparation containing plasma membranes and mitochondria results in activation of mitochondrial pyruvate dehydrogenase (149–151). Insulin does not have a direct effect on isolated mitochondria, therefore, this finding indicated that the interaction of insulin with the plasma membrane generated a substrate(s) that activates pyruvate dehydrogenase. Indeed, the substrate(s), after being freed from the plasma membranes and insulin, were able to stimulate mitochondrial pyruvate dehydrogenase (152,153) as well as glycogen synthase (154), low $K_m$ cyclic AMP-phosphodiesterase (155), and to inhibit cyclic AMP-dependent protein kinase (154). There were also reports on the insulin-dependent generation of a substance that stimulates RNA polymerase II in isolated nuclei (156). Thus, still-undefined soluble substances are produced by the interaction of insulin with the plasma membranes, the latter seem to modulate many insulin-dependent enzymatic systems in the right direction, and further studies should be made to determine the nature and the precise role of these substrates in transmembrane signaling.

Calcium ions and calmodulin or another $Ca^{2+}$-regulatory binding protein may also participate in transmembrane signaling. Many studies suggest that one of the first events after insulin binding to intact cells is a rise in cytoplasmic $Ca^{2+}$, most likely originating from internal calcium pools [reviewed in Czech (121)]. The latter event was hypothesized to modulate many insulin-dependent biochemical pathways (121). A rise in calcium by itself, however, does not seem to be sufficient, since conditions that increase the level of cytoplasmic $Ca^{2+}$ do not mimic insulin (157,158). In recent years it has become apparent that many of the biological effects of $Ca^{2+}$ are directed to specific regions or to specific activities within the cell by calmodulin or related proteins [reviewed in Means and Dedman (159) and Chueng (160)].

Two recent studies suggest that indeed such a protein participates in insulin action. The anticalmodulin drug trifluperazine inhibits the action of insulin on hexose transport and glucose metabolism but does not affect the ability of insulin to inhibit lipolysis (161). Direct addition of insulin to adipocytic plasma membrane increases their binding capacity toward labeled calmodulin (162). Therefore, one of the initial events in insulin action may be the recruitment of calmodulin from the cytoplasm to the internal side of the plasma membrane. The latter may participate in phosphorylation–dephosphorylation reactions by altering the activity of relevant protein kinases (163,164). Alternatively, calmodulin may participate more directly in the insulin-dependent recruitment of glucose-transport activity from the storage sites to the plasma membrane (165–168). This calcium binding regulatory protein may decrease the calcium threshold concentration which is required to induce the exocytotic processes most likely involved in the translocation of glucose-transport activity to the plasma membranes.

## ACKNOWLEDGMENTS

I wish to thank Professor M. Yatvin from the University of Wisconsin, Madison, Wisconsin, and Professor I. R. Cohen from the Department of Cell Biology, The Weizmann Institute of Science, for critically reading and commenting on the initial draft of this chapter.

## REFERENCES

1. Cuatrecasas, P. (1971). Insulin receptor interactions in adipose tissue cells: direct measure and properties. *Proc. Natl. Acad. Sci. U.S.A.* **68**, 1264–1268.
2. Freychet, P., Roth, J., and Neville, D. M. (1971). Insulin receptors in liver, specific binding of $^{125}$I-insulin to plasma membranes and its relation to bioactivity. *Proc. Natl. Acad. Sci. U.S.A.* **68**, 1833–1837.
3. Cuatrecasas, P. (1972). Isolation of the insulin receptor of liver and fat cell membranes. *Proc. Natl. Acad. Sci. U.S.A.* **69**, 318–322.
4. Jacobs, S., Shechter, Y., Bissell, K., and Cuatrecasas, P. (1977). Purification and properties of insulin receptors from rat liver membranes. *Biochem. Biophys. Res. Commun.* **77**, 981–988.
5. Williams, P. E., and Turkle, J. R. (1979). Purification of the insulin receptor from human placental membranes. *Biochim. Biophys. Acta* **579**, 367–374.
6. Harrison, L. C., and Itin, A. (1980). Purification of the insulin receptor from human placenta by chromatography on immobilized wheat germ lectin and receptor antibody. *J. Biol. Chem.* **255**, 12066–12072.
7. Siegel, T. W., Ganguly, S., Jacobs, S., Rosen, O. M., and Rubin, C. S. (1981). Purification and properties of the human placental insulin receptor. *J. Biol. Chem.* **256**, 9266–9273.
8. Fujita-Yamaguchi, Y., Choi, S., Sakamoto, Y., and Itakura, K. (1983). Purification of the insulin receptor with full binding activity. *J. Biol. Chem.* **258**, 5045–5049.
9. Jacobs, S., Hazum, E., Shechter, Y., and Cuatrecasas, P. (1979). Insulin receptor: covalent labeling and identification of subunits. *Proc. Natl. Acad. Sci. U.S.A.* **76**, 4918–4921.
10. Massague, J., and Czech, M. P. (1980). Multiple redox forms of the insulin receptor in native liver membranes. *Diabetes* **29**, 945–947.
11. Yip, C. C., Yeung, C. W. T., and Moule, M. L. (1978). Photoaffinity labeling of insulin receptor of rat adipocytes plasma membrane. *J. Biol. Chem.* **253**, 1743–1745.
12. Yeung, C. W. T., Moule, M. L., and Yip, C. C. (1980). Photoaffinity labeling of insulin receptor with an insulin analogue selectively modified at the amino terminal of the B chain. *Biochemistry* **19**, 2196–2203.
13. Yip, C. C., Yeung, C. W. T., and Moule, M. L. (1980). Photoaffinity labeling of insulin receptor proteins of liver plasma membrane preparations. *Biochemistry* **19**, 70–76.
14. Yip, C. C., Moule, M. L., and Yeung, C. W. T. (1982). Subunit structure of insulin receptor of rat adipocytes as demonstrated by photoaffinity labeling. *Biochemistry* **21**, 2940–2945.
15. Hofman, C., Ji, T. H., Miller, G., and Steiner, D. F. (1981). Photoaffinity labeling of the insulin receptor in H4 hepatoma cells: Lack of cellular receptor processing. *J. Supramol. Struct. Cell. Biochem.* **15**, 1–13.

16. Pilch, P. E., and Czech, M. P. (1979). Interaction of cross-linking agents with the insulin effector system of isolated fat cells. *J. Biol. Chem.* **254,** 3375–3381.
17. Pilch, P. E., and Czech, M. P. (1980). The subunit structure of the high-affinity insulin receptor. Evidence of disulfide linked receptor complex in fat cells and liver plasma membranes. *J. Biol. Chem.* **255,** 1722–1721.
18. Massague, J., Pilch, P. F., and Czech, M. P. (1980). Electrophoretic resolution of three major insulin receptor structures with unique subunit stoichiometries. *Proc. Natl. Acad. Sci. U.S.A.* **77,** 7137–7141.
19. Shechter, Y., Schlessinger, J., Jacobs, S., Chang, K.-J., and Cuatrecasas, P. (1978). Fluorescent labeling of hormone receptors in viable cells: Preparation and properties of highly fluorescent derivatives of epidermal growth factor and insulin. *Proc. Natl. Acad. Sci. U.S.A.* **75,** 2135–2139.
20. Schlessinger, J., Shechter, Y., Willingham, M. C., and Pastan, I. (1978). Direct visualization of binding, aggregation, and internalization of insulin and epidermal growth factor on living fibroblastic cells. *Proc. Natl. Acad. Sci. U.S.A.* **75,** 2659–2663.
21. Schlessinger, J., Van Obberghen, E., and Kahn, C. R. (1980). Insulin and antibodies against insulin receptor cap on membrane of cultured human lymphocytes. *Nature (London)* **286,** 729–731.
22. Reed, B. C., and Lane, M. D. (1980). Insulin receptor synthesis and turnover in differentiating 3T3-L1 preadipocytes. *Proc. Natl. Acad. Sci. U.S.A.* **77,** 285–289.
23. Krupp, M., and Lane, M. D. (1981). On the mechanism of ligand induced down regulation of insulin receptor level in the liver cell. *J. Biol. Chem.* **256,** 1689–1694.
24. Reed, B. C., Ronnett, G. V., Clements, P. R., and Lane, M. D. (1981). Regulation of insulin receptor metabolism, differentiation-induced alteration of receptor synthesis and degradation. *J. Biol. Chem.* **256,** 3917–3925.
25. Kahn, C. R., and Baird, K. (1978). The fate of insulin bound to adipocytes. *J. Biol. Chem.* **253,** 4900–4906.
26. Hammons, G. T., and Jarett, L. (1980). Lysozomal degradation of receptor bound $^{125}$I-labeled insulin by intact adipocytes. *Diabetes* **29,** 475–486.
27. Maxfield, F. R., Schlessinger, J., Shechter, Y., Pastan, I., and Willingham, M. C. (1978). Collection of insulin, EGF, and $\alpha_2$-macroglobulin in the same patches on the surface of cultured fibroblasts and common internalization. *Cell* **14,** 805–810.
28. Suzuki, K., and Kono, T. (1979). Internalization and degradation of fat cell bound insulin: Separation and partial characterization of subcellular vesicles associated with iodoinsulin. *J. Biol. Chem.* **254,** 9786–9794.
29. Marshall, S., Green, A., and Olefsky, J. M. (1981). Evidence of recycling of insulin receptors in isolated rat adipocytes. *J. Biol. Chem.* **256,** 11464–11470.
30. Green, A., and Olefsky, J. M. (1982). Evidence for insulin-induced internalization and degradation of insulin receptors in rat adipocytes. *Proc. Natl. Acad. Sci. U.S.A.* **79,** 427–431.
31. Kasuga, M., Karlsson, F. A., and Kahn, C. R. (1981). Insulin stimulates the phosphorylation of the 95,000 dalton subunit of its own receptor. *Science (Washington, D.C.)* **215,** 185–187.
32. Kasuga, M., Zick, Y., Blithe, D. L., Karlsson, F. A., Hüring, H. V., and Kahn, C. R. (1982). Insulin stimulation of phosphorylation of the subunit of the insulin receptor formation of both phosphoserine and phosphotyrosine. *J. Biol. Chem.* **257,** 9891–9894.
33. Avruch, J., Nemenoff, R. A., Blackshear, P. J., Pierce, M. W., and Osathanondh, R. (1982). Insulin stimulated tyrosine phosphorylation of the insulin receptor in detergent extracts of human placental membranes. *J. Biol. Chem.* **257,** 15162–15166.
34. Kasuga, M., Zick, Y., Blithe, D. L., Crettaz, M., and Kahn, C. R. (1982). Insulin stimulates tyrosine phosphorylation of the insulin receptor in a cell-free system. *Nature (London)* **298,** 667–669.

35. Van Obberghen, E., Rassi, B., Kowalski, A., Gazzano, H., and Ponzio, G. (1983). Receptor-mediated phosphorylation of the hepatic insulin receptor. Evidence that the $M_r$ 95,000 receptor subunit is its own kinase. *Proc. Natl. Acad. Sci. U.S.A.* **80,** 945–949.

36. Massague, J., Guillette, B. J., and Czech, M. P. (1981). Affinity labeling of multiplication stimulating activity receptors in membranes from rat and human tissues. *J. Biol. Chem.* **256,** 2122–2125.

37. Kasuga, M., Van Obberghen, E., Nissley, S. P., and Rechler, M. M. (1981). Demonstration of two subtypes of insulin-like growth factor receptors by affinity cross-linking. *J. Biol. Chem.* **256,** 5305–5308.

38. Bhaumick, B., Bala, R. M., and Hollenberg, M. D. (1981). Somatomedin receptor of human placenta: Solubilization, photolabeling, partial purification, and comparison with insulin receptor. *Proc. Natl. Acad. Sci. U.S.A.* **78,** 4279–4283.

39. Chernausek, S. D., Jacobs, S., and Van Wyk, J. J. (1981). Structural similarities between human receptors for somatomedin C and insulin analysis by affinity labeling. *Biochemistry* **20,** 7345–7350.

40. Kasuga, M., Van Obberghen, E., Nissley, S. P., and Rechler, M. M. (1982). Structure of the insulin-like growth factor receptor in chicken embryo fibroblasts. *Proc. Natl. Acad. Sci. U.S.A.* **79,** 1864–1868.

41. Cuatrecasas, P., and Tell, G. P. E. (1973). Insulin-like activity of concanavalin A and wheat germ agglutinin, direct interactions with insulin receptors. *Proc. Natl. Acad. Sci. U.S.A.* **70,** 485–489.

42. Cuatrecasas, P. (1972). Affinity chromatography and purification of the insulin receptor of liver cell membranes. *Proc. Natl. Acad. Sci. U.S.A.* **69,** 1277–1281.

43. Jacobs, S., Hazum, E., and Cuatrecasas, P. (1980). The subunit structure of rat liver insulin receptor. *J. Biol. Chem.* **255,** 6937–6940.

44. Jacobs, S., and Cuatrecasas, P. (1981). Insulin receptor: Structure and function. *Endocr. Rev.* **2,** 251–263.

45. Massague, J., and Czech, M. P. (1982). The subunit structures of two distinct receptors for insulin-like growth factors I and II and their relationship to the insulin receptor. *J. Biol. Chem.* **257,** 5038–5045.

46. Massague, J., Pilch, P. F., and Czech, M. P. (1981). A unique proteolytic cleavage site on the β subunit of the insulin receptor. *J. Biol. Chem.* **256,** 3182–3190.

47. Cuatrecasas, P. (1972). Properties of insulin receptor isolated from liver and fat cell membranes. *J. Biol. Chem.* **247,** 1980–1991.

48. Baron, M. D., Wisher, M. H., Thamm, P. M., Saunders, D. J., Brandenburg, D., and Sonksen, P. H. (1981). Hydrodynamic characterization of the photoaffinity-labeled insulin receptor solubilized in Triton X-100. *Biochemistry* **20,** 4156–4161.

49. Pollet, R. J., Haase, B. A., and Standaert, M. L. (1981). Characterization of detergent-solubilized membrane proteins. *J. Biol. Chem.* **256,** 12118–12126.

50. Cuatrecasas, P., and Illiano, G. (1971). Membrane sialic acid and the mechanism of insulin action in adipose tissue cells. *J. Biol. Chem.* **246,** 4938–4946.

51. Rosenthal, J. W., and Fain, J. N. (1971). Insulin-like effect of clostridial phospholipase C, neuraminidase, and other bacterial factors on brown fat cells. *J. Biol. Chem.* **246,** 5888–5895.

52. Yip, C. C., and Moule, M. L. (1983). Structure of the insulin receptor of rat adipocytes: The three interconvertible redox forms. *Diabetes* **32,** 760–767.

53. Cuatrecasas, P. (1973). Interaction of concanavalin A and wheat germ agglutinin with the insulin receptor of fat cells and liver. *J. Biol. Chem.* **248,** 3528–3534.

54. Rosen, O. M., Chia, G. H., Fung, C., and Rubin, C. S. (1979). Tunicamycin-mediated depletion of insulin receptors in 3T3-L1 adipocytes. *J. Cell. Physiol.* **99,** 37–42.

55. Ronnett, G. V., and Lane, M. D. (1981). Post translation glycosylation-induced activation of

aglycoinsulin receptor accumulation during tunicamycin treatment. *J. Biol. Chem.* **256,** 4704–4707.

56. Hedo, J. R., Kasuga, M., Van Oberghen, E., Roth, J., and Kahn, C. R. (1981). Direct demonstration of glycosylation of insulin receptor subunits by biosynthetic and external labeling: Evidence for heterogeneity. *Proc. Natl. Acad. Sci. U.S.A.* **78,** 4791–4795.

57. Jacobs, S., Hazum, E., and Cuatrecasas, P. (1980). Digestion of insulin receptors with proteolytic and glycosidic enzymes—effect on purified and membrane-associated receptor subunits. *Biochem. Biophys. Res. Commun.* **94,** 1066–1073.

58. Van Obberghen, E., Kasuga, M., LeCam, A., Hedo, J. A., and Harrison, L. C. (1981). Biosynthetic labeling of insulin receptor: Studies of subunits in cultured human IM-9 lymphocytes. *Proc. Natl. Acad. Sci. U.S.A.* **78,** 1052–1056.

59. Shechter, Y., and Sela, B.-A. (1981). Insulin-like effects of wax bean agglutinin in rat adipocytes. *Biochem. Biophys. Res. Commun.* **98,** 367–373.

60. Katzen, H. M., Vicario, P. P., Mumford, R. A., and Green, B. G. (1981). Evidence that the insulin-like activities of Concanavalin A and insulin are mediated by a common insulin receptor linked effector system. *Biochemistry* **20,** 5800–5809.

61. Bordier, C. (1981). Phase separation of integral membrane proteins in Triton X-114 solution. *J. Biol. Chem.* **256,** 1604–1607.

62. Meyer, H. E., Kopp, F., Bücher, V., Pablock, W., Sessiz, T., and Reimauer, H. (1983). Separation of insulin receptor proteins in the Triton X-114 two-phase system. Evidence for two distinct receptor populations. *Int. Symp. Insulin Recept., 2nd, Rome,* Abstr., p. 12.

63. Petruzzelli, L. M., Ganguly, S., Smith, C. J., Cobb, M. H., Rubin, C. S., and Rosen, O. (1982). Insulin activates a tyrosine-specific protein kinase in extracts of 3T3-L1 adipocytes and human placenta. *Proc. Natl. Acad. Sci. U.S.A.* **79,** 6792–6796.

64. Roth, R. A., and Cassel, D. J. (1983). Insulin receptor: Evidence that it is a protein kinase. *Science (Washington, D.C.)* **219,** 299–301.

65. Shia, M. A., and Pilch, P. F. (1983). The subunit of the insulin receptor is an insulin-activated protein kinase. *Biochemistry* **22,** 717–721.

66. Kasuga, M., Fujita-Yamaguchi, Y., Blithe, D. L., and Kahn, C. R. (1983). Tyrosine-specific protein kinase activity is associated with the purified insulin receptors. *Proc. Natl. Acad. Sci. U.S.A.* **80,** 2137–2141.

67. Eckhart, W., Hutchinson, M. A., and Hunter, T. (1979). An activity phosphorylating tyrosine in polyoma T antigen immunoprecipitates. *Cell* **18,** 925–933.

68. Hunter, T., and Sefton, B. M. (1980). Transforming gene product of Rous Sarcoma virus phosphorylates tyrosine. *Proc. Natl. Acad. Sci. U.S.A.* **77,** 1311–1315.

69. Carpenter, G., King, L., Jr., and Cohen, S. (1979). Rapid enhancement of protein phosphorylation in A-431 cell membrane preparations by epidermal growth factor. *J. Biol. Chem.* **254,** 4884–4891.

70. Ek, B., and Heldin, B.-H. (1982). Characterization of a tyrosine-specific kinase activity in human fibroblast membranes stimulated by platelet-derived growth factor. *J. Biol. Chem.* **257,** 10486–10492.

71. Jacobs, S., Kull, F. C., Jr., Earp, H. S., Svoboda, M. E., Van Wyk, J. J., and Cuatrecasas, P. (1983). Somatomedin-C stimulates the phosphorylation of the β-subunit of its own receptor. *J. Biol. Chem.* **258,** 9581–9584.

72. Shechter, Y., and Karlish, S. J. D. (1980). Insulin-like stimulation of glucose oxidation in rat adipocytes by vanadyl (IV) ions. *Nature (London)* **284,** 556–558.

73. Dubyak, G. R., and Kleinzeller, A. (1980). The insulin-mimetic effects of vanadate in isolated rat adipocytes. Dissociation from effects of vanadate as $(NA^+-K^+)ATPase$ inhibitors. *J. Biol. Chem.* **255,** 5306–5312.

74. Degani, H., Gochin, M., Karlish, S. J. D., and Shechter, Y. (1981). Electron paramagnetic

resonance studies and insulin-like effects of vanadium in rat adipocytes. *Biochemistry* **20**, 5795–5799.

75. Swarp, G., Speeg, K. V., Jr., Cohen, S., and Garbes, D. L. (1982). Phosphotyrosyl-protein phosphatase of TCRC-2 cells. *J. Biol. Chem.* **257**, 7298–7301.

76. Blundell, T. L., and Humbel, R. E. (1980). Hormone families: Pancreatic hormones and homologous growth factors. *Nature (London)* **287**, 781–787.

77. Rinder Knecht, E., and Humbel, R. E. (1978). The amino acid sequence of human insulin-like growth factor I and its structural homology with proinsulin. *J. Biol. Chem.* **253**, 2769–2776.

78. Das, M., Miyakawa, T., Fox, C. F., Pruss, R. M., Aharonou, A., and Hershman, H. R. (1977). Specific radiolabeling of a cell surface receptor for epidermal growth factor. *Proc. Natl. Acad. Sci. U.S.A.* **74**, 2790–2796.

79. Glenn, K., Bowen-Pope, D. F., and Ross, R. (1982). Platelet-derived growth factor. III. Identification of a platelet-derived growth factor by affinity labeling. *J. Biol. Chem.* **257**, 5172–5176.

80. Massague, J., Czech, M. P., Iwata, K., De Larco, J. E., and Todaro, G. J. (1982). Affinity labeling of a transforming growth factor receptor that does not interact with epidermal growth factor. *Proc. Natl. Acad. Sci. U.S.A.* **79**, 6822–6826.

81. Massague, J., Guillette, B. J., Czech, M. P., Morgan, C. J., and Bradshaw, R. A. (1981). Identification of a nerve growth factor receptor protein in sympathetic ganglia membranes by affinity labeling. *J. Biol. Chem.* **256**, 9419–9424.

82. Massague, J., and Czech, M. P. (1982). Role of disulfides in the subunit structure of the insulin receptor. *J. Biol. Chem.* **257**, 6729–6738.

83. Massague, J., Yu, K.-T., Heinrich, J., Mattola, C., and Czech, M. P. (1983). Structural and functional homologies among the receptors for insulin and the insulin-like growth factors. *Int. Symp. Insulin Recept., Rome,* Abstr., p. 13.

84. Jacobs, S., Kull, C. F., Jr., and Cuatrecasas, P. (1983). Monensin blocks the maturation of receptors for insulin and somatomedin C: Identification of receptor precursors. *Proc. Natl. Acad. Sci. U.S.A.* **80**, 1228–1231.

85. Hintz, R. L., Clemmons, D. R., Underwood, L. E., and Van Wyk, J. J. (1972). Competitive binding of somatomedin to insulin receptors of adipocytes. chondrocytes and liver membranes. *Proc. Natl. Acad. Sci. U.S.A.* **69**, 2351–2353.

86. Koonts, J. W., and Iwahashi, M. (1981). Insulin as a potent specific growth factor in a rat hepatoma cell line. *Science (Washington, D.C.)* **211**, 947–949.

87. Massague, J., Blinderman, L. A., and Czech, M. P. (1982). The high affinity insulin receptor mediates growth stimulation in rat hepatoma cells. *J. Biol. Chem.* **257**, 13958–13963.

88. Rechler, M. M., Kasuga, M., Sasaki, N., De Vroede, M. A., Romanus, J. A., and Nissley, S. P. (1983). Insulin-like growth factors. *In* "Somatomedins: Basic Chemistry, Biology and Clinical Importance" (E. M. Spencer, ed.), de Gruyter, Berlin.

89. Terris, S., and Steiner, D. (1975). Binding and degradation of $^{125}$I-insulin by rat hepatocytes. *J. Biol. Chem.* **250**, 8389–8398.

90. Terris, S., and Steiner, D. F. (1976). Detention and degradation of $^{125}$I-insulin by perfused livers from diabetic rats. *J. Clin. Invest.* **57**, 855–896.

91. Baldwin, D., Jr., Terris, S., and Steiner, D. F. (1980). Characterization of insulin-like actions of anti-insulin–receptor antibodies. *J. Biol. Chem.* **255**, 402–407.

92. Sonne, O., and Glieman, J. (1980). Insulin receptors in cultured human lymphocytes (IM-9): Lack of receptor mediated degradation. *J. Biol. Chem.* **255**, 7449–7454.

93. Gliemann, J., and Sonne, O. (1978). Binding and receptor-mediated degradation of insulin in adipocytes. *J. Biol. Chem.* **253**, 7857–7863.

94. Hofmann, C., Marsh, J. W., Miller, B., and Steiner, D. F.(1980). Cultured hepatoma cells as a model system for studying insulin processing and biologic responsiveness. *Diabetes* **29**, 865–874.

95. Bergeron, J. J. M., Levine, G., Sikstrom, R., O'Shaughnessy, D., Kopriwa, B., *et al.* (1978). Polypeptide hormone binding sites *in vivo:* Initial localization of $^{125}$I-labeled insulin to hepatocyte plasmalemma as visualized by electron microscope radioautography. *Proc. Natl. Acad. Sci. U.S.A.* **74**, 5051–5055.

96. Carpenter, J.-L., Gordon, P., Amherdt, M., Van Obberghen, E., Kahn, C.-R., and Orci, L. (1978). $^{125}$I-insulin binding to cultural human lymphocytes: Initial localization and fate of hormone determined by quantitative electron microscopic autoradiography. *J. Clin. Invest.* **61**, 1057–1070.

97. Gorden, P., Carpenter, J.-L., Freychet, P., Le Cam, A., and Orchi, L. (1978). Intracellular translocation of iodine-125-labeled insulin. Direct demonstration in isolated hepatocytes. *Science (Washington, D.C.)* **200**, 782–785.

98. Barazzone, P., Carpenter, J.-L., Gorden, P., Van Obberghen, E., and Orci, L. (1980). Polar redistribution of $^{125}$I-labelled insulin on the plasma membrane of cultured human lymphocytes. *Nature (London)* **286**, 401–403.

99. Carpenter, J.-L., Van Obberghen, E., Gordon, P., and Orci, L. (1981). Surface redistribution of $^{125}$I-insulin in cultured human lymphocytes. *J. Cell Biol.* **91**, 17–25.

100. Bergeron, J. J. M., Sikstrom, R., Hand, A. R., and Posner, B. I. (1979). Binding and uptake of $^{125}$I-insulin into rat liver hepatocytes and endothelium. *J. Cell Biol.* **80**, 427–443.

101. Tycko, B., and Maxfield, F. R. (1982). Rapid acidification of endocytic vesicles containing $\alpha_2$-macroglobulin. *Cell* **28**, 643–651.

102. Rennie, P., and Gliemann, J. (1981). Rapid down regulation of insulin receptors in adipocytes artifact of the incubation buffer. *Biochem. Biophys. Res. Commun.* **102**, 824–831.

103. Krupp, M. N., and Lane, M. D. (1982). Evidence for different pathways for the degradation of insulin and insulin receptor in the chick liver cells. *J. Biol. Chem.* **257**, 1372–1377.

104. Prince, M. J., Baldwin, D., Tsai, P., and Olefsky, J. M. (1981). Regulation of insulin receptors in cultured human fibroblasts. *Endocrinology (Baltimore)* **109**, 1754–1759.

105. Kahn, M. N., Posner, B. I., Kahn, R. J., and Bergeron, J. J. M. (1982). Internalization of insulin into rat liver Golgi elements. *J. Biol. Chem.* **257**, 5969–5976.

106. Kahn, M. N., Posner, B. I., Verma, A. K., Kahn, R. J., and Bergeron, J. J. M. (1981). Intracellular hormone receptors: Evidence for insulin and lactogen receptors in a unique vesicle sedimenting in lysosome fractions of rat liver. *Proc. Natl. Acad. Sci. U.S.A.* **78**, 4980–4984.

107. Posner, B. I., Patel, B. A., Khan, M. N., and Bergeron, J. J. M. (1982). Effect of chloroquine on the internalization of $^{125}$I-insulin into subcellular fractions of rat liver. *J. Biol. Chem.* **257**, 5789–5799.

108. Glieman, J., Gammeltoft, S., and Vinten, J. (1975). Time course of insulin receptor binding and insulin lipogenesis in isolated rat fat cells. *J. Biol. Chem.* **250**, 3368–3374.

109. Brandenburg, D., Diaconescu, C., Saunders, D., and Thamm, P. (1980). Covalent linking of photoreactive insulin to adipocyte produces a prolonged signal. *Nature (London)* **286**, 821–822.

110. Bar, R. S., Harrison, L. C., Muggeo, M., Gorden, P., Kahn, C. R., and Roth, J. (1979). Regulation of insulin receptor in normal and abnormal physiology in humans. *Adv. Intern. Med.* **24**, 23–52.

111. Gavin, J. R., III, Roth, J., Neville, D. M., Jr., De Meyts, P., and Buell, D. N. (1974). Insulin-dependent regulation of insulin receptor concentrations: A direct demonstration in cell culture. *Proc. Natl. Acad. Sci. U.S.A.* **71**, 84–88.

112. Blackard, W. G., Guzelian, P. S., and Small, M. E. (1978). Down regulation of insulin receptors in primary cultures of adult rat hepatocytes in monolayer. *Endocrinology (Baltimore)* **103**, 548–553.

113. Peterson, B., Beckner, S., and Blecker, M. (1978). Hormone receptors—characteristics of insulin receptors in a new line of cloned neonatal rat hepatocytes. *Biochim. Biophys. Acta* **542**, 470–485.

114. Marshall, S., and Olefsky, J. M. (1980). Effects of insulin incubation on insulin binding, glucose transport and degradation by isolated rat adipocytes. Evidence for hormone induced desensitization at the receptor and postreceptor level. *J. Clin. Invest.* **66**, 763–772.

115. Mott, D. M., Howard, B. V., and Bennett, P. H. (1979). Stoichiometric binding and regulation of insulin receptors on human diploid fibroblasts using physiological insulin levels. *J. Biol. Chem.* **254**, 8762–8767.

116. Livingston, J. N., Purvis, B. J., and Lockwood, D. H. (1978). Insulin-dependent regulation of the insulin-sensitivity of adipocytes. *Nature (London)* **273**, 394–396.

117. Ronnett, G. V., Knutson, V. P., and Lane, M. D. (1982). Insulin-induced down regulation of insulin receptors in 3T3-L1 adipocytes, altered rate of receptor inactivation. *J. Biol. Chem.* **257**, 4285–4291.

118. Kosmacos, F. C., and Roth, J. (1980). Insulin-induced loss of the insulin receptor in IM-9 lymphocytes: A biological process mediated through the insulin receptor. *J. Biol. Chem.* **255**, 9860–9869.

119. Carpenter, G., and Cohen, S. (1976). $^{125}$I-labeled epidermal growth factor: Binding internalization and degradation in human fibroblasts. *J. Cell Biol.* **71**, 159–171.

120. Das, M., and Fox, C. F. (1978). Molecular mechanism of mitogen action, processing of receptor induced by epidermal growth factor. *Proc. Natl. Acad. Sci. U.S.A.* **75**, 2644–2648.

121. Czech, M. P. (1977). Molecular basis of insulin action. *Annu. Rev. Biochem.* **46**, 359–384.

122. Zor, U. (1983). Role of cytoskeletal organization in the regulation of adenylate cyclase-cyclic adenosine monophosphate by hormones. *Endocr. Rev.* **4**, 1–21.

123. Charry, R. J., (1979). Rotational and lateral diffusion of membrane proteins. *Biochem. Biophys. Acta* **559**, 289–327.

124. Saffman, P. G., and Delbrück, M. (1975). Brownian motion in biological membranes. *Proc. Natl. Acad. Sci. U.S.A.* **72**, 3111–3113.

125. Sheets, M. P., Schindler, M., and Koppel, D. E. (1980). Lateral mobility of integral membrane proteins is increased in spherocytic erythrocytes. *Nature (London)* **285**, 510–512.

126. Wey, C.-L., Cone, R. A., and Edidin, M. A. (1981). Lateral diffusion of rhodopsin in photoreceptor cells measured by fluorescence photobleaching and recovery. *Biophys. J.* **33**, 225–232.

127. Tank, D. W., Wu, E. S., and Webb, W. W. (1982). Enhanced molecular diffusibility in muscle membrane blebs: release of lateral constraints. *J. Cell Biol.* **92**, 207–212.

128. Harrison, L. C., Flier, J. S., Roth, J., Karlsson, F. A., and Kahn, C. R. (1979). Immunoprecipitation of the insulin receptor: A sensitive assay for receptor antibodies and a specific technique for receptor purification. *J. Clin. Endocrinol. Metab.* **48**, 59–65.

129. De Meyts, P., Bianco, A. R., and Roth, J. (1976). Site–site interactions among insulin receptors: Characterization of the negative cooperativity. *J. Biol. Chem.* **251**, 1877–1888.

130. Gammeltoft, S., Lars, Q. K., and Sestoft, L. (1978). Insulin receptors in isolated rat hepatocytes. *J. Biol. Chem.* **253**, 8406–8413.

131. Gavin, J. R., III, Gorden, P., Roth, J., Archer, J. A., and Buell, D. N. (1973). Characteristics of the human lymphocyte insulin receptor. *J. Biol. Chem.* **248**, 2202–2207.

132. De Meyts, P., Roth, J., Neville, D. M., Jr., Gavin, J. R., III, and Lesniak, M. A. (1973). Insulin interaction with its receptors: Experimental evidence for negative cooperativity. *Biochem. Biophys. Res. Commun.* **55**, 154–161.

133. Harmon, J. T., Kahn, C. R., Kempner, E. S., and Schlegel, W. (1980). Characterization of the receptor in its membrane environment by radiation inactivation. *J. Biol. Chem.* **255**, 3412–3419.

134. Maturo, J. M., III, and Hollenberg, M. D. (1978). Insulin receptor: Interaction with non receptor glycoprotein from liver cell membranes. *Proc. Natl. Acad. Sci. U.S.A.* **75**, 3070–3074.

135. Feutrell, C., and Metzger, H. (1981). Larger oligomers of IgE are more effective than dimers in stimulating rat basophilic leukemia cells. *J. Immunol.* **125**, 701–710.

136. Shechter, Y., Hernaez, L., Schlessinger, J., and Cuatrecasas, P. (1979). Local aggregation of hormone-receptor complexes is required for activation by epidermal growth factor. *Nature (London)* **278**, 835–838.

137. Conn, P. M., Rogers, D. C., Stewart, J. M., Neidel, J., and Sheffield, T. (1982). Conversion of a gonadotropin-releasing hormone antagonist to an agonist. *Nature (London)* **296**, 653–656.

138. Kahn, C. R., Baird, K., Flier, J. S., and Jarrett, D. B. (1977). Effects of autoantibodies to the insulin receptor on isolated adipocytes. Studies of insulin binding and insulin action. *J. Clin. Invest.* **60**, 1094–1106.

139. Jacobs, S., Chang, K. J., and Cuatrecasas, P. (1978). Antibodies for purified insulin receptors have insulin-like activity. *Science (Washington, D.C.)* **200**, 1283–1284.

140. Pillion, D. J., Grantham, J. R., and Czech, M. P. (1979). Biological properties of antibodies against rat adipocytes intrinsic membrane proteins. Dependence on multivalency for insulin-like activity. *J. Biol. Chem.* **254**, 3211–3220.

141. Kahn, C. R., Baird, K. L., Jarrett, D. B., and Flier, J. S. (1978). Direct demonstration that receptor cross-linking or aggregation is important in insulin action. *Proc. Natl. Acad. Sci. U.S.A.* **75**, 4209–4213.

142. Shechter, Y., Chang, K. J., Jacobs, S., and Cuatrecasas, P. (1979). Modulation of binding and bioactivity of insulin by anti-insulin antibody: Relation to possible role of receptor self-aggregation in hormone action. *Proc. Natl. Acad. Sci. U.S.A.* **76**, 2720–2724.

143. Shechter, Y. (1983). Bound lectins that mimic insulin produce persistent insulin-like activities. *Endocrinology* **113**, 1921–1926.

144. Jacobs, S., and Cuatrecasas, P. (1976). The mobile receptor hypothesis and "cooperativity" of hormone binding. *Biochim. Biophys. Acta* **433**, 482–495.

145. Minton, A. P. (1981). The bivalent ligand hypothesis. A quantitative model for hormone action. *Mol. Pharmacol.* **19**, 1–14.

146. Mrotek, J. J., and Hall, P. F. (1977). Response of adrenal tumor cells to ACTH: Site of inhibition by cytochalasin B. *Biochemistry* **16**, 3177–3181.

147. Hall, P. F., Carponnier, C., Nakamura, M., and Gabbiani, G. (1979). The role of microfilaments in the response of adrenal tumor cells to ACTH. *J. Biol. Chem.* **254**, 9080–9084.

148. Hall, P. F., Charponnier, C., Nakamura, M., and Gabbiani, G. (1979). The role of microfilaments in the response of Leydig cells to LH. *J. Steroid Biochem.* **11**, 1361–1366.

149. Seals, J. R., and Jarett, L. (1980). Activation of pyruvate dehydrogenase by direct addition of insulin to an isolated plasma membrane/mitochondria mixture. Evidence for generation of insulin's second messenger in a subcellular system. *Proc. Natl. Acad. Sci. U.S.A.* **77**, 77–81.

150. Seals, J. R., McDonald, J. M., and Jarett, L. (1979). Insulin effect on protein phosphorylation of plasma membranes and mitochondria in a subcellular system from rat adipocytes. I. Identification of insulin-sensitive phosphoproteins. *J. Biol. Chem.* **254**, 6991–6996.

151. Seals, J. R., McDonald, J. M., and Jarett, L. (1979). Insulin effect on protein phosphorylation of plasma membranes and mitochondria in a subcellular system from rat adipocytes. II. Characterization of insulin-sensitive phosphoproteins and conditions for observation of the effect. *J. Biol. Chem.* **254**, 6997–7001.

152. Jarett, L., Kiechle, F. L., Popp, D. A., Kotagal, N., and Gavin, J. R., III (1980). Differences in the effect of insulin on the generation by adipocytes and IM-9 lymphocytes of a chemical mediator which stimulates the action of insulin on pyruvate dehydrogenase. *Biochem. Biophys. Res. Commun.* **96**, 735–741.

153. Seals, J. R., and Czech, M. P. (1981). Characterization of a pyruvate dehydrogenase activator released by adipocyte plasma membranes in response to insulin. *J. Biol. Chem.* **256**, 2894–2899.

154. Larner, J., Galasko, G., Cheng, K., De Paoli-Roach, A. A., Huang, L., *et al.* (1979). Generation by insulin of chemical mediator that controls protein phosphorylation and dephosphorylation. *Science (Washington, D.C.)* **206,** 1408–1410.
155. Parker, J. C., Kiechle, F. L., and Jarett, L. (1982). Partial purification from hepatoma cells of an intracellular which mediates the effects of insulin on pyruvate dehydrogenase and low $K_m$ cyclic AMP phosphodiesterase. *Arch. Biochem. Biophys.* **215,** 339–344.
156. Horvat, A. (1980). Stimulation of RNA synthesis in isolated nuclei by an insulin-induced factor in liver. *Nature (London)* **286,** 906–908.
157. Bonn, D., Belhadge, O., and Cohen, P. (1977). Modulation by calcium of the insulin action and of the insulin-like effect of oxytoxin on isolated rat lipocytes. *Eur. J. Biochem.* **75,** 101–105.
158. Grinstein, S., and Erlig, D. (1976). Action of insulin and cell calcium: Effect of ionophore A23187. *J. Membr. Biol.* **29,** 313–328.
159. Means, A. R., and Dedman, J. R. (1980). Calmodulin—An intracellular calcium receptor. *Nature (London)* **285,** 73–77.
160. Chueng, W. Y. (1980). Calmodulin plays a pivotal role in cellular regulation. *Science (Washington, D.C.)* **207,** 19–27.
161. Shechter, Y. (1984). Trifluperazine inhibits insulin action on glucose metabolism in fat cells without affecting inhibition of lipolysis. *Proc. Natl. Acad. Sci. U.S.A.* **81,** 327–331.
162. Goewert, R. R., Klaven, N. B., and McDonald, J. M. (1983). Direct effect of insulin on the binding of calmodulin to rat adipocyte plasma membranes. *J. Biol. Chem.* **258,** 9995–9999.
163. Schulman, H., and Greengard, P. (1978). Stimulation of brain membrane protein phosphorylation by calcium and an endogenous heat-stable protein. *Nature (London)* **271,** 478–479.
164. Weisman, D. M., Singh, T. J., and Wang, J. H. (1978). The modulator-dependent protein kinase. A multifunctional protein kinase activatable by the $Ca^{2+}$-dependent modulator protein of the cyclic nucleotide system. *J. Biol. Chem.* **253,** 3387–3390.
165. Suzuki, K., and Kono, T. (1980). Evidence that insulin causes translocation of glucose transport activity to the plasma membrane from an intracellular storage site. *Proc. Natl. Acad. Sci. U.S.A.* **77,** 2542–2545.
166. Kono, T., Suzuki, K., Dansey, L. E., Robinson, F. W., and Blevins, T. L. (1981). Energy-dependent and protein synthesis-dependent recycling of the insulin-sensitive glucose transport mechanism in fat cells. *J. Biol. Chem.* **256,** 6400–6407.
167. Karnieli, E., Zarnowski, M. J., Hissin, P. J., Simpson, I. A., Salans, L. B., and Cushman, S. W. (1981). Insulin-stimulated translocation of glucose transport systems in the isolated rat adipose cell—Time course reversal insulin concentration dependency, and relationship to glucose transport activity. *J. Biol. Chem.* **256,** 4772–4777.
168. Cushman, S. W., and Warzala, L. J. (1980). Potential mechanism of insulin action on glucose transport in the isolated rat adipose cell. Apparent translocation of intracellular transport systems to the plasma membrane. *J. Biol. Chem.* **255,** 4759–4762.

# 7

# Size of Neurotransmitter Receptors as Determined by Radiation Inactivation–Target Size Analysis

**J. CRAIG VENTER**
Section on Receptor Biochemistry
Laboratory of Neurophysiology
NINCDS
National Institutes of Health
Bethesda, Maryland

## I. INTRODUCTION

The principal receptors of the autonomic nervous system are the adrenergic and cholinergic receptors. The physiology and pharmacology associated with these receptors have been the subject of countless thousands of articles and books since the first decade of this century. Until the 1970s, receptors were discussed more as metaphysical than real entities. In fact, researchers generally could not even agree as to what part of the cell would contain the receptor molecules. Adrenergic and cholinergic responses have been the most extensively studied, in part, because of their overall physiological importance. β-Adrenergic receptors, which modulate a diverse array of cellular and organ functions including heart rate and vascular and airway diameter, have received extensive attention, in part, because of their demonstrated role in the activation of adenylate cyclase activity (Robison *et al.*, 1971). Nicotinic acetylcholine receptors, which control skeletal

**245**

muscle contraction and certain aspects of ganglionic function, are the most extensively studied receptors and consequently the receptor about which the most is known. The genes for nicotinic acetylcholine receptors have been cloned and the complete amino acid sequence has been determined (Noda *et al.*, 1983). The purification and structural elucidation of the nicotinic acetylcholine receptor was aided tremendously, particularly in the early stages, by having a rich source of the receptor (electric eels) and specific receptor probes (cobra toxins) (Lee, 1979).

The resolution of the structure of adrenergic (α and β), muscarinic cholinergic, and dopaminergic receptors is clearly underway in a variety of laboratories throughout the world. Structural determinations have generally depended on the application of new technologies and the development of affinity ligands (Venter and Fraser, 1983a,b; Venter, 1982). However, this chapter will discuss the application of target size analysis, a reemerging technique, for obtaining structural information for a number of receptors.

## II. RADIATION INACTIVATION–TARGET SIZE ANALYSIS

### A. Historical Perspective

The principle of radiation inactivation is based upon the observation that a relationship exists between a dose-dependent inactivation of functional macromolecules by ionizing radiation and the size of the macromolecule or protein (Pollard, 1953; Lea, 1947; Kampner and Schlegel, 1979; Jung, 1984). Target size analysis is not a new method of molecular size analysis. The concept of a target size was introduced by Crowther (1924), and the first attempt to measure the molecular size of a macromolecule (hemocyanin) was reported by Svedberg and Broholt (1939).

In the 1940s, the relationship between ionizing radiation dose and molecular size was established and applied to various biomolecules (Lea, 1947). Lea *et al.* (1944) used X-rays to determine the volumes of myosin and ribonuclease. The size reported in 1944 for myosin was 470,000 daltons and 30,000 for ribonuclease (Lea *et al.*, 1944). The target size of 30,000 for ribonuclease has since been reproduced by others (Pollard *et al.*, 1955; Fluke, 1966). The molecular mass for ribonuclease was subsequently reported to be 29,000 daltons, with two subunits of 14,000 daltons each (D'Alessio *et al.*, 1972), and for myosin 468,000 daltons (Holtzer and Lowey, 1959; Mueller, 1964; Weeds and Lowey, 1971).

In a Kempner and Schlegel (1979) review of the field of radiation inactivation of enzymes, it was reported that interest in radiation inactivation as a procedure

**TABLE I**

**Comparison of Target-Size Volumes of Enzymes to Known Structures**

| Enzyme | Molecular weight | Subunits | Subunit molecular weight | Target size |
|---|---|---|---|---|
| Malate dehydrogenase[a] | 74,000 | 2 | 36,000 | 73,000 |
| Succinic dehydrogenase[a] | 87,000 | 2 | 60,000 + 27,000 | 80,000 |
| Glutamic dehydrogenase[a] | 320,000 | 6 | 57,000 | 300,000 |
| Catalase[a] | 232,000 | 4 | 58,000 | 230,000 |
| Creatine kinase[a] | 84,000 | 2 | 44,000 | 73,000 |
| Alkaline phosphatase[a] | 140,000 | 2 | 69,000 | 140,000 |
| Ribonuclease I[a] | 29,000 | 2 | 14,000 | 27,000–30,000 |
| Lactate dehydrogenase[b] | 140,000 | 4 | 35,000 | 34,000 |
| Xanthine oxidase[b] | 300,000 | 2 | 150,000 | 125,000 |
| Glucose 6-phosphatase[b] | 130,000 | 2 | 63,000 | 70,000 |
| β-Glucuronidase[b] | 280,000 | 4 | 75,000 | 81,000 |

[a] Enzymes for which the target size detects the oligomeric structure of multisubunit enzymes.

[b] Enzymes for which the target size has detected the monomer. Data adapted from Kempner and Schlegel (1979).

to determine the molecular size of enzymes dissipated during the 1950s when the results from target size data did not apparently agree well with molecular weight values determined by other means. However, Kempner and Schlegel compared target size values of enzymes to known values for enzyme molecular weights and subunit composition (Table I). This reevaluation of target size data helped to support the validity of the technique (Kempner and Schlegel, 1979).

## B. Target Theory

According to target theory (Pollard *et al.*, 1955), the biological activity of a protein (target) can be destroyed by a single "hit" of high-energy radiation occurring with its molecular volume. This is often referred to as the "one target–one hit theory of radiation inactivation." When tissues are irradiated with ionizing radiation such as X-rays, $\gamma$-rays, or high energy electrons, ionization occurs randomly throughout the tissue. The direct action of the ionizing radiation is localized to a spherical radius of 20 Å (Jung, 1984). If this volume is occupied by a protein, each primary ionization releases 66 eV of energy or 1518 kcal/mole. This energy is clearly sufficient to break several structural bonds ($C_2H_5$—H, 98 kcal/mole) ($C_2H_5$—$CH_3$, 85 kcal/mole), which can lead to a loss of protein function. However, it is not necessary for the polypeptide backbone of proteins to be broken in order to disrupt the biological function of protein mac-romolecules, a simple derangement in the tertiary structure should be more than adequate.

Due to the small spherical radius (20 Å) of the direct ionization and the fact that ionizations are produced randomly throughout the total biological sample, the actual probability of a given protein being hit can be determined according to the Poisson formula in which the probability of 0, 1, 2, . . . ionizations occur-ring within a unit volume is given by

$$P(n) = e^{-x}(X)^n/n! \tag{1}$$

in which $x$ is a function of the amount of radiation exposure (Kempner and Schlegel, 1979). Because the average number of hits per target protein is a product of the target-protein volume and the number of hits per cubic centimeter, it should be clear that the larger the protein (volume), the greater the chance of it being hit with any given dosage of radiation. This inverse relationship between radiation dose and molecular size provides the basis of molecular weight determination.

$$P(0) = e^{-x} \tag{2}$$

As predicted by Eq. (2) and from numerous experiments, it was found that

$$A = A_o e^{-VD} \tag{3}$$

in which $A_o$ is the initial activity and $A$ the surviving activity after a high-energy "hit" occurring within its molecular volume $V$ from a radiation dose $D$ (in units of ionization per $cm^3$). Equation (3) states that surviving activities decrease as a simple exponential of the radiation dose $D$. The molecular weight ($M_r$) of the target can be obtained by

$$M_r = (N_o V)/\bar{v} \tag{4}$$

in which $N_o$ is Avogadro's number and $\bar{v}$ is the partial specific volume of the protein.

In most experiments $D$ is expressed in rads, (1 rad is a dose of radiation which results in the adsorption of 100 ergs of energy by 1 gram of sample). However, $D$ is difficult to determine in such units. One method to overcome this roadblock is to use the empirical relationship

$$M_r = (6.4 \times 10^{11})/D_{37} \text{ (rads)} \tag{5}$$

derived by Kempner and Macy (1968), for which at a given temperature (25°C), $D_{37}$ is the radiation dose in rads required to reduce the activity to 37% of the control value.

It was subsequently discovered that at substantially reduced temperatures, a further empirical correction factor of 280% is required (Kempner and Schlegel, 1979; Schlegel et al., 1979).

## C. Enzyme Calibration Procedure

The use of protein standards of known molecular size as an alternative to the empirical approach was published by Lo *et al.* (1982) and rapidly adopted by this laboratory (Venter, 1983; Venter *et al.*, 1983a; Lilly *et al.*, 1983).

The application of the enzyme-calibration procedure to frozen samples cooled with liquid nitrogen satisfies the key criteria for target size analysis: (1) that radiation inactivation occurs at a temperature within the specimen well below that in which any thermal inactivation commences; (2) that empirical correction factors for temperature be eliminated; and (3) that the determination should not depend upon knowledge of the absolute dose of radiation (Lo *et al.*, 1982).

In our studies, criterion (1) was satisfied by maintenance of the samples between $-45°$ and $-52°C$ throughout irradiation procedures. In all cases, activities of the frozen samples were within 90–100% of the control nonfrozen samples (Venter, 1983; Lilly *et al.*, 1983; Venter *et al.*, 1983a).

Criteria (2) and (3) were satisfied by the use of enzymes of known molecular weight as internal standards. Enzyme molecular weight and subunit composition were analyzed by SDS–polyacrylamide gel electrophoresis with results essentially identical with those reported (Lo *et al.*, 1982).

When the standard enzymes (see later) were subjected to a high-energy electron bombardment, enzymatic activity declined as a simple exponential of the radiation dosage (Fig. 1). Enzymes simultaneously irradiated had their inactivation ratios (Lo *et al.*, 1982; Venter, 1983; Lilly *et al.*, 1983; Venter *et al.*, 1983a) $S_X/S_G$ determined by the ratio of the slopes of the semilogarithmic plots of each enzyme ($S_X$) to the slope of either β-galactosidase ($S_G$) or horse liver alcohol dehydrogenase ($S_{ADH}$).

The ratio $S_X/S_G$ for each standard (X) was related to its molecular weight ($M_X$), normalized to the molecular weight of β-galactosidase ($M_G$) (Fig. 7), as shown by Lo *et al.* (1982) based on a transformation of radiation target theory.

$$-2.3 \log A/A_0 = \bar{v}/N_0(MD) \qquad (6)$$

in which $A$ is the enzyme activity upon irradiation at a dose $D$, $A_0$ is the initial control activity, $N_0$ is Avogadro's number, $\bar{v}$ is the partial specific volume of the protein, and $M$ is the protein molecular weight.

The slope of $\log A/A_0$ versus $D$ is proportional to $M$. Therefore,

$$S_X/S_G = M_X/M_G \qquad (7)$$

The requirement for the absolute value of $D$ is eliminated and, along with it, any temperature correction factors.

Once calibration data are established, the inactivation ratio for an unknown protein X can be related to the standard plot using a single enzyme standard, e.g., ADH by

$$S_X/S_G = (S_X/S_{ADH})(S_{ADH}/S_G) \qquad (8)$$

## 1. Enzyme Standards

The standard enzymes, horse liver alcohol dehydrogenase (84,000 daltons), yeast alcohol dehydrogenase (160,000 daltons), pyruvate kinase (224,000 daltons), and *Escherichia coli* β-galactosidase (464,000 daltons) originally used by Lo *et al.* (1982) are multisubunit enzymes (Table II). These enzymes all have two or more subunits under standard conditions and they inactivate as the oligomeric structure and, therefore, are useful as models for multisubunit proteins. The oligomeric size is obtained even with active enzyme monomers, strongly supporting the idea of energy transfer between subunits (Lo *et al.*, 1982). The subunit size and oligomeric molecular weight of the commercially available (Sigma Chemical Co.) enzymes were reconfirmed by Lo *et al.* (1982) and by this laboratory, e.g., Venter *et al.* (1983a), using SDS–polyacrylamide gel electrophoresis and polyacrylamide gel electrophoresis. Enzyme activity is routinely measured on a recording spectrophotometer (as described in Lo *et al.*, 1982; Venter, 1983; Venter *et al.*, 1983a).

For radiation inactivation, the enzymes are frozen in trays (see later) at a concentration of 0.4–4.0 mg/ml either in combination with membranes or individually. In the case of pyruvate kinase, 10 mg/ml of bovine serum albumin are added to prevent enzyme aggregation upon freezing and thawing. Enzymes included with membranes or other enzymes for irradiation have radiation-inactivation profiles identical with the purified enzyme irradiated alone. Samples are stored at −80°C both prior to and subsequent to irradiation procedures.

## 2. Receptor Standards

While the above enzymes have provided an excellent calibration system, none of the proteins are integral membrane proteins. In retrospect, we have not found

TABLE II

Molecular Properties of Calibrating Proteins[a]

| Protein | Molecular weight | Subunit | Subunit molecular weight |
|---|---|---|---|
| Horse liver alcohol dehydrogenase | 84,000 | 2 | 40,000[b] |
| Yeast alcohol dehydrogenase | 160,000 | 4 | 40,000 (37,000)[c] |
| Rabbit muscle pyruvate kinase | 224,000 | 4 | 57,000[d] |
| E. coli β-galactosidase | 464,000 | 4 | 116,000 |
| Rat brain muscarinic cholinergic receptor | 80,000 | 1 | 80,000 |

[a] Data from Lo *et al.* (1982) and Venter (1983) unless otherwise noted.
[b] Weber and Osborn (1969); Green and Mekay (1969).
[c] Harris (1964); Pfleiderer and Auricchio (1964).
[d] Morawiecki (1960); Steinmetz and Deal (1966).

any data which might suggest that membrane proteins have in anyway an altered inactivation profile when compared to $H_2O$-soluble proteins: however, we wanted to have a control integral membrane protein for study. Fortunately, this was provided by a study on the muscarinic cholinergic receptor (Venter, 1983). The muscarinic cholinergic receptor is a monomeric integral membrane protein with a molecular weight of 80,000 as determined by SDS–polyacrylamide gel electrophoresis (Venter, 1983; Venter et al., 1984a). Target size analysis of the muscarinic receptor using high-energy electrons and the above enzyme calibration procedure also indicated a molecular weight of 80,000 for this protein (see later). A similar value (76,000) was reported for the same receptor using cobalt as a source of radiation (Uchida et al., 1982). So, in addition to the standard enzymes, we frequently check our data by utilizing the muscarinic receptor as an additional control.

## 3. Sample Temperature and Conditions

**a. Indirect Radiation Effect.** The presence of liquid $H_2O$ is not compatible with target size analysis procedures, as $H_2O$ molecules will undergo ionization to produce the oxidant $H_2O_2$ and the highly reactive free radicals $H^+$ and $OH^-$. The free radicals can diffuse an average distince of 30 Å during their lifetime. Proteins which encounter these radicals can be inactivated. In aqueous solution, this indirect action is the primary effect seen (Jung, 1984).

Two basic methods exist by which one can limit or eliminate the indirect radiation effects, sample lyophilization and the use of frozen samples.

**b. Lyophilization.** By freezing samples and irradiating the dried material at room temperature, the $H_2O$ effect can be effectively eliminated. This has been the method of choice for many proteins because there are no requirements for temperature-correction factors required with Eq. (5) at reduced temperatures. However, with membrane proteins in a lipid environment, the use of the lyophilization approach may present problems, particularly when protein–protein interactions are understudy. In addition, for this technique (lyophilization) to be generally applicable, the membrane proteins which are the subject of investigation would need to be completely stable to freeze-drying and rethawing and to be heat-stable while lyophilized. While this was the case for the nicotinic acetylcholine receptor studied by Lo et al. (1982), we found that the slow inward calcium channel (Venter et al., 1983a), the $D_2$-dopamine receptor (Lilly et al., 1983), the muscarinic cholinergic receptor (Venter, 1983), and adrenergic receptors (Fraser and Venter, 1982; Venter et al., 1984b; Fraser et al., 1983) were extremely unstable under these conditions. It is highly likely that the majority of functional proteins (receptors, ion channels, etc.) will fall into this latter category. Therefore, we extended the enzyme-calibration procedure of Lo et al. (1982) applied to lyophilized samples, to the same enzymes in the frozen state (Venter, 1983; Venter et al., 1983a; Lilly et al., 1983).

**c. Frozen Samples.**   In that the enzyme-calibration procedure eliminates the need for empirical correction factors for temperature (Venter *et al.*, 1983a), we reasoned that the procedure should be equally applicable to frozen as well as lyophilized samples (Venter, 1983; Venter *et al.*, 1983a). Freezing samples effectively eliminates the indirect effects of radiation by preventing the diffusion of any formed free radicals. The criterion that radiation inactivation occurs at a temperature within the sample well below that in which any thermal inactivation commences [criterion (1)] was satisfied by our maintenance of the samples between $-45°$ and $-52°C$ throughout irradiation procedures. In all of the cases, activities of our frozen samples were within 90–100% of our control nonfrozen samples. The inactivation ratios and standard calibration curve for the enzyme standards under frozen conditions agreed well with the data by Lo *et al.* (1982) using lyophilized samples (Venter *et al.*, 1983a). The least-squares slope of the standard curve (theoretical slope = 1) was 0.97 for frozen samples (Venter, 1983) and 0.90 for lyophilized samples (Lo *et al.*, 1982).

## 4. Radiation Procedure

**a. Source of Radiation.**   For target size analysis, high-energy electrons, γ rays, and X-rays are most commonly used. These forms of radiation produce ionizing radiation randomly throughout the sample volume, a requirement for the application of target theory (see earlier).

In practice, one is generally limited by the type of facility available to the user, e.g., Van de Graaff generator, linear electron accelerator, etc. The radiation source and range must be matched with the actual sample thickness to ensure radiation penetration of the entire sample. For samples up to 2 mm in thickness, 1 MeV should be used. This electron beam can be provided by Van de Graaff electron generators, by γ rays from cobalt-60, or 1000-kVcp X-rays (Jung, 1984). Linear electron accelerators can produce higher MeV electron beams and are, therefore, required for thicker samples (Harmon *et al.*, 1980).

**b. Irradiation Times.**   For our receptor studies, we often needed to exceed radiation doses of 10 Mrads. Smaller proteins require up to 30 Mrads (Jung, 1984) or 50 Mrads (Kempner and Schlegel, 1979). The three most commonly used radiation sources are summarized in Table III. From this table it is clear that in order to complete target irradiation in a reasonable time (minutes to hours), either a Van de Graaff generator or a linear electron accelerator should be utilized. For a more complete discussion of the biophysics of radiation inactivation see Jung (1984), Pollard (1953), and Lea (1947).

Because the radiation absorbed by the sample generates heat, the sample temperature must be controlled. With high-radiation doses provided by the linear electron accelerator, heat generation can be a problem (Lo *et al.*, 1982). With

**TABLE III**

**Commonly Used Sources of Radiation for Target Size Analysis**

| Radiation source | Type of ionizing radiation energy (MeV) | Dose rate (Mrad/min) | Sample thickness | Time to inactivate a 64,000-dalton protein by 50% ($D_{37} = 40$ Mrad) |
|---|---|---|---|---|
| Linear electron accelerator | High-energy electron | 10–15 | ~2 | ≤1 cm | 5 min |
| Van de Graaff electron generator | High-energy electron | 1.5 | ~0.5 | ≤2 mm | 20 min |
| Cobalt-60 | γ-ray | | ~0.01 | ≤2 mm | 16.7 hr |

lyophilized samples and 2 Mrad/min, Lo *et al.* (1982) reported temperatures of 30°C in control tubes and 25°–35°C in sample tubes. In an attempt to control thermal inactivation, Lo *et al.* (1982) studied the thermal stability of freeze-dried nicotinic acetylcholine receptor between 5° and 100°C. They found a sharp drop in toxin binding at temperatures above 50°C and, therefore, stressed the need to keep sample temperatures below this level.

With frozen samples (temperatures −45°– −120°C) and with lower energy systems (Table III), heat dissipation from the sample is more easily controlled with liquid $N_2$ (Venter, 1983; Lilly *et al.*, 1983; Venter *et al.*, 1983a).

**c. Sample Preparation and Irradiation.** In our studies, purified cell membranes and/or the calibrating enzymes are layered at a depth of 0.5 mm in open aluminum trays. The membranes and/or standards are quickly frozen by immersion of the trays in liquid nitrogen. The aluminum trays containing the thin film of frozen samples are placed in an aluminum chamber (target chamber). The target chamber contains an aluminum foil window that provides an airtight seal. The target chamber is flushed for at least 3 min with liquid $N_2$ to maintain the sample temperature and to replace the air with a nitrogen atmosphere. The target chamber is continuously cooled with flowing liquid $N_2$ throughout the irradiation procedure to constantly maintain the sample temperature in the range of −45°– −52°C. The irradiation room is maintained at low humidity and has a high-velocity ventilation system to minimize frost and ozone accumulation. Our samples are routinely irradiated with 0.5 mA beam of 1.5 MeV electrons produced by a Van de Graaff electron generator (Table III). Samples are stored at −80°C both prior to and subsequent to irradiation procedures. For further details see Jung (1984).

## D. Target Size Analysis of Membrane Proteins

### 1. Enzyme Standards

When the standard enzymes are subjected to high-energy electron bombardment, enzyme activity declines as a simple exponential of the radiation dose (Fig. 1). The enzymes simultaneously irradiated had their inactivation ratio $S_X/S_G$ determined by the ratio of the slopes of the semilogarithmic plots of each enzyme ($S_X$) (e.g., Fig. 1) to the slope of β-galactosidase ($S_G$). The inverse ratio $S_X/S_{ADH}$ was also determined by comparison to horse liver alcohol dehydrogenase ($S_{ADH}$).

The ratio $S_X/S_G$ for each standard (X) was related to its molecular weight ($M_X$), normalized to the molecular weight ($M_G$) of β-galactosidase according to Eq. (7). Figure 2 is representative of such data. The theoretical slope of the relationship in Eq. 7 should be 1.0. The least-squares slope of the line in Fig. 2 is 0.97, indicating an excellent agreement between theory and practice. Illustrated on the same standard curve are the inactivation ratios for a number of neurotransmitter receptors.

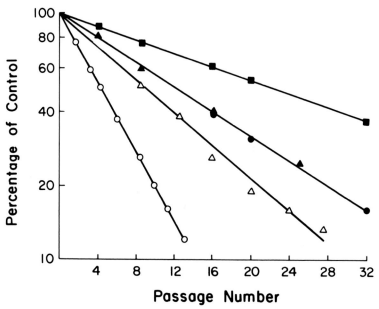

**Fig. 1.** Target size analysis of enzyme standards. Radiation inactivation of standard enzymes horse liver alcohol dehydrogenase (84,000 daltons) (■), yeast alcohol dehydrogenase (160,000 daltons) (▲), pyruvate kinase (224,000 daltons) (△), and *E. coli* β-galactosidase (464,000 daltons) (○). Surviving enzyme activity plotted as a function of the radiation dosage. (Data from Venter *et al.*, 1984a,b.)

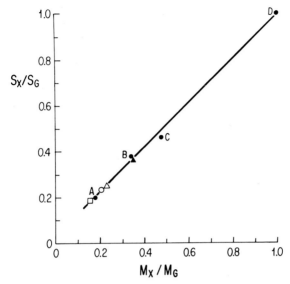

**Fig. 2.** Radiation inactivation molecular weight calibration curve. Standard enzymes horse liver ADH (A), yeast ADH (B), pyruvate kinase, (C), and *E. coli* β-galactosidase (D) were inactivated individually, in groups, and together with membranes with identical results and their inactivation ratios determined. The least-squares slope of the line is 0.97. The receptors were related to the standard plot by determining the inactivation ratio $S_{\alpha\text{-receptor}}/S_{\text{yeast ADH}}$ or $S_{\alpha\text{-receptor}}/S_G$ which permits a direct comparison to the enzyme standard plot of the relationship between the inactivation ratio $S_X/S_G$ and molecular weight ratio $M_X/M_G$. Symbols: $\alpha_1$-adrenergic receptor (▲), $D_2$-dopamine receptor (△), $\beta_2$-adrenergic receptor (○), and muscarinic cholinergic receptor (□).

The absolute slope of the enzyme standards varies from each inactivation experiment, regardless of whether passage number (Jung, 1984) or radiation dose in Mrads is used to plot the data, illustrating the uncertainty of the "radiation dose" measurement (Lo *et al.*, 1982). However, the slope ratios have remained constant throughout all of our studies, strongly supporting the validity of this procedure.

## 2. Neurotransmitter Receptors

**a. Muscarinic Cholinergic Receptor.**  The muscarinic cholinergic receptor mediates various central nervous system activities as well as the function of the parasympathetic side of the autonomic nervous system, including heart rate attentuation and smooth muscle contraction in the eye, intestine, and airways (Venter *et al.*, 1984a).

This laboratory has investigated the structure of the muscarinic cholinergic receptor and found it to be a monomeric protein with a molecular mass of 80,000 daltons (Venter, 1983). The SDS–polyacrylamide gel electrophoretic pattern of

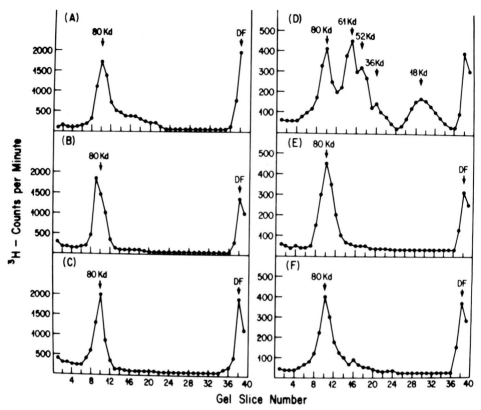

**Fig. 3.** SDS–polyacrylamide gel electrophoretic analysis of muscarinic receptors from a variety of tissues and species. Membranes were purified from various tissues and labeled with [$^3$H]-propylbenzylyl choline mustard ([$^3$H]PrBCM). Tissues included human brain (A), canine brain (B), rat brain (C), guinea pig ileum smooth muscle (D), rat heart (E), and canine heart (F). [$^3$H]PrBCM–muscarinic receptor complexes were solubilized with 2% SDS and analyzed on 10% SDS gels. Samples were treated at 100°C for 5 min in the presence of 5% 2-mercaptoethanol. Tissues in (A), (B), (C), (E), and (F), were isolated in the presence of phenylmethylsulfonyl fluoride (100 µM). Gels were sliced into 2.0-mm slices and counted in a scintillation counter subsequent to overnight incubation in Econofluor [New England Nuclear (NEN)] containing 3% Protosol (NEN). The data in each panel are representative of data from at least three experiments. SDS gels in the presence of 1 µM atropine show radioactivity only at the dye front. Standard proteins were analyzed in each gel. Protein standards are (1) phosphorylase b (94,000 daltons); (2) albumin (67,000 daltons); (3) ovalbumin (43,000 daltons); (4) carbonic anhydrase (30,000 daltons); and (5) trypsin inhibitor (20,000 daltons) (Pharmacia electrophoresis calibration kit 11-C021-01). (Data from Venter, 1983.)

muscarinic receptors isolated from a number of sources is illustrated in Fig. 3. All of the tissues show only the 80,000-dalton protein, except for guinea pig ileum smooth muscle, a preparation in which it has been difficult to inhibit protease activity (Venter, 1983). Topology studies indicate that over 50% of the muscarinic receptor protrudes into the extracellular aqueous space, with 14% of the receptor possibly on the cytoplasmic side of the membrane (Venter, 1983). A model of the structure of the muscarinic receptor is shown in Fig. 4.

Due to the fact that the muscarinic receptor is a highly conserved protein from human brain to *Drosophila* heads (Venter *et al.*, 1984a) and that it is a stable, easily assayed protein, it seemed ideal to use for target size studies (Venter, 1983). We also wanted to answer the question of whether the 80,000-dalton protein represented the intact receptor as it exists in the membrane or only a subunit or tryptic fragment of it. The multiple fragments seen with the ileum smooth muscle (Fig. 3D) also prompted further study (Venter, 1983). When rat brain membranes are frozen in thin layers and subjected to high-energy electron bombardment, there is an exponential loss of muscarinic receptors as a function of the radiation dose (Fig. 5). The loss of muscarinic receptors was assessed by the loss of [$^3$H]QNB binding over a wide range of ligand concentrations both above and below the $K_d$ for [$^3$H]-quinuclidinyl benzilate (QNB) binding (Venter,

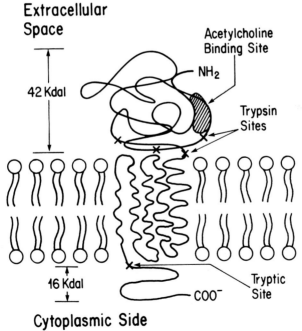

**Fig. 4.** Proposed model of the muscarinic cholinergic receptor membrane structure.

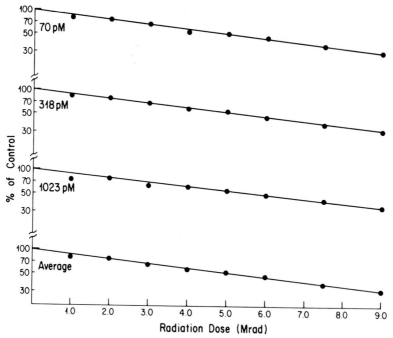

**Fig. 5.** Radiation inactivation–target size analysis of rat brain muscarinic acetylcholine receptor. Purified rat brain membranes were frozen in thin layers in aluminum trays along with enzyme standards and subjected to high-energy electron bombardment. Samples were thawed and the radiation-induced loss of muscarinic receptors was assessed by measuring [³H]QNB-specific binding over a range of ligand concentrations. Values represent the average of triplicate determinations from four separate experiments. The graph marked "average" is an average of all of the data from all of the experiments. Lines were drawn by least-squares linear regression. The radiation-induced loss of muscarinic receptors is linear over a wide range of radiation doses. The average molecular weight is 82,000, a value essentially identical with that found by SDS-gel analysis. (Data from Venter, 1983.)

1983). As can be seen in Fig. 5, the loss of the receptor was linear over a broad dose of radiation, indicating a single homogeneous class of sites. Scatchard analysis of saturation isotherms of QNB binding to rat membranes indicated a change in receptor number with no change in receptor affinity (of the remaining sites), consistent with the one target–one hit theory of radiation inactivation.

The membrane molecular weight of the muscarinic receptor was determined by the inclusion of enzyme standards with the receptor preparation (Venter, 1983). Figure 6 illustrates the target size analysis of the guinea pig ileum muscarinic receptor together with the target size of the enzyme standard pyruvate

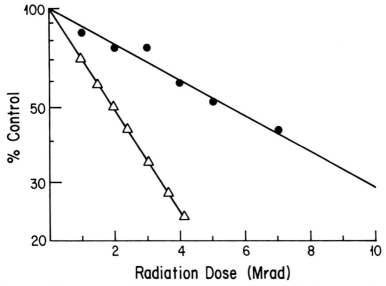

**Fig. 6.**   Radiation inactivation–target size analysis of guinea pig ileum smooth muscle muscarinic acetylcholine receptor. Guinea pig ileum longitudinal smooth muscle cell membranes were purified and frozen in thin layers in aluminum trays along with enzyme standards and subjected to high-energy electron bombardment. Radiation-induced loss of the muscarinic receptor was assessed by measuring [³H]QNB binding. The figure represents the survival of specific [³H]QNB binding. The data points represent the average of triplicate determinations with three ligand concentrations from the irradiation experiments. The radiation inactivation of the muscarinic receptor (●) is compared on the same plot to the inactivation of pyruvate kinase (△) from the same radiation experiment. Lines were drawn by least-squares linear regression. The molecular weight of the muscarinic receptor was calculated by determining the inactivation ratio $S_{muscarinic\ receptor}/S_{pyruvate\ kinase}$, which permits a direct comparison to enzyme standards on a linear plot (Fig. 7) of the relationship between the inactivation ratio $S_X/S_G$ and molecular weight ratio $M_X/M_G$ determined in a series of radiation inactivation experiments. The inactivation ratio of the muscarinic receptor was related to the standard plot by $S_X/S_G = (S_X\ S_{PK})/(S_{PK}\ S_G)$. These data indicate a molecular mass of 78,000 daltons for the ileum muscarinic receptor.

kinase. The inactivation ratio $S_{muscarinic\ receptor}/S_{pyruvate\ kinase}$ (PK) was related to the standard curve (Fig. 7) by

$$S_X/S_G = (S_X/S_{PK})\ (S_{PK}/S_G) \tag{9}$$

The slope ratio indicates a molecular mass of 82,000 daltons for the brain muscarinic receptor, a value in excellent agreement with the SDS gel data (Venter, 1983). Furthermore, a molecular mass value of 78,000 daltons was obtained for the ileum muscarinic receptor, supporting the contention that the 80,000-dalton peak (Fig. 3D) represents the intact receptor protein (Venter, 1983).

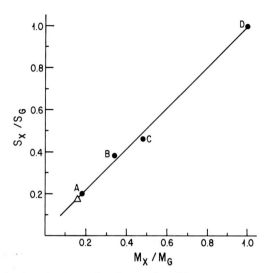

**Fig. 7.** Radiation inactivation molecular weight calibration curve. Standard enzymes horse liver alcohol dehydrogenase (A); yeast alcohol dehydrogenase (B); pyruvate kinase (C); and *Escherichia coli* β-galactosidase (D) were inactivated individually, in groups, and with membrane samples with identical results and their inactivation ratios determined. The least-squares slope of the line is 0.97. The muscarinic receptor (△) was related to the standard plot as described in the legend to Fig. 6. (Data from Venter, 1983.)

A target size of 76,000 has also been reported for the smooth muscle muscarinic cholinergic receptor by Uchida *et al.* (1982) using cobalt-60 as a radiation source.

As summarized in Table IV, the molecular size of the muscarinic cholinergic receptor determined by target size analysis agrees closely with the molecular weight of the receptor determined by other means from a wide variety of species and tissues.

**b. $\alpha_1$-Adrenergic Receptors.**  α-Adrenergic receptors ($\alpha_1$ and $\alpha_2$) are also principal neurotransmitter receptors of the autonomic and central nervous systems. α-Adrenergic receptors are found in most tissues, including brain, blood vessels, airway smooth muscle, cardiac muscle, and the liver. The rat liver $\alpha_1$-adrenergic receptor has undergone extensive physiological and pharmacological characterizations (see, e.g., Dehaye *et al.*, 1980; Blackmore *et al.*, 1983) and biochemical analysis (Kunos *et al.*, 1983; Venter *et al.*, 1983b; Fraser *et al.*, 1983; Venter and Fraser, 1983a).

In structural studies on $\alpha_1$-adrenergic receptors, we utilized the high-specific-activity [³H]phenoxybenzamine (POB) as an affinity reagent (Kunos *et al.*, 1983). [³H]POB was found to be highly selective for $\alpha_1$-receptors and can

TABLE IV

**Muscarinic Cholinergic Receptor Molecular Weights from Various Techniques, Tissues, and Species**

| Species tissue | Method | Molecular weight | Reference |
|---|---|---|---|
| Human | | | |
| Brain | SDS–PAGE[a] | 78,000 ± 1200 | Venter (1983) |
| | | 80,000 | Venter et al. (1984a) |
| Brain | Radiation inactivation | 82,000 | Venter (1983) |
| Monkey | | | |
| Ciliary muscle | SDS–PAGE | 80,000 | Venter et al. (1984a) |
| Canine | | | |
| Brain | SDS–PAGE | 82,000 ± 1800 | Venter (1983) |
| Heart | SDS–PAGE | 81,000 ± 3000 | |
| Rat | | | |
| Brain | Radiation inactivation | 82,000 | Venter (1983) |
| Brain | Hydrodynamic | 86,000 | Haga (1980) |
| Brain | SDS–PAGE | 80,000 ± 2000 | Venter (1983) |
| Brain | SDS–PAGE | 83,200 ± 2500 | Birdsall et al. (1979) |
| Heart | SDS–PAGE | 78,000 ± 1800 | Venter (1983) |
| Guinea pig | | | |
| Ileum smooth muscle | Radiation inactivation | 78,000 | Venter (1983) |
| Ileum smooth muscle | Radiation inactivation | 76,000 ± 4000 | Uchida et al. (1982) |
| Ileum smooth muscle | SDS–PAGE | 79,000 ± 4200 | Venter (1983) |
| Ileum smooth muscle | SDS–PAGE | 77,600 ± 2000 | Birdsall et al. (1979) |
| Brain | SDS–PAGE | 83,200 ± 6000 | Birdsall et al. (1979) |
| Frog | | | |
| Brain | SDS–PAGE | 80,000 | Birdsall et al. (1979) |
| Drosophila | | | |
| Head | SDS–PAGE | 80,000 | Venter et al. (1984a) |

[a] SDS–PAGE, sodium dodecyl sulfate–polyacrylamide gel electrophoresis.

readily be utilized in the 0.5–1.0 n$M$ range to specifically and covalently label the $\alpha_1$-adrenergic receptor (Kunos et al., 1983). SDS–polyacrylamide gel electrophoretic analysis of the $\alpha_1$-adrenergic receptor indicates a molecular weight of 85,000 (Venter et al., 1983b; Fraser et al., 1983; Venter and Fraser, 1983a) (Fig. 8).

We utilized target size analysis as a means of identifying the membrane form of the $\alpha_1$-adrenergic receptor and to be certain we had not isolated a tryptic fragment of the receptor despite our extensive use of protease inhibitors. When

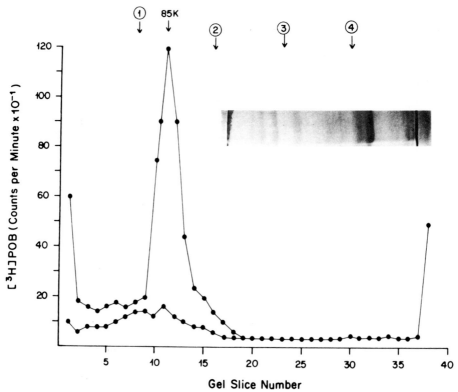

**Fig. 8.** SDS–polyacrylamide gel electrophoretic analysis of the rat liver $\alpha_1$-adre-
nergic receptor. Rat liver membranes were purified and affinity labeled with [³H]POB.
[³H]POB–$\alpha_1$-receptor complexes were solubilized with 2% SDS, treated at 100°C for 5
min in the presence of 5% β-mercaptoethanol, and analyzed on 10% SDS–polyacrylamide gels. Gels were sliced into 2.0-mm slices and counted in a scintillation counter subsequent to overnight incubation in Econofluor (NEN) containing 3% Protosol
(NEN). Data shown are representative of data from 18 experiments. The background
nonspecific binding (lower line) is representative of [³H]POB binding in the presence of
prazosin (0.1 μM). Standards (1)–(4) are as in Fig. 3. (Data from Venter et al., 1984b.)

rat liver membranes are frozen in thin layers and subjected to high-energy electron bombardment, there is an exponential loss of $\alpha_1$-adrenergic receptors, as a
function of the radiation dosage (Fig. 9 and 10). The loss of $\alpha_1$-receptors was
assessed by the loss of [³H]POB- and [³H]prazosin-specific binding over a wide
range of ligand concentrations both above and below the $K_d$ for ligand binding
(Fig. 9). The loss of receptor was linear over a broad dose of radiation, indicating
a single homogeneous class of sites (Figs. 9 and 10). Scatchard analysis of
[³H]prazosin-binding saturation isotherms indicated that the radiation produced a

loss of binding sites with no change in receptor affinity, consistent with the one target–one hit theory of radiation inactivation (Fig. 11). The target size of the liver $\alpha_1$-receptor was also assessed by the loss of the 85,000-dalton protein on SDS gels (Fig. 10).

The functional molecular size of the $\alpha_1$ receptor was determined by the inclusion of enzyme standards with the membrane preparations (Figs. 10 and 12). As can be seen in Figs. 10 and 12, the radiation-induced loss of the $\alpha_1$ receptor coincides with the loss of yeast ADH (molecular weight 160,000). The slope ratios clearly yield a molecular weight of 160,000 (Fig. 13) for the membrane-associated $\alpha_1$-adrenergic receptor. These findings contrast to those with the muscarinic receptor (see earlier), for which the SDS-gel data and the target size agree directly.

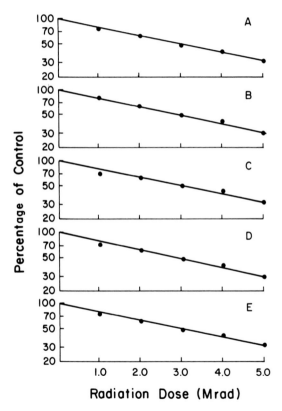

**Fig. 9.** Radiation inactivation of rat liver $\alpha_1$-adrenergic receptors. The effect of radiation on the survival of $\alpha_1$ receptors was assessed by measuring [³H]prazosin-specific binding at five concentrations (A–E), both above and below the $K_d$ value for prazos in binding. (Data from Venter et al., 1984b.)

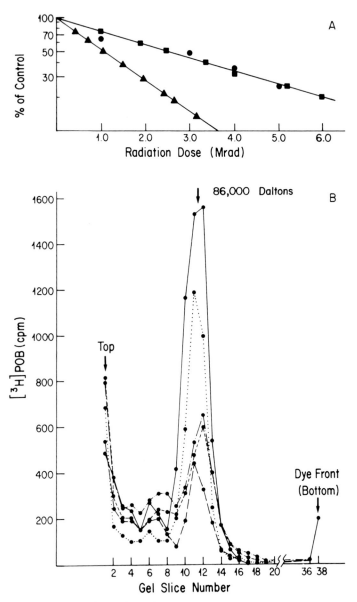

**Fig. 10.** Target size analysis of $\alpha_1$-adrenergic receptors as measured by the loss of the 85,000-dalton $\alpha$-receptor monomer on SDS–PAGE. Radiation inactivation of rat liver membranes was performed as described in Fig. 12. $\alpha_1$ Receptors were affinity labeled with [$^3$H]POB, solubilized with 2% SDS, and subjected to SDS–polyacrylamide gel electrophoresis. (B) The superimposed SDS-gel patterns from control membranes (●——●) and membranes subjected to inactivation with 1.0 Mrad (●·····●), 3.0 Mrad (●·——·●), 4.0 Mrad (●- - -●), and 5.0 Mrad (●— —●) are illustrated. (A) is a semilogarithmic plot of the loss of the 85,000-dalton $\alpha_1$-receptor monomer (●) from SDS gels. Compared on the same plot are the standard enzymes yeast ADH, 160,000 daltons (■) and β-galactosidase, 464,000 daltons (▲). These data show that the 85,000-dalton protein derived from a 160,000-dalton complex in the membrane. (Data from Venter et al., 1984b.)

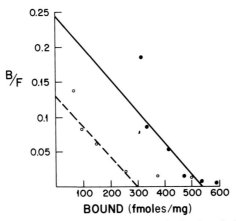

**Fig. 11.** Scatchard analysis of [³H]prazosin-specific binding before and subsequent to radiation inactivation. The solid line represents the control prazosin-specific binding to rat liver plasma membranes. The dashed line is the binding remaining after irradiation with ~3.0 Mrad. Receptor loss is due to a loss of binding sites not an affinity change. (Data from Venter *et al.*, 1984b.)

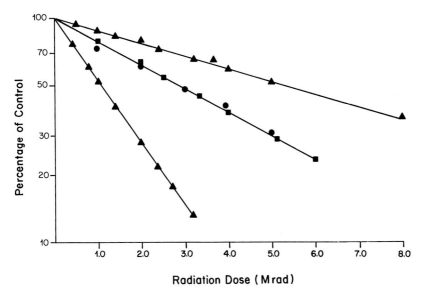

**Fig. 12.** Radiation inactivation–target size analysis of $\alpha_1$-adrenergic receptors and standard enzymes. Rat liver membranes were purified and frozen in thin layers in aluminum trays along with enzyme standards and subjected to high-energy electron bombardment. Radiation-induced loss of $\alpha_1$ receptors was assessed by measuring [³H]POB and [³H]prazosin-specific binding over a wide range of ligand concentrations, both above and below the $K_d$ values or apparent $K_d$ values of each ligand. The figure represents the survival of $\alpha_1$-receptor-specific binding (●). The data points represent the averages of triplicate determinations with the two ligands and five ligand concentrations (Fig. 10) from three radiation inactivation experiments. The inactivation of the α receptor is compared on the same plot to the inactivation of horse liver ADH (top curve, ▲), yeast ADH (■), and β-galactosidase (lower curve, ▲). Lines were drawn by least-squares linear regression and molecular weight values calculated as shown in Fig. 13.

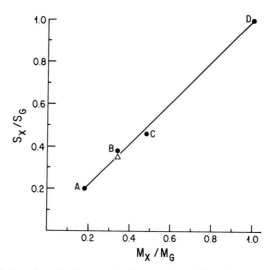

**Fig. 13.** Radiation inactivation molecular weight calibration curve. Standard enzymes as in Fig. 1; horse liver ADH (A), yeast ADH (B), pyruvate kinase (C), and $E.$ $coli$ $\beta$-galactosidase (D). The least-squares slope of the line is 0.97. The $\alpha_1$-adrenergic receptor ($\triangle$) was related to the standard plot by determining the inactivation ratio $S_{\alpha\ receptor}/S_{yeast\ ADH}$ or $S_{\alpha\ receptor}/S_G$ which permits a direct comparison to the enzyme standard plot of the relationship between the inactivation ratio $S_X/S_G$ and molecular weight ratio $M_X/M_G$. The inactivation ratio of the $\alpha$ receptor was related to the standard plot by $S_X/S_G = (S_X/S_{ADH})(S_{ADH}/S_G)$. These data indicate a molecular weight of 160,000 for the rat liver $\alpha_1$-adrenergic receptor.

Numerous studies have shown that target size analysis can provide information concerning the oligomeric structure of a protein (Table I). In the case of the nicotinic acetylcholine receptor, the target-size molecular mass of 300,000 daltons was close to the oligomeric molecular mass of 270,000 ± 20,000 for the five receptor polypeptide subunits, as determined by SDS–PAGE (Lo $et$ $al.$, 1982).

It is presently not known how the action of the $\alpha_1$-adrenergic receptor is mediated. There are indications that the slow inward calcium channel is activated in some tissues by this receptor (see later). The simplest model for the $\alpha_1$-receptor structure is illustrated in Fig. 14. The target size of 160,000 makes it intriguing to propose a dimeric model (Fig. 14), although it is equally possible that there exists a second yet to be identified subunit (Venter $et$ $al.$, 1983b; Venter and Fraser, 1983a; Fraser $et$ $al.$, 1983).

c. $\alpha_2$-**Adrenergic Receptors.** The $\alpha_2$-adrenergic receptor has been distinguished pharmacologically from the $\alpha_1$-adrenergic receptor by the interaction of selective antagonists (Berthelsen and Pettinger, 1977; Hoffman *et al.*, 1979). $\alpha_2$-Adrenergic receptors have been associated with the attenuation of adenylate cyclase activity (Jakobs, 1979) and the activation of the slow inward calcium channel (Andersson, 1973).

We have recently utilized target size analysis to provide the first structural data on the $\alpha_2$ receptor (Venter *et al.*, 1983c). In this study, we utilized human platelets which have been extensively characterized as a model system for $\alpha_2$-adrenergic receptor binding (Motulsky *et al.*, 1980) and coupling to adenylate cyclase (Jakobs *et al.*, 1978; Limbird *et al.*, 1982).

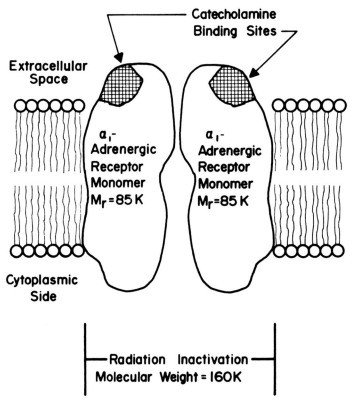

**Fig. 14.** Proposed dimeric model for $\alpha_1$-adrenergic receptor. (Data from Venter and Fraser, 1983b.)

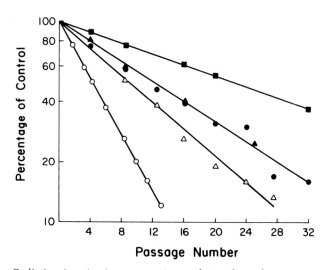

**Fig. 15.** Radiation inactivation–target size analysis of $\alpha_2$-adrenergic receptors and standard enzymes. Human platelet membranes were isolated and frozen in thin layers in aluminum trays along with enzyme standards and subjected to high-energy electron bombardment. Radiation-induced loss of $\alpha_2$ receptors (●) was assessed by measuring [³H]yohimbine binding at a number of ligand concentrations, below and above the $K_d$ value for yohimbine binding. The figure represents the survival of specific [³H]yohimbine binding (●) expressed on a semilogarithmic plot as a function of control [³H]yohimbine-specific binding and of increasing doses of radiation (passage number). Each point is the average of triplicate samples from seven separate determinations. The radiation inactivation of the $\alpha_2$ receptor is compared on the same plot to the inactivation of horse liver alcohol dehydrogenase (■), yeast alcohol dehydrogenase (▲), pyruvate kinase (△), and *E. coli* β-galactosidase (○). Lines were drawn by least-squares linear regression. Samples were irradiated with a 0.5-ma beam of 1.5-MeV electrons produced with a Van de Graaff generator. The cumulative dose was a direct function of the number of passes of the beam over the sample. (Data from Venter *et al.*, 1983c.)

When human platelet membranes were subjected to high-energy electron bombardment, $\alpha_2$-adrenergic receptors, as measured by [³H]yohimbine-specific binding, were inactivated as a simple exponential of the radiation dosage (Fig. 15). [³H]Yohimbine binding was assessed over a concentration range of 1–4 n*M*, that is, both above and below the $K_d$ for yohimbine binding to the platelet $\alpha_2$-adrenergic receptor (Motulsky *et al.*, 1980; Daiguji *et al.*, 1981). The calculated target size was independent of either ligand or receptor concentration (Venter *et al.*, 1983c). The molecular size of the $\alpha_2$-adrenergic receptor was determined again by the slope ratio method with enzyme standards (Fig. 15). The calculated molecular weight calibration curve (Fig. 16) indicates a molecular weight of 160,000 for the $\alpha_2$ receptor (Venter *et al.*, 1983c). These data indicate at least a superficial structural similarity between the $\alpha_1$- and $\alpha_2$-adrenergic receptors (Venter *et al.*, 1983b), a finding strongly supported by monoclonal antibody data and a subunit size of 85,000 for the $\alpha_2$-receptor (Shreeve *et al.*, 1984).

**d. $D_2$-Dopamine Receptors.** The $D_2$-dopamine receptor is the dopamine-sensitive site in the brain and pituitary to which neuroleptics bind with high (nanomolar) affinity (Lilly *et al.*, 1983). The $D_2$-dopamine receptor is thought to be involved in a variety of dopaminergic behaviors, such as rotation and stereotypy in rats and emesis in dogs and humans. As with the $\alpha_2$-adrenergic receptor, little biochemical information is available on the $D_2$-dopamine receptor (Lilly *et al.*, 1983). Our study (Lilly *et al.*, 1983) using target size analysis provided the first estimate of the molecular size of the $D_2$-dopamine receptor.

The loss of $D_2$-dopamine receptors resulting from irradiation was quantitated using [$^3$H]spiperone. Specific spiperone binding was assessed at four ligand concentrations which bracketed the $K_d$ value for [$^3$H]spiperone binding (Lilly *et al.*, 1983). The functional molecular size of the $D_2$-dopamine receptor was determined by the slope ratio method (Fig. 17). The calculated molecular weight

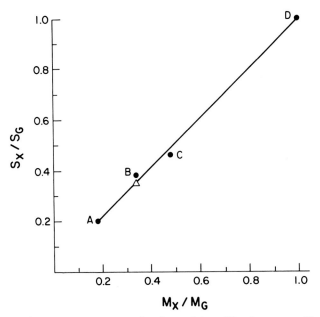

**Fig. 16.** Radiation inactivation molecular weight calibration curve. The molecular size of the $\alpha_2$-adrenergic receptor ($\triangle$) is 160,000. See Fig. 13 for details of method. (Data from Venter *et al.*, 1983c.)

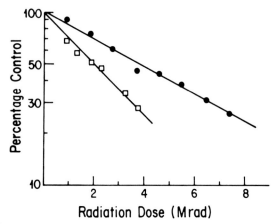

**Fig. 17.** Radiation inactivation–target size analysis of canine striatal $D_2$-dopamine receptors and internal enzyme standard. $D_2$-Dopamine receptor-containing canine striatal membranes were frozen in thin layers in aluminum trays with enzyme standards and subjected to high-energy electron bombardment, as described in the text. Radiation-induced loss of the $D_2$-dopamine receptor was assessed by measuring [³H]spiperone binding at four concentrations of ligand ranging from 0.2 to 1.7 nM. The survival of specific [³H]spiperone binding is represented by (●); each point represents the mean of triplicate determinations for four ligand concentrations. The degree of inactivation measured is independent of the ligand concentration used. The radiation inactivation of the $D_2$-dopamine receptor is compared on the same plot with the inactivation of yeast ADH (□) from the same radiation experiment. Lines were drawn by least-squares linear regression. The molecular size of the $D_2$-dopamine receptor was calculated as described in the text by determining the inactivation ratio $S_{\text{dopamine receptor}}/S_{\text{yeast ADH}}$, which permits a direct comparison to enzyme standards on a linear plot (Fig. 2) of the relationship between the inactivation ratio $S_X/S_G$ and molecular weight ratio $M_X/M_G$ determined in a series of radiation inactivation experiments. The inactivation ratio of the $D_2$-dopamine receptor was related to the standard plot by $S_{\text{dopamine receptor}}/S_G = (S_{\text{dopamine receptor}}/S_{\text{ADH}}/(S_{\text{ADH}}/S_G)$. (Data from Lilly et al., 1983.)

calibration curve (Fig. 18) indicates a molecular weight of 123,000 for the canine $D_2$ receptor.

The percentage decrease in [³H]spiperone binding caused by a given radiation dose was independent of the ligand concentration used to assess that change. Scatchard analysis of these data (Fig. 19) indicates that only receptor number, not ligand affinity, changed as a result of radiation treatment (Lilly *et al.*, 1983).

Further studies (Fig. 20) compared the target size of the human versus dog $D_2$-receptor. As illustrated in Fig. 20, the receptors from both species have essentially identical target sizes (Lilly *et al.*, 1983). (See Table V for a comparison of the target size of membrane proteins to values from other methods.)

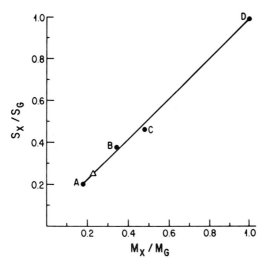

**Fig. 18.** Radiation inactivation molecular weight calibration curve. Standard enzymes are as in Figs. 1 and 13. The least-squares slope of the line is 0.97. The $D_2$-dopamine receptor ($\triangle$) was related to the standard plot as described in Fig. 1, and these data indicate a molecular weight of 123,000 for the canine $D_2$-dopamine receptor. (Data from Lilly *et al.*, 1983.)

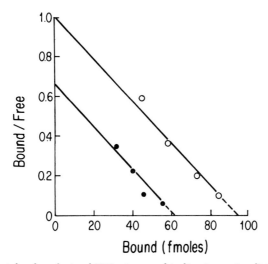

**Fig. 19.** Scatchard analysis of [$^3$H]spiperone binding in nonirradiated and irradiated canine striatal membranes. Illustrated are representative Scatchard analyses of specific [$^3$H]spiperone binding to canine $D_2$-dopamine receptors. Nonirradiated samples ($\bigcirc$) had a $K_d$ value of 0.74 n*M*; samples that received a radiation dose which reduced the mean amount of specific binding to 69% of control values had a $K_d$ value of 0.76 n*M* ($\bullet$). Determinations were made at identical protein concentrations. The parallel shift in the curves reflects a reduction in the bound value to about 65% of the control. (Data from Lilly *et al.*, 1983.)

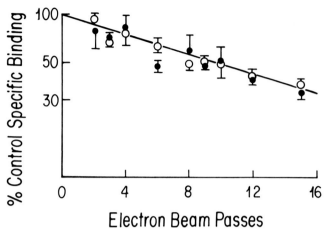

**Fig. 20.** Comparison of $D_2$-dopamine receptor radiation inactivation profiles from canine and human brain striata. Human striatal membranes were irradiated along with canine striatal membranes. The semilogarithmic plots of the decrease in specific [³H]spiperone binding with increasing radiation dose have identical slopes for canine (○) and human (●) striatal membranes. This illustrates that the $D_2$-dopamine receptors from the two species have identical functional sizes. Samples were irradiated with a 0.5-ma beam of 1.5-MeV electrons produced with a Van de Graaff generator. The cumulative dose was a linear function of the number of passes of the beam over the sample. (Data from Lilly et al., 1983.)

TABLE V

**Comparison of the Target Size of Membrane Proteins to Values from Other Methods**

| Receptor or membrane protein | Molecular size | Reference |
|---|---|---|
| Nicotinic acetylcholine receptor (5 subunits $\alpha_2$, $\beta$, $\gamma$, $\delta$) | | |
| Target size | 230,000 | Levinson and Ellory (1973) |
| | 300,000 | Lo et al. (1982) |
| Sequence analysis | 267,757 | Noda et al. (1983) |
| SDS–PAGE | 270,000 | Lo et al. (1982) |
| | 250,000 | Lindstrom et al. (1979) |
| | 260,000 | Biesecker (1973) |
| | 280,000 | Agnew et al. (1980) |
| Hydrodynamic | 250,000 | Reynolds and Karlin (1978) |
| | 250,000–300,000 | Ruchel et al. (1981) |
| Muscarinic cholinergic receptor (see Table IV for further details) | | |
| Target size | 82,000 | Venter (1983) |
| | 78,000 | Venter (1983) |
| | 76,000 | Uchida et al. (1982) |
| SDS–PAGE | 80,000 | Venter (1983); Venter et al. (1984a); Birdsall et al. (1979) |
| Hydrodynamic | 86,000 | Haga (1980) |
| $\alpha_1$-Adrenergic receptor | | |
| Target size | 160,000 | Venter et al. (1983b); Fraser et al. (1983) |

**TABLE V** (*Continued*)

| Receptor or membrane protein | Molecular size | Reference |
|---|---|---|
| SDS–PAGE | 80,000 | Kunos *et al.* (1983) |
| | 85,000 | Venter *et al.* (1984a) |
| Hydrodynamic | 96,000 | Guellan *et al.* (1979) |
| $\alpha_2$-Adrenergic receptor | | |
| Target size | 160,000 | Venter *et al.* (1983c) |
| SDS–PAGE | 85,000 | Shreeve *et al.* (1985) |
| $\beta_1$-Adrenergic receptor | | |
| Target size | 96,000 | Neilson *et al.* (1981) |
| | 65,000 | Fraser and Venter (1983) |
| | 130,000 | Fraser and Venter (1983) |
| SDS–PAGE | 65,000–70,000 | Fraser and Venter (1980) |
| | 65,000 | Schorr *et al.* (1982) |
| Hydrodynamic | 65,000 | Venter (1981); Fraser (1984); |
| $\beta_2$-Adrenergic receptor | | Strauss *et al.* (1979) |
| Target size | 109,000 | Fraser and Venter (1982) |
| | 96,000 | Nielson *et al.* (1981) |
| SDS–PAGE | 59,000 | Soiefer and Venter (1980) |
| | 58,000 | Schorr *et al.* (1982) |
| Hydrodynamic | 75,000 | Haga *et al.* (1977) |
| | 90,000 | Venter (1981); Strauss *et al.* (1979) |
| $D_2$-Dopamine receptor | | |
| Target size | 123,000 | Lilly *et al.* (1983) |
| Hydrodynamic | 136,000 | Lilly *et al.* (1984) |
| Insulin receptor | | |
| Target size | 87,000 | Harmon *et al.* (1980) (combined |
| SDS | 260,000–275,000 | subunits) Massague *et al.* (1980); |
| | | Jacobs and Cuatrecasas (1981) |
| Reduced component | 125,000–135,000 | Jacobs *et al.* (1977) |
| | 90,000 | Jacobs *et al.* (1979) |
| Hydrodynamic | ~300,000 | Cuatrecasas (1972a,b); Jacobs *et al.* (1977); Harrison *et al.* (1978); Ginsberg *et al.* (1978); Pollet *et al.* (1981); Seigel *et al.* (1981) |
| Slow inward calcium channel | | |
| Target size | | |
| Smooth muscle | 278,000 | Venter *et al.* (1983a) |
| Cortex and skeletal muscle | 210,000 ± 20,000 | Norman *et al.* (1983) |
| Skeletal muscle | 136,000 ± 12,000 | Goll *et al.* (1983) |
| SDS–PAGE | 45,000 (subunit) | Venter *et al.* (1983a) |
| Benzodiazepine receptor | | |
| Target size | 220,000 | Chang and Bernard (1982) |
| | 216,000 | Chang *et al.* (1982) |
| | 90,000–100,000 | Doble and Iversen (1982) |
| Hydrodynamic | 240,000 | Asano and Ogasawara (1981) |
| $H^+,K^+$-ATPase | | |
| Target size | 250,000–270,000 | Saccomani *et al.* (1981) |
| SDS–PAGE | 100,000 | Saccomani *et al.* (1981) |

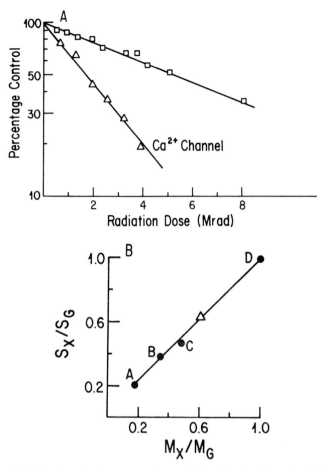

**Fig. 21.** Radiation inactivation–target size analysis of the slow inward calcium channel and standard enzymes. Guinea pig ileum longitudinal smooth muscle cell membranes were purified and frozen in thin layers in aluminum trays along with enzyme standards and subjected to high-energy electron bombardment, as described in Section II,C. Radiation-induced inactivation of the calcium channel was assessed by measuring [³H]nitrendipine-specific binding at a number of ligand concentrations, both below and above the equilibrium dissociation constant. (A) The graph represents the survival of specific [³H]nitrendipine binding (△) expressed on a semilogarithmic plot as a function of control [³H]nitrendipine-specific binding and of increasing doses of radiation. Each point is the average of triplicate samples from five separate determinations. The radiation inactivation of the calcium channel is compared on the same plot to the inactivation of horse liver alcohol dehydrogenase (□) from the same radiation experiment. Lines were drawn by least-squares linear regression. (B) The molecular weight of the calcium channel was determined by determining the inactivation ratio $S_{Ca^+ channel}/S_{liver ADH}$, which permits direct comparison to enzyme standards on a linear plot of the relationship between the inactivation ratio $S_X/S_G$ and molecular weight ratio $M_X/M_G$ determined in a series of radiation inactivation experiments. Standard enzymes were as in Figs. 1 and 13. The inactivation ratio of the calcium channel (△) was related to the standard plot by $S_X/S_G = (S_X/S_{ADH})(S_{ADH}/S_G)$. These data indicate a molecular weight of 278,000 for the calcium channel. (Data from Venter *et al.*, 1983a.)

## 3. *The Slow Inward Calcium Channel*

The slow inward calcium channel plays an important role in the regulation of both smooth muscle and cardiac muscle contraction (Janis and Triggle, 1983). This calcium channel is responsible for the transmembrane movement of calcium ions during the plateau of the cardiac action potential and can be modulated by β-adrenergic receptors. Similar calcium channels exist in ileum and airway smooth muscle where they are under the control of specific hormone and neurotransmitter receptors such as the muscarinic cholinergic receptor (Venter *et al.*, 1983a).

In 1982 we investigated the target size of the calcium channel of smooth muscle utilizing [$^3$H]nitrendipine, a specific calcium channel antagonist (Venter *et al.*, 1983a). As with the above receptors, when smooth muscle membranes isolated from the longitudinal smooth muscle layer of the guinea pig ileum were subjected to high-energy electron bombardment, the slow inward calcium channel, as measured by [$^3$H]nitrendipine-specific binding, was inactivated as a simple exponential of the radiation dose (Fig. 21). [$^3$H]Nitrendipine binding was assessed over a wide range (55–500 pM) of ligand concentrations. As earlier, the slope ratio method was utilized to assess the molecular size of the calcium channel (Fig. 21). The average calcium channel molecular weight from all of the radiation inactivation studies is 278,000 ± 21,000 (Venter *et al.*, 1983a).

## III. CONCLUSIONS AND AVENUES FOR FURTHER STUDY

Target size analysis is a technique that is finding increasing application in the field of membrane biochemistry, particularly with regard to receptors. Molecular size information can be obtained, as demonstrated earlier, with only a simple functional assay. In fact, information not necessarily obtainable with protein isolation techniques (e.g., functional parameters) can be obtained with radiation inactivation.

Future areas for investigation concern the molecular basis for inactivation of oligomeric structures. It is clear from some studies (Saccomani *et al.*, 1981; Lo *et al.*, 1982) that energy transfer can readily occur between subunits of an oligomeric protein. It is equally clear from Table I that some multisubunit proteins inactivate as the subunit monomer. These combined observations suggest that any generalization at this time concerning energy transfer based on any one model would clearly lead to erroneous conclusions with many other proteins. It seems reasonable that the closeness and type of interactions occurring between subunits would be a major determinant in whether energy transfer could occur between proteins. In this vein, the hydrophobic interactions which can occur between membrane proteins may lead to enhanced opportunity for the passage of inactivating energy between subunits. Another remaining question is to what extent the sugar portion of a glycoprotein can contribute to the overall target size of the molecule.

Another potential use for radiation inactivation would be in the quantitation of the amount or percentage of a membrane protein (receptor) present in different aggregation states (monomer, dimer, etc.). This would permit the study of the allosteric modulation of receptor function resulting from subunit interactions or from receptors interacting with membrane effector proteins including $Ca^{2+}$ channels (Venter *et al.*, 1983a) and the guanine nucleotide regulatory protein of adenylate cyclase (Nielson *et al.*, 1981).

## REFERENCES

Agnew, W. S., Moore, A. C., Levinson, S. R., and Raftery, M. A. (1980). Identification of a large molecular weight peptide associated with a tetradotoxin binding protein from the electroplax of Electrophorus electricus. *Biochem. Biophys. Res. Commun.* **92**, 860–866.

Andersson, R. (1973). Role of cyclic AMP and $Ca^{++}$ in mechanical and metabolic events in isometrically contracting vascular smooth muscle. *Acta Physiol. Scand.* **87**, 84–95.

Asano, T., and Ogasawara, N. (1981). Soluble gamma-amino butyric acid and benzodiazepine receptors from rat cerebral cortex. *Life Sci.* **29**, 193–200.

Berthelsen, S., and Pettinger, W. A. (1977). A functional basis for classification of α-adrenergic receptors. *Life Sci.* **21**, 595–606.

Biesecker, G. (1973). Molecular properties of the cholinergic receptor from electrophorus electricus. *Biochemistry* **12**, 4403–4409.

Birdsall, N. J. M., Burgen, A. S. V., and Hulme, E. (1979). A study of the muscarinic receptor by gel electrophoresis. *Br. J. Pharmacol.* **66**, 337–342.

Blackmore, P. F., Hughes, B. P., Charest, R., Shuman, E. A., and Exton, J. A. (1983). Time course of $\alpha_1$-adrenergic and vasopressin actions on phosphorylase activation, calcium efflux, pyridine nucleotide reduction, and respiration in hepatocytes. *J. Biol. Chem.* **258**, 10488.

Chang, L.-R., and Barnard, E. A. (1982). The benzodiazepine GABA receptor complex: Molecular size in brain synaptic membranes and in solution. *J. Neurochem.* **39**, 1507–1518.

Chang, L.-R., Barnard, E. A., Lo, M. M. S., and Dolly, J. O. (1982). Molecular sizes of benzodiazepine receptors and the interacting GABA receptors in the membrane are identical. *FEBS Lett.* **126**, 309–312.

Crowther, J. A. (1924). Actions of X-rays on tissue cells. *Proc. R. Soc. London, Ser. B* **96**, 207–211.

Cuatrecasas, P. (1972a). Isolation of the insulin receptor of liver and fat cell membranes. *Proc. Natl. Acad. Sci. U.S.A.* **69**, 318–322.

Cuatrecasas, P. (1972b). Properties of the insulin receptor isolated from liver and fat cell membranes. *J. Biol. Chem.* **247**, 1980–1991.

Daiguji, M., Meltzer, H. Y., and U'Prichard, D. C. (1981). Human platelet $\alpha_2$-adrenergic receptors: labelling with $^3$H-yohimbine, a selective antagonist ligand. *Life Sci.* **28**, 2705–2717.

D'Alessio, G. D., Parente, A., Guida, C., and Leone, E. (1972). Dimeric structure of seminal ribonuclease. *FEBS Lett.* **27**, 285–288.

Dehayne, J.-P., Blackmore, P. F., Venter, J. C., and Exton, J. H. (1980). Studies on the α-adrenergic activation of hepatic glucose output. α-Adrenergic activation of phosphorylase by immobilized epinephrine. *J. Biol. Chem.* **255**, 3905–3910.

Doble, A., and Iversen, L. L. (1982). Molecular size of benzodiazepine receptor in rat brain *in situ:* evidence for a functional dimer *Nature (London)* **295**, 522–523.

Fluke, D. J. (1966). Temperature dependence of ionizing radiation effect on dry lysozyme and ribonuclease. *Radiat. Res.* **28**, 677–693.

Fraser, C. M. (1984). Anti-β-adrenergic receptor monoclonal antibodies and β-receptor structure. *In* "Monoclonal and Anti-Idiotypic Antibodies: Probes for Receptor Structure and Function" (J. C. Venter, C. M. Fraser, and J. M. Lindstrom, eds.) pp. 69–84. Alan R. Liss, New York.

Fraser, C. M., and Venter, J. C. (1980). Monoclonal antibodies to β-adrenergic receptors: Use in purification and molecular characterization of β-receptors. *Proc. Natl. Acad. Sci. U.S.A.* **77**, 7034–7038.

Fraser, C. M., and Venter, J. C. (1982). The size of the mammalian lung $\beta_2$-adrenergic receptor as determined by target size analysis and immunoaffinity chromatography. *Biochem. Biophys. Res. Commun.* **109**, 21–29.

Fraser, C. M., Greguski, R., Eddy, B., and Venter, J. C. (1983). Autoantibodies and monoclonal antibodies in the purification and molecular characterization of neurotransmitter receptors. *J. Cell. Biochem.* **21**, 219–231.

Ginsberg, B. H., Cohen, R. M., Kahn, C. R., and Roth, J. (1978). Properties and partial purification of the detergent solubilized insulin receptor. *Biochim. Biophys. Acta* **542**, 88.

Goll, A., Ferry, D. R., and Glossman, H. (1983). Target size analysis of skeletal muscle $CA^{2+}$ channels. *FEBS Lett.* **157**, 63–69.

Green, R. W., and Mekay, R. H. (1969). The behavior of horse liver alcohol dehydrogenose in guanidine hydrocloride. *J. Biol. Chem.* **244**, 5034–5043.

Guellan, G., Aggerbeck, M., and Hanoune, J. (1979). Characterisation and solubilization of the α-adrenoceptor of rat liver plasma membranes labeled with [$^3$H]-phenoxybenzamine. *J. Biol. Chem.* **254**, 10761–10768.

Haga, T., (1980). Molecular size of muscarinic acetylcholine receptors of rat brain. *FEBS Lett.* **113**, 68–72.

Haga, T., Haga, K., and Gilman, A. G. (1977). Hydrodynamic properties of the β-adrenergic receptor and adenylate cyclase from wild type and variant S49 lymphoma cells. *J. Biol. Chem.* **252**, 5776–5782.

Harmon, J. T., Kahn, C. R., Kempner, E. S., and Schlegel, W. (1980). Characterization of the insulin receptor in its membrane environment by radiation inactivation. *J. Biol. Chem.* **255**, 3412–3419.

Harris, I. (1964). Structure and catalytic activity of alcohol dehydrogenase. *Nature (London)* **203**, 30–34.

Harrison, L. C., Billington, T., East, I. J., Nicholas, R. J., and Clark, S. (1978). The effect of solubilization on the properties of the insulin receptor of human placental membrane. *Endocrinology (Baltimore)* **102**, 1485.

Hoffman, B. B., DeLean, A., Wood, C. L., Shocken, D. D., and Lefkowitz, R. J. (1979). Alpha-adrenergic receptor subtypes: quantitative assessment by ligand binding. *Life Sci.* **24**, 1739–1746.

Holtzer, A., and Lowey, S. (1959). The molecular weight, size and shape of the myosin molecule. *J. Am. Chem. Soc.* **81**, 1370–1377.

Jacobs, S., and Cuatrecasas, P. (1981). Insulin receptor: Structure and function. *Endocr. Rev.* **2**, 251–263.

Jacobs, S., Schechter, Y., Bissell, K., and Cuatrecasas, P. (1977). Purification and properties of insulin receptors from rat liver membranes. *Biochem. Biophys. Res. Commun.* **77**, 981–988.

Jacobs, S., Hazum, E., Shechter, Y., and Cuatrecasas, P. (1979). Insulin receptor: Covalent labelling and identification of subunits. *Proc. Natl. Acad. Sci. U.S.A.* **6**, 4918–4921.

Jakobs, K. H. (1979). Inhibition of adenylate cyclase by hormones and neurotransmitters. *Mol. Cell. Endocrinol.* **16**, 147–156.

Jakobs, K. H., Saur, W., and Schultz, G. (1978). Metal and metal-ATP interactions with human platelet adenylate cyclase: effects of *alpha* adrenergic inhibition. *Mol. Pharmacol.* **14,** 1073–1078.

Janis, R. A., and Triggle, D. J. (1983). New developments in $Ca^{2+}$ channel antagonists. *J. Med. Chem.* **26,** 775–785.

Jung, C. Y. (1984). Molecular weight determination by radiation inactivation. *In* "Molecular and Chemical Characterization of Membrane Receptors—Receptor Biochemistry and Methodology" (J. C. Venter and L. C. Harrison, eds.), Vol. 3, pp. 193–208. Alan R. Liss, New York.

Kempner, G. R., and Macey, R. L. (1968). Membrane enzyme systems molecular size determination by radiation inactivation. *Biochim. Biophys. Acta* **163,** 188–203.

Kempner, E. S., and Schlegel, W. (1979). Size determination of enzymes by radiation inactivation. *Anal. Biochem.* **92,** 2–10.

Kunos, G., Kan, W. H., Greguski, R., and Venter, J. C. (1983). Affinity labeling and molecular weight determination of hepatic $\alpha_1$-adrenergic receptors with [$^3$H] phenoxybenzamine. *J. Biol. Chem.* **258,** 326–332.

Lea, D. E. (1947). "Actions of Radiations on Living Cells." Cambridge Univ. Press, Cambridge, England.

Lea, D. E., Smith, K. M., Holmes, B., and Markham, R. (1944). Direct and indirect actions of radiation on viruses and enzymes. *Parasitology* **36,** 110–118.

Lee, C. Y., (1979). Recent advances in the chemistry and pharmacology of snake toxins. *Adv. Cytopharmacol.* **3.**

Levinson, S. R., and Ellory, J. C. (1973). Molecular size of the tetrodotoxin binding site estimated by irradiation inactivation. *Nature (London) New Biol.* **245,** 122–123.

Lilly, L., Fraser, C. M., Jung, C. Y., Seeman, P., and Venter, J. C. (1983). Molecular size of the canine and human brain D2 dopamine receptor as determined by radiation inactivation. *Mol. Pharmacol.* **24,** 10–14.

Lilly, L., Davis, A., Fraser, C. M., Seeman, P., and Venter, J. C. (1985). Molecular characterization of brain $D_2$-dopamine receptors. Submitted.

Limbird, L. E., Speck, J. L., and Smith, S. K. (1982). Sodium ion modulates agonist and antagonist interactions with the human platelet alpha$_2$-adrenergic receptor in membrane and solubilized preparations. *Mol. Pharmacol.* **21,** 607–619.

Lindstrom, J., Merlie, J., and Yogeeswaran, G. (1979). Biochemical properties of acetylcholine receptor subunits from Torpedo californica. *Biochemistry* **18,** 4465–4470.

Lo, M. M. S., Barnard, E. A., and Dolly, J. O. (1982). Size of acetylcholine receptors in the membrane. An improved version of the radiation inactivation method. *Biochemistry* **21,** 2210–2217.

Massague, J., Pilch, P. F., and Czech, M. P. (1980). Electrophoretic resolution of three major insulin receptor structures with unique subunit stoichiometries. *Proc. Natl. Acad. Sci. U.S.A.* **77,** 7137–7141.

Morawiecki, A. (1960). Dissociation of pyruvic kinase in urea solutions. *Biochim. Biophys. Acta* **44,** 604–605.

Motulsky, H. T., Shattil, S. J., and Insel, P. A. (1980). Characterization of $\alpha_2$-adrenergic receptors on human platelets using [$^3$H]Yohimbine. *Biochem. Biophys. Res. Commun.* **97,** 1562–1570.

Mueller, H. (1964). Molecular weight of myosin and meromyosins by archibald experiments preformed with increasing speed of rotation. *J. Biol. Chem.* **239,** 797–804.

Nielson, T. B., Lad, P. M., Preston, M. S., Kempner, E., Schlegel, W., and Rodbell, M. (1981). Structure of the turkey erythrocyte adenylate cyclase system. *Proc. Natl. Acad. Sci. U.S.A.* **78,** 722–726.

Noda, M., Takahashi, H., Tanabo, T., Toyosato, M., Kikyotani, S., Furutani, Y., Hirose, T., Takashima, H., Imayama, S., Miyata, T., and Numa, S. (1983). Structural homology of Torpedo californica acetylcholine receptor subunits. *Nature (London)* **302,** 528–532.

Norman, R. I., Borsotto, M., Fosset, M., and Lazdanski, M. (1983). Determination of the molecular size of the nitrendipine-sensitive CA$^{2+}$ channel by radiation inactivation. *Biochem. Biophys. Res. Commun.* **111**, 878–883.

Pfleiderer, G., and Auricchio, F. (1964). The DPNH-binding capacity of various dehydrogenases. *Biochem. Biophys. Res. Commun.* **16**, 53–59.

Pollard, E. C. (1953). Primary ionization as a test of molecular organization. *Adv. Biol. Med. Phys.* **3**, 153–189.

Pollard, E. C., Guild, W. R., Hutchinson, F., and Setlow, R. B. (1955). The direct action of ionizing radiation on enzymes and antigens. *Prog. Biophys. Biophys. Chem.* **5**, 72–108.

Pollet, F. J., Haase, B. A., and Standaert, M. L. (1981). Characterisation of detergent-solubilized membrane proteins. *J. Biol. Chem.* **256**, 12118–12126.

Reynolds, J., and Karlin, A. (1978). Molecular weight in detergent solution of acetylcholine receptor from Torpedo californica. *Biochemistry* **17**, 2035–2038.

Robison, G. A., Butcher, R. W., and Sutherland, E. W. (1971). "Cyclic AMP." Academic Press, New York.

Ruchel, R., Watters, D., and Maelicke (1981). Molecular forms and hydrodynamic properties of acetylcholine receptor from electric tissue. *Eur. J. Biochem.* **119**, 215–223.

Saccomani, G., Sachs, G., Cuppoletti, J., and Jung, C.-Y. (1981). Target molecular weight of the gastric (H$^+$K$^+$)-ATPase: Functional and structural molecular size. *J. Biol. Chem.* **256**, 7727–7729.

Schlegel, W., Kempner, E. S., and Rodbell, M. (1979). Activation of adenylate cyclase in hepatic membranes involves interactions of the catalytic unit with multimeric complexes of regulatory proteins. *J. Biol. Chem.* **254**, 5168–5176.

Schorr, R. G. L., Heald, S. L., Jeffs, P. W., Lavin, T. N., Strohsacker, M. W., Lefkowitz, R. J., and Caron, M. G. (1982). The β-adrenergic receptor: Rapid purification and covalent labeling by photoaffinity cross-linking *Proc. Natl. Acad. Sci. U.S.A.* **79**, 2778–2782.

Seigel, T. W., Ganguly, S., Jacobs, S., Rosen, O. M., and Rubin, C. S. (1981). Purification and properties of the human placental insulin receptor. *J. Biol. Chem.* **256**, 9266–9273.

Shreeve, S. M., Fraser, C. M., and Venter, J. C. (1985). Molecular Comparison of α$_1$- and α$_2$-adrenergic receptors suggests that these proteins are structurally related "iso-receptors." *Proc. Natl. Acad. Sci. U.S.A.* (in press).

Soiefer, A. I., and Venter, J. C. (1980). Mammalian lung β$_2$-adrenergic receptor purification utilizing an affinity reagent. *Fed. Proc., Fed. Am. Soc. Exp. Biol.* **39**, 313.

Steinmetz, M. A., and Deal, W. C., Jr. (1966). Metabolic control and structure of glycolytic enzymes. III. Dissociation and subunit structure of rabbit muscle pyruvate kinase. *Biochemistry* **5**, 1399–1405.

Strauss, W. L., Ghai, G., Fraser, C. M., and Venter, J. C. (1979). Hydrodynamic properties and sulfhydryl reagent sensitivity of β$_1$ (cardiac) and β$_2$ (lung and liver) adrenergic receptors: Molecular evidence for isoreceptors. *Fed. Proc. Fed. Am. Soc. Exp. Biol.* **38**, 843.

Svedberg, T., and Broholt, S. (1939). Splitting of protein molecules by ultra-violet light and α-rays. *Nature (London)* **143**, 938–939.

Uchida, S., Matsumoto, K., Takeyasu, K., Higuchi, H., and Yoshida, H. (1982). Molecular mechanism of the effects of guanine nucleotide and sulfhydryl reagent on muscarinic receptors in smooth muscle studied by radiation inactivation. *Life Sci.* **31**, 201–209.

Venter, J. C. (1981). β-Adrenoceptors, adenylate cyclase, and the adrenergic control of cardiac contractility. *In* "Adrenoceptors and Catecholinamine Actions" (G. Kunos, ed.), pp. 213–245. Wiley (Interscience), New York.

Venter, J. C. (1982). Monoclonal antibodies and autoantibodies in the isolation and characterization of neurotransmitters: The future of receptor research. *J. Mol. Cell. Cardiol.* **14**, 687–693.

Venter, J. C. (1983). Muscarinic cholinergic receptor structure. 1. Receptor size, membrane orientation and absence of major phylogenetic structural diversity. *J. Biol. Chem.* **258**, 4842–4848.

Venter, J. C., and Fraser, C. M. (1983a). The structure of α- and β-adrenergic receptors. *Trends Pharmacol. Sci.* **4,** 256–258.

Venter, J. C., and Fraser, C. M. (1983b). β-Adrenergic receptor isolation and characterization with immobilized drugs and monoclonal antibodies. *Fed. Proc., Fed. Am. Soc. Exp. Biol.* **42,** 273–278.

Venter, J. C., Fraser, C. M., Schaber, J. S., Jung, C. Y., Bolger, G., and Triggle, D. J. (1983a). Molecular properties of the slow inward calcium channel: Molecular weight determinations by radiation inactivation and covalent affinity labelling. *J. Biol. Chem.* **258,** 9344–9348.

Venter, J. C., Fraser, C. M., Lilly, L., Seeman, P., Eddy, B., and Schaber, J. (1983b). The structure of neurotransmitter receptors (adrenergic, dopaminergic and muscarinic cholinergic). *In* "Catecholamines, Part A: Basic and Peripheral Mechanisms," (E. Usdin, A. Carlsson, A. Dählstrom, and J. Engel, eds.), pp. 293–302. Alan R. Liss, Inc., New York.

Venter, J. C., Schaber, J. S., U'Prichard, D. C., and Fraser, C. M. (1983c). Molecular size of the human platelet $\alpha_2$-adrenergic receptor as determined by radiation inactivation. *Biochem. Biophys. Res. Commun.* **116,** 1070–1075.

Venter, J. C., Eddy, B., Hall, L. M., and Fraser, C. M. (1984a). Monoclonal antibodies detect the conservation of muscarinic cholinergic receptor structure from drosophila to human brain and possible structural homology with $\alpha_1$-adrenergic receptors. *Proc. Natl. Acad. Sci. U.S.A.* **81,** 272–276.

Venter, J. C., Horne, P., Eddy, B., Greguski, R., and Fraser, C. M. (1984b). Alpha$_1$-adrenergic receptor structure. *Mol. Pharmacol.* **26,** 196–205.

Weber, K., and Osborn, M. (1969). The reliability of molecular weight determinations by dodecyl sufate-polyacerylamide gel electrophoresis. *J. Biol. Chem.* **244,** 4406–4412.

Weeds, A. G., and Lowey, S. (1971). Substructure of the myosin molecule. II. The light chains of myosin. *J. Mol. Biol.* **61,** 701–725.

# 8

# $\alpha_2$-Adrenergic Receptors: Apparent Interaction with Multiple Effector Systems

## L. E. LIMBIRD and J. D. SWEATT
Department of Pharmacology
School of Medicine
Vanderbilt University
Nashville, Tennessee

## I. HISTORICAL PERSPECTIVE AND INTRODUCTION

Adrenergic receptors for the native catecholamines epinephrine and nor-epinephrine were first classified into distinct receptor populations, termed $\alpha$- and $\beta$-adrenergic receptors, based on the relative order of potency of the catecholamines in eliciting physiological effects. Subsequently, specific antagonists for $\beta$-adrenergic (e.g., propranolol) and $\alpha$-adrenergic (e.g., phentolamine) receptors were developed which confirmed the original postulate of Ahlquist (1948) that distinct populations of adrenergic receptors exist.

More recently, pharmacological as well as biochemical studies have demonstrated that $\alpha$- and $\beta$-adrenergic receptors can be further subdivided into receptor

THE RECEPTORS, VOL. II

subtypes. Thus, $\beta_1$- and $\beta_2$-adrenergic receptors can be distinguished by the relative potency of agonists and selective antagonists in interacting with the receptor binding sites (Lands *et al.*, 1967). Both $\beta$-adrenergic subtypes are linked to the activation of adenylate cyclase activity and appear to elicit their physiological effect, at least in part, by the elevation of intracellular cAMP concentrations.

$\alpha$-Adrenergic receptors have also been categorized into subtypes. The early distinction between $\alpha$-adrenergic subtypes was anatomical in nature: the archetypal $\alpha_2$-adrenergic receptor was presynaptic and mediated inhibition of norepinephrine release in the central nervous system, whereas $\alpha_1$-adrenergic receptors were thought to be solely postsynaptic and mediated, for example, the contraction of smooth muscle (Doxey *et al.*, 1977; Langer, 1974, 1976; Starke, 1977; Westfall, 1977). Clonidine and epinephrine were potent agonists at "central $\alpha$ receptors," whereas methoxamine and norepinephrine were potent at so-called "peripheral $\alpha$-receptors." However, as research has continued, the anatomical description of $\alpha$-adrenergic receptors has proven inappropriate. Thus, the $\alpha$-adrenergic receptor which mediates epinephrine-induced platelet aggregation is clearly not "presynaptic," but nonetheless demonstrates a specificity for $\alpha$-adrenergic agents that resembles that observed for central nervous system presynaptic receptors. Consequently, it is now regarded as more appropriate to refer to what were previously called "postsynaptic receptors" as $\alpha_1$-adrenergic receptors and to previously named "presynaptic receptors" as $\alpha_2$-receptors (Berthelson and Pettinger, 1977; Borowski *et al.*, 1977; see Hoffman and Lefkowitz, 1980b, for a review).

Unlike the situation for $\beta$-adrenergic receptors, in which both $\beta_1$- and $\beta_2$-adrenergic receptors are linked to the activation of adenylate cyclase, it appears at present that each $\alpha$-receptor subtype may be linked to discrete effector systems for eliciting its ultimate physiological effect. Thus, $\alpha_1$-adrenergic receptors have been shown to be linked to phosphatidylinositol turnover and the gating of $Ca^{2+}$. In contrast, $\alpha_2$-adrenergic receptors are typically described as being linked to inhibition of adenylate cyclase activity (Fain and Garcia-Sainz, 1980).

This chapter will focus solely on $\alpha_2$-adrenergic receptor systems. The classification and direct identification of $\alpha_2$-adrenergic receptors using radioligand binding techniques will only be summarized briefly, as this has been extensively reviewed elsewhere (Hoffman and Lefkowitz, 1980b). The biochemical properties of $\alpha_2$ receptors in target tissues where these receptors are linked to attenuation of adenylate cyclase and the postulated mechanism by which this inhibition is effected will then be described. Finally, the remainder of this chapter will summarize the extant data that suggest that only in a limited number of situations can physiological functions elicited via $\alpha_2$-adrenergic receptors be entirely accounted for by decreases in cAMP levels intracellularly. Thus, examples will be given in which physiological functions elicited by $\alpha_2$-adrenergic receptors appear to be entirely independent of changes in cellular cAMP accumulation, whereas other examples will indicate that $\alpha_2$-adrenergic receptor-mediated decreases in cAMP levels may parallel changes in cell function but that concomitant changes in intracellular "messengers" distal to changes in cAMP content must also be involved in eliciting $\alpha_2$-adrenergic effects.

## II. IDENTIFICATION OF α-ADRENERGIC RECEPTORS

The direct identification of $\alpha$-adrenergic binding sites in a number of target tissues has relied on the availability of radioligands of appropriate selectivity and sufficient specific radioactivity. Table I summarizes the currently available radioligands for the identification of $\alpha_2$-adrenergic receptors. As indicated in Table I, [³H]dihydroergocryptine (DHE) or [³H]phentolamine are so-called "nonsubtype selective" $\alpha$-adrenergic antagonists in that they are equally potent at both the $\alpha_1$- and $\alpha_2$-adrenergic receptors. However, one can selectively label $\alpha_2$-adrenergic receptors in tissues possessing both subtypes by incubating with the nonsubtype-selective radiolabeled antagonist in the presence of a concentration of prazosin, usually $10^{-7} M$, that occupies all of the $\alpha_1$-adrenergic receptors present (Hoffman *et al.*, 1979, 1980). This approach relies on the virtual specificity of the antagonist prazosin for $\alpha_1$-adrenergic receptors. Alternatively, $\alpha_2$-adrenergic receptors can be identified using the $\alpha_2$-adrenergic receptor-selective antagonist [³H]yohimbine or its stereoisomer [³H]rauwolscine. These ligands have been particularly useful in the identification of $\alpha_2$ receptors in human platelets, because yohimbine and its congeners have an anomalously higher potency at platelet receptor binding sites than at $\alpha_2$ receptors in other tissues. Investigators have also used radioactively labeled agonists, e.g., [³H]epinephrine and [³H]norepinephrine, or partial agonists, e.g., [³H]clonidine and p-[³H]aminoclonidine, to identify $\alpha_2$-adrenergic receptors and especially to compare the properties of receptor–agonist versus receptor–antagonist interactions. The references cited in Table I are not intended to be exhaustive but serve only to direct the reader to methodological details. In all of the cases cited, however, assurance that the radioligand binding detected represents binding to the physiologically relevant receptor has been provided by data demonstrating the saturability, appropriate kinetics, and specificity of radioligand binding.

## III. PROPERTIES OF α₂-ADRENERGIC RECEPTORS LINKED TO INHIBITION OF ADENYLATE CYCLASE

$\alpha_2$-Adrenergic receptors are classically described as one of a number of receptor populations that are linked to inhibition of adenylate cyclase activity (Jakobs, 1979). Consequently, a number of experimental avenues have been taken to understand the precise molecular mechanism by which $\alpha_2$-adrenergic agents

**TABLE I**

**Radioligands Used for the Identification of $\alpha_2$-Adrenergic Receptors**

| Type of radioligand | Subtype specificity | Comment | Reference |
|---|---|---|---|
| Antagonists | | | |
| [$^3$H]Dihydroergocryptine | Equal potency at $\alpha_1$ and $\alpha_2$ receptors | Labels entire $\alpha$-receptor population | Williams et al. (1976); Williams and Lefkowitz (1976); Newman et al. (1978); Tsai and Lefkowitz (1978); Wood et al. (1979); Jakobs and Raushek (1978) |
| [$^3$H]Dihydroergonine | Equal potency at $\alpha_1$ and $\alpha_2$ receptors | | Steer et al. (1979) |
| [$^3$H]Phentolamine | Equal potency at $\alpha_1$ and $\alpha_2$ receptors | | |
| [$^3$H]Yohimbine | $\alpha_2$ Selective | May be particularly advantageous in platelet, because of 10-fold higher potency there | Smith and Limbird (1981) |
| Agonists | | | |
| [$^3$H]Epinephrine | Could theoretically interact with $\alpha_1$ or $\alpha_2$ receptors | Probably labels only the guanine nucleotide-sensitive high-affinity state, $\alpha_{2H}$ | Hoffman et al. (1982); Smith and Limbird (1981); Mac-Farlane et al. (1981) |
| [$^3$H]Norepinephrine | | May label only $\alpha_{2H}$, simply because at $\alpha_{2L}$ or $\alpha_1$, norepinephrine may have too low a potency to be detectable in direct binding assays | Lynch and Steer (1981); U'Prichard et al. (1977); Greenberg et al. (1976) |
| Partial Agonists | | | |
| [$^3$H]Clonidine | $\alpha_2$ Selective | It is possible that, because of the high potency of these agents, both affinity states of the $\alpha_2$ receptor ($\alpha_{2H}$ and $\alpha_{2L}$) may be labeled | Garcia-Sevilla et al. (1981) |
| p-[$^3$H]Aminoclonidine | $\alpha_2$ Selective | | Shattil et al. (1981) Mooney et al. (1982) |

inhibit this enzyme. In many cases, these studies have focused on the different properties of receptor–agonist and receptor–antagonist interactions. The underlying rationale of such studies is that those molecular properties that uniquely characterize agonist occupancy of the receptors are those properties that may be crucial for receptor coupling to the adenylate cyclase system. The human platelet has been a frequently exploited model system for evaluating the molecular consequences of agonist occupancy of the $\alpha_2$ receptor, primarily because platelets are easily isolated as a reasonably homogeneous preparation and contain α-adrenergic receptors solely of the $\alpha_2$-adrenergic subtype which can be easily demonstrated to elicit inhibition of adenylate cyclase activity.

One approach for comparing agonist versus antagonist interactions with human platelet $\alpha_2$-adrenergic receptors has been to compare the properties of agonist and antagonist competition for radiolabeled-antagonist binding to human platelet membranes. Although both [³H]DHE and [³H]yohimbine have been used to study human platelet $\alpha_2$ receptors in this manner, [³H]yohimbine, because of its reduced interaction with nonspecific sites, has been relied on more frequently for quantitative analysis of receptor–agonist versus receptor–antagonist interactions. [³H]Yohimbine itself appears to interact with a single population of saturable binding sites possessing nanomolar affinity for this ligand (Hoffman et al., 1982; Smith and Limbird, 1981). Competition for [³H]yohimbine binding by α-adrenergic antagonists, such as yohimbine itself or phentolamine, demonstrates a competition profile characteristic of ligand binding to a single population of recognition sites possessing a discrete affinity for each antagonist. Thus, 10–90% of antagonist competition for the radioligand occurs over an 81-fold range of competitor (Koshland, 1969) and pseudo-Hill slopes $(n_H)$ of these plots are equal to 1.0 (Hoffman et al., 1982). Thus, the curves are said to be of "normal steepness." In contrast, agonist competition curves are "shallow," i.e., 10–90% competition by agonists for [³H]yohimbine binding occurs over a greater than 81-fold range of competitor and pseudo-Hill coefficients are less than 1.0 (Hoffman et al., 1982). These data reflect a heterogeneity of receptor–agonist interactions that could be accounted for by multiple and distinct receptor binding sites possessing differing affinities for the agonist or by multiple but interconvertible affinity states of the $\alpha_2$ receptor for the agonist agents. The data are also consistent with agonist-promoted negative cooperativity among the binding sites, such that the affinity of the overall receptor population of agonist decreases as receptor occupancy by agonist increases.

A number of lines of experimental evidence suggest that agonists interact with human platelet $\alpha_2$-adrenergic receptors in two interconvertible affinity states, termed the "high-affinity state" $(\alpha_2 R_H)$ and the "low-affinity state" $(\alpha_2 R_L)$ (Hoffman et al., 1982). Interconversion between these two affinity states has been demonstrated to be mediated by guanine nucleotides. Thus, in a manner analogous to the regulation of β-adrenergic receptors linked to the stimulation of adenylate cyclase, guanine nucleotides decrease $\alpha_2$-receptor affinity for agonist agents and simultaneously remove the apparent heterogeneity of agonist binding, as evidenced by a return of the shape of the agonist competition curve to "normal steepness" (pseudo $n_H = 1.0$) (Tsai and Lefkowitz, 1979; Hoffman et al., 1982; Smith and Limbird, 1981). The ability of guanine nucleotides to decrease agonist potency at $\alpha_2$ receptors correlates with the intrinsic activity of the agonist in attenuating prostaglandin $E_1$- ($PGE_1$) stimulated human platelet adenylate cyclase activity (Tsai and Lefkowitz, 1979). Thus, a greater fraction of the $\alpha_2$-adrenergic population is in the high-affinity state, $\alpha_2 R_H$, when the receptor population is occupied by a full agonist, such as epinephrine, than when the receptor population is occupied by a partial agonist, such as clonidine. Computer-assisted analysis of the competition profiles permits a quantitative assessment of the fraction of the total receptor population in the high-affinity state under particular experimental conditions as well as the apparent $K_D$ of the receptor for the ligand in the high ($K_{DH}$) and low- ($K_{DL}$) affinity states (DeLean et al., 1978, 1980; Munson and Rodbard, 1980).

The molecular interpretation of the multiple affinity states of human platelet $\alpha_2$-adrenergic receptors linked to the inhibition of adenylate cyclase has relied heavily on analogy with the previously characterized $\beta$-adrenergic system linked to the activation of adenylate cyclase (see Limbird, 1981, and Ross and Gilman, 1980, for reviews). For $\beta$-adrenergic receptors, a number of lines of experimental evidence had suggested that agonist occupancy of the receptor stabilized receptor interaction with the guanine nucleotide regulatory protein that mediates stimulation of the adenylate cyclase enzyme ($N_s$). As a result of this interaction, agonists facilitated exchange of the GDP for GTP into the guanine nucleotide binding site of $N_s$ (Cassel and Selinger, 1978). GTP, or a triphosphate analogue of GTP, is essential for activation of the enzyme. In fact, the naked catalytic subunit of the adenylate cyclase system does not possess catalytic activity in the presence of the physiological substrate $Mg^{2+}ATP$, unless it is in association with the GTP-liganded $N_s$ (Ross et al., 1978). Consequently, the ability of $\beta$-adrenergic agonists to increase adenylate cyclase activity can be visualized as an agonist-enhanced rate of interaction of GTP with the holoadenylate cyclase system. Adenylate cyclase activity persists until GTP is hydrolyzed to GDP via a GTPase activity presumably inherent in the guanine nucleotide-binding protein (Cassel et al., 1977).

In the above scheme for the $\beta$-adrenergic stimulation of adenylate cyclase, the agonist–receptor–$N_s$ "ternary complex," presumably formed transiently in vivo, is felt to represent the high-affinity state of the receptor for agonist (DeLean et al., 1980). The binding of GTP to $N_s$, however, dissociates the receptor–$N_s$ interaction (Limbird et al., 1979), presumably concomitant with stabilization of the $N_s$–catalytic subunit interaction. The receptor, once dissociated from $N_s$, is felt to represent the low-affinity state for agonist (Kent et al., 1979; DeLean et *al.*, 1980). Thus, in the presence of GTP, all of the receptors possess a lower affinity for agonists and consequently display binding properties consistent with a homogeneous population of receptors possessing a single, and lower, affinity for agonist.

As noted earlier, the similarity of agonist interactions with β-adrenergic receptors and agonist interactions with α₂-adrenergic receptors suggested the possibility that agonist occupancy of the α₂ receptor resulted in the formation of a ternary complex of agonist–receptor–GTP-binding protein and that this complex represented the high-affinity state of the α₂ receptor for agonist detected in competition binding studies. Additional biochemical studies beyond the analysis of competition binding provided evidence that agonist occupancy of the human platelet α₂-adrenergic receptor indeed stabilized the association of the receptor with the GTP-binding protein modulating receptor affinity for agonist. Exposure of human platelet membranes to the biological detergent digitonin resulted in the solubilization of α₂ receptors that retained the identical recognition properties of α-adrenergic antagonists that had been seen in intact membranes. In contrast, detergent solubilization resulted in a loss of higher affinity receptor–agonist interactions and their sensitivity to guanine nucleotides (Smith and Limbird, 1981). These data were interpreted to suggest that digitonin had disrupted receptor interaction with membrane component(s) responsible for guanine nucleotide-sensitive, high-affinity receptor–agonist interaction. Thus, digitonin had presumably disrupted α₂-receptor interactions with the GTP-binding protein modulating receptor affinity for agonist. In contrast, if human platelet α₂-adrenergic receptors were occupied with the radiolabeled agonist [³H]epinephrine prior to solubilization, the receptor retained its higher affinity for agonists as well as its sensitivity to guanine nucleotides, manifested by the ability of the hydrolysis-resistant analogue of GTP, guanyl-5′-yl imidodiphosphate [Gpp(NH)p], to facilitate the dissociation of prelabeled [³H]epinephrine binding from the solubilized receptor complex (Smith and Limbird, 1981). Further evidence that the solubilized, prelabeled [³H]epinephrine–α₂ receptor complex might contain a GTP-binding protein, or at least an additional membrane component, came from a comparison of the molecular properties of solubilized agonist–α₂ receptor complexes with those of antagonist–α₂ receptor complexes or with those of receptors solubilized in an unoccupied state. Thus, sucrose gradient centrifugation demonstrated that agonist occupancy of the human platelet α₂-adrenergic receptor resulted in an increase in apparent receptor size (faster sedimentation) that was not mimicked by antagonist occupancy of the receptor (Michel *et al.*, 1981; Smith and Limbird, 1981). This increase in apparent receptor size was reversed or prevented by guanine nucleotides (Smith and Limbird, 1981). Similar molecular changes had been noted previously for agonist occupancy of β-adrenergic receptors linked to the stimulation of adenylate cyclase in a number of model systems (Limbird and Lefkowitz, 1978; Limbird *et al.*, 1979). In studies with β-adrenergic receptors, the complex of larger size was demonstrated to contain the $N_s$ protein, identified as a substrate for ADP ribosylation by cholera toxin (Limbird et al., 1980), and this protein was also demonstrated in reconstitution studies to be effective in eliciting the activation of adenylate cyclase (Stadel et al., 1981), providing direct evidence that a single GTP-binding component of the adenylate cyclase architecture could modulate both catalytic activity and receptor affinity for agonist agents. In contrast, however, the agonist-stabilized $\alpha_2$-adrenergic receptor complex of larger size was demonstrated not to contain the cholera toxin substrate (Smith and Limbird, 1982). This observation provided evidence, as had other extant indirect data (Cooper, 1982), that distinct GTP-binding proteins were involved in mediating the activation and inhibition of adenylate cyclase. (Further discussion of this conclusion is provided in the following section.)

The functional communication of $\alpha_2$-adrenergic receptors with a GTP-binding regulatory protein was demonstrated using two additional experimental approaches. As suggested earlier, one predicted consequence of the interaction of $\alpha_2$ receptors with a GTP-binding protein was an agonist-facilitated rate of exchange of guanine nucleotides at the GTP binding site on the protein, by analogy with findings in the $\beta$-adrenergic receptor system. This prediction was documented to be true by the observation that $\alpha_2$-adrenergic agonists, but not antagonists, provoke the release of prebound [$^3$H]Gpp(NH)p from human platelet membranes and that partial agonists facilitate the release of [$^3$H]Gpp(NH)p to a lesser extent. $PGE_1$, an agent which activates adenylate cyclase activity in the human platelet, also promotes the release of [$^3$H]Gpp(NH)p from human platelet membranes, but the quantity of [$^3$H]Gpp(NH)p released in response to $PGE_1$ is additive to that released by epinephrine, which is what would be expected if "stimulatory" (i.e., $PGE_1$) and "inhibitory" (i.e., $\alpha_2$) receptors were associated with distinct populations of GTP-binding proteins (Michel and Lefkowitz, 1982).

A final line of experimentation that suggests that human platelet $\alpha_2$-adrenergic receptors functionally couple to a GTP-binding regulatory protein is the demonstration of $\alpha_2$-adrenergic-stimulated GTP hydrolysis in the human platelet membranes (Aktories et al., 1982). It had previously been demonstrated that $\beta$-adrenergic agents, as well as other stimuli of adenylate cyclase activity, activated GTP hydrolysis and that this hormone-stimulated hydrolysis was specifically inhibited by cholera toxin (Cassel and Pfeuffer, 1978). Since GTP hydrolysis is felt to terminate the stimulation of adenylate cyclase activity, it might seem anomalous that agents which stimulate cAMP synthesis might also stimulate termination of this synthesis. However, since one consequence of the agonist-facilitated exchange of guanine nucleotide at the GTP-binding protein is a more rapid replenishment of GTP at the binding site, investigators have interpreted the stimulation of GTPase activity by agents that activate adenylate cyclase as a manifestation of agonist-promoted substrate replenishment at the GTPase-substrate binding site, rather than an increase in the velocity of the hydrolysis reaction per se.

The activation of GTP hydrolysis by agents that inhibit adenylate cyclase activity is not altered by cholera toxin but is inhibited by islet-activating protein (IAP) (Aktories *et al.*, 1983), a toxin secreted by *Bordetella pertussis* which ADP-ribosylates the GTP-binding subunit of the membrane component felt to elicit the inhibition of adenylate cyclase activity ($N_i$). This latter finding suggests that the GTPase activity being stimulated by $\alpha_2$-adrenergic agents is at $N_i$ and not $N_s$. However, the data are still inconclusive as to whether or not $\alpha_2$-adrenergic-stimulated GTP hydrolysis results from agonist-promoted substrate replenishment at the GTP binding site or agonist-activated hydrolysis per se. As mentioned earlier, $\alpha_2$-adrenergic agents have been demonstrated to facilitate guanine nucleotide exchange at a GTP-binding site, monitored as agonist-promoted release of prebound [$^3$H]Gpp(NH)p. This observation indicates that $\alpha_2$-adrenergic agonists would likely facilitate replenishment of GTP at the substrate binding site, once hydrolysis had occurred. Furthermore, the observation that IAP can inhibit $\alpha_2$-adrenergic GTPase activity may result from the known ability of IAP to destabilize $\alpha_2$ receptor–GTP-binding protein interactions (Kurose *et al.*, 1983). Thus, although it would be a reasonable postulate that $\alpha_2$-adrenergic-mediated inhibition of adenylate cyclase activity might result from a direct acceleration of the ''turn off'' mechanism of the adenylate cyclase system, i.e., GTP hydrolysis, this remains to be documented. Nonetheless, $\alpha_2$-adrenergic-stimulated GTPase activity is another measurable manifestation of a functional association between the agonist-occupied $\alpha_2$-adrenergic receptor and a GTP-binding regulatory protein. (Figure 1 summarizes the several experimental approaches which suggest that the $\alpha_2$ receptor interacts with a GTP-binding protein.)

Another modifier of $\alpha_2$-adrenergic receptor interactions is the sodium ion. The interest in evaluating the effects of sodium on $\alpha_2$-receptor properties resulted from the observation that sodium was required for or enhanced $\alpha_2$-adrenergic-mediated inhibition of adenylate cyclase activity in broken cell incubations (Jakobs, 1979). Sodium was demonstrated to decrease receptor affinity for agonists 10- to 20-fold (Tsai and Lefkowitz, 1978; Limbird *et al.*, 1982), while slightly (<2-fold) increasing receptor affinity for antagonists (Limbird *et al.*, 1982). The specificity for the effect of sodium is $Na^+ > Li^+ \gg K^+$, and the $EC_{50}$ for sodium is approximately 5–15 m$M$ in $\alpha_2$-adrenergic systems. Sodium, like GTP, does not decrease receptor affinity for antagonist agents and, if anything, causes a slight increase in receptor affinity for antagonists over the same concentration range that $Na^+$ mediates decreases in $\alpha_2$-receptor affinity for agonists (Limbird *et al.*, 1982). The addition of GTP together with $Na^+$ further reduces $\alpha_2$-

1. Multiple Receptor "Affinity States" for Agonists

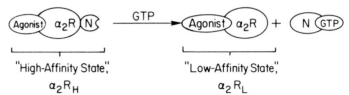

"High-Affinity State",                "Low-Affinity State",

$\alpha_2 R_H$                       $\alpha_2 R_L$

2. Guanine Nucleotide-Enhanced-Rate of $[^3H]$ Agonist Dissociation

$$\text{GTP} \rightarrow \uparrow \text{Receptor Affinity} \equiv \uparrow K_D \equiv \uparrow k_{off}/k_{on}$$
$$\text{for Agonist}$$

3. Agonist–Facilitated Increase in Apparent Receptor "Size"

$$\text{Agonist} + (\alpha_2 R) + (N \rightarrow (Agonist) R \ N$$
"Ternary Complex"

4. Agonist–Facilitated Guanine Nucleotide Exchange

$$(\alpha_2 R) + (N \ ^3H\ GTP \underset{GTP}{\overset{Agonist}{\rightleftharpoons}} (Agonist)\ R ) + (N\ GTP) + \ ^3HGDP$$

5. Agonist-Stimulated Rate of GTP Hydrolysis (GTP → GDP)

Due to facilitated guanine nucleotide exchange (#4),
agonists facilitate the replenishment of GTP substrate

**Fig. 1.** Manifestations of $\alpha_2$-adrenergic receptor–GTP-binding protein interactions. Symbols: $\alpha_2R$, $\alpha_2$-adrenergic receptor; N, guanine nucleotide-binding protein ($?N_i$).

receptor affinity for agonists (Michel *et al.*, 1980; Limbird *et al.*, 1982), suggesting that there is either an additive or a synergistic effect of sodium and GTP on receptor–agonist interactions (see Table II).

The phenomenological similarity of the effects of sodium and guanine nucleotides on human platelet receptor–agonist interactions and the synergistic effects of sodium and GTP in mediating the inhibition of adenylate cyclase has suggested to some investigators that the effects of sodium may be mediated via a monovalent cation binding site that exists on the GTP-binding protein which "couples" the $\alpha_2$-adrenergic receptor to the adenylate cyclase complex (Aktories *et al.*, 1981; Mooney *et al.*, 1982). However, a number of perturbants are able to differentially resolve the effects of GTP from those of sodium on receptor–agonist interactions in the human platelet system (Limbird and Speck, 1983). Thus, incubation of intact platelets with the sulfhydryl-directed reagent *N*-ethylmaleimide (NEM) results in an uncoupling of $\alpha_2$-adrenergic-mediated receptors from the inhibition of adenylate cyclase (Jakobs *et al.*, 1982). In membranes subsequently prepared from NEM-treated human platelets, there is no change in receptor–antagonist interactions but there is a decrease in receptor affinity for agonists. This decrease in receptor affinity for agonists is paralleled by a loss in the ability of guanine nucleotides to modulate receptor–agonist interactions. Taken together, these findings suggest that NEM treatment of the intact platelet destabilizes receptor interactions with the GTP-binding protein responsible both for higher affinity receptor–agonist interactions and for the effects of guanine nucleotides on these interactions. In contrast, the ability of sodium to decrease receptor affinity for agonists is not altered in membranes derived from NEM-treated platelets. Another perturbant which eliminates the effects of GTP on receptor–agonist interactions without affecting sodium-promoted decreases in receptor affinity for agonists is elevated temperature (Limbird and Speck, 1983). Thus, incubation of human platelet membranes at 45°C for 30 min (or 60°C for 5 min) causes no alterations in [³H]yohimbine binding but causes a selective decrease in receptor affinity for agonists, which again is paralleled by a loss in the ability by GTP to modulate receptor–agonist interactions. However, as with NEM, elevated temperature does not modify the ability of the sodium ion to decrease receptor affinity for agonist agents. These two findings provide indirect data consistent with the hypothesis that distinct components in the human platelet membrane are mediating the effects of GTP and sodium on receptor–agonist interactions. However, these data cannot rule out the possibility that the same molecular component mediates the effects of GTP and Na⁺ but via different domains and that these domains are differentially sensitive

**TABLE II**

**Na⁺ and α₂-Adrenergic Receptors**

| Effects of Na⁺ on α₂-Adrenergic Receptors |
|---|
| $Na^+ > Li^+ >> K^+$ |
| $EC_{50} \simeq 5$–15 mM $Na^+$ |
| ↓ Receptor affinity for agonists (10- to 50-fold) |
| ↑ Receptor affinity for antagonists (~2-fold) |

| Differential Effects of Perturbants on Na⁺- and GTP-Modulation of Receptor–Agonist Interactions | Receptor–agonist interactions | |
|---|---|---|
| Perturbation | Na⁺ effects | GTP effects |
| Exposure of intact platelets to N-ethylmaleimide (1 mM) | Retained | Lost |
| Exposure of platelet membranes to elevated temperature (45°C, 20 min; 60°C, 5 min) | Retained | Lost |
| Solubilization of the platelet membrane with digitonin | Retained | Lost |

to alkylation or elevated temperature. The observation that sodium still affects $\alpha_2$ receptor–agonist interactions in detergent-solubilized extracts of human platelets has provided more convincing evidence that distinct molecular components mediate the effects of GTP and sodium on the human platelet $\alpha_2$-adrenergic receptor system, since previous studies have suggested that solubilization of the $\alpha_2$ receptor from the human platelet membrane in an unoccupied state resolves the receptor from the GTP-binding protein subunit modulating receptor affinity for agonist (Smith and Limbird, 1981). The conclusion that $Na^+$ and GTP elicit their effects via distinct molecular mechanisms was suggested previously for $\alpha_2$ receptors in rabbit platelet membranes, since it was observed that, unlike GTP, the addition of $Na^+$ to agonist competition studies did not cause the transition of the shape of the competition curve to one of normal steepness (Michel *et al.*, 1980). Similar conclusions that distinct molecular mechanisms, and presumably distinct components, elicit the effects of $Na^+$ and GTP have also been drawn for muscarinic receptor systems that mediate the inhibition of adenylate cyclase (McMahon and Hosey, 1983). Thus, it may be a general property of inhibitory adenylate cyclase systems that the effects of sodium are mediated by a component distinct from the GTP-binding subunit that modulates receptor–agonist interactions. It has not been demonstrated by any laboratory at present, however, whether or not the effects of sodium result from interaction with a monovalent cation binding site existing on the receptors themselves, on an entirely unique peptide that is part of an already identified component, or on a not yet characterized component within the adenylate cyclase system architecture.

## IV. POSTULATED MECHANISMS
## BY WHICH $\alpha_2$-ADRENERGIC AGENTS
## INHIBIT ADENYLATE CYCLASE ACTIVITY

Inherent in the previous discussion is the important role of a guanine-nucleotide binding protein in mediating $\alpha_2$-adrenergic receptor-evoked inhibition of adenylate cyclase. Thus, in a manner analogous to stimulatory adenylate cyclase systems, GTP is required for $\alpha_2$-adrenergic inhibition of adenylate cyclase and also modulates $\alpha_2$-receptor affinity for agonists. As alluded to earlier, despite the important role of GTP in both the stimulation and inhibition of adenylate cyclase, a substantial amount of indirect evidence has suggested that distinct GTP-binding proteins elicit the phenomenologically similar effects of GTP in stimulatory and inhibitory adenylate cyclase systems (Cooper, 1982). Thus, GTP-mediated activation and attenuation of adenylate cyclase manifest differential sensitivity to proteases, $Mn^{2+}$, radiation activation, and sulfhydryl-directed reagents. The discovery that islet-activating protein catalyzes the ADP ribosylation of an $M_r$ 41,000 GTP-binding subunit, for which covalent modification paralleling the loss of $\alpha_2$-adrenergic receptor-mediated inhibition of adenylate cyclase activity is observed, has provided a method for the selective identification of the so-called "inhibitory GTP-binding protein," termed $N_i$ (Katada and Ui, 1982).

The IAP substrate has been purified from rat liver (Bokoch *et al.*, 1983, 1984) and human erythrocytes (Codina *et al.*, 1983) and contains at least two subunits: the $M_r$ 41,000 GTP-binding subunit ADP-ribosylated by IAP and an associated $M_r$ 35,000 subunit that dissociates from the $M_r$ 41,000 subunit upon the binding of guanine nucleotides. A smaller peptide of approximately 10,000 $M_r$ is also detected in the presumably pure $N_i$ preparation, but its function at present is obscure. The $M_r$ 35,000 subunit of $N_i$ appears to be structurally indistinguishable from that which is associated with the $M_r$ 45,000 GTP-binding subunit of the regulatory protein responsible for the activation of adenylate cyclase, $N_s$ (Manning and Gilman, 1983). It has been demonstrated in reconstitution experiments that the concentration of the $M_r$ 35,000 subunit can regulate the extent of the activation of adenylate cyclase (Northrup *et al.*, 1982; Bokoch *et al.*, 1984; Katada *et al.*, 1984a,b,c). Thus, it is the "free" GTP-liganded $M_r$ 45,000 subunit that activates adenylate cyclase, and the addition of excess $M_r$ 35,000 subunit facilitates the reassociation of the 45,000/35,000 heterodimer and results in deactivation of the adenylate cyclase system. Since binding of GTP to either $N_s$ or $N_i$ provokes the dissociation of the 45,000/35,000 or 41,000/35,000 heterodimer, and since there is typically 10 times more IAP substrate ($M_r$ 41,000 GTP-binding subunit of $N_i$) than cholera toxin substrate ($M_r$ 45,000 GTP-binding subunit of $N_s$) in a given target tissue, it has been postulated that $N_i$-mediated inhibition of adenylate cyclase, rather than necessarily occurring via direct inhibition of the catalytic subunit, may primarily occur "indirectly," via provision of excess free $M_r$-35,000 subunit, which in turn decreases the quantity of free $M_r$-45,000 subunit available for activation of the catalytic subunit (Katada *et al.*,

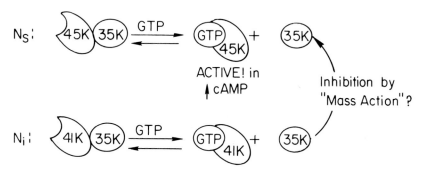

**Fig. 2.** Postulated interplay between the stimulatory ($N_s$) and inhibitory ($N_i$) GTP-binding regulatory proteins. Agonists facilitate the rate of GTP interaction with $N_s$, $N_i$. See text for explanation.

1984a,b,c; Gilman, 1984). Figure 2 is a schematic diagram of the postulated interplay between stimulatory and inhibitory GTP-binding proteins.

## V. POSTULATED MECHANISMS BY WHICH $\alpha_2$-ADRENERGIC AGENTS ELICIT PHYSIOLOGICAL EFFECTS

The intense investigation of the molecular mechanism by which $\alpha_2$-adrenergic agents inhibit adenylate cyclase activity has to some extent overshadowed studies of the molecular basis by which adrenergic agents working via $\alpha_2$-adrenergic receptors elicit their ultimate physiological effects. As indicated at the outset of this chapter, although lowering intracellular cAMP concentrations may be sufficient to evoke a physiological effect in some target tissues, in other tissues such a signal may be neither necessary nor sufficient to elicit the physiological function typically attributed to $\alpha_2$-adrenergic agonists (see Table III).

The one system in which $\alpha_2$-adrenergic receptor-mediated effects may be entirely accounted for by the ability of catecholamines to attenuate cAMP accumulation is the regulation of lipid metabolism in the fat cell. Work in this field has firmly established that adipocyte cAMP levels regulate the breakdown of intracellular triglyceride stores (Fain, 1982; Hales *et al.*, 1978). Thus, stimulation of cAMP accumulation by catecholamines (acting through $\beta$-adrenergic receptors), thyroid hormone, glucagon, and adrenocorticotropic hormone

TABLE III

$\alpha_2$-Adrenergic Receptor-Elicited Physiological Effects:
Relationship to Decreases in cAMP

| Tissue | Response | $\downarrow$ cAMP detected | Physiological mechanism |
|---|---|---|---|
| Adipose | Antilipolytic effect | Yes | May be accounted for solely by $\downarrow$ cAMP |
| Human platelet | Primary aggregation | Yes | May be accounted for by $\downarrow$ cAMP |
| | Secretion, "secondary aggregation" | | Events distal to or independent of $\downarrow$ cAMP |
| Endothelial cells | $Cl^-$, $H_2O$ secretion | Yes (by some) | Events distal to or independent of $\downarrow$ cAMP |
| Rabbit ileum (heterogeneous cell populations) | $Cl^-$, $H_2O$ secretion | Yes | Events distal to or independent of $\downarrow$ cAMP |

(ACTH) results in activation of triacyl glycerol lipase, presumably via a cAMP-dependent phosphorylation of this enzyme by a specific cAMP-dependent protein kinase. Enhancement of cAMP accumulation appears to result from the activation of the adenylate cyclase enzyme (Birnbaumer and Rodbell, 1969; Butcher et al., 1965, 1968) rather than from the inhibition of phosphodiesterase activity, and hormonal stimulation of adenylate cyclase activity in fat cells correlates with the level of stimulated protein-kinase activity in the same cells (Soderling et al., 1973). Exogenous cAMP can also mediate lipolysis, as can cholera toxin (Vaughn et al., 1970), an agent that results in the persistent accumulation of cAMP as a result of the inhibition of the GTPase "turn off" mechanism subsequent to ADP ribosylation of the $N_s$ protein. The large amount of experimental evidence supporting the hypothesis that cAMP is the sole messenger regulating the rate of lipolysis in the adipocyte (Fain, 1982; Hales et al., 1978; Steinberg et al., 1975) has made inhibition of cAMP accumulation a likely locus for physiological attenuation of triglyceride breakdown. Consequently, once epinephrine was found to inhibit adenylate cyclase activity in adipocyte membranes (Burns and Langley, 1975), this action was postulated to represent a mechanism for regulation of adipocyte metabolism. Subsequent studies have demonstrated that the α₂-adrenergic receptor mediates this inhibition of adenylate cyclase by epinephrine. In addition, clonidine has also been demonstrated to decrease lipolysis in stimulated adipocytes (Burns et al., 1981). Finally, α₂-adrenergic agents are not unique in their ability to decrease adenylate cyclase activity and lipolysis in fat cells. Thus, free-fatty acids (Fain and Shepherd, 1975), prostaglandin $E_2$ ($PGE_2$) (Garcia-Sainz et al., 1981), and adenosine (Dole, 1961; Fain et al., 1972) also decrease adenylate cyclase activity and the rate of adipocyte lipolysis.

In an attempt to manipulate the ability of adenosine, $PGE_2$, and epinephrine to lower cAMP levels and thus evaluate the importance of this property in receptor-mediated attenuation of lipolysis, isolated hamster adipocytes were studied in the presence and absence of extracellular $Na^+$ (Garcia-Sainz et al., 1981). As indicated earlier in this chapter, $Na^+$ has been demonstrated to be required for or to enhance hormone-mediated attenuation of adenylate cyclase activity in broken cell preparations. Furthermore, Lichtshtein et al. (1979) had demonstrated that opiate-mediated attenuation of cAMP in cultured neuroblastoma × glioma hybrid cells was not observed when $Na^+$ was replaced by choline in the extracellular medium. Thus, it was postulated that receptor-mediated inhibition of adenylate cyclase involved a $Na^+$-sensitive step and that removal of extracellular $Na^+$ might effectively uncouple receptor occupancy from inhibition of adenylate cyclase in intact cells. However, Garcia-Sainz et al. (1981) demonstrated that the extent of inhibition of isoproterenol or ACTH-stimulated cAMP accumulation and of lipolysis by adenosine, $PGE_2$, and clonidine was virtually indistinguishable in control hamster adipocytes and in adipocytes suspended in medium in which Na$^+$ (and its accompanying anion) were substituted by iso-molar sucrose. Consequently, these findings argue against a general requirement for Na$^+$, at least extracellular Na$^+$, in receptor-mediated attenuation of adenylate cyclase activity in an intact cell environment and also further demonstrate the parallelism between $\alpha_2$ receptor-mediated attenuation of cAMP accumulation and receptor-mediated function in the adipocyte.

The effect of removal of extracellular Na$^+$ was also evaluated in the human platelet, since epinephrine, acting via $\alpha_2$-adrenergic receptors, is known to both decrease cAMP accumulation and elicit aggregation and secretion in human platelets. In a manner analogous to findings in the hamster adipocyte, the removal of extraplatelet Na$^+$ did not alter the ability of epinephrine to attenuate PGE$_1$-stimulated cAMP accumulation in the human platelet nor did the removal of extraplatelet Na$^+$ interfere with epinephrine-induced "primary" aggregation, the reversible platelet–platelet interaction that results from stimulus-provoked exposure of platelet surface receptors for fibrinogen, a physiological platelet "crosslinking" agent (Connolly and Limbird, 1983a). Thus, epinephrine-promoted decreases in cAMP accumulation may be sufficient to explain the effects of epinephrine on primary aggregation and fibrinogen–receptor exposure, a response that is known to be blocked by increases in cAMP and felt to be sensitive to the prevailing intraplatelet cAMP levels (Graber and Hawiger, 1982).

In contrast to the effects of removal of extraplatelet Na$^+$ on epinephrine-stimulated primary aggregation, however, removal of extraplatelet Na$^+$ blocked human platelet secretion and the so-called "secondary" aggregation elicited by epinephrine, ADP, and lower concentrations of thrombin, all agents which are known to attenuate cAMP accumulation but which also require the release of arachidonic acid and the production of the arachidonic acid metabolite thromboxane A$_2$ to elicit their secretory and irreversible aggregatory effects on human platelets (Connolly and Limbird, 1983b). Consequently, it appears that at least part of the effects of $\alpha_2$-adrenergic agents on platelets, i.e., those related to platelet secretion, are mediated by signal(s) independent of or distal to alterations in cAMP accumulation, since removal of extraplatelet Na$^+$ does not alter epinephrine-induced inhibition of cAMP accumulation but does eliminate epinephrine-induced secretion in human platelets. Second, the observation that $\alpha_2$-adrenergic-mediated cAMP attenuation is unaltered by changes in extracellular Na$^+$ in human platelets (Connolly and Limbird, 1983a) as well as in hamster adipocytes (Garcia-Sainz et al., 1981) may suggest that intracellular Na$^+$ concentrations, although lowered by removal of extracellular Na$^+$, are nonetheless sufficient to drive the hormone-mediated inhibition of adenylate cyclase which appears sensitive to Na$^+$ in broken cell incubations. In fact, intraplatelet Na$^+$ concentrations have been demonstrated to be responsible for modulating the affinity of human platelet $\alpha_2$-receptor affinity for agonist agents (Motulsky, 1983).

Another system in which $\alpha_2$-adrenergic effects on physiological processes appear to result from effects independent of or distal to decreases in cAMP accumulation is the $\alpha_2$-adrenergic inhibition of glucose-stimulated insulin release from the pancreas. Thus, glucose-stimulated insulin release can be demonstrated to be inhibited by epinephrine, via $\alpha_2$-adrenergic receptors, without any changes in islet cell cAMP levels (Wollheim and Sharp, 1981; Ullrich and Wollheim, 1984). Epinephrine, however, can be detected to lower islet cAMP concentrations in the presence of phosphodiesterase inhibitors (Wollheim and Sharp, 1981), a condition which favors cAMP accumulation. These latter findings confirm that $\alpha_2$-adrenergic agents can attenuate adenylate cyclase activity in intact cells in a manner analogous to effects observed in broken cell preparations of pancreatic islets (Katada and Ui, 1981a). Nonetheless, two additional findings further dissociate the effects of $\alpha_2$-adrenergic agents on glucose-stimulated insulin secretion from those on inhibition of adenylate cyclase activity. First, epinephrine (acting via $\alpha_2$ receptors) can transiently augment stimulation of cAMP accumulation by forskolin, a diterpene activator of cyclase that appears to activate the catalytic moiety of the adenylate cyclase architecture, perhaps via another component distinct from $N_s$ (Ullrich and Wollheim, 1984). This $\alpha_2$-adrenergic-stimulated transient increase in cAMP is nonetheless accompanied by the characteristic $\alpha_2$-adrenergic decrease in insulin secretion. After longer incubation times (i.e., 15 min), $\alpha_2$-adrenergic agents attenuate forskolin-stimulated cAMP accumulation in rat islet cells, as they consistently do in homogenates prepared from the same tissue. These findings indicate that epinephrine effects on insulin secretion can be dissociated from the ability of epinephrine to attenuate cyclic AMP accumulation. One potential caveat to the interpretation of these studies, however, is that a crucial $\alpha_2$ receptor-mediated decrease in cAMP signal may be occurring in the microenvironment of the islet essential for regulation of insulin secretion but may not be detected or may be overshadowed by a measurement of total cellular cAMP accumulation, even over the short time courses ($<3$ min) that were evaluated in the previously discussed studies (Ullrich and Wollheim, 1984). However, further evidence that $\alpha_2$-adrenergic agents attenuate glucose-stimulated insulin release via a mechanism distinct from or distal to the inhibition of adenylate cyclase activity is the observation that $\alpha_2$-adrenergic agents also block insulin release stimulated by exogenous dibutyryl-cAMP (Malaisse et al., 1970; Kato and Nakaki, 1983; Nakaki et al., 1983b; Ullrich and Wollheim, 1984).

The mechanism by which epinephrine inhibits insulin release distal to cAMP accumulation is not known. An early hypothesis that epinephrine may work by inhibiting glucose-stimulated $Ca^{2+}$ uptake (Brisson and Malaisse, 1973) is less compelling when one realizes that inhibition of $Ca^{2+}$ uptake is small compared to the effects on insulin release (Wollheim et al., 1977) and, furthermore, that clonidine, a partial agonist at $\alpha_2$-adrenergic receptors, blocks insulin release without altering $Ca^{2+}$ uptake (Wollheim and Sharp, 1981). Nonetheless, epinephrine may alter the intracellular handling of $Ca^{2+}$ or islet cells' sensitivity to $Ca^{2+}$ once mobilized, since $Ca^{2+}$ is a crucial component of the islet cell secretory mechanism. Interestingly, the important relationship between the adenylate cyclase-linked $\alpha_2$-adrenergic receptor and $Ca^{2+}$ in pancreatic islet secretion was also manifested in early studies evaluating the molecular basis for the hyperinsulinemia induced by islet-activating protein, the toxin secreted by *Bordetella pertussis* that is now known to ADP-ribosylate $N_i$ and block $\alpha_2$-mediated inhibition of cAMP accumulation. Thus, IAP treatment of rats 3 days prior to sacrifice enhanced glucose-stimulated insulin release in the presence of extracellular $Ca^{2+}$ but actually decreased glucose-stimulated insulin release when $Ca^{2+}$ was omitted (and EGTA was added) to the perfusate of subsequently isolated islets (Katada and Ui, 1978). IAP was later demonstrated to block the effects of epinephrine not only on cAMP accumulation and insulin release but also on $^{45}Ca^{2+}$ uptake in isolated islets. Thus, Ui and colleagues have concluded that IAP not only blocks $\alpha_2$-adrenergic-mediated inhibition of cAMP accumulation but also may cause a sustained opening of the native cation "ionophore" located on the surface membrane of the $\beta$ cell that is responsible for the $Ca^{2+}$ translocation postulated to evoke pancreatic secretion (Katada and Ui, 1979). Their additional observation that the cellular level of cAMP was only higher in islets exposed to IAP compared to control islets when breakdown of cAMP was prevented by inclusion of a phosphodiesterase inhibitor argues that a further modification of cellular function elicited by IAP may be the rate of degradation of cAMP. One interpretation of these findings is that the $M_r$ 41,000 GTP-binding subunit of $N_i$ may be pivotal in not only the attenuation of adenylate cyclase activity but also in the stimulation of phosphodiesterase activity and/or in the alteration of intracellular $Ca^{2+}$ disposition. Whether or not this hypothesis is correct, the extant data regarding the mechanism for $\alpha_2$-adrenergic inhibition of insulin release argue that at least part of the effects of catecholamines occur distal to or independent of changes in cAMP accumulation.

Another physiological system in which $\alpha_2$-adrenergic regulation of physiological functions does not appear to be adequately explained by decreases in cAMP accumulation is the regulation of secretory diarrhea in the intestinal epithelial cell. Cyclic AMP is felt to mediate two important processes in the normal physiology of intestinal epithelial cells: NaCl cotransport (absorption) and $Na^+$-dependent $Cl^-$ secretion (Field, 1980). Sodium chloride is transported across the brush border and the $Na^+$ thus absorbed is extruded from the contralumenal surface via a $Na^+$, $K^+$-ATPase. The choride ion passively diffuses across the contralumenal border. Thus, NaCl is effectively resorbed from the intestine in an electroneutral but energy-dependent process. Cyclic cAMP appears to block this overall process by blocking absorption across the brush border (Nellans et al., 1973). However, simply blocking absorption cannot account for the net secretion found in secretory diarrhea. This latter effect can be explained by the putative effects of cAMP and/or $Ca^{2+}$ to directly stimulate secretion from the crypt cell of the intestinal epithelium. In the crypt cell, NaCl cotransport occurs across the contralumenal surface, increasing intracellular NaCl concentrations. The $Na^+$ is then actively extruded back across the contralateral surface, resulting in no net intracellular $Na^+$ increase but a high-intracellular $Cl^-$ content. This $Cl^-$ then diffuses across the lumenal surface into the intestinal lumen. Apical conductance of $Cl^-$ is increased in the stimulated cell, leading to net secretion. The postulated effect of cAMP to activate this $Na^+$-dependent $Cl^-$ secretion is strengthened by the ability of cholera toxin to cause a persistent secretion from these cells in parallel with a persistent elevation of intracellular cAMP levels (Nakaki *et al.*, 1982). However, the ability of α₂-adrenergic agents to block cAMP-mediated secretion is not adequately explained by α₂-adrenergic-mediated attenuation of cAMP accumulation. Thus, although rat intestinal epithelial cells isolated from the mucosal layer contain α₂-adrenergic receptors, the ability of α₂ agonists to modulate cAMP levels is not always observed. For example, Laburthe *et al.* (1982) demonstrated that epinephrine attenuated basal and VIP-stimulated cAMP accumulation in rat intestinal epithelial cells, whereas Nakaki *et al.* (1983a) were unable to observe a clonidine-promoted basal, $PGE_1$- or vasoactive intestinal peptide (VIP)-stimulated reduction in cAMP levels in these cells, despite the ability of clonidine to block cAMP-mediated secretion in the same preparations. Epinephrine can partially inhibit cholera toxin- and $PGE_1$-stimulated cAMP accumulation in the rabbit ileum, but these decreases in cAMP levels are not associated with a significant reduction in toxin-induced secretion (Field *et al.*, 1975). Finally, perhaps more important is the finding that epinephrine and clonidine inhibit intestinal secretion elicited by dibutyryl-cAMP in rat intestinal epithelial cells, and norepinephrine inhibits dibutyryl-cAMP-induced secretion in rabbit ileum (Field *et al.*, 1975; Field and McColl, 1973). Thus, at least part of the physiological effects elicited by α₂-adrenergic agents must result from the regulation of a process distal to the formation of cAMP.


## VI. SUMMARY

α-Adrenergic receptors of the α₂ subtype have been categorized by the order of potency of agonists (clonidine > epinephrine > norepinephrine) in eliciting physiological effect and the greater sensitivity of these effects to blockade by the antagonist yohimbine than by prazosin. The receptor binding sites can be confidently identified using a number of commercially available agonist, partial agonist, and antagonist radioligands. α₂-Adrenergic receptors have been demonstrated to be linked to inhibition of adenylate cyclase activity via a GTP-binding protein ($N_i$) that is distinct from that GTP-binding protein involved in mediating hormonal stimulation of adenylate cyclase activity ($N_s$). However, it is interesting that only in a limited number of circumstances can the ultimate physiological effects mediated via $\alpha_2$-adrenergic receptors be solely attributed to the lowering of intracellular cAMP concentrations. More frequently, it appears that at least part of the effects elicited by $\alpha_2$-adrenergic agents occur distal to (or independent of) changes in cAMP accumulation. These latter observations are provocative and suggest that the $\alpha_2$-adrenergic receptor in a single target cell may be linked to multiple potential effector systems.

## REFERENCES

Ahlquist, R. P. (1948). A study of the adrenotropic receptors. *Am. J. Physiol.* **153**, 586–600.

Aktories, K., Schultz, G., and Jakobs, K.-H. (1981). The hamster adipocyte adenylate cyclase system II. Regulation of enzyme stimulation and inhibition by monovalent cations. *Biochim. Biophys. Acta* **676**, 59–67.

Aktories, K., Schultz, G., and Jakobs, K.-H. (1982). Cholera toxin inhibits prostaglandin E₁ but not adrenaline-induced stimulation of GTP hydrolysis in human platelet membranes. *FEBS Lett.* **146**, 65–68.

Aktories, K., Schultz, G., and Jakobs, K.-H. (1983). Islet-activating protein prevents nicotinic acid-induced GTPase stimulation and GTP but not GTPase-induced adenylate cyclase inhibition in rat adipocytes. *FEBS Lett.* **156**, 88–92.

Berthelson, S., and Pettinger, W. A. (1977). A functional basis for classification of α-adrenergic receptors. *Life Sci.* **21**, 595–606.

Birnbaumer, L., and Rodbell, M. (1969). Adenylate cyclase in fat cells II. Hormone receptors. *J. Biol. Chem.* **244**, 3477–3482.

Bokoch, G. M., Katada, T., Northup, J. K., Hewlett, E. L., and Gilman, A. G. (1983). Identification of the predominant substrate for ADP-ribosylation by islet activating protein. *J. Biol. Chem.* **258**, 2072–2075.

Bokoch, G. M., Katada, T., Northrup, J. K., Ui, M., and Gilman, A. G. (1984). Purification and properties of the inhibiting guanine nucleotide-binding regulatory component of adenylate cyclase. *J. Biol. Chem.* **259**, 3560–3567.

Borowski, E., Starke, K., Ehrl, H., and Endo, T. (1977). A comparison of pre- and post-synaptic effects of β-adrenolytic drugs in the pulmonary artery of the rabbit. *Neuroscience* **2**, 285–296.

Brisson, G. R., and Malaisse, W. J. (1973). The stimulus–secretion coupling of glucose-induced insulin release. XI Effects of theophylline and epinephrine on ⁴⁵Ca⁺⁺ efflux from perifused islets. *Metab. Clin. Exp.* **22**, 455–465.

Burns, T. W., and Langley, P. E. (1975). The effect of alpha- and beta-adrenergic receptor stimulation on the adenylate cyclase activity of human adpiocytes. *J. Cyclic Nucleotide Res.* **1**, 321–328.

Burns, T. W., Langley, P. E., Terry, B. E., Bylard, D. B., Hoffman, B. B., Tharp, M. D., Lefkowitz, R. J., Garcia-Sainz, J. A., and Fain, J. N. (1981). Pharmacological characterization of adrenergic receptors in human adpiocytes. *J. Clin. Invest.* **67**, 467–475.

Butcher, R. W., Ho, R. J., Meng, H. C., and Sutherland, E. W. (1965). Adenosine 3′,5′-monophosphate in biological materials. II the measurement of adenosine 3′,5′-monophosphate in tissues and the role of the cyclic nucleotide in the lipolytic response of fat cells to epinephrine. *J. Biol. Chem.* **240**, 4515–4523.

Butcher, R. W., Baird, C. E., and Sutherland, E. W. (1968). Effects of lipolytic and antilipolytic substances on adenosine 3′,5′-monophosphate levels in isolated fat cells. *J. Biol. Chem.* **243**, 1705–1712.

Cassel, D., and Pfeuffer, T. (1978). Mechanism of cholera toxin action: Covalent modification of the guanylnucleotide-binding protein of the adenylate cyclase system. *Proc. Natl. Acad. Sci. U.S.A.* **75**, 2669–2673.

Cassel, D., and Selinger, Z. (1978). Mechanism of adenylate cyclase activation through the β-adrenergic receptor: Catecholamine-induced displacement of bound GDP by GTP. *Proc. Natl. Acad. Sci. U.S.A.* **75**, 4155–4159.

Cassel, D., Levkovitz, H., and Selinger, Z. (1977). The regulatory GTPase cyclase of turkey erythrocyte adenylate cyclase. *J. Cyclic Nucleotide Res.* **3**, 393–406.

Codina, J., Hildebrandt, J., Iyengar, R., Birnbaumer, L., Sekura, R. D., and Manclark, C. R. (1983). Pertussis toxin substrate, the putative Ni component of adenylyl cyclases is an αβ heterodimer regulated by guanine nucleotide and magnesium. *Proc. Natl. Acad. Sci. U.S.A.* **80**, 4276–4280.

Connolly, T. M., and Limbird, L. E. (1983a). The influence of Na⁺ on the alpha₂-adrenergic receptor-adenylate cyclase system of human platelets. I. A method for removal of extraplatelet Na⁺. Effect of Na⁺ removal on aggregation, secretion and cAMP accumulation. *J. Biol. Chem.* **258**, 3907–3912.

Connolly, T. M., and Limbird, L. E. (1983b). Removal of extraplatelet Na⁺ eliminates indomethacin-sensitive secretion from human platelets stimulated by epinephrine, ADP and thrombin. *Proc. Natl. Acad. Sci. U.S.A.* **80**, 5320–5324.

Cooper, D. M. F. (1982). Biomodal regulation of adenylate cyclase. *FEBS Lett.* **138**, 157–163.

DeLean, A. L., Munson, P. J., and Rodbard, D. (1978). Simultaneous analysis of families of *sigmoidal* curves: application to bioassay, radioligand assay and physiological dose-response curves. *Am. J. Physiol.* **235**, E97–E102.

DeLean, A., Stadel, J. M., and Lefkowitz, R. J. (1980). A ternary complex model explains the agonist-specific binding properties of the adenylate cyclase-coupled β-adrenergic receptor. *J. Biol. Chem.* **255**, 7108–7117.

Dole, V. P. (1961). Effect of nucleic acid metabolites on lipolysis in adipose tissue. *J. Biol. Chem.* **236**, 3125–3130.

Doxey, J. C., Smith, C. F. C., and Walker, J. M. (1977). Selectivity of blocking agents for pre- and post-synaptic α-adrenoreceptors. *Br. J. Pharmacol.* **60**, 91–96.

Fain, J. N. (1982). Regulation of lipid metabolism by cyclic nucleotides. In "Cyclic Nucleotides II" (J. W. Kebabian and J. A. Nathanson, eds.), Handbook of Experimental Pharmacology, Vol. 58/II, pp. 89–150. Springer-Verlag, Berlin and New York.

Fain, J. N., and Garcia-Sainz, J. A. (1980). Role of phosphatidylinositol turnover in alpha₁ and of adenylate cyclase inhibition in alpha₂ effects of catecholamines. *Life Sci.* **26**, 1183–1194.

Fain, J. N., and Shepherd, R. E. (1975). Free fatty acids as feedback regulators of adenylate cyclase and cyclic AMP accumulation in rat fat cells. *J. Biol. Chem.* **250**, 6586–6592.

Fain, J. N., Panter, R. H., and Ward, W. F. (1972). Effects of adenosine nucleotides on adenylate cyclase, phosphodiesterase, cyclic adenosine monophosphate accumulation and lipolysis in fat cells. *J. Biol. Chem.* **247**, 6866–6872.

Field, M. (1980). Regulation of small intestinal ion transport by cyclic nucleotides and calcium. In "Secretory Diarrhea" (M. Field, J. S. Fordtran, and S. G. Schultz, eds), pp. 21–30. Am. Physiol. Soc., Bethesda, Maryland.

Field, M., and McColl, I. (1973). Ionic transport in rabbit ileal mucosa III. Effects of catecholamines. *Am. J. Physiol.* **225**, 852–857.

Field, M., Sheerin, H. E, Henderson, A., and Smith, P. L. (1975). Catecholamine effects on cyclic AMP levels and ion secretion in rabbit ileal mucosa. *Am. J. Physiol.* **229**, 86–92.

Garcia-Sainz, J. A., Li, S.-Y., and Fain, J. N. (1981). Alpha₂-adrenergic amines, adenosine and prostaglandins inhibit lipolysis and cAMP accumulation in hamster adipocytes in the absence of extracellular sodium. *Life Sci.* **28**, 401–406.

Garcia-Sevilla, J. A., Hollingsworth, P. J., and Smith, C. B. (1981). $\alpha_2$-Adrenergic receptors in human platelets: Selective labeling by [³H]clonidine and [³H]-yohimbine and competitive inhibition of antidepressant drugs. *Eur. J. Pharmacol.* **74**, 329–341.

Gilman, A. G. (1984). G proteins and dual control of adenylate cyclase. *Cell* **36**, 577–579.

Graber, S., and Hawiger, J. (1982). Evidence that changes in platelet cAMP levels regulate the fibrinogen receptor on human platelets. *J. Biol. Chem.* **257**, 14606–14609.

Greenberg, D. A., U'Prichard, D. C., and Snyder, S. H. (1976). Alpha-noradrenergic receptor binding in mammalian brain: differing labeling of agonist and antagonist states. *Life Sci.* **19**, 69–76.

Hales, L. N., Luzio, J. P., and Siddle, K. (1978). Hormonal control of adipose tissue lipolysis. *Biochem. Soc. Symp. No.* 43, 97–135.

Hoffman, B. B., and Lefkowitz, R. J. (1980a). An assay for alpha-adrenergic subtypes using [³H]-dehydroergocryptine. *Biochem. Pharmacol.* **29**, 452–454.

Hoffman, B. B., and Lefkowitz, R. J. (1980b). Alpha-adrenergic receptor subtypes. *N. Engl. J. Med.* **25**, 1390–1396.

Hoffman, B. B., DeLean, A., Wood, C. L., Schocken, D. D., and Lefkowitz, R. J. (1979). Alpha-adrenergic receptor subtypes: Quantitative assessment by ligand binding. *Life Sci.* **24**, 1739–1745.

Hoffman, B. B., Michel, T., Mullikin-Kilpatrick, D., Lefkowitz, R. J., Tolbert, M. E. M., Gilman, H., and Fain, J. N. (1980). Agonist versus antagonist binding to α-adrenergic receptors. *Proc. Natl. Acad. Sci. U.S.A.* **77**, 4569–4573.

Hoffman, B. B., Michel, T., Brenneman, T. B., and Lefkowitz, R. J. (1982). Interactions of agonists with platelet alpha₂-adrenergic receptors. *Endocrinology (Baltimore)* **119**, 926–932.

Jakobs, K.-H. (1979). Inhibition of adenylate cyclase by hormones and neurotransmitters. *Mol. Cell. Endocrinol.* **16**, 147–156.

Jakobs, K.-H., and Raushek, R. (1978). [³H]dehydroergonine binding to α-adrenergic receptors in human platelets. *Klin. Wochenschr.* **56**, 139–145.

Jakobs, K.-H., Lasch, P., Minuth, M., Aktories, K., and Schultz, G. C. (1982). Uncoupling of α-adrenoreceptor-mediated inhibition of human platelet adenylate cyclase by N-ethylmaleimide. *J. Biol. Chem.* **257**, 2829–2833.

Katada, T., and Ui, M. (1978). Effects of *in vivo* pretreatment of rats with a new protein purified from *Bordetella pertussis* on *in vitro* secretion of insulin: Role of calcium. *Endocrinology (Baltimore)* **104**, 1822–1827.

Katada, T., and Ui, M. (1979). Islet activating protein. Enhanced insulin secretion and cyclic AMP accumulation in pancreatic islets due to activation of native ionophores. *J. Biol. Chem.* **254**, 469–479.

Katada, T., and Ui, M. (1981a). Islet-activating protein. A modifier of receptor-mediated regulation of rat islet adenylate cyclase. *J. Biol. Chem.* **256**, 8310–8317.

Katada, T., and Ui, M. (1981b). *In vitro* effects of islet-activating protein on cultured rat pancreatic islets. Enhancement of insulin secretion, adenosine $3',5'$-monophosphate accumulation and $^{45}Ca^{++}$ flux. *J. Biochem. (Tokyo)* **89**, 979–990.

Katada, T., and Ui, M. (1982). ADP-ribosylation of the specific membrane protein of C6 glioma cells by islet-activating protein. Association with modification of adenylate cyclase activity. *J. Biol. Chem.* **257**, 7210–7216.

Katada, T., Bokoch, G. M., Northrup, J. K., Ui, M., and Gilman, A. G. (1984a). The inhibitory guanine nucleotide-binding regulatory component of adenylate cyclase. Properties and function of the purified protein. *J. Biol. Chem.* **259**, 3568–3577.

Katada, T., Northup, J. K., Bokoch, G. M., Ui, M., and Gilman, A. G. (1984b). The inhibitory guanine nucleotide-binding regulatory component of adenylate cyclase. Subunit dissociation and guanine-nucleotide-dependent hormonal inhibition. *J. Biol. Chem.* **25**, 3578–3585.

Katada, T., Bokoch, G. M., Smigel, M. D., Ui, M., and Gilman, A. G. (1984c). The inhibitory guanine nucleotide-binding regulatory component of adenylate cyclase. Subunit dissociation and the inhibition of adenylate cyclase in S46 Lymphoma cyc-and wild type membranes. *J. Biol. Chem.* **259**, 3586–3595.

Kato, R., and Nakaki, T. (1983). Alpha₂-adrenoceptors beyond cAMP generation: Islets of Langerhans and intestinal epithelium. *Trends Pharmacol. Sci.* **4**, 34–36.

Kent, R. S., DeLean, A., and Lefkowitz, R. J. (1979). A quantitative analysis of beta-adrenergic receptor interactions. Resolution of high and low affinity states of the receptor by computer modeling of ligand binding data. *Mol. Pharmacol.* **17**, 14–23.

Koshland, D. E., Jr. (1969). Conformational aspects of enzyme regulation. *Cur. Top. Cell. Regul.* **1**, 1–27.

Kurose, H., Katada, T., Amano, T., and Ui, M. (1983). Specific uncoupling by islet activating protein, Pertussis Toxin, of negative signal transduction via α-adrenergic, cholinergic and opiate receptors in neuroblastoma x glioma hybrid cells. *J. Biol. Chem.* **258**, 4870–4875.

Laburthe, M., Amiranoff, S., and Boissard, C. (1982). Adrenergic inhibition of cyclic AMP accumulation in epithelial cells isolated from rat small intestine. *Biochim. Biophys. Acta* **721**, 101–108.

Lands, A. M., Arnold, A., McAuliff, J. P., Luduena, F. P., and Brown, T. G., Jr. (1967). Differentiation of receptor systems activated by sympathomimetic amines. *Nature (London)* **214**, 597–598.

Langer, S. Z. (1974). Presynaptic regulation of catecholamine release. *Biochem. Pharmacol.* **23**, 1793–1800.

Langer, S. Z. (1976). The role of α and β-presynaptic receptors in the regulation of noradrenaline release elicited by nerve stimulation. *Clin. Sci. Mol. Med.* **51**, 4235–4265.

Lichtshtein, D., Boone, G., and Blume, A. J. (1979). A physiological requirement of Na⁺ for the regulation of cAMP levels in intact NG 108-15 cells. *Life Sci.* **25**, 985–992.

Limbird, L. E. (1981). Activation and attenuation of adenylate cyclase: GTP-binding proteins as macromolecular messengers in receptor–cyclase coupling. *Biochem. J.* **195**, 1–13.

Limbird, L. E., and Lefkowtiz, R. J. (1978). Beta-adrenergic receptors: Agonist induced increase in apparent molecular size. *Proc. Natl. Acad. Sci. U.S.A.* **75**, 228–232.

Limbird, L. E., and Speck, J. L. (1983). N-ethylmaleimide elevated temperature and digitonin solubilization eliminate guanine nucleotide but not Na⁺ effects of human platelet alpha₂-adrenergic receptor–agonist interactions. *J. Cyclic Nucleotide Protein Phosphorylation Res.* **9**, 191–202.

Limbird, L. E., Gill, D. M., Stadel, J. M., Hickey, A. R., and Lefkowitz, R. J. (1979). Loss of β-adrenergic receptor-guanine nucleotide regulatory protein interactions accompanies decline in catecholamine responsiveness of adenylate cyclase in maturing rat erythrocytes. *J. Biol. Chem.* **255**, 1854–1861.

Limbird, L. E., Gill, D. M., and Lefkowitz, R. J. (1980). Agonist promoted coupling of the beta-adrenergic receptors with the guanine nucleotide regulatory protein of the adenylate cyclase system. *Proc. Natl. Acad. Sci. U.S.A.* **77**, 775–779.

Limbird, L. E., Speck, J. L., and Smith, S. K. (1982). Sodium ion modulates agonist and antagonist interactions with the human platelet α₂-adrenergic receptors in membrane and solubilized preparations. *Mol. Pharmacol.* **41**, 607–619.

Lynch, C. J., and Steer, M. L. (1981). Evidence for high and low affinity alpha₂-receptors. Comparison of [³H] -norepinephrine and [³H]-phentolamine binding to human platelet membranes. *J. Biol. Chem.* **256**, 3298–3303.

MacFarlane, D. E., Wright, B. L., and Stump, D. C. (1981). Use of methyl-[³H]-yohimbine as a radioligand for alpha$_2$-adrenergic receptors on intact platelets. Comparison with de-hydroergocryptine. *Thromb. Res.* **24**, 31–43.

McMahon, K. K., and Hosey, M. M. (1983). Potentiation of monovalent cation effects on ligand binding to cardiac muscarinic receptors in N-ethylmaleimide treated membranes. *Biochem. Biophys. Res. Commun.* **111**, 41–46.

Malaisse, W. J., Brisson, G., and Malaisse-Lagae, F. (1970). The stimulus-secretion coupling of glucose-induced insulin release. I. Interaction of epinephrine and alkaline earth cations. *J. Lab. Clin. Med.* **76**, 895–902.

Manning, D. R., and Gilman, A. G. (1983). The regulatory components of adenylate cyclase and transducin. A family of structurally homologous guanine nucleotide-binding proteins. *J. Biol. Chem.* **258**, 7059–7063.

Michel, T. M., and Lefkowitz, R. J. (1982). Hormonal inhibition of adenylate cyclase. $\alpha_2$-Adrenergic receptors promote release of [$^3$H]-Gpp(NH)p from platelet membranes. *J. Biol. Chem.* **257**, 13557–13563.

Michel, T. M., Hoffman, B. B., and Lefkowitz, R. J. (1980). Differential regulation of the alpha$_2$-adrenergic receptors by Na$^+$ and guanine nucleotides. *Nature (London)* **288**, 709–711.

Michel, T. M., Hoffman, B. B., Lefkowitz, R. J., and Caron, M. G. (1981). Different sedimentation of properties of agonist- and antagonist-labelled platelet alpha$_2$-adrenergic receptors. *Biochem. Biophys. Res. Commun.* **100**, 1131–1136.

Mooney, J. J., Horne, N. C., Handin, R. I., Schildkraut, J. J., and Alexander, R. W. (1982). Sodium inhibits both adenylate cyclase and high affinity [$^3$H]-p-aminoclonidine binding to $\alpha_2$-adrenergic receptors in purified human platelet membranes. *Mol. Pharmacol.* **21**, 600–608.

Motulsky, H. J. (1983). Influence of sodium on the $\alpha_2$-adrenergic receptor system of human platelets. Role for intraplatelet sodium in receptor binding. *J. Biol. Chem.* **258**, 3913–3919.

Munson, P. J., and Rodbard, D. (1980). LIGAND: A versatile computerized approach for characterization of Ligand-binding systems. *Anal. Biochem.* **107**, 220–239.

Nakaki, T., Nakadate, T., Yamamoto, S., and Kato, R. (1982). $\alpha_2$-Adrenoceptors inhibit the cholera-toxin-induced intestinal fluid accumulation. *Naunyn-Schmiedeberg's Arch. Pharmacol.* **318**, 181–184.

Nakaki, T., Nakadate, T., Yamamoto, S., and Kato, R. (1983a). Alpha$_2$-adrenergic receptor in intestinal epithelial cells: identification by [$^3$H]-yohimbine and failure to inhibit cyclic AMP accumulation. *Mol. Pharmacol.* **23**, 228–234.

Nakaki, T., Nakadate, T., Yamamato, S., and Kato, R. (1983b). Inhibition of dibutyryl cyclic AMP-induced insulin release by alpha$_2$-adrenergic stimulation. *Life Sci.* **32**, 191–195.

Nellans, H. N., Frizzell, R. A., and Schultz, S. G. (1973). Coupled sodium–chloride influxes across the brush border of rabbit ileum. *Am. J. Physiol.* **225**, 467–475.

Newman, K. D., Williams, L. T., Bishopric, N. M., and Lefkowitz, R. J. (1978). Identification of $\alpha$-adrenergic receptors by [$^3$H]-dihydroergocriptine binding. *J. Clin. Invest.* **61**, 395–402.

Northrup, J. K., Smigel, M. D., and Gilman, A. G. (1982). The guanine nucleotide activating site of the regulatory component of adenylate cyclase. Identification of ligand binding. *J. Biol. Chem.* **257**, 11416–11423.

Ross, E. M., and Gilman, A. G. (1980). Biochemical properties of hormone-sensitive adenylate cyclase. *Ann. Rev. Biochem.* **49**, 553–564.

Ross, E. M., Howlett, A. C., Ferguson, K. M., and Gilman, A. C. (1978). Reconstruction of hormone-sensitive adenylate cyclase activity with resolved components of the enzyme. *J. Biol. Chem.* **253**, 6401–6412.

Shattil, S. J., McDonough, M., Turnbull, J., and Insel, P. A. (1981). Characterization of alpha-adrenergic receptors in human platelets using [$^3$H]-clonidine. *Mol. Pharmacol.* **19**, 179–183.

Smith, S. K., and Limbird, L. E. (1981). Solubilization of human platelet $\alpha$-adrenergic receptors:

Evidence that agonist occupancy of the receptors stabilizes receptor–effector interactions. *Proc. Natl.Acad. Sci. U.S.A.* **78**, 4026–4030.

Smith, S. K., and Limbird, L. E. (1982). Apparent independence of the alpha-adrenergic system of the human platelet from the cholera toxin-catalyzed ADP-ribosylated 42,000 Mr subunit of the adenylate cyclase system. *J. Biol. Chem.* **257**, 10471–10478.

Soderling, T. R., Corbin, J. A., and Park, C. R. (1973). Regulation of adenosine 3′,5′-monophosphate-dependent protein kinase II Hormonal regulation of the adipose tissue enzyme. *J. Biol. Chem.* **248**, 1822–1829.

Stadel, J. M., Schorr, R. G. L., Limbird, L. E., and Lefkowitz, R. J. (1981). Evidence that a beta-adrenergic receptor-associated guanine nucleotide regulatory protein conveys GTP-γ-s dependent adenylate cyclase activity. *J. Biol. Chem.* **256**, 8718–8723.

Starke, K., (1977). Regulation of noradrenaline release by presynaptic receptor systems. *Rev. Physiol., Biochem. Pharmacol.* **77**, 1–124.

Steer, M. L., Khorana, J., and Galgoci, B. (1979). Quantitation and characterization of human platelet alpha-adrenergic receptors using [³H]-phentolamine. *Mol. Pharmacol.* **16**, 719–728.

Steinberg, D., Mayer, S. E., Khoo, J. C., Miller, E. A., Miller, R. E., Fredholm, B., and Eichner, R. (1975). Hormonal regulation of lipase, phosphorylase, and glycogen synthase in adipose tissue. *Adv. Cyclic Nucleotide Res.* **5**, 549–568.

Tsai, B.-S., and Lefkowitz, R. J. (1978). Agonist-specific effects of monovalent and divalent cations in adenylate cyclase-coupled alpha-adrenergic receptors in rabbit platelets. *Mol. Pharmacol.* **14**, 540–548.

Tsai, B.-S., and Lefkowitz, R. J. (1979). Agonist-specific effects of guanine nucleotides on alpha-adrenergic receptors in human platelets. *Mol. Pharmacol.* **16**, 61–68.

Ullrich, S., and Wollheim, C. B. (1984). Islet cyclic AMP levels are not lowered during α₂-adrenergic inhibition of insulin release studies with epinephrine and forskolin. *J. Biol. Chem.* **259**, 4111–4115.

U'Prichard, D. C., Greenberg, D. A., and Snyder, S. H. (1977). Binding characteristics of a radiolabeled agonist and antagonist at central nervous system alpha-noradrenergic receptors. *Mol. Pharmacol.* **13**, 454–473.

Vaughn, M., Pierce, N. F., and Greenbough, W. B., II (1970). Stimulation of glycerol production in fat cells by cholera toxin. *Nature (London)* **226**, 658–659.

Westfall, T. C. (1977). Local regulation of adrenergic neurotransmission. *Physiol. Rev.* **57**, 659–728.

Williams, L. T., and Lefkowitz, R. J. (1976). Alpha-adrenergic receptor identification by [³H]-dehydroergocryptine binding. *Science (Washington, D.C.)* **192**, 791–793.

Williams, L. T., Mullikin, D., and Lefkowitz, R. J. (1976). Identification of α-adrenergic receptors in uterine smooth muscle membranes by [³H]-dehydroergocryptine binding. *J. Biol. Chem.* **251**, 6915–6923.

Wollheim, C. B., and Sharp, G. W. G. (1981). Regulation of insulin release by calcium. *Physiol. Rev.* **61**, 941–973.

Wollheim, C. B., Kikuchi, M., Renold, A. E., and Sharp, G. W. G. (1977). Somatostatin- and epinephrine-induced modifications of ⁴⁵Ca⁺⁺ fluxes and insulin release in rat pancreatic islets maintained in tissue culture. *J. Clin. Invest.* **60**, 1165–1173.

Wood, C. L., Arnett, C. D., Clarke, W. R., Tsai, B. S., and Lefkowitz, R. J. (1979). Subclassification of alpha-adrenergic receptors by direct binding studies. *Biochem. Pharmacol.* **28**, 1277–1282.

# 9

# Protein Glycosylation and Receptor–Ligand Interactions

## M. R. SAIRAM

Reproduction Research Laboratory
Clinical Research Institute of Montreal
Montreal, Quebec, Canada

## I. INTRODUCTION

Communication between cells usually accomplished by the generation of specific signals for message transmission requires the participation of integral components of the membrane. Cell membranes perform the important task of

**307**

conveying the chemical message contained in the signal (ligand) into the interior of the cell. The mechanism by which the cell membrane accomplishes this function in most biological systems is generally well defined. Macromolecules called receptors located on the cell surface serve to discriminate, with a very high degree of specificity and efficacy, the ligand(s) present in the close environment surrounding the cell. The high affinity of the receptor–ligand interactions assures binding of the ligands present at low concentrations (usually $10^{-9}–10^{-12}\,M$) in biological systems. The coupling of the ligand to the receptor activates an intramembrane component having the net effect of transducing the message to activate an effector system in the interior of the cell membrane generating a second messenger to amplify the initial signal culminating in the biological response typical of the ligand and the cell.

Many receptor systems have been recognized in modern biology describing the interactions of drugs, toxins, antigens, antibodies, hormones, neurotransmitters, etc. Receptor–ligand interactions are complex reactions involving all the components of the membrane, namely, protein, carbohydrate, and lipid. The receptor composed of an assembly of glycoproteins is embedded in the lipid bilayer. The carbohydrate units are usually clustered on the periphery of the molecule so that this hydrophilic area is exposed to the external side of the cell membrane. Some of the glycoproteins may extend their contact to the cytoplasmic side of the membrane, providing a means of conveying an external stimulus such as a hormone or neurotransmitter.

In glycoproteins, the carbohydrate units are covalently attached to the polypeptide backbone. These prosthetic groups can be part of the receptor or ligand or both. There is now extensive evidence documenting the fact that the carbohydrate moieties of glycoproteins fulfill important biological functions. In addition to the well-recognized phenomena such as stabilization of protein conformation, regulation of metabolic half-life in circulation in animals, and uptake by cells, most recent evidence suggests that the carbohydrate units in certain ligands such as the glycoprotein hormones are intimately involved in message transmission, the coupling of the receptor to the adenylate cyclase system. The aim of this chapter is to discuss the experimental approaches presently available to evaluate the role of protein glycosylation in receptor–ligand interactions and their implications for the study of the mechanism of hormone action.

## II. CARBOHYDRATES IN DETERMINING SPECIFICITY

Monosaccharide units or oligosaccharides on cell surfaces serve as attachment sites for many biological agents, such as antibodies, plant lectins, bacterial toxins, and viruses. In addition, they are now recognized as integral components of receptors or ligands in a more complex form. Some of the receptors, such as the acetylcholine receptor (1,2), human choriogonadotropin receptor (3), β-adrenergic receptor (4), and hepatic binding protein (5,6) have been obtained in highly homogeneous form. Direct evidence for their glycoprotein nature comes from analysis of their composition. However, the role of these sugar residues in receptor function is not completely clear. Carbohydrate units have two unique features that render them highly suitable in processes linked with cellular recognition. They can be linked in various ways, giving rise to structures which can be arranged in much greater combinations than possible with the linear ordering of amino acids in peptides. The second most important feature is to impart a hydrophilic character to a membrane component which might protrude into the external environment and provide access to an incoming signal.

Among hormones, the glycoprotein hormones provide the best examples of ligands which are large glycoproteins serving important biological functions.

## III. GENERAL FEATURES OF GLYCOPROTEINS

Glycoproteins are ubiquitous macromolecules present in almost all living cells, with the exception of bacteria; they function as antibodies, receptors, lectins, enzymes, and hormones. The sugar residues in glycoproteins are covalently attached to the polypeptide backbone. Of the five different types of carbohydrate–polypeptide linkages present in biological systems, two are most commonly found in secreted proteins. These are the N-glycosidic β-$N$-acetyl-glucosaminylasparagine (GlcNAc-Asn) and O-glycosidic α-$N$-acetylgalac-tosaminylserine-threonine (GalNAc-Ser/Thr). Both kinds of glycosidic linkages may be found in the same protein as in the placental gonadotropins human choriogonadotropin (hCG) and equine choriogonadotropin (eCG) which have a high percentage of carbohydrate. Glycoproteins in general are more heterogeneous than simple proteins, as the length of the carbohydrate chain may vary to a greater degree than the polypeptide backbone. This is because of the nature of biosynthetic mechanisms. While the polypeptide backbone is a direct gene product, the covalently attached carbohydrate moiety is not, because it is assembled by a series of enzymatic reactions involving specific glycosyltransferases which add on or remove sugar units at specific stages. Such an assembly is generally less precise, resulting in their typical microheterogeneity. This behavior can often be frustrating if the glycoprotein is being isolated for the first time, as considerable effort would have to be devoted to the analysis of various components. There can be no doubt that microheterogeneity presents special problems in their isolation and characterization.

## IV. STRUCTURAL ORGANIZATION
   OF THE GLYCOPROTEIN HORMONES*

As a considerable part of this chapter will deal with glycoprotein hormones, a general description of their structural features will be most appropriate to appreciate the importance of the role of carbohydrates in biological function. The glycoprotein hormones are secreted into the general circulation in animals to control the growth and function of the gonads and thyroid. These hormones, of which five have been well characterized (Table I), contain a quaternary structure, a feature unique to this family of hormones. Their two subunits, designated α and β, are noncovalently linked by hydrophobic–electrostatic and other interactions. The isolated subunits are virtually inactive, and their specific combination is essential for binding to specific receptors in target tissues and cellular activation. The α subunit is more or less common to all the hormones in a given species, and the β subunit imparts hormonal specificity to the α–β complex. It is believed that both subunits contribute binding sites for receptor interaction, with the common α subunit playing a more critical role in the formation of an active α–β complex.

The hormones interact with receptors of high affinity and low capacity in specific cells of the ovary and testis for gonadotropins and thyroid for thyrotropin. A number of attempts are underway to isolate and characterize these receptor proteins. A soluble form of LH receptor from bovine corpus luteum has been isolated recently (3) and details of its structure are awaited. The structures of all of the hormones are known. They are 28,300–45,000 molecular weight proteins with both subunits containing covalently linked carbohydrates. In the hormones of pituitary origin, the α subunit generally carries about twice as much carbohydrate as the β, but in the placental gonadotropins which may contain up to 35–40% sugars, the larger β subunit carries a much greater proportion of the carbohydrate in its carboxyl terminus. The α and β subunits of the pituitary hormones contain N-glycosidically linked sugars[†] of which the terminal hexosamines appear to be sulfated. The placental β subunits have both N-glycosidic and O-glycosidic (predominant) linkages, none of which are sulfated.

The molecular biology of these hormones is fairly well advanced, resulting in

---

*There are five glycoprotein hormones of mammals which have been well characterized. Follicle-stimulating hormone or follitropin (FSH), luteinizing hormone or lutropin (LH), and thyroid stimulating hormone or thyrotropin (TSH) are secreted by the anterior pituitary. Human choriogonadotropin (hCG) and equine choriogonadotropin (eCG) are secreted in large quantities during pregnancy and are usually extracted from the urine of pregnant women and blood of pregnant mares, respectively. The hCG is an LH-like hormone, both in structure and function, binding to LH receptors. The eCG is a hybrid hormone having both LH and FSH activities and can bind to both LH and FSH receptors. The placental hormones have a greater biological activity than the pituitary counterparts.

[†]An O-glycosidic linkage has been found in the α-subunit of bovine pituitary (7).

TABLE I

**Properties of Glycoprotein Hormones**

---

Secreted by the pituitary and placenta
High-biological specificity
Are glycoproteins with 15–40% carbohydrate. Placental hormones have a
    greater content
Molecular weight 28,300–45,000
Consist of two nonidentical and noncovalently linked subunits $\alpha$ and $\beta$ forming
    a hormone dimer ($\alpha\beta$)
Isolated subunits are inactive—show no receptor binding or cause adenylate
    cyclase activation
Recombination restores all aspects of biological activity
$\alpha$ Subunits are virtually identical; $\beta$ subunits are hormone specific
Both subunits contain carbohydrate moiety(ies). O-Glycosidically linked carbo-
    hydrates are generally found in the placental hormones
Structures of the polypeptide and carbohydrate moieties are known

---

the isolation of cDNA clones for some of the subunits and the identification of their genes. The hCG $\alpha$- and $\beta$-subunit genes have been recently expressed in bovine papilloma virus, resulting in biologically active hormone.

The most interesting and unique aspect of the study of gonadotropin–thyrotropin hormone action is that both the ligand and receptor are glycoproteins. At present, this is the only instance in which the carbohydrates of the ligand have been definitely shown to be essential in cellular activation. These aspects are discussed in Section VIII.

## V. BIOLOGICAL PROPERTIES
## OF THE GLYCOPROTEIN HORMONES

The gonadotropic hormones are essential for the normal growth and secretory activity of the gonads and consequently regulate the reproductive functions in the male and female. The coordinated and sequential actions of FSH and LH on the ovary and testis lead to the production of steroids in the appropriate cellular compartments and gametogenesis. Thyrotropin, on the other hand, stimulates the activity of the thyroid, causing the secretion of thyroidal hormones such as thyroxine.

The individual subunits of the glycoprotein hormones are virtually inactive but full biological activity, including appropriate cellular responses, can be regained by reconstitution of the $\alpha$ and $\beta$ subunits. The basic concepts of protein–peptide hormone action are applicable to these systems also and are depicted in Fig. 1 so as to allow a discussion of the various events in which the carbohydrate moiety of the ligands appears to play a critical role.

GLYCOPROTEIN HORMONE

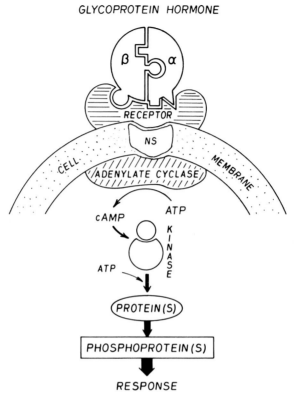

**Fig. 1.** Model of the action of glycoprotein hormones on target cells. The hormone is shown as having three binding sites (two in the $\alpha$ and one in the $\beta$) in its interaction with the receptor (8). Binding of the hormone to the receptor facilitates GTP binding to the $N_s$ (a regulatory protein) component in the intramembrane compartment, allowing coupling to the adenylate cyclase on the inner side of the plasma membrane. The complete integrity of the carbohydrate moiety is now known to be essential for effecting this coupling (Section VIII). The increased size of the arrows signifies amplification of the initial hormonal signal at each step. [Modified from Sairam (8).]

Both subunits contribute binding sites for interaction with the receptor, as revealed by structure–function studies (8,9). The high affinity and specific interaction cause the coupling of the receptor via the GTP-dependent binding protein to the adenylate cyclase in the inner membrane of the cell. The generation of cyclic AMP activates further cellular events, producing typical responses such as steroidogenesis. The widely studied phenomena, such as down-regulation, internalization, and desensitization, are also applicable to the gonadotropins. These hormones provide one of the most interesting models to investigate the role of the carbohydrate at each one of these steps.

## VI. EXPERIMENTAL APPROACHES TO STUDY THE ROLE OF CARBOHYDRATE UNITS

As mentioned earlier, glycoproteins are universally distributed, are common components of the cell surface in animals, and are found in many products exported by the cell, although it is now generally accepted that glycosylation is not essential for secretion. The glycoproteins of the cell membrane have been shown to play an important role in differentiation (10,11), intercellular recognition and aggregation (10,12–15), malignancy (10,16), and pinocytosis (5,17), and they also function as receptors for hormones and viruses (10) and as determinants of immunological specificity (10). Among the secreted glycoproteins are biologically important molecules, such as immunoglobulins, enzymes, and hormones.

Despite the ubiquitous nature of glycoproteins in living cells, the precise biological roles of the carbohydrate units remain to be clearly defined. Since cells expend great amounts of energy in carrying out the various reactions required for the glycosylation of proteins (the general structure of which is highly conserved throughout evolution), the carbohydrate units must be important in unique but still universal physiological process(es).

Research on glycoproteins has been slow, because of difficulties associated with their preparation and handling and because suitable microanalytical techniques were not available. This situation is, however, changing with the result that a pattern of the general structure of the carbohydrate moieties in glycoproteins is now available (18). Specific methods for the stripping of the carbohydrate residues from the glycoproteins would aid in exploring their functional role in the protein. One can conceive of several approaches to accomplish this task of selective removal: (i) by generation of glycosylation-deficient mutants of cells; (ii) by using inhibitors of glycosylation such as tunicamycin; (iii) treatment of glycoproteins with exoglycosidases which sequentially cleave specific sugar residues in the oligosaccharide chain; (iv) digestion with endoglycosidases which tend to cleave near the site of attachment of the sugar moiety to the protein core; and (v) chemical treatment of the glycoproteins with reagents such as anhydrous hydrogen fluoride or trifluoromethanesulfonic acid. All of these methods, except the first for obvious reasons, have been applied in investigating the role of carbohydrates in either receptors or ligands or both. Since this author has been personally involved in studies on chemical deglycosylation, emphasis will be placed on this approach.

## A. Inhibition of Glycosylation by Tunicamycin

Tunicamycin is a glucosamine-containing antibiotic which inhibits the synthesis of $N$-acetylglucosaminylpyrophosphoryl polyisoprenol (19–22). Because the formation of this intermediate is required for the eventual synthesis of N-glycosidically linked oligosaccharides, culturing of cells *in vitro* with tunicamycin results in the synthesis of glycoproteins deficient in asparagine-linked oligosaccharides. Therefore, this drug is very useful in analyzing the deficiency of N-glycosylation in biological function of receptors and ligands.

Different cells vary in their susceptibility to the action of tunicamycin, and for this reason the effects of the antibiotic on inhibition of glycosylation as compared to inhibition of general protein synthesis must be very carefully evaluated before extensive use. A clear dissociation between the two events has been reported in numerous instances to make this reagent very useful (e.g., Table II) in the study of biosynthesis and the assembly of membrane receptors. Generally the presence of a small quantity ($\mu$g/ml) is sufficient to cause greater than 95% inhibition of N-glycosylation without significantly reducing protein synthesis. Examples of receptor populations biosynthesized in the presence of tunicamycin are considered in Section IX.

There are other techniques which might fall into the category of altering glycosylation during biosynthesis which have not been as extensively employed as tunicamycin. The monovalent ionophore monensin inhibits the secretion of macromolecules from several cell types. Monensin reduces sialylation of lymphoid cell surface glycoproteins (24) and lowers the incorporation of labeled galactose and fucose into secreted immunoglobulin M molecules (25). Monensin apparently released incompletely processed forms of fibronectin from human skin fibroblasts in culture (26).

Other methods of inhibition of glycosylation employing glucose starvation or 2-deoxyglucose have not been widely used and are complicated because of the associate effects on general cellular metabolism.

**TABLE II**

**Effects of Tunicamycin on Protein Synthesis and Glycosylation in Cultured Muscle Cells**[a]

| Tunicamycin ($\mu$g/ml) | Protein synthesis [$^{14}$C]leucine (%) | Glycosylation [2-$^{3}$H]mannose (%) |
|---|---|---|
| 0 | 100 | 100 |
| 0.05 | 83 ± 8 | 7 ± 2 |
| 0.10 | 67 ± 10 | 5 ± 2 |
| 0.20 | 47 ± 4 | 2 ± 1 |

[a] Adapted from Prives and Olden (23). Protein synthesis and glycosylation have been assessed by measuring the incorporation of L-[U-$^{14}$C]leucine and D-[2-$^{3}$H]mannose, respectively, into TCA-precipitable protein in cultured muscle cells after a 6-hr exposure to the antibiotic tunicamycin.

## B. Exoglycosidase Digestion

Exoglycosidases such as neuraminidase, α-fucosidase, β-D-galactosidase, β-D-N-acetylglucosaminidase, and α-D-mannosidase obtained from bacteria, fungi, plants, and animal tissues are valuable in the selective removal of sugar residues from the oligosaccharide moiety. Until these enzymes become more easily commercially available in highly purified forms free of proteolytic activity, their use remains restricted to a few laboratories that are engaged in purifying them. Neuraminidase is the most widely known in this class and has been extensively used for a number of glycoproteins, including receptors and glycoprotein hormones because it is relatively quick in its action of quantitatively removing terminal sialic acid from the carbohydrate moiety. The removal of sialic acid residues changes the molecule in many ways. In many instances, charge heterogeneity is considerably reduced, rendering the molecule less acidic in character. Several membrane receptor preparations, for example, those for gonadotropins from the testis and ovary (25), show a significant increase in the ability to bind the respective radioligand. In most instances, the asialoglycoproteins lose their biological activity because of quick elimination by hepatic binding sites (receptors) which now recognize the exposed galactose residues on the protein. Hence, sialic acid is essential for maintaining such ligands in circulation. The hepatic receptor–asialoglycoprotein interactions are discussed further in Section VIII. While causing the destruction of biological activity *in vivo,* most desialylated ligands retain full biological activity *in vitro,* including the ability to bind to receptors and to cause cellular activation.

The other enzymes which are required for the subsequent removal of additional sugar residues act very slowly in comparison to neuraminidase, and the reactions require long incubations for completion, a situation which might affect the stability of the receptor and/or the ligands. This limitation has restricted the application of exoglycosidases, with the exception of neuraminidase, to the study of receptors. However, Bahl and colleagues have made detailed studies on the effects of these various enzymes on hCG (26). Such enzymes have no effect on the pituitary glycoprotein hormones because of the presence of sulfated hexosamines which render them resistant to their action (27).

## C. Application of Endoglycosidases

The discovery and isolation of endoglycosidases from plant and bacterial sources has provided a boon in the study of glycoproteins, including many ligands and receptors. These enzymes, usually referred to as endo-β-N-acetyl-glucosaminidase H or D designating their source, now provide a major step forward, are particularly suited for glycoproteins with complex sugar chains, and offer possibilities that are nonexistent with the exoglycosidases. In contrast with the exoglycosidases, the endoglycosidases act on the core of the sugar moiety and cleave most of the oligosaccharide units as a block, thus essentially stripping most of the carbohydrate from the protein. The action of endo-β-$N$-acetylglucos-aminidase D from diplococcal pneumoniae is depicted below.

$$X\text{-Man}\alpha1 \diagdown \atop \text{Man}\alpha1 \diagup} {6' \atop 3} \text{Man}\beta1\text{-4GlcNAc}\beta1\text{-4GlcNAc}\beta1\text{-4GlcN Ac}\beta1\text{-Asn} \qquad (1)$$

with $X$ above Asn and protein below, arrow ↑ under penultimate GlcNAc.

$X$ represents monosaccharides or oligosaccharides containing mannose, fucose, galactose, and/or sialic acid.

The enzymes are commercially available in a highly purified form. While being attractive in principle, the enzymes suffer from the disadvantages of requiring long incubations and the incomplete reaction with various glycoproteins despite the fact that sugar moieties in glycoproteins are most likely found on the periphery. Cleavage occurs at the penultimate GlcNAc linkage, leaving one sugar residue still linked to the Asn at each site of attachment in the glycoprotein. They have been successfully applied to prepare hCG subunits essentially free of carbohydrate (28). There are no reports of the application of the endoglyosidases to the study of highly purified receptors and the consequent effects on function. This may indeed pose a problem unless some means are devised to maintain the stability of the purified receptor under the conditions of incubation required for enzyme action.

## D. Chemical Deglycosylation

Chemical methods of removing carbohydrates from glycoproteins have attracted increasing attention since it was shown that anhydrous HF (a reagent routinely used in the chemical synthesis of peptides for deblocking of protecting groups for amino acids) could effectively remove part of the carbohydrate moiety from glycoproteins (29). Its applicability to the deglycosylation of glycoprotein hormones and subunits has been studied in detail in the authors' laboratory. Treatment of lyophilized and moisture-free hormone or subunit preparations with anhydrous liquid HF at 0°C under vacuum for 30–60 min results in the disappearance of 90% of the accessible sugar residues without altering the structure of the polypeptide backbone. The overall loss of carbohydrate is about 75%. The quaternary structure of the hormone is fully maintained following carbohydrate removal. On the other hand, it may be rendered even more stable after the treatment. The biochemical properties of the deglycosylated gonadotropins are summarized in Table III. There can be no doubt that the effects on function discussed in Section VIII are solely the result of the removal of the carbohydrate

TABLE III

Biochemical Characteristics of Deglycosylated Hormones[a]

---

The loss of carbohydrate is approximately 75%
Chemical deglycosylation does not affect the polypeptide moiety
There is no detectable change in conformation
Quaternary structure of the hormone is fully preserved
Chromatographic properties are different from the native hormones
DG Hormones are excluded on lectin columns such as concanavalin A–Sepharose
DG Hormones show less electrophoretic heterogeneity
DG Hormones are more stable than native hormones

---

[a] The properties listed are for deglycosylated preparations obtained by treatment of purified LH, hCG, or FSH with anhydrous HF for 1 hr at 0°C. Adapted from Sairam [8].

residues. The general methods of chemical deglycosylation which could be applicable to other glycoprotein ligands have been described in detail elsewhere (30).

Based on detailed kinetic analysis, it has been concluded that deglycosylation of ovine LH, a protein containing about 16% carbohydrate, produces alterations in the following manner. Each of the three carbohydrate moieties now have two N-acetylglucosaminyl linkages left intact (31); a situation very similar to the action of endo-$\beta$-$N$-acetylglucosaminidases on other glycoproteins. It should be noted that such enzymes are ineffective on pituitary glycoprotein hormones (27), because of the presence of blocked (sulfated) $N$-acetylglucosamine residues at their termini.

The rapidity of the reaction, the ease of removing the reagent (HF) by evaporation, and the subsequent high recovery of the deglycosylated hormone are attractive features of the application of the HF method which could be easily extended to the study of other glycoprotein ligands. As discussed elsewhere (30), other alternative reagents such as trifluoromethanesulfonic acid are cumbersome and have not been as successful as the anhydrous HF method.

The HF treatment has not been tested with receptors. It is doubtful if the HF-deglycosylation procedure can be applied to strip carbohydrate from receptors in the form of crude plasma membranes. While anhydrous HF, which is an excellent solvent for most proteins, might achieve the desired result, the multitude of deglycosylated protein components of the membrane could lead to an insoluble product after removal of the reagent. This is most likely due to the disorganization of the membrane structure and the large increase in the net hydrophobicity of the proteins following stripping of the hydrophilic part of the molecules. However, if a receptor glycoprotein can be obtained in a stable lyophilizable form, chemical deglycosylation becomes feasible. Until such time, producing non-glycosylated receptors by biosynthetic means, that is, by employing suitable inhibitors of glycosylation such as tunicamycin, remains the only alternative.

## VII. ROLE OF CARBOHYDRATE IN THE RECOGNITION AND UPTAKE OF GLYCOPROTEIN BY CELLS

Carbohydrate structures serve as important and unique recognition signals or markers on glycoproteins that are in the soluble form as well as those situated on the cell surface. An impressive array of evidence now suggests the involvement of carbohydrate residues in the extent of survival of important glycoproteins in circulation and their uptake by hepatic cells, in directing the uptake and intracellular translocation of lysosomal enzymes in fibroblasts, and more recently in the coupling of the membrane receptor to adenylate cyclase in the glycoprotein hormone-responsive tissues. It would be most appropriate to consider well-defined examples to illustrate the generalities of these phenomena.

### A. Cell–Cell Interactions

Most if not all of the acidic glycoproteins, including those on the cell surface, contain the acidic sugar $N$-acetylneuraminic acid (sialic acid) at the nonreducing terminus, which could serve as a marker for the interaction of two proteins. The first evidence which showed that sugars in cell surface components acted as binding sites (receptors) for the attachment of influenza virus came from the classical experiments of Burnet (32). They showed the absolute requirement of terminal sialic acid residues on the erythrocyte membrane for the attachment of virus particles. The attachment of virus particles to the cell surface was accompanied by the agglutination of erythrocytes. Treatment of human erythrocytes with the enzyme sialidase (neuraminidase) which effectively splits the terminal sialic acids from the erythrocyte surface abolished the agglutination reaction. Agglutination was also inhibited by the low concentration of other sialoglycoproteins which also subsequently lost their ability to do so upon treatment with neuraminidase.

The major sialoglycoprotein of the erythrocyte membrane which is responsible for virus attachment has been identified as glycophorin. The attachment of virus to the glycophorin complex on the cell surface is a prerequisite for infection of the host (33). Human erythrocyte glycophorin is a glycoprotein of 31,000 daltons with 60% carbohydrate content. The complex structure of glycophorin reveals the presence of two types of oligosaccharide moieties. Fifteen of these units are in O-glycosidic linkage to Thr/Ser with one linked by the N-glycosidic bond to an Asn residue.

Enzymatic attachment of sialic acid, by employing specific sialyltransferases from pig submaxillary glands (34) to asialoglycophorin (asialoerythrocytes), restores the ability of these red cells to be agglutinated by different viruses. Such studies provide strong evidence, suggesting that the linkage of sialic acid in a specific configuration at the termini of oligosaccharide units is very important in receptor–ligand interactions.

## B. Pinocytosis of Circulating Glycoproteins

Specific receptors on the surface of eukaryotic cells recognize a large variety of molecules of different chemical nature. In several instances, binding to the receptor(s) is followed by an invagination of the cell surface, leading to the entry of the interacting ligand within a vesicular structure of the plasma membrane. This mechanism, termed receptor-mediated pinocytosis, is responsible for the internalization of several macromolecules, such as proteins, toxins, hormones, and viruses.

Carbohydrate residues of glycoproteins serve an important function in regulating their survival in the circulation of animals. In some of the best documented series of experiments, Ashwell and colleagues demonstrated that mammalian hepatocytes (liver cells) contain a unique binding protein (receptor) that promotes the rapid uptake and clearance of serum glycoproteins (5). Their important observation revealed that the presence of sialic acid in the proteins was critical to their survival in circulation. Only those serum glycoproteins with all of their sialic acids intact remained in cirulation with a half-life of more than 48 hr. This was dramatically reduced to a few minutes upon removal of sialic acid. Their initial studies (5), conducted with the copper-containing serum glycoprotein ceruloplasmin, were later extended to many others including hCG (35). The biological activity of the hormone is closely related to the extent of its sialylation. In ceruloplasmin, only 2 out of the 10 sialic acids need to be removed to reduce the half-life from 60 hr to less than 5 min.

The mechanism responsible for the disappearance of asialoglycoproteins has been identified to be a binding protein of the liver membrane which recognizes exposed galactosyl residues of the oligosaccharide moiety. At present, the studies on the hepatic binding protein and serum glycoproteins represent the best examples in which both the receptor and ligand are glycoproteins which have been well characterized. The hepatic binding protein has been isolated from the liver plasma membranes (6) and also shown to be present in other hepatic structures, including the Golgi apparatus, smooth microsomes, and lysosomes. The hepatic receptor for asialoglycoproteins has been identified as a macromolecule of 88,000 daltons composed of two subunits (40,000 and 48,000) which requires $Ca^{2+}$ for binding (36). As mentioned earlier, this receptor protein is also a glycoprotein which has the carbohydrate residues linked N-glycosidically in a manner found in most other members of this family (37). The binding protein is highly specific to glycoproteins with exposed galactose residues. The hepatic receptor protein recycles between the cell surface and lysosomes, allowing continuous clearance of asialoglycoproteins from the general circulation (38).

Metabolic clearance systems in which sugar residues other than galactose serve as recognition markers have also been identified in liver and other cells of the reticuloendothelial system. These binding proteins recognizing terminal mannose or *N*-acetylglucosamine residues have been isolated from rabbit and rat livers (39,40). These receptor proteins also require $Ca^{2+}$ for binding. It is likely that other binding proteins involved in the clearance of glycoprotein ligands, with specificities different from those recognizing the above mentioned determinants, exist in mammalian cells and particularly in the liver.

Many of the properties of the hepatic binding proteins discussed earlier are reminiscent of the lectinlike substances. These substances, thought to be found in plants (seeds), are widely known to recognize and bind specific carbohydrate residues, causing reactions such as hemagglutination and mitogenic response. In view of the many similarities in properties, it has been proposed that the lectin terminology be extended to mammalian biological systems also in which they may be involved in generalized functions, such as recognition, differentiation, growth, and metabolism (41).

Active pinocytosis of several lysosomal enzymes, such as glycosidases, by fibroblasts is mediated by the presence of phosphorylated mannose (mannose 6-phosphate) as a recognition marker (42). Large glycoproteins (215,000 daltons) which recognize such determinants have been isolated from bovine liver membranes and also demonstrated to be present in several mammalian cells, such as rat hepatocytes, human fibroblasts, and Chinese hamster ovary cells (43).

This receptor–ligand system has great importance in the manifestation of a hereditary defect of glycoprotein metabolism called I-cell disease. In this instance, the lysosomal glycosidases are not taken up by the fibroblasts by active pinocytosis unlike normal cells (44). The lack of enzymes catalyzing the formation of mannose 6-phosphate, which can serve as the recognition signal for uptake, is a basic defect in I-cell disease.

A present view visualizes a major role for phosphomannosyl residues on newly synthesized acid hydrolases and of phosphomannose receptors on some membranes, which would allow segregation in the Golgi complex. In this organelle, the products that need to be contained within the cell (directed to lysosomes) are sorted out from those that do not have the recognition signals for retention and are destined to be exported (secreted) (44). In such an instance, the receptor–ligand system functions to prevent the loss of key enzymes from the lysosomes.

## VIII. GLYCOPROTEIN HORMONE–RECEPTOR INTERACTIONS

Among hormonal systems, the glycoproteins represent the most complex structures, characterized by the presence of two unique structural features. First, by the presence of high amounts of covalently linked carbohydrates and second, by a quaternary structure composed of an equimolar complex of $\alpha$–$\beta$ subunits, which are attached noncovalently. The subunits can be dissociated by a variety of conditions, such as acidic pH and denaturants such as 6 $M$ urea or guanidine hydrochloride. The union of both subunits is required for biological activity, because the individual subunits are inactive. Their specific interaction creates a unique conformation that is distinctly different from either of the component subunits and which can efficiently interact with the receptor. Many studies involving the modification of the nine or so reactive amino acid side chains in the polypeptide moiety in either or both of the subunits and selected peptide bonds, have conclusively demonstrated their essential nature for biological activity. Such studies have been very useful in mapping the intersubunit and hormone–receptor interaction sites (8,9).

Until recently, it was believed that the major role of carbohydrates in the glycoprotein hormones was only to maintain their stability in circulation. This has been clearly established in studies using asialohormones such as hCG, hLH, FSH, and eCG which were shown to lose their biological activity *in vivo*. The molecular mechanisms underlying this rapid clearance from circulation are mediated by the presence of galactose-specific receptors in the hepatocytes, as was discussed in the previous section. It was also shown that the asialohormones were active *in vitro,* suggesting that sialic acid residues were not essential for the induction of hormone response per se. [For review, see Sairam (8).]

The application of exoglycosidases to sequentially remove carbohydrate residues from a hormone such as hCG (26) revealed that additional sugars were not also required for binding to the receptor. The treatment also had some variable effects on cellular activation, depending upon the experimental model, and were thus inconclusive [reviewed in Sairam (8)]. In 1979, it was first demonstrated by us that ovine LH (45) could be stripped of most of its carbohydrate moiety by treatment with anhydrous HF (see Section VI,D), resulting in a modified hormone, the protein core of which remains unaltered. In the subsequent discussion, these modified preparations will be referred to as DG hormones* (DG-LH, DG-FSH, DG-hCG).

Treatment of gonadotropins with HF for about 60 min at 0°C is adequate for stripping approximately 75% of the carbohydrate moiety from the intact hormone and the subunits (Table IV). The chemical deglycosylation procedure is specific to the carbohydrate moiety and permits the isolation of the modified hormone in a lyophilized form and in a state completely free of the native hormone. Although these preparations are designated deglycosylated gonadotropins, it should be borne in mind that they are not completely free of carbohydrate residues. The *N*-acetylglucosamine residues still remain attached to the asparagine residues of the protein core. As none of the currently available methods can achieve sugar-

---

*Abbreviations: DG-LH, deglycosylated lutropin; DG-FSH, deglycosylated follitropin; and DG-hCG, deglycosylated human choriogonadotropin.

TABLE IV

**Carbohydrate Composition of Chemically Deglycosylated Gonadotropins**[a]

| Carbohydrate | Percentage remaining | | | |
| --- | --- | --- | --- | --- |
| | Sheep LH | Bovine LH | Sheep FSH | hCG |
| Hexoses | <3 | <5 | 14 | 13 |
| Glucosamine | 66 | 66 | 44 | 52 |
| Galactosamine | 0 | 0 | 0 | 95 |
| Sialic acid | 0 | 0 | 0 | 6 |
| Fucose | 0 | 0 | 0 | 0 |

[a] Values shown as percentage of sugar residue remaining after chemical deglycosylation by anhydrous HF. Adapted from Sairam (8).

free glycoprotein hormones, it is predicted that the more recent techniques of genetic engineering will permit the design of systems which can produce unglycosylated forms of the α and β subunits.

Most of our recent knowledge on the molecular mechanisms of hormone action in which carbohydrate has been shown to be critical arises from the use of DG hormones. The hormone action consists of a multitude of events in which an initial signal is amplified, producing the appropriate cellular response (Fig. 1). In gonadotropin-responsive cells, the parameters that can be easily and quickly measured by *in vitro* incubation procedures without the complicating factors of metabolism are (a) binding to the membrane receptors; (b) activation of adenylate cyclase, causing the accumulation of cyclic AMP; and (c) secretion of steroids, a response typical of the gonadotropins. This does not imply that other metabolic parameters are not influenced by the action of gonadotropins.

By comparing the effects of intact and deglycosylated gonadotropins in testicular or ovarian model systems, the requirement of carbohydrate in various steps of hormone action have been analyzed (8).

## A. Receptor Binding

All deglycosylated gonadotropins retain excellent binding ability to their specific receptors (Fig. 2). Our initial observations on ovine LH (oLH) in this regard have been confirmed and extended to other hormones like sheep FSH (46) and hCG in a number of laboratories (28,47–49). Deglycosylated hormones such as DG-hCG and DG-FSH even exhibit a higher degree of binding to the receptor as shown by their enhanced ability to compete with respective [125]I-labeled native hormones for binding to membrane receptors. This is clearly due to a 2- to 3-fold increase in the affinity of the DG hormones to receptors.

In addition to retaining good binding to their receptors, the deglycosylated hormones do not show any loss of receptor specificity as shown by the failure of excess unlabeled DG-hCG, DG-oLH, or DG-bLH, to compete with $^{125}$I-labeled oFSH for its own testicular receptor or vice versa. An example of this is illustrated in Fig. 3 in which it is shown that the same membrane preparation contains two specific and separate receptors for LH and FSH. These data provide conclusive evidence, showing that the complete integrity of the carbohydrate moiety in the native hormones is not critical for interaction with the receptor on gonadal membranes.

Removal of most of the bulky carbohydrate units from the subunits of gonadotropins does not impede recombination or receptor binding as was first

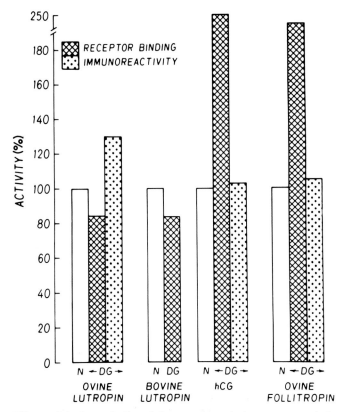

**Fig. 2.** Effects of deglycosylation of the gonadotropic hormones on their receptor-binding activity and immunoreactivity. These activities were measured in respective assays using the $^{125}$I-labeled native hormone as the tracer. The activity of the native hormone preparation is taken as 100%. The hormones tested were deglycosylated by anhydrous HF at 0°C for 60 min and purified. Abbreviations: N, native; DG, deglycosylated. [From Sairam (8).]

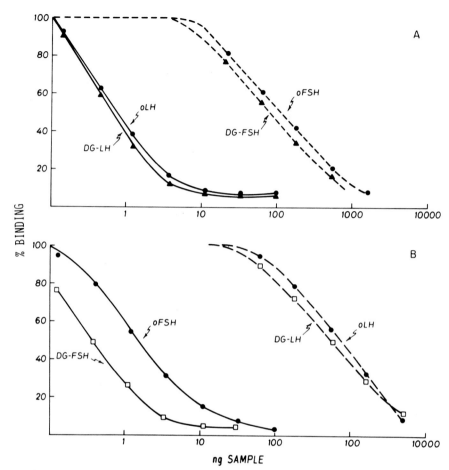

**Fig. 3.** Receptor specificity of the deglycosylated hormones. Using the same rat testicular membrane fraction which contains LH and FSH receptors (but on different cell types), competition assays were set up with $^{125}I$-labeled native hormone as tracers to verify the binding activity of the DG-LH and DG-FSH. The data show that there is no change in receptor-binding specificity following carbohydrate removal, i.e., (A) DG-FSH like FSH does not actively bind to the LH receptor or (B) DG-LH like LH fails to interact with the FSH receptor. The small degree of displacement (<0.1%), when compared to the active hormone, is judged to be due to cross contamination. Similar data have been obtained with mature ovarian granulosa cells—an instance in which both FSH and LH receptors are on the same cell. Such data demonstrate that the carbohydrate does not play a significant role in specifying receptor-binding specificity. This information appears, therefore, to be the exclusive domain of the polypeptide moiety of the two interacting α and β subunits.

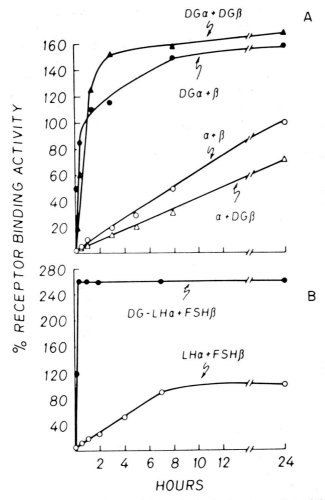

**Fig. 4.** Effect of deglycosylation of the individual subunits on their ability to reassociate. (A) Deglycosylated LH subunits were allowed to recombine with each other and the native counterparts in the following combinations: α + DGβ, DGα + β, α + β, DGα + DGβ. Preparations were incubated in 0.05 M phosphate buffer, pH 7.5 at 37°C. At different time intervals, aliquots were removed and stored at −20°C until activity determination in binding assay with LH receptor in testis membranes. Receptor-binding activity is normalized to the native hormone. (B) The LHα and DG-LHα were incubated with oFSHβ subunit and binding activity determined in specific FSH receptor assay. In both experiments, it is evident that deglycosylation of the α subunit greatly enhances its recombining ability with the native β subunit (M. R. Sairam and B. Prasad, unpublished observations).

shown with the subunits of ovine LH (50). The recombinants in which both subunits are deglycosylated have full receptor binding activity. Similar data are now available for hCG subunits (8,51). Our recent data on this aspect are even more revealing and discriminate the contributions of the carbohydrate in the α and β subunits. When the kinetics of recombination using the various combinations of the native and deglycosylated α and β subunits were examined, it was found that whenever the LHα was deglycosylated, its recombination with the β subunit was almost instantaneous, generating full (perhaps greater) receptor binding activity. Using the same DG-LH α subunit to recombine with FSHβ an identical pattern of data were obtained, but this time the complex bound specifically to the FSH receptor, because of the choice of the hormone-(FSH)-specific β subunit (Fig. 4).

Such studies prove beyond doubt that the carbohydrate residues of the ligand, in this instance hormones, do not directly participate in receptor-binding events.

## B. Coupling of Receptor–Adenylate Cyclase Unit

In the functioning cell or intact membrane unit, the receptor on the external surface is coupled to the adenylate cyclase complex in the inner membrane. Having established that carbohydrate residues of the glycoprotein hormones are not required for binding to the receptor, we examined their effect on adenylate cyclase or cyclic AMP accumulation.

All deglycosylated hormones, despite being fully active in binding to the receptor, have virtually lost all of their ability to activate the membrane-bound enzyme. Data depicted in Fig. 5 demonstrate that the three deglycosylated hormones fail to cause the accumulation of cyclic AMP in the medium even in the presence of effective phosphodiesterase inhibitors such as isobutylmethyl xanthine (8). Based on its close structural resemblance to gonadotropins and mechanism of action, it can be inferred that deglycosylated thyrotropin (TSH) may produce similar effects. Although this has not yet been achieved by chemical means, data derived from naturally occurring altered forms of TSH varying in their extent of glycosylation (52) lend support to the above conclusions. The observations on the lack of cyclic AMP accumulation have been confirmed by further studies on the actual assessment of adenylate cyclase by DG-hCG in rat testis (48) and DG-FSH in human and rat testis (53).

It was indicated earlier that deglycosylation of the α subunit enhanced its recombining ability with the β subunit as well its affinity to the receptor (Fig. 4). Analysis of the effects of these recombinants on cells shows that receptor binding and activation of adenylate cyclase can be clearly dissociated. A recombinant in which the α subunit is preferentially deglycosylated loses its ability to initiate intracellular events (Fig. 6). These data (Figs. 2–6) considered together provide conclusive evidence establishing that removal of the carbohydrate moiety of the

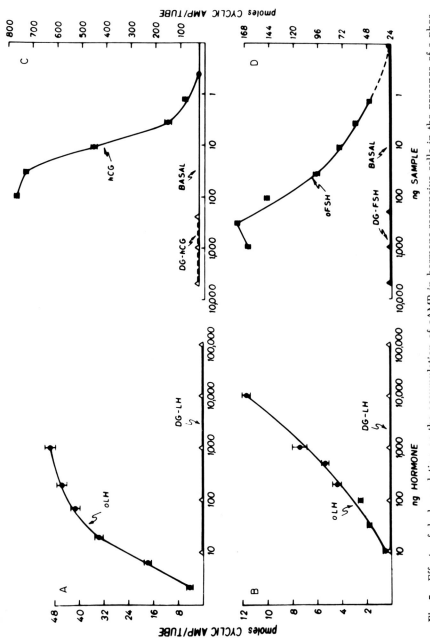

**Fig. 5.** Effects of deglycosylation on the accumulation of cAMP in hormone-responsive cells in the presence of a phosphodiesterase inhibitor. (A) Ovine LH and DG-LH in rat testicular cells, (B) same as (A) in ovarian cells, (C) hCG and DG-hCG in rat testicular cells, and (D) ovine FSH and DG-FSH in immature rat seminiferous tubular suspensions. [From Sairam (8).]

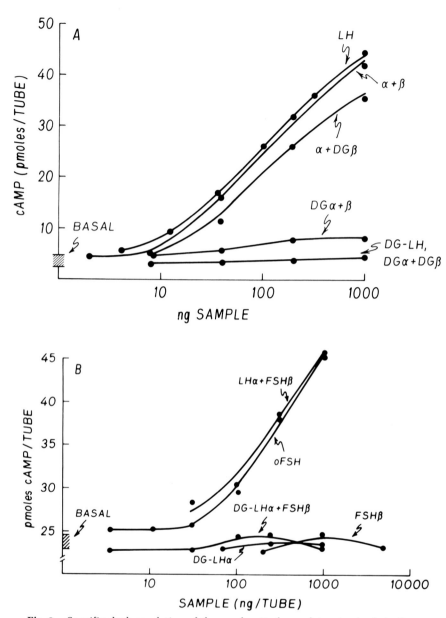

**Fig. 6.** Specific deglycosylation of the α subunit of gonadotropins leads to the uncoupling of the receptor–adenylate cyclase unit in testicular cells. The 24-hr recombinants of DG-LHα + LHβ (Fig. 4A) and DG-LHα + FSHβ (Fig. 4B) which had shown more than 100% receptor-binding activity were incubated with (A) LH- and (B) FSH-responsive testicular cells, respectively, and cyclic AMP accumulation after 30 min of incubation was measured. In both instances, the recombinant in which the α subunit is fully glycosylated can activate the adenylate cyclase. When about 75% carbohydrate residues are removed, as in chemically deglycosylated α, the recombinant virtually loses all such activity.

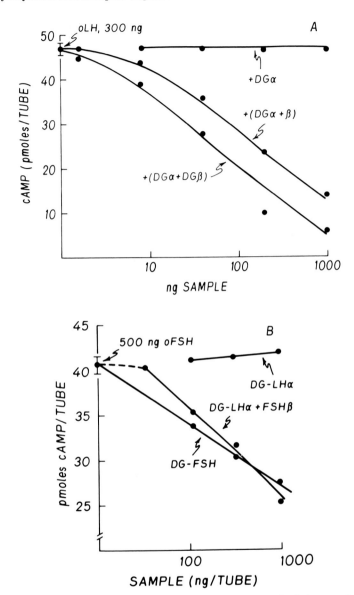

**Fig. 7.** Inhibition of hormone response by recombinants in which the gonadotropin α subunit is deglycosylated. Recombinants DG-LHα + LHβ and DG-LHα + FSHβ (see Figs. 6A and B) were tested for their influence on cyclic AMP accumulation, induced by the respective native hormones in testicular cells. (A) Cyclic AMP produced by 200 ng of ovine LH is considered 100% hormone response. (B) Ovine FSH (500 ng) was used as the challenging dose to test the effect of DG-LHα + FSHβ. Note that in both experiments the recombinant in which the α was deglycosylated inhibited the hormone response.

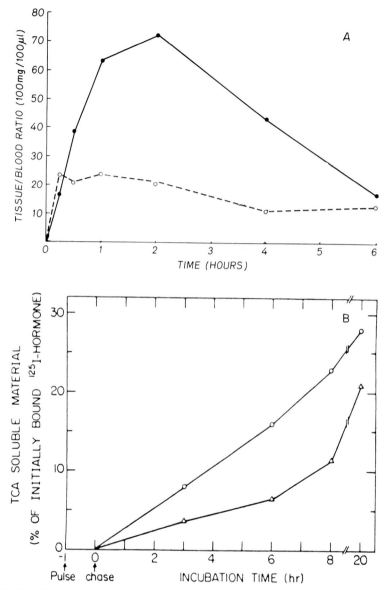

**Fig. 8.** (A) Comparison of the ovarian uptake (binding) of [125]I-hCG and DG-hCG in pseudopregnant rats. In this condition, the ovaries acquire the capacity to bind LH (hCG) in a specific manner. The labeled hormones were administered by the intracardiac route and at different intervals the radioactivity in the ovary and blood was measured and expressed as a ratio. The significant difference observed in the early uptake (2 hr) of hCG (solid lines) and DG-hCG (broken lines) is due to the loss of sialic acid which has a protective role in hCG. Despite the reduced uptake, it is important to note that a greater

common α subunit leads to uncoupling of the receptor–adenylate cyclase system. The mechanism by which this is effected is not understood at present and will no doubt be the object of intensive investigations.

By virtue of their discordant effects on receptor binding and the activation of adenylate cyclase, all of the deglycosylated hormones are effective antagonists of the action of native hormones in the respective systems, as reviewed recently (8). There is no loss of receptor-binding specificity after deglycosylation of the hormones and consequently inhibition can be obtained only with the appropriate antagonist, e.g., DG-hCG inhibits hCG or LH action and not FSH action and DG-FSH will antagonize only FSH action.

Since the DG hormones do not affect cyclic AMP accumulation in gonadal cells *in vitro* caused by various nonhormonal stimuli, such as choleragen (8), prostaglandin $E^2$, isoproterenol (54), forskolin, and fluoride (53) some of which act by nonreceptor-mediated mechanisms, there can be no doubt that antagonism is caused by a specific blockade at the receptor on the cell surface.

Recent data from our studies reveal that loss of carbohydrate from the common α subunit alone is adequate to induce an uncoupling of the receptor–adenylate cyclase system (Figs. 4 and 6). Consequently, the recombinant of DG α plus native β in each case is an antagonist of the native hormone (Fig. 7). The same preparation of deglycosylated α subunit (e.g., DG-LHα as in Fig. 7), when recombined with fully glycosylated LHβ or FSHβ, acquires the ability to antagonize LH and FSH action, respectively, in specific cells. Because of their structural and functional similarities, it may be expected that the same DG-LHα, when associated with the native TSHβ, would generate a specific TSH antagonist. This demonstrates that the carbohydrate moiety on the α subunit is of critical importance in transduction of the hormonal signal.

## C. Other Biological Effects of Deglycosylation of Hormones

In addition to exhibiting antagonistic properties *in vitro,* two deglycosylated gonadotropins have been shown to be active *in vivo,* as shown by their ability to inhibit ovulation (8,55) and terminate pregnancy in the rat (56). Despite possibly

---

percentage of radioactive DG-hCG remains bound to the receptor (compare 15 min and 6 hr). In [125]I-hCG, the disappearance of radioactivity from 2 to 6 hr is faster, indicating higher metabolic disposal at the target cell. (B) Differences in the degradation of hCG (○) and DG-hCG (△) by granulosa cells *in vitro*. The experiment examines the time course of degradation of receptor bound [125]I-hCG or [125]I-DG-hCG. Granulosa cells in cultures were pulse labeled with the radioactive ligands for 1 hr and then chased in a hormone-free medium. Medium collected at noted time intervals was subjected to 10% trichloroacetic acid precipitation. Radioactivity in the TCA soluble fraction which is an indication of hormone degradation was determined. At 8 hr, the rate and degradation of DG-hCG was about half of that of hCG. [From Zor et al. (58).]

enhanced clearance from circulation, as in the case of DG-hCG, the native form of which is heavily glycosylated enough of the antagonist must reach the receptor *in vivo* to exert a blockade of hormone action. Evidence gathered from experiments in which the uptake of labeled oLH/DG-LH and hCG/DG-hCG were compared in pseudopregnant rats demonstrated that the deglycosylated hormones that interact with the receptor remains bound to the receptor longer than the labeled native hormone (Fig. 8A), giving rise to the speculation that the metabolic fate of the DG hormones may differ from that of the native counterparts. There is now direct experimental evidence supporting this (Fig. 8B). The intracellular fate of $^{125}$I-labeled ligands such as hCG can be studied by high-resolution autoradiography (57). Granulosa cells incubated with $^{125}$I-hCG or $^{125}$I-DG-hCG handled the two ligands quite differently, with $^{125}$I-DG-hCG about 61% of the specifically bound ligand was retained on the cell membrane in 60 min. In contrast, only 28% of the labeled hormone was retained when the cells were treated with $^{125}$I-hCG. This was also reflected in the intracellular accumulation of labeled ligand in the lysosomes, where much less $^{125}$I-DG-hCG was found compared to $^{125}$I-hCG (58). Similar results have been reported with mouse Leydig tumor cells (59).

There can be no doubt that the deglycosylated hormones have provided an excellent tool in the form of antagonists to investigate the mechanism of action of the glycoprotein hormones. DG gonadotropins have now been used to differentiate receptor occupancy and such phenomena as down-regulation and desensitization (54,58). DG-LH or DG-hCG cause only moderate (58) or no desensitization (59) in responsive cells. The deglycosylated gonadotropins acting in a highly specific manner prevent desensitization induced by the homologous hormones.

Deglycosylation studies performed during the last 4 years have conclusively demonstrated the importance of the carbohydrate moiety in the biological function of the hormone at the cellular level. Whereas most alterations of the protein core of these hormones [see Sairam (8) for review] destroy even the receptor binding which is the first step in hormone action, major modifications in carbohydrates (reduction by 80%) induces discriminatory effects on the various events, as discussed earlier. While the precise mechanisms are not yet known, several speculations are fertile ground for further studies. It is highly unlikely that the loss of a major portion of the carbohydrate causes drastic alterations in the overall conformation of the hormones, as revealed by biophysical studies (8). This is also supported by the fact that a highly conformation-dependent event such as recognition by the receptor is also fully preserved (and increased in several instances). Carbohydrate removal from the hormone (or for that matter from the α subunit alone) could alter the conformational changes that occur in the hormone–receptor complex in such a manner as to affect the lateral mobility of the complex along the membrane (60) which might couple with the adenylate cyclase unit in the inner membrane. Another possibility is that the receptor–GTP regulatory protein complex is not formed in the presence of the deglycosylated hormone. If this were so, the increase in the binding of guanyl nucleotides to the plasma membranes that is observed in the presence of LH (61) either does not occur or is suboptimal when the DG gonadotropin is present, so as to induce the coupling.

## IX. ROLE OF CARBOHYDRATE IN RECEPTOR ASSEMBLY AND FUNCTION

As noted earlier, the plasma membrane of eukaryotic cells serves several important roles, including receptor function. Most, if not all, cell surface receptor proteins are believed to contain covalently bound carbohydrates. Many of these have also been shown to consist of subunits, a proper assembly of which may be required for appropriate function and transmission of the ligand signal. The association of carbohydrate with receptors has been demonstrated by one or more of the following approaches: (i) by isolation of the receptor and subunits and subsequent chemical analysis; (ii) by incorporation of labeled sugars into membrane protein in cultures; and (iii) employing compounds which preferentially inhibit glycosylation reactions of cells in culture.

Because of their complexity and the fact that the receptors have to be coupled with other membrane proteins to achieve the biological purpose of signal transmission, the functional role of the carbohydrate moiety cannot be directly assessed by performing experiments with isolated receptor-protein preparations. Such questions can at the present time be best answered by employing inhibitors of glycosylation such as tunicamycin (see Section VI,A). Indeed, several elegant studies already performed with receptor systems such as acetylcholine and insulin have revealed the important role of carbohydrates in receptor assembly as well as function. It may be useful to review some of these observations which serve as models for studies on other systems.

The acetylcholine receptor (Ach-R) of the skeletal muscle is a membrane glycoprotein that mediates the reception of neural impulses by muscle cells at vertebrate neuromuscular junctions. Its biochemical and pharmacological properties have been extensively investigated (1,62). The Ach-R which contains approximately 5% carbohydrate (by weight) is an oligomer of four polypeptide chains ($\alpha$, $\beta$, $\gamma$, and $\delta$) assembled into a complex of $\alpha_2-\beta-\gamma-\delta$ (63). At least two of the four polypeptides of the Ach-R contain N-glycosidically linked oligosaccharide chains (64).

The role of glycosylation in the biogenesis of Ach-R has been studied using tunicamycin (23,64,65), an antibiotic which inhibits the N-glycosylation of proteins. Treatment of muscle cells which normally synthesize Ach-R with tunicamycin produces several important effects. The glycosylation of protein is almost completely inhibited (see Table II), and the accumulation of Ach-R on the cell surface, as determined by specific binding of ligands such as $^{125}$I-bungarotoxin to intact muscle cell is reduced (64). In addition, Ach-R that can be extracted by detergents is also reduced significantly (23). Glycosylation of the protein may protect the newly synthesized Ach-R from cellular degradative processes, permitting the accumulation of such receptor molecules on the cell surface of responsive cells. In the absence of glycosylation, the Ach-R accumulated on the cell surface apparently gets degraded faster by proteolytic enzymes, a process which can be inhibited by common protease inhibitors such as leupeptin (23). While it may be generally true that nonglycosylated proteins may be more susceptible to proteolytic digestion, more recent studies on Ach-R assembly question this view (64) and propose that this can only be a partial explanation.

A dramatic reduction in the Ach-R on cells caused by incomplete glycosylation has now been attributed to the failure of nonglycosylated subunits of the receptor to assemble into a functional unit which can bind the ligand (64). When glycosylation is inhibited, the $\alpha$ subunit of the Ach-R does not acquire the toxin binding sites. Furthermore, inhibition of glycosylation by tunicamycin is thought to prevent the transport of sub-units to the Golgi where they are normally assembled into matrix Ach-R (66). Yet another differing view suggests that impairment of protein glycosylation results in the expression of Ach-R with an altered conformation which now displays different metabolic and functional properties (65). While there may be differences in the interpretation of the data, there can be little doubt as to the importance of glycosylation in the cells' ability to acquire a functional Ach-R.

Glycosylation also appears to be one of the important processing steps in the synthesis of the insulin receptor. Culture of 3T3-L1 adipocytes with tunicamycin causes rapid depletion of cell receptors (67), presumably by reducing the half-life. Now there is evidence to suggest that inhibition of N-glycosylation is accompanied by the depletion of cell surface and total detergent-extractable insulin receptors, owing to the reduction in the formation of newly synthesized insulin receptor. Glycosylation of the receptor appears to be essential for its activation after translation (68).

In cultured IM-9 human lymphocytes, tunicamycin causes a time- and dose-dependent decrease in the expression of insulin and growth hormone receptors (69). A deficiency of glycosylation caused a decrease in receptor numbers without altering the affinity.

Tunicamycin should prove useful in many other studies involving receptors, as it could provide an approach by which nonglycosylated receptor populations can be produced and investigated. However, in such studies care should be taken to account for the possible decrease in the amount of receptors which could arise as a result of a true decrease in the assembly of new receptor or the enhanced degradation of susceptible naked receptor molecules or both.

## X. CONCLUDING REMARKS

In most cellular systems, proteins undergo extensive and often complex series of cotranslational and posttranslational events which may include enzymatic processing at specific sites, covalent modification, or compartmentalization into discrete sites within or outside of the cell. For most receptor proteins localized on the cell membrane and several secreted proteins (ligands), glycosylation represents an important covalent modification that could significantly affect biological function, as discussed in this chapter.

The precise role(s) of carbohydrate moiety in glycoproteins has not been completely understood, but there is evidence supporting each one of the following points of view: (a) the carbohydrate residues serve as recognition markers both on the cell surface and on ligands; (b) targetting of macromolecules, e.g., some enzymes to organelles such as lysosomes; (c) conformational stabilization of the protein; (d) assembly of receptor oligomeric structures; and (e) biological activity.

While there are numerous examples showing that the carbohydrate moiety of many glycoprotein enzymes, such as deoxyribonuclease, ribonuclease, carboxypeptidase Y, or interferon, is not essential for biological activity, most recent evidence gathered from our own studies as well as those of others and highlighted in this article conclusively show for the first time the critical role of the carbohydrate moiety not in the recognition of the receptor but in the transduction of the biological signal contained in the glycoprotein hormone molecule into the interior of the cell. It appears to be involved in the coupling of the glycoprotein hormone receptor to the adenylate cyclase. Because of this failure in inducing coupling, the deglycosylated hormones are biologically inactive. Thus, in these hormones we have an example to show that complete glycosylation is not essential for receptor interaction. The carbohydrate moiety of the $\alpha$ subunit but not that of the $\beta$ subunit plays a critical role in effecting the coupling. Such differential characteristics render the deglycosylated hormones to be antagonists of the action of native hormones. Since deglycosylation can alter the biological profile of the glycoprotein hormones not only in circulation but also at the cellular level, it may be speculated that such phenomena could be involved in the regulation of hormone activity and cell function, e.g., down-regulation and desensitization.

Aberrant glycosylation, or the lack of it, can interfere with either the assembly or the expression and/or stability of the plasma membrane (or intracellular) receptors, as noted with examples such as acetylcholine and insulin receptors. There can be no doubt that extension of these studies to other model systems will generate new and interesting data, expanding our knowledge on the role of glycosylation in receptor–ligand interactions. The ability to clone receptors and ligands provides a new approach of tailoring altered structures which could theoretically eliminate glycosylation sites. Tumor cells which are spontaneous or induced, having altered forms of receptors or secreting ligands modified in the carbohydrate moiety, should also provide useful models in delineating the role of glycosylation.

NOTE ADDED IN PROOF

Consistent with the hypothesis that lateral mobility of the hormone receptor-gonadotropin complex may be involved in coupling to cyclase (60) Bahl *et al.* have recently found (70) that deglycosylated hCG (which is an antagonist of hCG action) does not cause aggregation of receptors following binding to responsive cells in the rat corpus luteum. The conformational alterations caused by deglycosylation of hCG which induce a change in its biological profile can be partially restored by the addition of an antibody to the hormone (71).

**ACKNOWLEDGMENTS**

The collaboration of Dr. Bhargavi Prasad in some of the experiments noted in this chapter is greatly appreciated. Part of the work in the author's laboratory was aided by the MRC of Canada.

**REFERENCES**

1. Heidmann, T., and Changeux, J. P. (1978). Structural and functional properties of the acetylcholine receptor protein in its purified and membrane-bound states. *Annu. Rev. Biochem.* **47**, 317–357.
2. Weill, C. L., McNamee, M. G. and Karlin, A. (1974). Affinity-labeling of purified acetylcholine receptor from *Torpedo carlifornica. Biochem. Biophys. Res. Commun.* **61**, 997–1003.
3. Dattatreyamurthy, B., Rathnam, B., and Saxena, B. B. (1983). Isolation of the luteinizing hormone–chorionic gonadotropin receptor in high yield from bovine corpora lutea. *J. Biol. Chem.* **258**, 3140–3158.
4. Shorr, R. G. L., Lefkowitz, R. J., and Caron, M. G. (1981). Purification of the β-adrenergic receptor. *J. Biol. Chem.* **256**, 5820–5826.
5. Ashwell, G., and Morell, A. G. (1974). The role of surface carbohydrates in the hepatic recognition and transport of circulating glycoproteins. *Adv. Enzymol. Relat. Areas Mol. Biol.* **41**, 99–128.
6. Pricer, W. E., Jr., and Ashwell, G. (1976). Subcellular distribution of a mammalian hepatic binding protein specific for asialoglycoproteins. *J. Biol. Chem.* **251**, 7539–7544.
7. Parsons, T. F., Bloomfield, F. A., and Pierce, J. G. (1983). Purification of an alternate form of the α subunit of the glycoprotein hormones from bovine pituitaries and identification of its O-linked oligosaccharide. *J. Biol. Chem.* **258**, 240–244.
8. Sairam, M. R. (1983). Gonadotropic hormones: Relationship between structure and function with emphasis on antagonists. *In* "Hormonal Proteins and Peptides" (C. H. Li, ed.), Vol. XI, pp. 1–79. Academic Press, New York.
9. Pierce, J. G., and Parsons, T. F. (1981). Glycoprotein hormones: structure and function. *Annu. Rev. Biochem.* **50**, 465–495.

10. Hughes, R. C. (1976). "Membrane glycoproteins," pp. 6–27, 269–284. Butterworth, London.
11. Subtelny, S., and Wessells, N. K. (1980). "The Cell Surface: Mediator of Developmental Processes." Academic Press, New York.
12. Hynes, R. O. (1976). Cell surface proteins and malignant transformation. *Biochim. Biophys. Acta* **458**, 73–107.
13. Yamada, K. M., and Olden, K. (1978). Fibronectins—adhesive glycoproteins of cell surface and blood. *Nature (London)* **275**, 179–184.
14. Yamada, K. M., Yamada, S. S., and Pastan, I. (1976). Cell surface protein partially restores morphology, adhesiveness and contact inhibition of movement to transformed fibroblasts. *Proc. Natl. Acad. Sci. U.S.A.* **73**, 1217–1221.
15. Carter, W. G., Fukuda, M., and Hakomori, S. (1978). Chemical composition, gross structure and organization of transformation-sensitive glycoproteins. *Ann. N.Y. Acad. Sci.* **312**, 160–177.
16. Hynes, R. O., and Fox, C. F. (1980). "Tumor Cell Surfaces and Malignancy." Liss, New York.
17. Neufeld, E. F., and Ashwell, G. (1980). Carbohydrate recognition systems for receptor-mediated pinocytosis. *In* "The Biochemistry of Glycoproteins and Proteoglycans" (W. J. Lennard, ed.), pp. 241–266. Plenum, New York.
18. Kornfeld, R., and Kornfeld, S. (1976). Comparative aspects of glycoprotein structure. *Annu. Rev. Biochem.* **45**, 217–237.
19. Takatsuki, A., Arima, K., and Tamura, G. (1971). Tunicamycin, a new antibiotic. I. Isolation and characterization of tunicamycin. *J. Antibiot.* **24**, 215–223.
20. Tkacz, J. S., and Lampen, J. O. (1975). Tunicamycin inhibition of polyisoprenyl N-Acetyl glucosaminyl pyrophosphate formation in calf-liver microsomes. *Biochem. Biophys. Res. Commun.* **65**, 248–257.
21. Kuo, S. C., and Lampen, J. O. (1976). Tunicamycin inhibition of [$^3$H]glucosamine incorporation into yeast glycoproteins: binding of tunicamycin and interaction with phospholipids. *Arch. Biochem. Biophys.* **172**, 574–581.
22. Struck, D. K., and Lennarz, W. J. (1977). Evidence for the participation of saccharide-lipids in the synthesis of the oligosaccharide chain of ovalbumin. *J. Biol. Chem.* **252**, 1007–1013.
23. Prives, J. M., and Olden, K. (1980). Carbohydrate requirement for expression and stability of acetylcholine receptor on the surface of embryonic muscle cells in culture. *Proc. Natl. Acad. Sci. U.S.A.* **77**, 5263–5267.
24. Taratakoff, A., Hoessli, D., and Vassalli, P. (1981). Intracellular transport of lymphoid surface glycoproteins. Role of the Golgi complex. *J. Mol. Biol.* **150**, 525–535.
25. Reichert, L. E., Jr., and Abou-Issa, H. (1976). Properties of FSH receptors in rat gonadal tissue. *In* "Hormone-Receptor Interaction Molecular Aspects" (G. Levey, ed.), pp. 153–170. Dekker, New York.
26. Moyle, W. R., Bahl, O. P., and Marz, L. (1975). Role of the carbohydrate of human chorionic gonadotropin in the mechanism of hormone action. *J. Biol. Chem.* **250**, 9163–9169.
27. Parsons, T. F., and Pierce, J. G. (1980). Oligosaccharide moieties of glycoprotein hormones: bovine lutropin resists enzymatic deglycosylation because of terminal O-sulphated N-acetylhexosamines. *Proc. Natl. Acad. Sci. U.S.A.* **77**, 7089–7093.
28. Goverman, J. M., Parsons, T. F., and Pierce, J. G. (1982). Enzymatic deglycosylation of the subunits of chorionic gonadotropin. Effects on formation of tertiary structure and biological activity. *J. Biol. Chem.* **257**, 15059–15064.
29. Mort, A. J., and Lamport, D. T. A. (1977). Anhydrous hydrogen fluoride deglycosylates glycoproteins. *Anal. Biochem.* **82**, 289–309.
30. Manjunath, P., and Sairam, M. R. (1985). Chemical deglycosylation of glycoprotein hormones. *In* "Methods in Enzymology Hormone Action" (L. Birnbaumer and B. O'Malley, eds.), Vol. 109, 725–735. Academic Press, New York.

31. Manjunath, P., Sairam, M. R., and Schiller, P. (1982). Chemical deglycosylation of ovine pituitary lutropin. A study of the reaction conditions and effects on biochemical, biophysical and biological properties of the hormone. *Biochem. J.* **207**, 11–19.
32. Burnet, F. M. (1951). Mucoproteins in relation to virus action. *Physiol. Rev.* **31**, 131–150.
33. Tomita, M., and Marchesi, V. T. (1975). Amino acid sequence and oligosaccharide attachment sites of human erythrocyte glycophorin. *Proc. Natl. Acad. Sci. U.S.A.* **72**, 2964–2968.
34. Paulson, J. C., Sadler, J. E., and Hill, R. L. (1979). Restoration of specific myxovirus receptors to asialoerythrocytes by incorporation of sialic acid with pure sialyltransferases. *J. Biol. Chem.* **254**, 2120–2124.
35. Van Hall, E. V., Vaitukaitis, J. L., Ross, G. T., Hickmann, J., and Ashwell, G. (1971). Immunological and biological activity of hCG following progressive desialylation. *Endocrinology* **88**, 456–464.
36. Hudgin, R. L., Pricer, W. E., Jr., Ashwell, G., Stockert, R. J., and Morell, A. G. (1974). The isolation and properties of a rabbit liver binding protein specific for asialoglycoproteins. *J. Biol. Chem.* **249**, 5536–5543.
37. Kawasaki, T., and Ashwell, G. (1976). Chemical and physical properties of an hepatic membrane protein that specifically binds asialoglycoproteins. *J. Biol. Chem.* **251**, 1296–1302.
38. Tanabe, T., Pricer, W. E., and Ashwell, G. (1979). Subcellular membrane topology and turnover of a rat hepatic binding protein specific for asialoglycoproteins. *J. Biol. Chem.* **254**, 1038–1043.
39. Kawasaki, T., Eton, R., and Yamashina, I. (1978). Isolation and characterization of a mannan-binding protein from rabbit liver. *Biochem. Biophys. Res. Commun.* **81**, 1018–1024.
40. Townsend, R., and Stahl, P. (1981). Isolation and characterization of a mannose/N-acetyl-glucosamine/fucose-binding protein from rat liver. *Biochem. J.* **194**, 209–214.
41. Ashwell, G. (1977). A functional role for lectins. *Trends Biochem. Sci.* **2**, N186.
42. Kaplan, A., Achord, D. T., and Sly, W. S. (1977). Phosphohexosyl components of a lysosomal enzyme are recognized by pinocytosis receptors on human fibroblasts. *Proc. Natl. Acad. Sci. U.S.A.* **74**, 2026–2030.
43. Sahagian, G., Distler, J., and Jourdian, G. N. (1981). Characterization of a membrane-associated receptor from bovine liver that binds phosphomannosyl residues of bovine testicular β-galactosidase. *Proc. Natl. Acad. Sci. U.S.A.* **78**, 4289–4293.
44. Neufeld, E. F., Lim, T. W., and Shapiro, L. J. (1975). Inherited disorders of lysosomal metabolism. *Annu. Rev. Biochem.* **44**, 357–376.
45. Sairam, M. R., and Schiller, P. W. (1979). Receptor binding, biological and immunological properties of chemically deglycosylated pituitary lutropin. *Arch. Biochem. Biophys.* **197**, 294–301.
46. Manjunath, P., Sairam, M. R., and Sairam, J. (1982). Studies on pituitary follitropin. X. Biochemical, receptor binding and immunological properties of deglycosylated ovine hormone. *Mol. Cell. Endocrinol.* **28**, 125–138.
47. Manjunath, P., and Sairam, M. R. (1982). Biochemical, biological and immunological properties of chemically deglycosylated human choriogonadotropin. *J. Biol. Chem.* **257**, 7109–7115.
48. Chen, H. C., Shimohigashi, Y., Dufau, M. L., and Catt, K. J. (1982). Characterization and biological properties of chemically deglycosylated human chorionic gonadotropin. Role of carbohydrate moieties in adenylate cyclase activation. *J. Biol. Chem.* **257**, 14446–14452.
49. Kalyan, N. K., and Bahl, O. P. (1983). Role of carbohydrate in human chorionic gonadotropin. Effect of deglycosylation on the subunit interaction and on its *in vitro* and *in vivo* biological properties. *J. Biol. Chem.* **258**, 67–74.
50. Sairam, M. R. (1980). Deglycosylation of ovine pituitary lutropin subunits. Effects on subunit interaction and hormone activity. *Arch. Biochem. Biophys.* **204**, 199–206.
51. Kalyan, N. K., and Bahl, O. P. (1981). Effect of deglycosylation on the subunit interactions and

receptor binding activity of human chorionic gonadotropin. *Biochem. Biophys. Res. Commun.* **102**, 1246–1253.

52. Joshi, L. R., and Weintraub, B. D. (1983). Naturally occurring forms of thyrotropin with low bioactivity and altered carbohydrate content act as competitive antagonists to more bioactive forms. *Endocrinology* **113**, 2145–2154.

53. Berman, M. I., Srivastava, M. A., and Sairam, M. R. (1985). Characterization of gonadotropin-sensitive adenylate cyclase activity in human testis: uncoupling of the receptor-cyclase complex by specific hormonal antagonist. *Mol. Cell Endocrinol.* (in press).

54. Zor, U., Shentzer, P., Azrad, A., Sairam, M. R., and Amsterdam, A. (1984). Deglycosylated luteinizing hormone (LH) prevents desensitization of cyclic adenosine monophosphate response by LH. Dissociation between receptor uncoupling and down regulation. *Endocrinology* **114**, 1954–1959.

55. Kato, K., and Sairam, M. R. (1983). Inhibition of ovulation in the rat by a hCG antagonist. *Contraception* **27**, 515–520.

56. Kato, K., Sairam, M. R., and Manjunath, P. (1983). Inhibition of implantation and termination of pregnancy in the rat by a human chorionic gonadotropin antagonist. *Endocrinology* **113**, 195–199.

57. Amsterdam, A., Nimrod, A., Lamprecht, S. A., Burstein, Y., and Linder, H. R. (1979) Internalization and degradation of receptor-bound hCG in granulosa cell cultures. *Am J. Physiol.* **236**, E129–138.

58. Zor, U., Sairam, M. R., Shentzer, P., Azrad, A., and Amsterdam, A. (1984). The role of carbohydrate moiety of the gonadotropin molecule in transduction of biological signal. *In* "Hypothalamo-Pituitary-Gonadal Axis" (K. W. McKerns and Z. Naor, eds.), pp. 235–248. Plenum, New York.

59. Rebois, R. V., and Fishman, P. H. (1983). Deglycosylated human chorionic gonadotropin. An antagonist to desensitization and down-regulation of the gonadotropin receptor-adenylate cyclase system. *J. Biol. Chem.* **258**, 12775–12778.

60. Amsterdam, A., Berkowitz, A., Nimrod, A., and Kohen, F. (1980). Aggregation of luteinizing hormone receptors in granulosa cells: a possible mechanism of desensitization to the hormone. *Proc. Natl. Acad. Sci. U.S.A.* **77**, 3440–3444.

61. Dufau, M. L., Baukal, A. J., and Catt, K. J. (1980). Hormone-induced guanyl nucleotide binding and activation of adenylate cyclase in the Leydig cell. *Proc. Natl. Acad. Sci. U.S.A.* **77**, 5837–5841.

62. Fambrough, D. M. (1979). Control of acetylcholine receptors in skeletal muscle. *Physiol. Rev.* **59**, 165–227.

63. Raftery, M. A., Hunkapiller, M. W., Strader, C. B. D., and Hood, L. E. (1980). Acetylcholine receptors: complex of homologous subunits. *Science* **208**, 1454–1456.

64. Merlier, J. P., Sebbane, R., Tzartos, S., and Lindstrom, J. (1982). Inhibition of glycosylation with tunicamycin blocks assembly of newly synthesized acetylcholine receptor subunits in muscle cells. *J. Biol. Chem.* **257**, 2694–2701.

65. Prives, J., and Sagi, D. B. (1983). Effect of tunicamycin, an inhibitor of protein glycosylation, on the biological properties of acetylcholine receptor in cultured muscle cells. *J. Biol. Chem.* **258**, 1775–1780.

66. Fambrough, D. M., and Devreotes, P. N. (1978). Newly synthesized acetylcholine receptors are located in the golgi apparatus. *J. Cell. Biol.* **76**, 237–244.

67. Rosen, O. M., Chia, G. H., Fung, C., and Rubin, C. S. (1979). Tunicamycin-mediated depletion of insulin receptors in 3T3-LI adipocytes. *J. Cell Physiol.* **99**, 37–42.

68. Reed, B. C., Ronnett, G. V., and Lane, M. D. (1981). Role of glycosylation and protein synthesis in insulin receptor metabolism by 3T3-LI mouse adipocytes. *Proc. Natl. Acad. Sci. U.S.A.* **78**, 2908–2912.

69. Keefer, L. M., and De Meyts, P. (1981). Glycosylation of cell surface receptors. Tunicamycin treatment decreases insulin and growth hormone binding to different levels in cultured lymphocytes. *Biochem. Biophys. Res. Commun.* **101,** 22–29.
70. Bahl, O. P., Thotakura, N. R., and Anumula, K. R. (1984). Biological role of carbohydrates in gonadotropins. *In* "Hormone Receptors in growth and reproduction" (B. B. Saxena, K. J. Catt, L. Birnbaumer, and L. Martini, eds.), pp. 165–183. Raven, New York.
71. Rebois, R. V., and Fishman, P. M. (1984). Antibodies against human chorionic gonadotropin convert the deglycosylated hormone from an antagonist to an agonist. *J. Biol. Chem.* **259,** 8087–8090.

# 10

# Role of Steroid Hormone Receptors in Development and Puberty

**B. D. GREENSTEIN AND I. M. ADCOCK**
Department of Pharmacology
St. Thomas's Hospital Medical School
London, United Kingdom

## I. INTRODUCTION

Steroid hormones play a critical role in the development and proper function of many organs, including the brain. The importance of the steroids in development has been recognized for a long time, but in recent years, there have been major advances in our knowledge of the processes which underlie the steroid influence in this respect. It is hoped here to provide a coherent review of the work done and to identify unifying threads in the studies carried out on different tissues. Earlier studies have been reviewed in a previous article (Greenstein, 1978a).

**341**

## Mechanism of Steroid Action

According to current dogma, steroids travel in the bloodstream either freely or largely bound to plasma proteins. In the latter state, they are physiologically inert and their half-life is prolonged (Westphal, 1972). The steroids dissociate from the plasma proteins and diffuse freely into the tissue. Within the cell, there are cytoplasmic receptor proteins with which the steroid reacts to form a tight but not irreversible bond. The act of union between steroid and receptor appears to alter the conformation of the protein, and the complex is able to enter the nucleus where it interacts with the genome. As a result, transcription is initiated and *de novo* protein synthesis occurs. The exact nature of the nuclear events is still unknown but may involve binding of the steroid–receptor complex to DNA "upstream" of the gene-promoter region (Parker, 1983). The scheme is shown diagrammatically in Fig. 1, and for a good critical review of the theory, the reader is referred to Gorski and Gannon (1976).

This "two-step" model of cytoplasmic–nuclear receptor action, originally proposed by Jensen *et al.* (1968), has recently been challenged in two studies which failed to detect cytoplasmic receptors (King and Greene, 1984: Welshons *et al.*, 1984) (Fig. 1). This controversy is probably a side issue in that the important aspects are the hormonal specificity of the initial reaction between

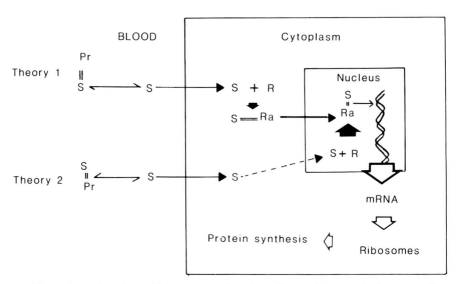

**Fig. 1.** Theories of steroid hormone action. According to theory 1, the hormone (S) dissociates from its plasma protein (Pr) and diffuses into the cell. The steroid binds to a cytoplasmic receptor protein (R) which is thereby activated (Ra), and the complex is translocated to the nucleus. According to more recent evidence (see text), the steroid hormone may diffuse into the nucleus where it binds to its receptor (theory 2).

steroid and receptor and subsequent gene expression. That the steroid–receptor complex plays a role in this process is assuredly beyond question.

Clearly, for the normal expression of steroid action, all the components of the system must be functional; a major problem facing the reviewer is in deciding whether the identification of the primary point of interaction, namely the receptor protein, necessarily constitutes definitive evidence that a developing cell is competent to respond normally to the steroid. Therefore, whenever possible, a functional link between the receptor and the cellular or whole-animal response has been aimed for here.

## II. GLUCOCORTICOID RECEPTORS

### Peripheral Tissues

The major natural adrenal glucocorticoids are cortisol, also known as hydrocortisone, corticosterone, cortisone, and 11α-deoxycorticosterone. Potent synthetic glucocorticoids include dexamethasone, prednisolone, and triamcinolone.

The glucocorticoids influence a wide range of metabolic functions (Schulster *et al.*, 1976), and it is hardly surprising that this class of steroids has a potent and sometimes adverse influence on the development of many tissues.

In the adult, glucocorticoids have an anabolic effect on the liver and enhance the activity of enzymes such as tyrosine aminotransferase and tryptophan dioxygenase (Beato *et al.*, 1972). Significantly, glucocorticoids do not induce tyrosine monooxygenase in the fetal liver (Sereni *et al.*, 1959; Holt and Oliver, 1968; Yeoh *et al.*, 1979). Nevertheless, the fetal liver does contain cytosolic glucocorticoid receptors (Giannopoulos *et al.*, 1974; Giannopoulos, 1975a,b), and these appear to be capable of being bound by the cell nucleus (Feldman, 1974; Parchman *et al.*, 1978). Yet even the presence of nuclear receptors is not proof that the receptor system is functional. Nuclear binding may occur without the normal (i.e., adult) patterns of genomic response. The situation is also complicated by the possibility that the cell might contain heterogeneous populations of glucocorticoid-binding proteins (Schmid and Grote, 1977), and these may differ qualitatively in fetal and adult liver (Kalimi and Gupta, 1982); this may express itself as a lower affinity for the hormone (Giannopoulos, 1975a,b; Parchmen *et al.*, 1978).

The picture seems equally complicated in other tissues. Sakly and Koch (1981) studied the ontogenesis of glucocorticoid receptors in the rat anterior pituitary gland and included a welcome attempt to relate nuclear binding of the steroid to pituitary function in terms of inhibition of adrenocorticotropic hormone (ACTH) secretion *in vitro*. In that study, nuclear uptake of [³H] dexamethasone in neonates was about one third of that in adults, with a correspondingly lower functional activity. Despite extensive perfusion of the whole animal prior to tissue preparation, they also detected in cytosols from neonates corticosterone-binding protein resembling the plasma steroid-binding corticosteroid-binding protein (CBG). In a subsequent study (Sakly and Koch, 1983), they postulated that this CBG-like protein might inhibit nuclear uptake of the steroid which would explain differences between neonates and adults in this respect. For other evidence of CBG-like proteins in rat pituitary cytosols, see de Kloet and McEwen (1976) and Koch et al. (1976).

The brain, too, possesses glucocorticoid receptors, particularly in the limbic system (McEwen et al., 1969, 1970; Stevens et al., 1971; Gerlach and McEwen, 1972; Rhees et al., 1975), and in this organ, too, there appears to be a CBG-like protein present in neonatal cytosols, although it is absent from cytosols prepared from brains of adult rats (Al-Khouri and Greenstein, 1980).

The neonatal receptors seem to have a lower affinity for glucocorticoids and display a broader range of binding selectivities (Clayton et al., 1977), but it remains to be seen whether these age-related changes influence the neural actions of the glucocorticoids.

In summary, therefore, it seems that some fetal or neonatal tissues may possess an immature glucocorticoid receptor system which binds the steroid with lower affinity and specificity than its adult counterpart(s). Nevertheless, these neonatal receptor systems do respond to glucocorticoid challenge.

Administration of corticosterone to immature mice stimulated activity of the enzyme tyrosine monooxygenase in the locus coeruleus of the brain (Markey et al., 1982). Glucocorticoids also accelerated surfactant production and lung maturation in developing animals (DeLemos et al., 1970; Kikkawa et al., 1971), and the immature lung does possess glucocorticoid receptors (Giannopoulos et al., 1972; Toft and Chytil, 1973). Glucocorticoids can also induce cleft-palate formation in experimental animals, and Yoneda and Pratt (1981) demonstrated the presence of glucocorticoid receptors in human embryonic palatal mesenchyme cells. These authors found that dexamethasone, a potent synthetic glucocorticoid, inhibited cell growth in culture, and they suggested that glucocorticoids might potentially be able to interfere with palate formation. The power of the glucocorticoids to suppress cellular function is also well illustrated by the inhibition of the growth of primary cultures of neonatal mouse dermal fibroblasts by corticosterone (Verbruggen and Solomon, 1980).

The above studies, while establishing a role for the glucocorticoid receptor, reveal nothing about the factors which determine its appearance and activity. But a novel and interesting avenue of research has been explored by Moscona and associates (Linser et al., 1982). They studied the precocious induction by glucocorticoids of glutamine synthetase in the Müller glia cells of the embryonic chick neural retina. The enzyme, which is involved in neurotransmitter metabolism, was inducible in cell cultures which also contained glucocorticoid receptors; but if the tissue was dissociated, the dispersed single cells were not inducible for glutamine synthetase and glucocorticoid receptors were reduced by about 75%. Reassociation partially restored enzyme inducibility and receptor concentrations.

The importance of cell–cell contacts in the induction of cellular function, including that of steroid receptor activity, is also underlined by similar studies of the prostate (see Section III,A); the identification of the factors which mediate intercellular correspondence are of considerable importance for the development not only of normal but also of pathological tissues and deserves more attention.

## III. ANDROGEN RECEPTORS

## A. Normal Development of Peripheral Tissues

Testosterone is the major circulating androgen of the male. The hormone is synthesised in the Leydig cells of the testis in response to luteinizing hormone (LH), an anterior pituitary gland gonadotropic hormone. In many species, including man, testosterone is converted in the target cell to a more active androgenic metabolite, $5\alpha$-dihydrotestosterone (DHT) by the $5\alpha$-reductase system (Wilson, 1975). Testosterone also serves as the substrate for target-organ synthesis of $17\beta$-estradiol (Fig. 2), and this dichotomy of testosterone metabolism subserves a divergence of roles, especially in developing tissues. The testis also secretes estradiol, although this constitutes only about 15% of the estrogen available to the target cell; the vast majority is generated by aromatization of testosterone (MacDonald et al., 1979).

Once in the bloodstream, testosterone is bound to the plasma protein sex hormone-binding globulin (SHBG) and to albumin, from which it dissociates and diffuses into the cell. Once intracellular, the hormone may bind to a specific receptor or first be metabolized before being bound (Fig. 3). This model will underpin the remainder of the section.

The functional significance of separate testosterone and DHT receptors is not fully understood, but it is believed that testosterone per se acts directly on the hypothalamo-hypophyseal axis to regulate LH secretion, on the testis to regulate spermatogenesis, and to stimulate the Wolffian duct during internal differentiation. DHT is thought to effect differentiation of the external genitalia and virilization at puberty (Wilson, 1978). Testosterone also causes sexual differentiation of the fetal brain, and the active metabolite of testosterone in this respect appears to be estradiol (see Section VI).

Receptors for androgens appear very early in cellular differentiation, and the influences which dictate their onset of synthesis are being clarified through work being done on the prostate gland. In the fetus, androgen of testicular origin is

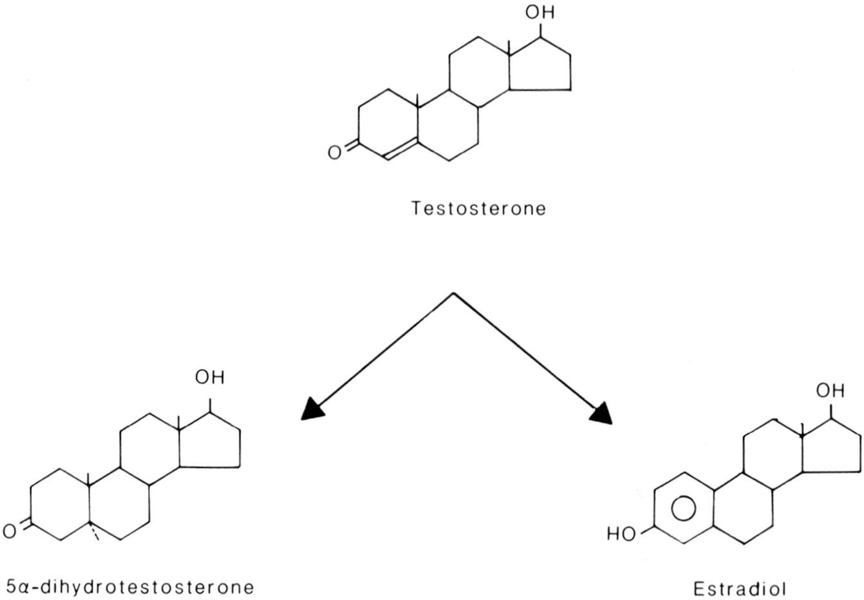

Testosterone

5α-dihydrotestosterone                                              Estradiol

**Fig. 2.** Metabolism of testosterone in target tissues. As well as being a hormone in its own right, testosterone serves as a prohormone and is metabolized both to 5α-dihydrotestosterone, a powerful androgen, and to estradiol.

taken up by the parenchymal cells which are then able to induce morphogenesis of the prostatic secretory epithelium (Cunha, 1976). The morphogenic induction by the mesenchyme of the epithelium almost certainly includes as one of its components the induction of prostatic epithelial androgen receptors. Cunha et al. (1980) showed that urogenital mesenchyme from embryonic mice was able to induce nuclear androgen receptors in the embryonic urinary bladder, which actually developed prostate-like acini when associated with the mesenchyme (Cunha and Lung, 1979). This form of tissue interaction and induction of steroid receptors may well prove to be a general feature of receptor development, and one wonders, for example, about possible glial influences on neuronal steroid-receptor induction and vice versa.

Once the androgen receptor in the prostate has been induced and provided sufficient testosterone is available, normal development and virilization ultimately occur. The ontogenesis of cytoplasmic androgen receptors has been studied in the rat ventral prostate gland (Rajfer et al., 1980), and there appeared to be a peak of available cytoplasmic receptors at about day 50 of age, i.e., around the time of male puberty in the rat.

Androgen target cells have also been described in the anterior pituitary and associated mesenchyme of the chick embryo (Gasc et al.,1979). These receptors

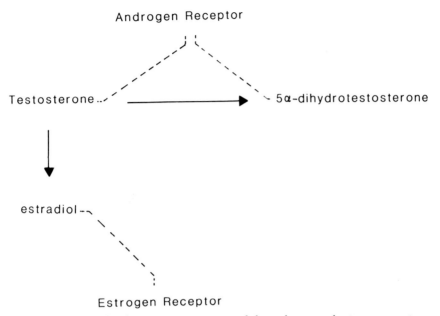

**Fig. 3.** Relationship between testosterone and the androgen and estrogen receptors. All of these primary sites of metabolism and tissue interaction must be functional for normal sexual differentiation of peripheral and neural tissues.

bound DHT and possibly testosterone but not estradiol. The possible role of androgens in the fetal pituitary gland, not only of birds but also of mammals, remains to be elucidated. It may be speculated, however, that apart from proliferative effects, the androgens (and possibly other steroids) may influence the eventual sensitivity of specific pituitary cells which respond to fluctuations in the concentrations of gonadal steroid and hypothalamo-hypophyseal hormones (see Section VI).

Studies on experimental animals point to the presence of androgen receptors in gonadal and accessory sex organs long before they are called into play to mediate the virilizing actions of testosterone and its androgenic metabolites at puberty. For example, Fichman *et al.* (1981) measured cytosolic and nuclear androgen receptors in human preputial skin samples obtained after elective circumcision. Receptors were predominantly localized to the nuclei in newborn subjects, during puberty, and also in adults when circulating androgens were high. In tissue samples from prepubertal boys, receptors were present virtually exclusively in the cytosol. Apart from providing further evidence for the theory of cytosolic–nuclear receptor interactions, that study by implication underscores the perils associated with the inappropriate absence or presence of androgens at critical stages of accessory sex organ development and during the prepubertal period.

## B. Androgen Receptors and Clinical Disorders

Disorders of sexual development have been usefully classified depending on the level of development at which the lesion occurs (Wilson and Goldstein, 1975). According to this classification, disorders are chromosomal, gonadal, or phenotypic. An example of a disorder of the chromosomes is Klinefelter's syndrome, which is primary testicular failure associated with an additional X chromosome, yielding an XXY karyotype. The problem is due to nondysjunction during meiosis. A gonadal sex disorder results from the failure or overactivity of the testes during development, and an example is cryptorchidism. But of primary interest here are the disorders of phenotypic sex, when the gonad appears to function normally but the target organs of 46, XY males do not respond to androgens. The condition was originally detected in patients with male pseudohermaphroditism, and three possible sites of the lesion have been identified so far. The 5α-reductase system which converts testosterone to DHT (see Fig. 2; Section III, A) may be deficient or absent and at birth these patients have a minimal masculinization of their external genitalia, although phallic enlargement does occur at puberty (Peterson et al., 1977).

Resistance may occur as a result of androgen receptor deficiency. Thus, Keenan et al. (1974) found a virtually complete absence of DHT receptors in fibroblasts cultured from patients with complete testicular feminization.

Alternatively, the androgen receptor population may be quantitatively normal but the receptor may be structurally deficient or unstable (Griffin, 1979). The receptor may lose affinity for the androgen. This was found in ventral prostate glands of aging rats (Greenstein, 1979) (Fig. 4). Nevertheless, androgen resistance may occur despite apparently normal 5α-reductase and androgen receptor systems and is presumed to involve a resistance of the genome to the androgen–receptor complex (Griffin and Wilson, 1980). In addition, there may be lesions at the hypothalamo-hypophyseal level. Whatever the level of the disorder, this field of clinical research provides an excellent illustration of the clinical usefulness of the androgen receptor and it is to be hoped that more general use will be made of this diagnostic tool.

## IV. ESTROGEN RECEPTORS

The modern theories of the mechanism of steroid action stem from the introduction of radioactive estrogens of high specific activity. The first report was that of Glascock and Hoekstra (1959) who used tritiated hexestrol, a synthetic estrogen, to show for the first time the localization of radioactivity in estrogen-responsive organs of lambs and goats. Jensen and Jacobsen (1962) used [$^3$H]estradiol to demonstrate the specific uptake and retention of radioactivity by

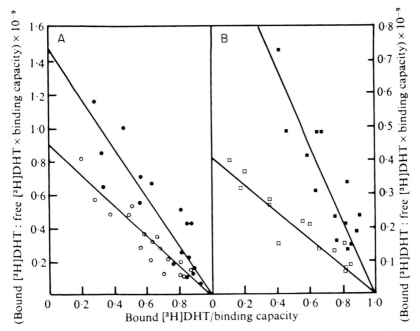

**Fig. 4.** Scatchard plots of the high-affinity binding of dihydrotestosterone (DHT) in (A) amygdaloid cytosol from 12-week-old rats orchidectomized 0.75 (●) or 7 (○) days previously and (B) in cytosol from the ventral prostate glands of 12-week-old (■) or 18-month-old (□) orchidectomized rats. The slope of the plot provides the molar association constant ($K_a$, liter/mole) and the abscissa describes the bound hormone as a fraction of the binding capacity. [From Greenstein (1979).]

the uterus and vagina after administration of the radioactive hormone to immature rats, and since then, a monumental amount of literature devoted to estrogen receptors has accumulated. For a careful and thorough review of the earlier work on estrogen receptors, see James and Fotherby (1970).

## Peripheral Tissues

The cytosolic estrogen receptor appears to be a useful index of the estrogenic responsiveness in developing tissues, although not as useful as a study of responsiveness itself. This distinction is well exemplified by studies of the onset of estrogenic responses in the embryonic chick liver (Lazier et al., 1981) and the Müllerian duct (Teng, 1980) where specific cellular products can be identified. From these studies, it appears that a major increase in responsiveness occurs from about the tenth day of embryonic development. Lazier et al. (1981) measured a marked increase in soluble nuclear estrogen receptors in embryonic chick liver from day 10 to 12 of development, and this increase corresponded nicely with the increased synthesis of vitellogenin and the apoproteins B and II of very low-density lipoproteins. The cytosol receptors increased more gradually from days 10 to 20 of development.

The factors which determine the development of estrogen receptors appear to vary with the tissue, since unlike the liver, chick Müllerian cytosolic estrogen receptors attain maximal levels at day 12 (Teng and Teng, 1975). Nevertheless, the Müllerian duct requires more prolonged estrogen treatment for cell differentiation, and this is probably because the liver is highly developed at an early stage with respect to estrogen-responsive parenchymal cells (Lazier et al., 1981).

Estrogen-concentrating nuclei have been localized in embryonic chick gonads (Gasc, 1980) and in the urogenital tract using the technique of autoradiography (Gasc and Stumpf, 1981). [$^3$H]Estradiol-concentrating cells were observed in the germinal epithelium of the left but not the right gonad in both sexes. This fits in with the known ovarian differentiation of the left gonad of birds which occurs as a result of estrogen-stimulated proliferation of the gonadal cortex (Wolff, 1950). In the autoradiographic studies of Gasc and Stumpf (1981), estrogen-concentrating cells were detected at $5\frac{1}{2}$ days of incubation, although the tissue may not have been responsive to estrogens at this stage. In this context, larvae of the amphibian *Xenopus laevis* contain adult-type estrogen receptors before the onset of responsiveness to estradiol (May and Knowland, 1981).

Earlier studies of the development of estrogen receptors in mammalian tissues such as the uterus have been reviewed (Eisenfeld, 1972; Greenstein, 1978a). Essentially, estrogen receptors in all tissues examined appeared to be present before birth. The factors governing normal uterine receptor development remain unknown. The importance of the mesenchyme in androgen receptor induction in epithelium has been mentioned (Section III,A) and a similar mechanism may operate in the uterus. Cunha et al. (1982), in an autoradiographic study of [$^3$H]estrogen uptake by mouse uterine nuclei during development, observed only a mesenchymal nuclear concentration of radioactivity from days 1 to 15 postpartum, while epithelial cells took up radioactivity from day 20 postpartum.

Normal estrogen receptor ontogenesis appears to occur independently of steroid hormones, but this development can be influenced by the gonads. Powell-Jones et al. (1981) reported a sexual differentiation of high-capacity estrogen-binding proteins in rat liver which could be abolished by neonatal castration of the male. Testosterone has been shown to inhibit estrogen receptors in the immature rat ovary (Saiduddin and Zassenhaus, 1978).

The neonatal and immature uterine estrogen receptor system and the uterus itself are apparently fully prepared to respond to estrogens. The preweanling rat has relatively high plasma concentrations of estradiol (MacKinnon et al., 1978) but the uterus is virtually vestigial in appearance. Yet, repeated doses of estradiol will produce adult-type uterine responses (Sheehan et al., 1981). This apparent paradox is explained by the presence in preweanling rat plasma of an estradiol-binding protein, α-fetoprotein (AFP) (Raynaud et al., 1971; Raynaud, 1973) which sequesters estradiol in a bound, inert form in the plasma. This estradiol is presumably of maternal origin.

In considering the ontogenesis of estrogen receptors, it might also be germaine to include more studies of the rate of recycling or regeneration of the receptor (Alvarez and Hancke, 1978). The recycling or "replenishment", as it is some-times called, can be adversely affected by progesterone, as has been shown in fetal guinea pig uterus (Sumida et al., 1981). These workers have also demonstrated the ready responsiveness of the fetal uterus to administered estradiol (Sumida and Pasqualini, 1980).

## V. PROGESTERONE RECEPTORS

The study of the progesterone interaction with mammalian target tissues has presented problems for three main reasons. First, there exists in the blood of most, if not all, mammals a protein [corticosteroid-binding globulin (CBG)] which binds progesterone at least as avidly as it does the corticosteroids (Westphal, 1972). This protein is, therefore, an interfering contaminant in cytosolic preparations. The problem is compounded by the presence, extravascularly and possibly even intracellularly, of a progesterone-binding CBG-like protein in the uterus, which appears to be distinct from the true progesterone receptor (Milgrom and Baulieu, 1970; Milgrom et al., 1970; Al-Khouri and Greenstein, 1980). The function of this protein is unknown, but there is evidence that it may serve as a "sink" for the hormone (B. D. Greenstein, unpublished observations).

A second problem is the central position occupied by progesterone in steroid metabolism in target tissues. Although there is very little evidence that progesterone is a "prohormone," the possibility cannot be ruled out. Third, progesterone receptors in neural and peripheral tissues form extremely labile bonds with the hormone in vitro (Philibert and Raynaud, 1973; Greenstein, 1978b). This problem and that of the interference by CBG were to some extent overcome by the introduction of a potent progestin, promegestone (R5020; Roussel-Uclaf) which in low concentrations appears not to bind CBG (Philibert and Raynaud, 1973).

The actions of progesterone and of its receptors cannot be considered in isolation, since in many instances these require the prior exposure of the tissue to estradiol. An important action of estradiol in this respect is its stimulant effect on progesterone receptor induction. In neural tissues this results in a facilitation of estrous behavior in laboratory animals (Boling and Blandau, 1939; Frank and Fraps, 1945), and in peripheral tissues in, for example, typical progesterone-induced changes in the uterine endometrium, and in the suppression of lactogenesis (Kuhn, 1969).

All the above factors must be considered when studying the development of progesterone-sensitive systems, and much remains to be done in this area.

## Peripheral Tissues

Undoubtedly, the tissue most studied in this respect has been the chick oviduct which produces well-defined proteins, such as ovalbumin, in response to estrogen priming and progesterone administration *in vivo* (O'Malley *et al.*, 1977). Estradiol stimulates progesterone receptor synthesis and, after entry into the cell by diffusion, progesterone binds to its receptor. The "activated" complex binds to the nuclear chromatin. The exact mechanism is unknown, but the net result is the appearance some 2 hr later of specific mRNA species (Spelsberg and Toft, 1976). The mechanism appears to be operational in the newly hatched chick, but a detailed investigation of the ontogenesis of the many components of this system appears to be lacking to date. Interestingly, there appears to be a circannual rhythm in steroid receptor levels and nuclear binding in the chick oviduct when receptor levels are depressed during winter and early spring (Spelsberg and Halberg, 1980).

An ontogenetic study of progestin-binding sites in guinea pig fetal uterus and ovary has been done by Pasqualini and Nguyen (1980). These putative progesterone receptor sites were undetectable before 37–42 days of gestation and gradually increased in number until parturition. Injection of estradiol into the mother resulted in a marked increase in these sites even at the earlier times when progesterone binding sites were normally undetectable. It would be interesting to know just how early in uterine development this effect of estradiol can be induced.

While estradiol stimulates progesterone receptor synthesis, progesterone appears to have an inhibitory effect on its own receptors. In the postpartum rat, uterine progesterone receptors are lower in number in suckling than in nonsuckling rats (Gomez *et al.*, 1977). Similarly, progestin receptors were present in mammary glands of virgin and pregnant mice but decreased as gestation progressed and were undetectable in mammary glands during the suckling period (Haslam and Shyamala, 1979).

Other steroids, notably the androgens, may influence progesterone receptor turnover. Schmidt and Katzenellenbogen (1979) were able to induce uterine growth with androgens which apparently were acting via androgen receptors but were also able to induce progesterone receptors. Relatively high doses of DHT were needed, however, and the androgen might also have been binding to the estrogen receptor in order to induce progesterone receptors (Schmidt and Katzenellenbogen, 1979).

Despite the large amount of work done on progesterone receptors in normal adult and neoplastic tissues, there is still scope for the study of the development of the progesterone receptor systems in peripheral tissues, particularly with respect to actions at the genomic level.

## VI. GONADAL STEROID RECEPTORS AND BRAIN DEVELOPMENT

In the developing brain, there appear to be many complex interactions between the various classes of gonadal hormones, and it is probably best if their receptors are considered together.

### Sexual Differentiation of the Brain

There has been intense interest in the role of testicular androgens in the sexual differentiation of those areas of the diencephalon which control male and female sexual behavior and which mediate gonadotropin release from the anterior pituitary gland (for reviews see Greenstein, 1978a; Thomas and Knight, 1978; McEwen, 1981a; MacLusky and Naftolin, 1981; Dudley, 1981; Arnold and Gorski, 1984).

From anatomical studies, there are grounds to believe that areas of the fetal brain associated with normal reproduction and sexual behavior in the adult are differentiated with respect to structural organization (Raisman and Field, 1973; Gorski et al., 1978). This differentiation, not only of synaptic volume, but also of certain enzyme systems, is to a large extent effected by androgen of testicular origin which enters the brain and then, curiously, is converted to estradiol which is an active metabolite in this respect (MacLusky and Naftolin, 1981). The mechanism of action of the steroid is unknown, but may involve protection of developing neuronal processes (Kolata, 1979). The area apparently most affected is the medial preoptic area anterior to the hypothalamus (Fig. 5). This area, in the rat at any rate, appears to be crucial in the control of cyclic release of luteinizing hormone from the anterior pituitary gland and in the mediation of mounting behavior in the adult male rat. Gorski et al. (1978) showed a sexual dimorphism of an area in the diencephalon, which they termed the sexually dimorphic nucleus of the preoptic area (SDN–POA). This group went on to show that after treatment of the neonatal female with testosterone, the SDN–POA subsequently resembled that of the male (Arnold and Gorski, 1984).

The current model of sexual differentiation of the brain is that the fetal testis secretes testosterone, prenatally in the primates and perinatally in the rat. The hormone travels to the brain where it is converted to estradiol (see Fig. 3) which binds to its receptors and as a result there are permanent changes in neuronal

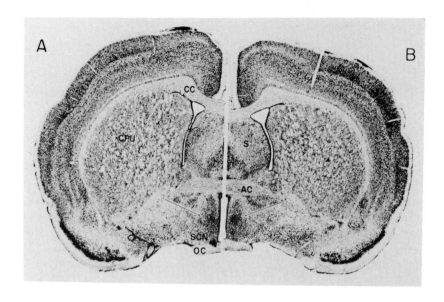

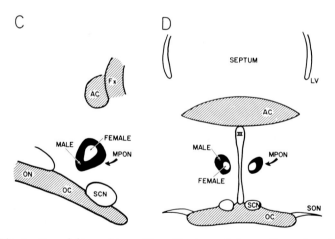

**Fig. 5.** Sexual dimorphism of the rat brain medial preoptic nucleus. Coronal sections through the adult female (A) and male (B) rat sacrificed 2 weeks after gonadectomy. Arrows indicate that portion of the medial preoptic nucleus that exhibits a marked sexual dimorphism (both at same magnification). The absence of the suprachiasmatic nucleus (SCN) in (B) is an artifact of the plane of tissue sectioning. Localization of the sexually dimorphic component of the medial preoptic nucleus (MPON) is shown in the sagittal (C) and coronal (D) planes. The nucleus of the female rat is drawn completely within the volume of the nucleus of the male. Drawn from magnified slides (×37). Abbreviations: AC, anterior commissure; CC, corpus callosum; CPU, caudate putamen; OC, optic chiasm; S, septum; Fx, fornix; LV, lateral ventricle; ON, optic nerve; SON, supraoptic nerve; and III, third ventricle. [From Gorski et al. (1978).]

growth and certain enzyme systems, possibly related to neurotransmitter turnover. Certainly estrogen receptors have developed in the diencephalon of both primates and rodents in time for the release of testosterone by the fetal testis (Barley *et al.*, 1974; MacLusky *et al.*, 1979), and these receptors are qualitatively similar to those in the adult brain (Salaman *et al.*, 1976). Also, the necessary enzymes are present in the diencephalon for the aromatization of testosterone to estradiol (MacLusky and Naftolin, 1981).

This intriguing action of testosterone on the developing brain is not confined to mammals but occurs in birds as well. In an elegant series of experiments (Konishi and Gurney, 1982), it was shown that there was a close relationship between sexual differentiation of the vocal system of the zebra finch (only the male sings) and cellular differentiation of neurones which control the vocal system. These studies also support the "aromatization" theory of sexual differentiation of the brain. The theory is neat and apparently self-contained but has undoubtedly oversimplified the situation.

The complexity of the sexual differentiation of the brain is hinted at by the finding that adult male rats are less sensitive than females to estrogen in eliciting female sexual behavior, while females are less sensitive than are males to estrogen in eliciting male sexual behavior (Pfaff, 1970; Pfaff and Zigmond, 1971; Whalen *et al.*, 1971; McEwen, 1981b). Thus, we can distinguish between defeminizing and masculinizing effects of testicular androgen on the fetal male brain (Whalen, 1974). Interestingly, it seems that mounting behavior is controlled in the preoptic area of the rat brain (Christensen and Clemens, 1974), while lordosis (female) sexual behavior in the rat is controlled in the medial basal hypothalamus (Christensen and Gorski, 1978). Both of these areas contain estrogen receptors in relatively great abundance, and in a recent study, Nordeen and Yahr (1983) measured nuclear estradiol receptors in these brain areas in rats. They found, essentially, that less estradiol was bound in male than in female medial basal hypothalamus, and that this brain region could be "defeminized" if the females received testosterone neonatally. In the preoptic area, the situation was more complex in that the sex difference was dependent on the time after initial exposure to estradiol. It remains to be seen whether the differences they observed are functionally related to the behavioral effects of the steroid.

Far less attention has been paid to the factors which might influence the postnatal development of estradiol receptors, and an illustration of the scope offered is provided by Barbanel and Assenmacher (1982) who reported that neonatal hypothyroidism markedly suppressed estrogen receptor development, especially in the pituitary gland, and this was independent of serum estrogen levels. The implications for the clinical situation are obvious. Clearly more work of this kind is needed.

Testosterone and other steroids have been overshadowed by estradiol in developmental studies of the brain, and it can be safely predicted that testosterone will be shown to play an important role as a hormone in its own right in brain development. Receptors which bind both testosterone and its powerful androgenic metabolite DHT do occur in the adult brain (Ginsburg *et al.*, 1974; Barley *et al.*, 1975; Kato, 1975), and these can alter their responsiveness to androgens with aging (Greenstein, 1979). But androgen receptors have also been detected in fetal tissues, for example, in the chick embryo pituitary gland (Gasc *et al.*, 1979) and in the neonatal rodent brain (Attardi and Ohno, 1976; Kato, 1976; Vito *et al.*, 1979; Lieberburg *et al.*, 1980). Their role(s) (if any) remain to be elucidated.

Unquestionably, the most difficult neural estrogen receptor to find and characterize has been that for progesterone, and there is still controversy about its identity. Progesterone is active in the brain in mediating neuronal electrical activity (Terasawa and Sawyer, 1970), sexual function and behavior (Morin and Feder, 1974), and protein synthesis in neural tissue (Wade and Feder, 1974). Yet, many attempts to find a neural progesterone receptor were frustrated or produced equivocal results. Eventually, it was discovered that progesterone forms a highly labile bond *in vitro* with its receptors (Moguilewsky and Raynaud, 1977; Greenstein, 1978b), and this problem was partially overcome by the introduction of a potent progestin R5020 (see Section V). The progestin does

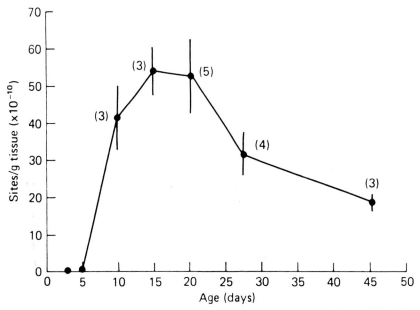

**Fig. 6.**  Ontogenesis of high-affinity progesterone binding in cytosols from brains of developing female rats (mean ± SEM). Figures in brackets indicate number of experiments. [From Greenstein, (1978a).]

not form as labile a bond with the receptor (Raynaud *et al.*, 1976). Using R5020, several different workers described progestin-binding systems in the rat hypothalamus, pituitary gland, and uterus which could be induced *in vivo* by estrogen priming (MacLusky and McEwen, 1978; Moguilewsky and Raynaud, 1979), and this effect of estrogen on progesterone receptors has now also been observed using as tracer the natural hormone (Al-Khouri and Greenstein, 1982).

Progesterone receptors have also been measured in the brains of developing rats (Fig. 6) (Greenstein, 1978a; MacLusky and McEwen, 1980), and in both studies these were found to increase with postnatal development, although they were detected earlier in development in the latter study.

Does progesterone play a role in the developing brain? It has been reported that progesterone "protects" the fetal brain from the effects of testosterone (Kincl and Maqueo, 1965). It has also been claimed that progesterone exposure in the fetus influences subsequent educational attainment during childhood (Dalton, 1976). At this stage it is too premature to assign a role for these progesterone-binding sites in these or other actions of progesterone during development. Progesterone receptors in the brain may lie dormant until the onset of puberty.

## VII. PUBERTY

It is perhaps fitting to end this review with a discussion of the role of steroid hormones and their receptors in arguably the most important developmental change, the onset of puberty. Generally, there has been a preoccupation with the unknown factors which occur around the time of puberty onset, but from more recent experiments (see later), it is perhaps more important to look at the more gradual changes both in the blood and in the gonads and accessory sex organs before puberty. We should also concentrate more on those manipulations that will delay, prevent, or precociously induce puberty.

### Theories of Puberty and the Prepubertal Internal Milieu

In both sexes, the gonads lie seemingly dormant during a period termed sexual immaturity and within this timespan they can be brought to maturity to precipitate precocious puberty. Therefore, the hypothalamo-hypophyseal–gonadal axis presumably is mature well before the onset of puberty but is held in check. The brake appears to be a lack of active sex hormones. Ramirez and Sawyer (1965) were able to induce precocious puberty in female rats by injecting them with estradiol.

The dominant theory for several years has been that in the prepubertal female rat the hypothalamo-hypophyseal axis is more sensitive to the inhibitory effect of estrogens on LH release from the pituitary gland. This theory is supported by experiments in which lower doses of estradiol were required to suppress the elevated circulating levels of LH after castration of prepubertal rats (see, e.g., Steele, 1977). This theory is known as the "altered gonadostat" theory, and its main proponents have been Ramirez and McCann (1963) and Donovan and Van Der Werff Ten Bosch (1965). There is remarkably little evidence to support the idea that the hypothalamus or anterior pituitary alter their intrinsic sensitivity to the inhibitory effects of circulating hormones around the time of puberty, and from more recent studies, there is evidence that any apparent change in sensitivity is simply a reflection of permissive changes on an already mature hypothalamo-hypophyseal axis allowed by the maturing ovary (Greenstein, 1978a; Ojeda *et al.*, 1981).

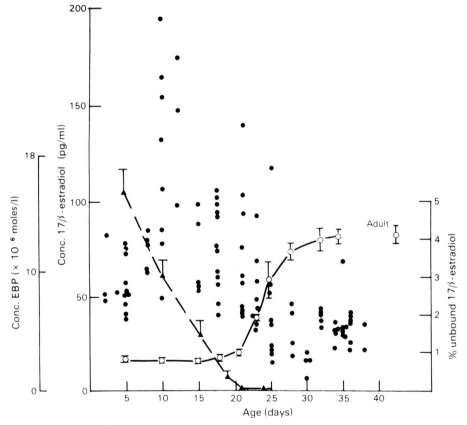

**Fig. 7.** Age-related changes in concentrations of serum estradiol-binding α-fetoprotein (▲), the free fraction of estradiol in serum (○), and of total serum estradiol concentrations (●) during development of the female rat (mean ± SEM). [From Puig-Duran *et al.* (1979).]

In the prepubertal rat, during the first 2 weeks of postnatal life, there is an apparently anomalous situation in that plasma levels of estradiol and of follicle-stimulating hormone (FSH) are raised, when it is known that the inhibitory effect of estradiol on pituitary release of FSH has already matured (MacKinnon *et al.*, 1978). But this anomaly is readily explained because the estradiol in these animals is bound to the plasma protein AFP (see Section IV) and is, therefore, unavailable to the tissues. When AFP levels in plasma decline, at about 21–23 days of age, the circulating estradiol enters the tissues and is metabolized. At the same time, the free and, therefore, physiologically active levels of estradiol rise dramatically in the plasma (Fig. 7) (Puig-Duran *et al.*, 1979) and exert their inhibitory effect on FSH secretion, and a fall in plasma FSH occurs as well at this time. In fact, if 21-day-old female rats are injected with AFP-rich serum, the free fraction of estradiol is depressed and FSH levels are kept elevated (Puig-Duran *et al.*, 1979).

Although there seems to be little evidence at present, it may be predicted that estradiol begins to exert its effects on the uterus and ovary from about 21–23 days of age in the rat. The rise in free estradiol which occurs in the plasma at this time may be the trigger which sets in train an inexorable series of ovarian events leading to puberty (Greenstein, 1978a; Ojeda *et al.*, 1981). As stated above, more evidence is needed, but it can be postulated that under the influence of raised FSH, coupled with the appearance of free estradiol in plasma, development of FSH and LH receptors is triggered in the ovary which in turn produces more estradiol. This estradiol acts upon the brain to entrain the neural rhythms required for the first preovulatory surge of LH from the anterior pituitary gland.

In the meantime, between the appearance of free estradiol in the plasma at 21 and 23 days and the first ovulation about 19 days later, the ovarian follicle has matured under a balanced milieu of estradiol, FSH, and probably prolactin (Richards and Midgley, 1976). The influences on follicle development are more complex than this summary suggests, and it appears that the theca cell produces androstenedione in response to LH and catecholamines. This androgen substrate diffuses into the granulosa cell which in response to LH and FSH aromatizes it to estradiol (Erickson, 1983). The theory is summarized in Fig. 8. The role of estradiol and its receptor in the ovary and the hypothalamo-hypophyseal axis is critical, but there is no need to postulate any changes in neural sensitivity to estradiol at puberty.

Extrapolation from rat to man needs care. For example, there is no estrogen-binding AFP in man, although human plasma does contain a sex hormone-binding globulin (Westphal, 1972). Nevertheless, it is likely that fundamentally there are similarities in the factors which influence ovarian development in both species, and the reader is doubtless aware that many of the endocrine processes now understood in man stem from studies carried out on the rat. The factors which determine the onset of puberty in the male are still unknown.

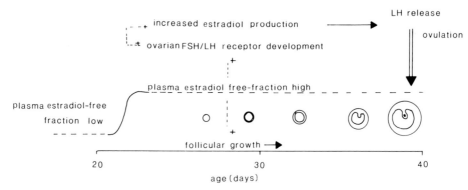

**Fig. 8.** A theory of the first ovulation in the rat. The initial stimulus is the rise in free, unbound circulating estradiol, which is biologically active. The estradiol stimulates follicular growth and synergizes with FSH to stimulate LH receptor development. Estrogen secretion from the granulosa cell is, therefore, increased, and shortly before the first ovulation, sufficient estrogen is released to trigger the critical series of events which occurs in the brain and anterior pituitary gland prior to the preovulatory surge of luteinizing hormone.

## ACKNOWLEDGMENT

This work was supported by the Mental Health Foundation, the Medical Research Council of Great Britain, and St. Thomas' Hospital Research (Endowments) Fund. I am grateful to Mrs. J. Andrews for typing the manuscript.

## REFERENCES

Al-Khouri, H., and Greenstein, B. D. (1980). Role of corticosteroid-binding globulin in interaction of corticosterone with uterine and brain progesterone receptors. *Nature (London)* **287,** 58–60.

Al-Khouri, H., and Greenstein, B. D. (1982). Effects of oestradiol benzoate injection on [$^3$H]progesterone-binding components in rat brain, uterus and serum. *J. Physiol. (London)* **332,** 110P–111P.

Alvarez, E. O., and Hancke, J. L. (1978). Binding capacity of estradiol and depletion-replenishment rate of cytoplasmic estrogen receptor in uterine tissue of maturing rats with hypothalamic lesions. *Acta Physiol. Latinoam.* **28,** 151–162.

Arnold, A. P., and Gorski, R. A. (1984). Gonadal steroid induction of structural sex differences in the central nervous system. *Annu. Rev. Neurosci.* **7,** 413–442.

Attardi, B., and Ohno, S. (1976). Androgen and estrogen receptors in the developing mouse brain. *Endocrinology (Baltimore)* **99,** 1279–1290.

Barbanel, G., and Assenmacher, I. (1982). Effects of thyroid hormones on the ontogeny of oestradiol-binding sites in the rat. *Mol. Cell. Endocrinol.* **28,** 247–261.

Barley, J., Ginsburg, M., Greenstein, B. D., MacLusky, N. J., Morris, I. D., and Thomas, P. J. (1974). A receptor mediating sexual differentiation? *Nature (London)* **252,** 259–260.

Barley, J., Ginsburg, M., Greenstein, B. D., MacLusky, N. J., and Thomas, P. J. (1975). An androgen receptor in rat brain and pituitary. *Brain Res.* **100,** 383–393.

Beato, M., Kalimi, M., and Feigelson, P. (1972). Correlation between glucocorticoid binding to specific liver cytosol receptors and enzyme induction *in vivo. Biochem. Biophys. Res. Commun.* **47,** 1464–1472.

Boling, J., and Blandau, R. (1939). The estrogen–progesterone induction of sexual receptivity in the spayed female rat. *Endocrinology (Baltimore)* **25,** 359–364.

Christensen, L. W., and Clemens, L. G. (1974). Intrahypothalamic implants of testosterone or estradiol and resumption of masculine sexual behavior in long term castrated male rats. *Endocrinology (Baltimore)* **95,** 984–990.

Christensen, L. W., and Gorski, R. A. (1978). Independent masculinization of neuroendocrine systems by intracerebral implants of testosterone or estradiol in the neonatal female rat. *Brain Res.* **146,** 325–340.

Clayton, C. J., Grosser, B. I., and Stevens, W. (1977). The ontogeny of corticosterone and dexamethasone receptors in rat brain. *Brain Res.* **134,** 445–453.

Cunha, G. R. (1976). Epithelial–stromal interactions in development of the urogenital tract. *Int. Rev. Cytol.* **47,** 137–194.

Cunha, G. R., and Lung, B. (1979). The importance of stroma in morphogenesis and functional activity of urogenital epithelium. *In Vitro* **15,** 50–71.

Cunha, G. R., Reese, B. A., and Sekkingstad, M. (1980). Induction of nuclear androgen-binding sites in epithelium of the embryonic urinary bladder by mesenchyme of the urinary sinus of embryonic mice. *Endocrinology (Baltimore)* **107,** 1767–1770.

Cunha, G. R., Shannon, J. M., Vanderslice, K. D., Sekkingstad, M., and Robboy, S. J. (1982). Autoradiographic analysis of nuclear estrogen binding sites during postnatal development of the genital tract of female mice. *J. Steroid Biochem.* **17,** 281–286.

Dalton, K. (1976). Prenatal progesterone and educational attainments. *Br. J. Psychol.* **129,** 438–442.

de Kloet, E. R., and McEwen, B. S. (1976). A putative glucocorticoid receptor and a transcortin-like macromolecule in pituitary cytosol. *Biochim. Biophys. Acta* **421,** 115–123.

De Lemos, R. A., Shermata, D. W., Knedson, T., Kotas, R., and Avery, M. E. (1970). Acceleration of appearance of pulmonary surfactant in fetal lung by administration of corticosteroids. *Am. Rev. Respir. Dis.* **102,** 459–461.

Donovan, B. T., and Van Der Werff Ten Bosch, J. J. (1965). "Physiology of Puberty," Monograph of the Physiological Society. Arnold, London.

Dudley, S. D. (1981). Prepubertal ontogeny of responsiveness to estradiol in female rat central nervous system. *Neurosci. Biobehav. Rev.* **5,** 421–435.

Eisenfeld, A. J. (1972). Ontogeny of estrogen and androgen receptors. *In* "The Control of the Onset of Puberty" (M. M. Grumbach, G. D. Grave, and F. E. Mayer, eds.), pp. 271–306. Wiley, New York.

Erickson, G. F. (1983). Primary cultures of ovarian cells in serum-free medium as models of hormones-dependent differentiation. *Mol. Cell. Endocrinol.* **24,** 21–49.

Feldman, D. (1974). Ontogeny of rat hepatic glucocorticoid receptors. *Endocrinology (Baltimore)* **95,** 1219–1227.

Fichman, K. R., Nyberg, L. M., Bujnovszky, P., Brown, T. R., and Walsh, P. C. (1981). The ontogeny of the androgen receptor in human foreskin. *J. Clin. Endocrinol. Metab.* **52,** 919–923.

Frank, A., and Fraps, R. (1945). Induction of estrus in the ovariectomised golden hamster. *Endocrinology (Baltimore)* **37,** 357–361.

Gasc, J.-M. (1980). Estrogen target cells in gonads of the chicken embryo during sexual development. *J. Embryol. Exp. Morphol.* **55,** 331–342.

Gasc, J.-M., and Stumpf, W. E. (1981). Sexual differentiation of the urogenital tract in the chicken embryo: autoradiographic localisation of sex-steroid target cells during development. *J. Embryol. Exp. Morphol.* **63,** 207–223.

Gasc, J.-M., Stumpf, W. E., and Sar, M. (1979). Androgen target cells in the pituitary of the chick embryo. *J. Steroid Biochem.* **11**, 1201–1203.

Gerlach, J. L., and McEwen, B. S. (1972). Rat brain binds adrenal steroid hormone: radioautography of hippocampus with corticosterone. *Science (Washington, D.C.)* **175**, 1133–1136.

Giannopoulos, G. (1975a). Ontogeny of glucocorticoid receptors in rat liver. *J. Biol. Chem.* **250**, 5847–5851.

Giannopoulos, G. (1975b). Early events in the action of glucocorticoids in developing tissues. *J. Steroid Biochem.* **6**, 623–631.

Giannopoulos, G., Mulay, S., and Solomon, S. (1972). Cortisol receptors in rabbit fetal lung. *Biochem. Biophys, Res. Commun.* **47**, 411–418.

Giannopoulos, G., Hassan, Z., and Solomon, S. (1974). Glucocorticoid receptors in fetal and adult rabbit tissues. *J. Biol. Chem.* **249**, 2424–2427.

Ginsburg, M., Greenstein, B. D., MacLusky, N. J., Morris, I. D., and Thomas, P. J. (1974). Dihydrotestosterone binding in brain and pituitary. *J. Endocrinol.* **61**, 24P.

Glascock, R. F., and Hoekstra, W. G. (1959). Selective accumulation of tritium-labeled hexoestrol by the reproductive organs of immature female goats and sheep. *Biochem. J.* **72**, 673–682.

Gomez, F., Bohnet, H. G., and Friesen, H. G. (1977). Changes in progesterone and estrogen receptors in the rat uterus during the estrous cycle and puerperium. *Prog. Cancer Res. Ther.* **4**, 245–259.

Gorski, J., and Gannon, F. (1976). Current models of steroid hormone action: a critique. *Annu. Rev. Physiol.* **38**, 425–450.

Gorski, R. A., Gordon, J. H., Shryne, J. E., and Southam, A. M. (1978). Evidence for a morphological sex difference within the medial preoptic area of the rat brain. *Brain Res.* **148**, 333–346.

Greenstein, B. D. (1978a). The role of hormone receptors in development and puberty. *J. Reprod. Fertil.* **52**, 419–426.

Greenstein, B. D. (1978b). Evidence for specific progesterone receptors in rat brain cytosol. *J. Endocrinol.* **79**, 327–338.

Greenstein, B. D. (1979). Androgen receptors in the rat brain, anterior pituitary gland and ventral prostate gland: effects of orchidectomy and ageing. *J. Endocrinol.* **81**, 75–81.

Griffin, J. E. (1979). Testicular feminization associated with a thermolabile androgen receptor in cultured human fibroblasts. *J. Clin. Invest.* **64**, 1624–1631.

Griffin, J. E., and Wilson, J. D. (1980). The syndrome of androgen resistance. *N. Engl. J. Med.* **302**, 198–209.

Haslam, S. Z., and Shyamala, G. (1979). Progesterone receptors in normal mammary glands of mice. Characterisation and relationship to development. *Endocrinology (Baltimore)* **105**, 786–795.

Holt, P. G., and Oliver, I. T. (1968). Plasma corticosterone concentrations in the perinatal rat. *Biochem. J.* **108**, 334–341.

James, F., and Fotherby, K. (1970). Interaction of sex hormones with target tissues. *In* "Steroid Biochemistry and Pharmacology" (M. H. Briggs, ed.), pp. 315–372. Academic Press, New York.

Jensen, E. V., and Jacobson, H. I. (1962). Basic guides to the mechanism of estrogen action. *Rec. Prog. Horm. Res.* **18**, 378–414.

Jensen, E. V., Suzuki, T., Kawashima, T., Stumpf, W. E., Jungblut, P. W., and De Sombre, E. R. (1968). A two step mechanism for the interaction of estradiol with rat uterus. *Proc. Natl. Acad. Sci. U.S.A.* **49**, 632–638.

Kalimi, M., and Gupta, S. (1982). Physiochemical characterisation of rat liver glucocorticoid receptor during development. *J. Biol. Chem.* **257**, 13324–13328.

Kato, J. (1975). The role of hypothalamic and hypophyseal 5α-dihydrotestosterone, estradiol and progesterone receptors in the mechanism of feedback action. *J. Steroid Biochem.* **6**, 979–987.

Kato, J. (1976). Ontogeny of 5α-dihydrotestosterone receptors in the hypothalamus of the rat. *Ann. Biol. Anim. Biochim. Biophys.* **16**, 467–469.

Keenan, B. S., Meyer, W. J., III, Hadjian, A. J., Jones, H. W., and Migeon, C. G. (1974). Syndrome of androgen insensitivity in man: absence of 5α-dihydrotestosterone binding protein in skin fibroblasts. *J. Clin. Endocrinol.* **38**, 1143–1146.

Kikkawa, Y., Kaibara, M., Motoyama, E. K., Orzalesi, M. M., and Cook, C. D. (1971). Morphologic development of the fetal rabbit lung and acceleration with cortisol. *Am. J. Pathol.* **64**, 423–442.

Kincl, F. A., and Maqueo, M. (1965). Prevention by progesterone of steroid-induced sterility in neonatal male and female rats. *Endocrinology (Baltimore)* **77**, 859–862.

King, W. J., and Greene, G. L. (1984). Monoclonal antibodies localize oestrogen receptor in the nuclei of target cells. *Nature (London)* **307**, 745–747.

Koch, B., Lutz, B., Briaud, B., and Mialhe, C. (1976). Heterogeneity of pituitary glucocorticoid binding. Evidence for a transcortin-like component. *Biochim. Biophys. Acta* **444**, 497–507.

Kolata, G. B. (1979). Sex hormones and brain development. *Science (Washington, D.C.)* **205**, 985–987.

Konishi, M., and Gurney, M. E. (1982). Sexual differentiation of brain and behavior. *Trends Neurosci.* **5**, 20–23.

Kuhn, N. J. (1969). Specificity of progesterone inhibition of lactogenesis. *J. Endocrinol.* **45**, 615–616.

Lazier, C. B., Nadin-Davis, S. A., Elbrecht, A., Blue, M.-L., and Williams, D. L. (1981). Estrogen receptor and the development of estrogenic responses in embryonic chick liver. *Adv. Exp. Med. Biol.* **138**, 19–38.

Lieberburg, I., MacLusky, N. J., and McEwen, B. S. (1980). Androgen receptors in the perinatal rat brain. *Brain Res.* **196**, 125–138.

Linser, P., Saad, A. D., Soh, B. M., and Moscona, A. A. (1982). Cell contact-dependent regulation of hormonal induction of glutamine synthetase in embryonic neural retina. *Prog. Clin. Biol. Res.* **85(B)**, 445–458.

MacDonald, P. C., Madden, J. C., Brenner, P. F., and Wilson, J. D. (1979). Origin of estrogen in normal men and in women with testicular feminization. *J. Clin. Endocrinol. Metab.* **49**, 905–916.

McEwen, B. S. (1981a). Sexual differentiation of the brain. *Nature (London)* **291**, 610.

McEwen, B. S. (1981b). Neural gonadal steroid actions. *Science (Washington, D.C.)* **211**, 1303–1311.

McEwen, B. S., Weiss, J. M., and Schwartz, L. S. (1969). Uptake of corticosterone by rat brain and its concentration by certain limbic structures. *Brain Res.* **16**, 227–241.

McEwen, B. S., Weiss, J. M., and Schwartz, L. S. (1970). Retention of corticosterone by cell nuclei from brain regions of adrenalectomised rats. *Brain Res.* **17**, 471–482.

MacKinnon, P. C. B., Puig-Duran, E., and Laynes, R. (1978). Reflections on the attainment of puberty in the rat: have circadian signals a role to play in its onset? *J. Reprod. Fertil.* **52**, 410–412.

MacLusky, N. J., and McEwen, B. S. (1978). Oestrogen modulates progestin receptor concentrations in some brain regions but not in others. *Nature (London)* **274**, 276–278.

MacLusky, N. J., and McEwen, B. S. (1980). Progestin receptors in the developing rat brain and pituitary. *Brain Res.* **189**, 262–268.

MacLusky, N. J., and Naftolin, F. (1981). Sexual differentiation of the central nervous system. *Science (Washington, D.C.)* **211**, 1294–1303.

MacLusky, N. J., Lieberburg, I., and McEwen, B. S. (1979). The development of estrogen receptor systems in the rat brain: Perinatal development. *Brain Res.* **178**, 129–142.

Markey, K. A., Towle, A. C., and Sze, P. Y. (1982). Glucocorticoid influence on tyrosine hydroxylase activity in mouse locus coeruleus during postnatal development. *Endocrinology (Baltimore)* **111**, 1519–1523.

May, F. E. B., and Knowland, J. (1981). Oestrogen receptor levels and vitellogenin synthesis during development of xenopus laevis. *Nature (London)* **292**, 853–855.

Milgrom, E., and Baulieu, E.-E. (1970). Progesterone in uterus and plasma. I. Binding in rat uterus 105,000 g supernatant. *Endocrinology (Baltimore)* **87**, 276–287.

Milgrom, E., Atger, M., and Baulieu, E.-E. (1970). Progesterone in uterus and plasma. IV. Progesterone receptor(s) in guinea pig uterus cytosol. *Steroids* **16**, 741–754.

Moguilewsky, M., and Raynaud, J.-P. (1977). Progestin-binding sites in the rat hypothalamus, pituitary and uterus. *Steroids* **30**, 99–109.

Moguilewksy, M. and Raynaud, J.-P. (1979). The relevance of hypothalamic and hypophyseal progestin receptor regulation in the induction and inhibition of sexual behavior in the female rat. *Endocrinology (Baltimore)* **105**, 516–522.

Morin, L. P., and Feder, H. H. (1974). Hypothalamic progesterone implants and facilitation of lordosis behavior in estrogen-primed ovariectomised guinea-pigs. *Brain Res.* **70**, 81–93.

Nordeen, E. J., and Yahr, P. (1983). A regional analysis of estrogen binding to hypothalamic cell nuclei in relation to masculinization and defeminization. *J. Neurosci.* **3**, 933–941.

Ojeda, S. R., Smith, S., White, J. P., and Andrews, W. W. (1981). Current studies on the mechanisms underlying the onset of female puberty. *Prog. Clin. Biol. Res.* **74**, 199–218.

O'Malley, B., Vedeckis, W. V., Birnbaumer, M.-E., and Schrader, W. T. (1977). Steroid receptor action: The role of receptors in regulating gene expression. *In* "Molecular Endocrinology" (I. MacIntyre and M. Szelke, eds.), pp. 135–150. Elsevier, Amsterdam.

Parchman, G. L., Cake, M. H., and Litwack, G. (1978). Functionality of the liver glucocorticoid receptor during the life cycle and development of a low affinity membrane binding site. *Mech. Ageing Dev.* **7**, 227–240.

Parker, M. (1983). Enhancer elements activated by steroid hormones? *Nature (London)* **304**, 687–688.

Pasqualini, J. R., and Nguyen, B. L. (1980). Progesterone receptors in the fetal uterus and ovary of the guinea pig: Evolution during fetal development and induction and stimulation in estradiol-primed animals. *Endocrinology (Baltimore)* **106**, 1160–1165.

Peterson, R. E., Imperato-McGinley, J., Gautier, T., and Sturba, E. (1977). Male pseudohermaphroditism due to the steroid 5α-reductase deficiency. *Am. J. Med.* **62**, 180–191.

Pfaff, D. W. (1970). Nature of sex hormones effect on rat sexual behavior: specificity of effects and individual patterns of response. *J. Comp. Physiol. Psychol.* **73**, 349–358.

Pfaff, D. W., and Zigmond, R. E. (1971). Neonatal androgen effects on sexual and non-sexual behavior of adult rats tested under various hormone regimes. *Neuroendocrinology* **7**, 129–145.

Philibert, D., and Raynaud, J.-P. (1973). Progesterone binding in the immature mouse and rat uterus. *Steroids* **22**, 89–98.

Powell-Jones, W., Thompson, C., Raeford, S., and Lucier, G. W. (1981). Effect of gonadectomy on the ontogeny of estrogen-binding components in rat liver cytosol. *Endocrinology (Baltimore)* **109**, 628–636.

Puig-Duran, E., Greenstein, B. D., and MacKinnon, P. C. B. (1979). The effects of serum oestrogen-binding components on the unbound oestradiol-17β fraction in the serum of developing female rats and on inhibition of [³H]oestradiol uptake by uterine tissue *in vitro*. *J. Reprod. Fertil.* **56**, 707–714.

Raisman, G., and Field, P. M. (1973). Sexual dimorphism in the neuropil of the preoptic area of the rat and its dependence on neonatal androgen. *Brain Res.* **54**, 1–29.

Rajfer, J., Namkung, P. C., and Petra, P. H. (1980). Ontogeny of the cytoplasmic androgen receptor in the rat ventral prostate gland. *Biol. Reprod.* **23**, 518–521.

Ramirez, D. V., and McCann, S. M. (1963). Comparison of the regulation of luteinizing hormone (LH) secretion in immature and adult rats. *Endocrinology (Baltimore)* **72**, 452–464.

Ramirez, D. V., and Sawyer, C. H. (1965). Advancement of puberty in the female rat by estrogen. *Endocrinology (Baltimore)* **76,** 1158–1168.

Raynaud, J.-P. (1973). Influence of rat estradiol binding plasma protein (EBP) on uterotropic activity. *Steroids* **21,** 249–258.

Raynaud, J.-P., Mercier-Bodard, C., and Baulieu, E.-E. (1971). Rat oestradiol binding plasma protein. *Steroids* **18,** 767–788.

Raynaud, J.-P., Bouton, M.-M., Philibert, D., and Vannier, B. (1976). Steroid binding in the hypothalamus and pituitary. *In* "Hypothalamus and Endocrine Functions" (F. Labrie, J. Meites, and G. Pelletier, eds.), pp. 171–189. Plenum, New York.

Rhees, R. W., Grosser, B. I., and Stevens, W. (1975). The autoradiographic localisation of [$^3$H]dexamethasone in the brain and pituitary of the rat. *Brain Res.* **100,** 151–156.

Richards, J. S., and Midgley, A. R. (1976). Protein hormone action: a key to understanding follicular and luteal cell development. *Biol. Reprod.* **14,** 82–94.

Saiduddin, S., and Zassenhaus, H. P. (1978). Effect of testosterone and progesterone on the estradiol receptor in the immature rat ovary. *Endocrinology (Baltimore)* **102,** 1069–1076.

Sakly, M., and Koch, B. (1981). Ontogenesis of glucocorticoid receptors in anterior pituitary gland: transient dissociation among cytoplasmic receptor density, nuclear uptake and regulation of corticotropic activity. *Endocrinology (Baltimore)* **108,** 591–596.

Sakly, M., and Koch, B. (1983). Ontogenetical variations of transcortin modulate glucocorticoid receptor function and corticotropic activity in the pituitary gland. *Horm. Metab. Res.* **15,** 92–96.

Salaman, D. F., Thomas, P. J., and Westley, B. R. (1976). Characterisation and properties of a high affinity estradiol receptor in neonatal rat brain. *Ann. Biol. Anim. Biochim. Biophys.* **16,** 479–490.

Schmid, W., and Grote, H. (1977). Multiple forms of glucocorticoid binding proteins in glucocorticoid target cells. *In* "Multiple Molecular Forms of Steroid Hormone Receptors" (M. K. Agarwal, ed.), pp. 35–48. Elsevier/North-Holland, New York.

Schmidt, W. N., and Katzenellenbogen, B. S. (1979). Androgen-induced interactions: an assessment of androgen interaction with the testosterone- and estradiol-receptor systems and stimulation of uterine growth and progesterone receptor synthesis. *Mol. Cell. Endocrinol.* **15,** 91–108.

Schulster, D., Burstein, S., and Cooke, B. A. (1976). "Molecular Endocrinology of the Steroid Hormones." Wiley, New York.

Sereni, F., Kenny, F. T., and Kretchmer, N. (1959). Factors influencing the development of tyrosine-α-ketoglutarate transaminase activity in rat liver. *J. Biol. Chem.* **234,** 609–612.

Sheehan, D. M., Branham, W. S., Medlock, K. L., Olsen, M. E., and Zehr, D. R. (1981). Uterine responses to estradiol in the neonatal rat. *Endocrinology (Baltimore)* **109,** 76–82.

Spelsberg, T. C., and Halberg, F. (1980). Circannual rhythms in steroid receptor concentration and nuclear binding in the chick oviduct. *Endocrinology (Baltimore)* **107,** 1234–1244.

Spelsberg, T. C., and Toft, D. O. (1976). The mechanism of action of progesterone. *In* "Receptors and the Mechanism of Action of Steroid Hormones," Part I (J. R. Pasqualini, ed.), pp. 261–309. Dekker, New York.

Steele, R. E. (1977). Role of the ovaries in maturation of the estradiol-luteinizing hormone negative feedback system of the pubertal rat. *Endocrinology (Baltimore)* **101,** 587–597.

Stevens, W., Grosser, B. I., and Reed, D. J. (1971). Corticosterone-binding molecules in rat brain cytosols: regional distribution. *Brain Res.* **35,** 602–607.

Sumida, C., and Pasqualini, J. R. (1980). Dynamic studies on estrogen response in fetal guinea pig uterus: effect of estradiol administration on estradiol receptor, progesterone receptor and uterine growth. *J. Recept. Res.* **1,** 439–457.

Sumida, C., Gelly, C., and Pasqualini, J. R. (1981). Progesterone antagonizes the effects of estradiol in the fetal uterus of guinea pig. *J. Recept. Res.* **2,** 221–232.

Teng, C. S. (1980). Ontogeny of the receptor and responsiveness to estrogen in the genital tract of the chick embryo. *Adv. Biosci.* **25,** 77–94.

Teng, C. S., and Teng, C. T. (1975). Studies on sex-organ development. Isolation and characterization of an oestrogen receptor from chick mullerian duct. *Biochem. J.* **150**, 183–190.

Terasawa, E., and Sawyer, C. H. (1970). Diurnal variation in the effects of progesterone on multiple unit activity in the rat hypothalamus. *Exp. Neurol.* **27**, 359–374.

Thomas, P. J., and Knight, A. (1978). Sexual differentiation of the brain. *In* "Current Studies of Hypothalamic Function" (K. Lederis and W. L. Veale, eds.), Vol. 1, pp. 192–203. Karger, Basel.

Toft, D., and Chytil, E. (1973). Receptors for glucocorticoids in lung tissue. *Arch. Biochem. Biophys.* **152**, 464–469.

Verbruggen, L. A., and Solomon, D. S. (1980). Glucocorticoid receptors and inhibition of neonatal mouse dermal fibroblast growth in primary culture. *Arch. Dermatol. Res.* **269**, 111–126.

Vito, C. C., Wieland, S. J., and Fox, T. O. (1979). Androgen receptors exist throughout the "critical period" of brain sexual differentiation. *Nature (London)* **282**, 308–310.

Wade, G. N., and Feder, H. H. (1974). Stimulation of [$^3$H]leucine incorporation into protein by estradiol-17β or progesterone in brain tissues of ovariectomized guinea pigs. *Brain Res.* **73**, 545–549.

Welshons, W. V., Lieberman, M. E., and Gorski, J. (1984). Nuclear localization of unoccupied oestrogen receptors. *Nature (London)* **307**, 747–749.

Westphal, U. (1972). Binding of hormones to serum proteins. *In* "Biochemical Actions of Hormones" (G. Litwack, ed.), Vol. 1, pp. 209–265. Academic Press, New York.

Whalen, R. E. (1974). Sexual differentiation: models, methods and mechanisms. *In* "Sex Differences in Behavior" (R. C. Friedman, R. M. Richart, and R. L. Vande Wiele, eds.), pp. 467–481. Wiley, New York.

Whalen, R. E., Luttge, W. G., and Gorzalka, B. B. (1971). Neonatal androgenization and the development of estrogen responsivitiy in male and female rats. *Horm. Behav.* **2**, 83–90.

Wilson, J. D. (1975). Metabolism of testicular androgens. *In* "Male Reproductive System. Handbook of Physiology, Sect. 7: Endocrinology" (R. D. Greep and E. B. Astwood, eds.), Vol. 5, pp. 491–508. Am. Physiol. Soc., Washington, D.C.

Wilson, J. D. (1978). Sexual differentiation. *Annu. Rev. Physiol.* **40**, 279–306.

Wilson, J. D., and Goldstein, J. L. (1975). Classification of hereditary disorders of sexual development. *Birth Defects, Orig. Artic. Ser.* **11**, 1–16.

Wolff, E. (1950). Le rôle des hormones embryonnaires dans la différenciation sexuelle des oiseaux. *Arch. Anat. Microsc. Morphol. Exp.* **39**, 426–450.

Yeoh, G. C., Arbuckle, T., and Oliver, I. T. (1979). Tyrosine aminotransferase induction in hepatocytes cultured from rat foetuses treated with dexamethasone *in utero*. *Biochem. J.* **180**, 545–549.

Yoneda, T., and Pratt, R. M. (1981). Glucocorticoid receptors in palatal mesenchymal cells from the human embryo: Relevance to human cleft palate formation. *J. Craniofacial Genet. Dev. Biol.* **1**, 411–423.

# 11

# Functions and Regulation
# of Cell Surface Receptors
# in Cultured Leydig Tumor Cells

**MARIO ASCOLI***
Division of Endocrinology
Departments of Medicine and Biochemistry
School of Medicine
Vanderbilt University
Nashville, Tennessee

*Present address: Center for Biomedical Research, The Population Council, 1230 York
Avenue, New York, New York 10021.

THE RECEPTORS, VOL. II

## I. INTRODUCTION

It is now well recognized that cell surface receptors play a pivotal role in the actions of polypeptide hormones and other regulatory proteins. The studies summarized in this chapter illustrate some aspects of the functions and regulation of the receptors for three distinct regulatory ligands: human choriogonadotropin (hCG), mouse epidermal growth factor (mEGF), and low-density lipoprotein (LDL) in a single cell type.

## II. DIFFERENTIATED FUNCTION OF MA-10 CELLS

The MA-10 cells are one of the clonal strains of Leydig tumor cells adapted to culture in this laboratory (Ascoli, 1981a). These cell lines originated from a transplantable tumor (designated M5480P) that originated spontaneously in a C57Bl/6 mouse (Ascoli and Puett, 1978a; Neaves, 1975).

Like their normal counterparts, MA-10 cells have the capacity to convert cholesterol into steroid hormones and to respond to LH/hCG with increased steroid biosynthesis. There are two major differences (besides the malignant

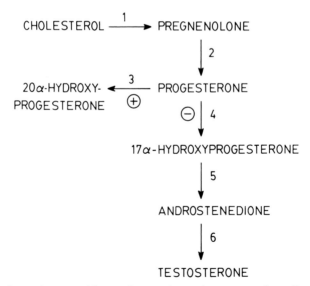

**Fig. 1.** Pathway for steroid biosynthesis in normal mouse Leydig cells and MA-10 cells. Normal mouse Leydig cells synthesize testosterone primarily via the $\Delta^4$ pathway. The MA-10 cells synthesize progesterone and 20α-dihydroxyprogesterone presumably because of reduced levels of 17α-hydroxylase (indicated by −) and elevated levels of 20α-hydroxylase (indicated by +) activities. The enzymes involved in this pathway are as follows: (1) cholesterol side chain cleavage; (2) $\Delta^5$-3β-hydroxysteroid dehydrogenase isomerase; (3) 20α-hydroxylase; (4) 17α-hydroxylase; (5) $^C$17−20 lyase; and (6) 17β-hydroxysteroid dehydrogenase. Data from Lacroix et al. (1979) and Ascoli (1981a).

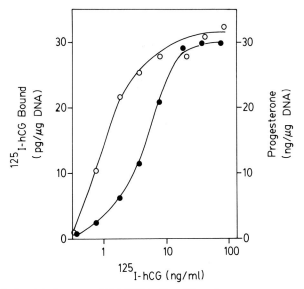

**Fig. 2.**   Relationship between $^{125}$I-hCG binding (●) and progesterone production (○) in MA-10 cells. Hormone binding and progesterone accumulation were measured after a 2 hr incubation at 37°C in serum-free medium.

origin) between MA-10 cells and normal Leydig cells. First, because of a decrease in the activity of one of the enzymes (17α-hydroxylase, see Fig. 1) involved in the conversion of progesterone to testosterone, the major steroid produced by MA-10 cells is progesterone rather than testosterone (Ascoli, 1981a; Ascoli and Puett, 1978a; Lacroix *et al.*, 1979; Payne *et al.*, 1985). In addition to this change that apparently occurred during transplantation (Moyle and Greep, 1974; Ascoli and Puett, 1978a), the cells gained the ability to metabolize progesterone to 20α-hydroxyprogesterone when adapted to culture (Ascoli, 1981a, 1982a). Thus, under basal conditions, MA-10 cells synthesize equivalent amounts of progesterone and 20α-hydroxyprogesterone; when exposed to hCG, however, the major steroid produced is progesterone (Ascoli, 1981a,b; Freeman and Ascoli, 1982a).

The second major difference between MA-10 cells and normal Leydig cells is in the coupling between hCG binding and the stimulation of steroid biosynthesis. In normal rat Leydig cells, maximal stimulation of steroid biosynthesis occurs when only about 1% of the LH/hCG receptors are occupied (Mendelson *et al.*, 1975), whereas in MA-10 cells, maximal stimulation of progesterone biosynthesis occurs when 60–70% of the receptors are occupied (Fig. 2). Both the affinity of the receptor for hCG ($ED_{50}$ for binding at 37°C $\cong 1 \times 10^{-10}$ $M$) and the number of LH/hCG receptors (10,000–20,000 per cell) are similar in MA-10 cells and in normal rat Leydig cells (Ascoli, 1981a,b; Catt *et al.*, 1980; Freeman and Ascoli, 1981; Lloyd and Ascoli, 1983; Tsuruhara *et al.*, 1977).

In spite of these differences, the main feature of the differentiated function of normal Leydig cells—the presence of LH/hCG receptors that are "coupled" to the steroidogenic pathway—has been retained by MA-10 cells. In addition, the "close coupling" between hCG binding and the stimulation of steroidogenesis in MA-10 cells offers the experimental advantage that changes in the number of hCG receptors can be easily correlated with changes in the steroidogenic response of the cells to the hormone (see later).

The ability of the receptor-bound hCG to stimulate steroid biosynthesis in MA-10 cells (as well as in normal Leydig and granulosa–luteal cells) is believed to be mediated by the classical cAMP pathway (for reviews see Ascoli, 1982a; Hunzicker-Dunn and Birnbaumer, 1985). Thus, the effects of LH/hCG on steroid biosynthesis can be mimicked with cAMP analogues or with other compounds that stimulate adenylate cyclase activity, such as cholera toxin and forskolin. During short-term incubations (~4hr), hCG and cAMP analogues always stimulate progesterone biosynthesis to the same extent. The response to cholera toxin, however, is rather variable, reaching 60–100% of the maximal response to hCG and cAMP analogues (Ascoli, 1978, 1981a,b; Freeman and Ascoli, 1981; Segaloff *et al.*, 1981a,b). The ability of these compounds to stimulate progesterone biosynthesis offers a useful experimental tool that can be used to study the steroidogenic potential of the cells independently of the hCG receptor.

The message that is initiated by the binding of hCG to its cell surface receptor and translated by the adenylate cyclase–protein kinase pathway eventually leads to an increased mobilization of cholesterol into the mitochondria and an increase in the conversion of cholesterol to pregnenolone (Fig. 1) (see Payne *et al.*, 1985, for a review). Pregnenolone is then converted to progesterone (the major product synthesized by MA-10 cells, see earlier) in the smooth endoplasmic reticulum. This step does not appear to be under hormonal control.

From the foregoing discussion, it is obvious that hCG plays an important role in the regulation of steroid biosynthesis. In addition, there are two other ligands, low-density lipoprotein and mouse epidermal growth factor, that directly, or indirectly, affect the actions of hCG on these cells. In the following sections, we review our data on the mechanisms by which these ligands affect the differentiated function of MA-10 cells.

## III. INTERACTION OF hCG WITH MA-10 CELLS

### A. General Properties of Binding, Internalization, and Degradation

The ability of cells to internalize and degrade receptor-bound ligands is a well recognized, ubiquitous phenomenon (Steinman *et al.*, 1983; Goldstein *et al.*, 1979; Brown *et al.*, 1983). The MA-10 cells are no exception; all the ligands studied to date that bind to specific cell surface receptors of MA-10 cells are

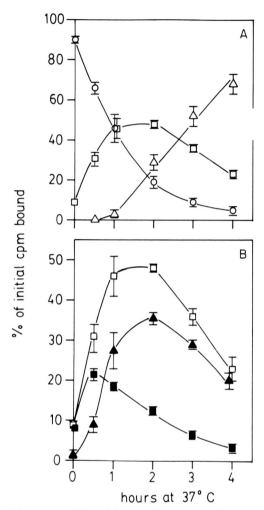

**Fig. 3.** Fate of the receptor-bound hCG. Cells were incubated with $^{125}$I-hCG (20 ng/ml) for 2 hr at 4°C and washed to remove the free hormone. At $t = 0$, the cells received warm medium and were incubated at 37°C for the times indicated. (A) The surface-bound (○) and internalized (□) radioactivity that remained cell associated and the release of mono[$^{125}$I]iodotyrosine into the medium (△) were measured as described elsewhere (Ascoli, 1982b). (B) After removing the surface-bound $^{125}$I-hCG, the proportions of internalized $^{125}$I-hCG that were free (▲) or receptor bound (■) were determined as described elsewhere (Dufau et al., 1973; Ascoli, 1983; Ascoli, 1984). The total radioactivity internalized (□) is reproduced from (A).

ultimately internalized. These include hCG (Ascoli and Puett, 1978b; Ascoli, 1982b), mEGF (Ascoli, 1981b; Lloyd and Ascoli, 1983), LDL (Freeman and Ascoli, 1982b, 1983), and transferrin (Ascoli, 1984a).

The internalization and degradation of hCG have received special attention in our laboratory because of their possible involvement in the regulation of the actions of hCG. Our understanding of these phenomena has improved because of the availability of (1) hCG derivatives that are exclusively labeled in either the $\alpha$ or $\beta$ subunits (Morgan et al., 1974; Ascoli, 1980, 1982b); (2) a biochemical method that removes the surface-bound hormone without affecting the viability and responsiveness of the cells (Ascoli, 1982b; Segaloff and Ascoli, 1981); and (3) a biochemical method that allows us to determine if the internalized hormone is free or receptor-bound (Dufau et al., 1973; Ascoli, 1983, 1984).

The process of internalization and degradation of hCG are best studied by first binding [125]I-hCG at 4°C to prevent internalization and by washing the cells to remove the free hormone. At this time, the cells are incubated at 37°C in hormone-free medium and the amount of surface-bound, internalized, and degraded hormone measured as a function of time after the temperature shift.

Figure 3 shows the fate of the receptor-bound [125]I-hCG. By measuring the

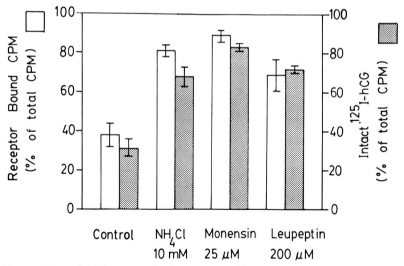

**Fig. 4.** Effect of inhibitors on the dissociation of the internalized [125]I-hCG–receptor complex and the degradation of the internalized [125]I-hCG. Cells were incubated with [125]I-hCG, as described in Fig. 3, and processed at $t = 1$ hr. After removing the surface-bound [125]I-hCG, the internalized receptor-bound and intact [125]I-hCG were determined by polyethylene glycol precipitation (Ascoli, 1983, 1984; Dufau et al., 1973) and sodium dodecyl sulfate–polyacrylamide gel electrophoresis (Ascoli, 1982b), respectively. The indicated inhibitors were presented throughout the experiment.

TABLE I

Distribution of Internalized, Receptor-Bound $^{125}I$-hCG in Endocytic Vesicles and Lysosomes[a]

| | Receptor-bound $^{125}I$-hCG (% of total cpm in each fraction) | |
| Inhibitor | Endocytic vesicles | Lysosomes |
| --- | --- | --- |
| None | 97 ± 3 | 27 ± 1 |
| $NH_4Cl$ (10 mM) | 95 ± 2 | 74 ± 5 |
| Monensin (25 μM) | 99 ± 1 | 97 ± 3 |
| Leupeptin (200 μM) | 99 ± 1 | 60 ± 4 |

[a] Cells were incubated with $^{125}I$-hCG for 2 hr at 4°C, washed to remove the free hormone, and then incubated at 37°C for 1 hr. The indicated inhibitors were present throughout the experiment. After removing the surface-bound hormone, the cells were homogenized and fractionated by the procedure of Merion and Sly (1983). The amount of receptor-bound $^{125}I$-hCG present in the endocytic vesicles and lysosomes was determined as described elsewhere (Ascoli, 1983, 1984; Dufau et al., 1973).

radioactivity only (Fig. 3A), one can determine the fate of the hormone and show that there is a precursor–product relationship whereby the surface-bound $^{125}I$-hCG is internalized, the internalized hormone is degraded, and the final degradation products appear in the medium. It should be noted that (1) this is a quantitative pathway; over 90% of the radioactivity that was initially bound to the cell surface is internalized and/or degraded within 4 hr of the temperature shift, the remainder appears back in the medium as intact $^{125}I$-hCG (Ascoli and Puett, 1978a,b; Ascoli, 1982b) and (2) the only degradation product detectable in the medium is monoiodotyrosine, which is the final product of hormone degradation (Ascoli and Puett, 1978a; Ascoli, 1980). The appearance of intermediate degradation products can be detected intracellularly and is discussed later.

By measuring how much of the internalized radioactivity is precipitable with polyethylene glycol (Dufau et al., 1973; Ascoli, 1983, 1984), it is possible to determine if the internalized $^{125}I$-hCG is free or receptor bound. The results presented in the Fig. 3B suggest a precursor–product relationship whereby the surface-bound hormone is internalized while bound to the receptor, the internalized hormone–receptor complex dissociates, and the free hormone is degraded. A more accurate, quantitative analysis of the internalization of the $^{125}I$-hCG–receptor complex can only be performed by using compounds that block the dissociation of the internalized complex. As show in Fig. 4 and Table I, there are at least three compounds ($NH_4Cl$, monensin, and leupeptin) that block the degradation of the internalized $^{125}I$-hCG and the intracellular dissociation of the hormone–receptor complex to about the same extent. The results (Table I) also

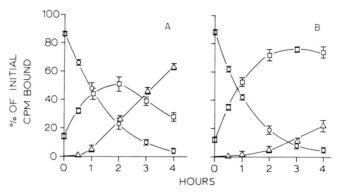

**Fig. 5.** Fate of the subunits of the cell-bound hCG. Cells were incubated with 20 ng/ml of $\alpha$-[125]I-hCG (A) or $\beta$-[125]I-hCG (B) for 2 hr at 4°C. After washing to remove the free hormone ($t = 0$), the incubation was continued at 37°C in hormone-free medium. The surface-bound (○) and internalized radioactivity (□) that remained associated with the cells and the mono[125]I]iodotyrosine released into the medium (△) were measured at the times indicated. (Reproduced with permission from Ascoli, 1982b.)

show that the internalized [125]I-hCG–receptor complex can be localized in endocytic vesicles and lysosomes and that the complex dissociates in the lysosomes. Intracellular degradation products of the hormone can also be detected in the lysosomes but not in the endocytic vesicles (Ascoli, 1984).

A more detailed analysis of the fate of the hormone was made possible by following the internalization and degradation of derivatives of hCG labeled exclusively in either the $\alpha$ or $\beta$ subunits (Ascoli, 1980, 1982b).* As shown in Fig. 5, both derivatives disappear from the cell surface at the same rate, but the accumulation of intracellular radioactivity and the rate of appearance of monoiodotyrosine in the medium are different. Additional experiments showed that the rates of internalization of either hormone derivative are very similar (Fig. 6) and that all the radioactivity that remained bound to the surface (at any of the time points shown in Fig. 5 and 6) is intact hCG, as opposed to the individual subunits (Ascoli, 1982b). These data, together with the results discussed earlier, show that the hormone is internalized in the intact form while bound to the receptor.

The difference in the rates of degradation of the two derivatives shown in Fig. 5, however, suggests that at some point the hormone subunits are processed differently. Since the hormone reaches the lysosomes while bound to the receptor

---

*Because of the distribution of reactive tyrosyl groups, the routine procedure used to label hCG with Na[125]I results in a derivative labeled exclusively in the $\alpha$ subunit (Ascoli, 1980, 1982a; Morgan et al., 1974). Thus, [125]I-hCG and $\alpha$-[125]I-hCG represent the same labeled species. The latter is used only in Figs. 5 and 6 to clarify the discussion.

(see Table I), it is safe to conclude that there is no subunit dissociation until the hormone–receptor complex reaches the lysosomes. At this point, however, the intracellular pathway for degradation depends on whether the α or β subunits are labeled.

A detailed analysis of the intracellular degradation products of the α- and β labeled derivatives is presented elsewhere (Ascoli, 1982b). Those results showed that subunit dissociation occurs during degradation and that the slower rate of release of monoiodotyrosine from the β-labeled hCG is due to the formation of several intermediate intracellular degradation products of the β subunit that are formed prior to the appearance of monoiodotyrosine. It is not known if this is simply due to differences in the position of the iodinated tyrosyl residues or to intrinsic differences in the susceptibility of each subunit to proteolysis. It is clear, however, that the observed differences are not due to differences in the location or mechanisms of degradation of each subunit, since (1) the only intracellular organelle that contains degradation products is the lysosome (Ascoli, 1984a) and

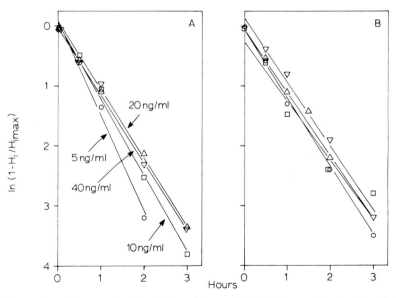

**Fig. 6.** Rate of internalization of the cell-bound hCG. Cells were incubated with the indicated concentrations of α-[125]I-hCG (A) or β-[125]I-hCG (B) for 2 hr at 4°C. After washing to remove the free hormone ($t = 0$), the cells were incubated in hormone-free medium at 37°C. At the indicated times the surface-bound, internalized, and degraded radioactivity were measured as described elsewhere (Ascoli, 1982b). The rates of internalization were calculated from the plot shown in which $H_{max}$ is the maximal amount of radioactivity processed (i.e., internalized and/or degraded) by the cells and $H_i$ is the radioactivity internalized at the indicated times (Schartz et al., 1982).

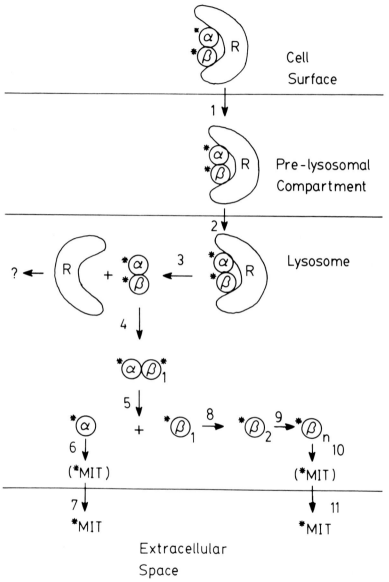

**Fig. 7.** A model for the intracellular pathway followed by the internalized receptor-bound hCG. See the text for details; asterisks indicate that the molecules are radioactively labeled. Abbreviations: *MIT, mono[$^{125}$I]iodotyrosine; R, receptor. (After Ascoli, 1980, 1982b, 1984).

(2) several inhibitors of hormone degradation block the degradation of either subunit to the same extent (Ascoli, 1980, 1982b, 1984).

These results are consistent with a model (Fig. 7) in which the hormone–receptor complex is internalized into endocytic vesicles (step 1) and transferred to the lysosomes (step 2). During these steps, both subunits of the hormone remain bound to the receptor. Once in the lysosomes, the hormone dissociates from the receptor but does not dissociate into subunits (step 3). The mechanisms responsible for dissociation of the hormone–receptor complex are not clearly understood. This phenomenon may occur as a consequence of the acidic pH that prevails in the lysosomes (Okhuma and Poole, 1978) and/or the effects of hydrolyases on the hormone and/or the receptor. In Fig. 7, hCG degradation is shown to begin after dissociation of the hormone–receptor complex (step 4) only for clarity. It is not known if this is indeed the case. The ultimate fate of the receptor is also unknown. Inasmuch as it is located in the lysosomes, it seems plausible that it is degraded, like the hormone is.

The first degradation product of hCG is composed of an intact α subunit and a partially degraded β subunit (step 4). This product dissociates into the intact α subunit and a partially degraded β subunit (step 5). The free α subunit is then quickly degraded to monoiodotyrosine, which leaves the cell and is detectable in the medium only (steps 6 and 7), while the free, partially degraded β subunit gives rise to several intracellular degradation products (steps 8 and 9), which, in turn, are degraded to monoiodotyrosine and released into the medium (steps 10 and 11). Thus, the preferential accumulation of intracellular radioactivity derived from the β subunit (see Fig. 5) is due to the formation of several labeled intracellular degradation products, while the lack of accumulation of intracellular radioactivity derived from the α subunit is due to the lack of intermediate, labeled degradation products. The only detectable degradation product of the α subunit is monoiodotyrosine, and this product is present in the medium only.

## B. Internalization and Degradation of hCG Are Not Needed for the Stimulation of cAMP and Steroid Biosynthesis

One obvious question that arises from the findings discussed earlier is whether the internalization and/or degradation of hCG are needed for the stimulation of cAMP and steroid biosynthesis. Two different approaches were used to test this possibility. First, because of the availability of a number of compounds that block either the internalization of the surface-bound hormone or the degradation of the internalized hormone, we tested the effects of these inhibitors on the ability of hCG to stimulate steroid biosynthesis. The data presented in Table II show the inhibitors of degradation (Ascoli, 1978, 1979, 1980, 1982b, 1984; Ascoli and Puett, 1978b,c). There are three classes of these inhibitors. Ammonium

**TABLE II**

**Inhibitors of the Degradation of Internalized hCG**[a]

| | Percentage of initial cpm bound | | | Progesterone (%) |
|---|---|---|---|---|
| Inhibitor | Surface | Internalized | Degraded | |
| None | 6 | 38 | 52 | 100 |
| NH$_4$Cl (10 mM) | 6 | 88 | 5 | 93 |
| Chloroquine (10 μM) | 6 | 89 | 5 | 88 |
| Chlorpromazine (10 μM) | 7 | 81 | 10 | 43 |
| Quinacrine (10 μM) | 6 | 92 | 1 | 75 |
| Monensin (25 μM) | 13 | 85 | 1 | 5 |
| Tosyl-Lys-CH$_2$Cl (200 μM) | 8 | 73 | 4 | 49 |
| Leupeptin (200 μM) | 7 | 88 | 4 | 96 |

[a] Cells were preincubated with [125]I-hCG (20 ng/ml) and the indicated inhibitors for 2 hr at 4°C. After washing, the cells were incubated for 3 hr at 37°C in hormone-free medium containing the indicated inhibitors. The surface-bound, internalized, and degraded [125]I-hCG and the amount of progesterone produced were determined as described elsewhere (Ascoli, 1982a). None of these inhibitors affect the binding of [125]I-hCG.

chloride, chloroquine, chlorpromazine, and quinacrine are weak bases that accumulate in acidic intracellular vesicles, such as endocytic vesicles and lysosomes, and inhibit their function by raising the intravesicular pH (De Duve *et al.*, 1974; Okhuma and Poole, 1978; Maxfield, 1982). Monensin also raises the intravesicular pH by exchanging H$^+$ for Na$^+$ (Pressman, 1976; Maxfield, 1982). Tosyl-Lys-CH$_2$Cl and leupeptin inhibit the action of some of the lysosomal proteases (Barret and Heath, 1977). As shown in Table II, these compounds have little or no effect on hormone internalization but are effective inhibitors of the degradation of the internalized hormone. These results also show that these compounds have variable effects on the ability of hCG to stimulate steroid biosynthesis. Three of them, NH$_4$Cl, chloroquine, and leupeptin, however, have little (if any) effect on hCG-stimulated steroidogenesis at concentrations that effectively inhibit hormone degradation. Thus, it was concluded that the degradation of the internalized hCG is not required for the stimulation of steroidogenesis. It should be noted that, although these results were obtained with the derivative of hCG labeled in the α subunit, we have also shown that these compounds are equally effective in blocking the degradation of the β subunit (Ascoli, 1980, 1982b). Moreover, recent studies have shown that all of these compounds prevent hCG degradation by blocking the function of lysosomes and also prevent the dissociation of the internalized hormone–receptor complex (see Fig. 4; Table I; Ascoli, 1984). Thus, in the presence of these inhibitors, the internalized hCG–receptor complex accumulates in the lysosomes and the hormone is not degraded.

The data presented in Table III show the inhibitors of hormone internalization (Ascoli and Puett, 1978b,c; Haigler *et al.*, 1980). There are two classes of these inhibitors. Sodium azide ($NaN_3$) and NaCN are representative of a number of metabolic inhibitors that have been shown to block internalization. The second class includes two "classic" chymotrypsin inhibitors and three other seemingly unrelated compounds. The mechanisms by which these inhibit internalization is not known, but interestingly, they have a single structural similarity: a carbonyl group adjacent to (or removed by at most one carbon atom) a benzene ring. All of these compounds prevent the internalization of the surface-bound hormone and thus inhibit hormone degradation. As shown in Table III, they are also effective inhibitors of hCG-stimulated steroid biosynthesis. These results do not necessarily mean that hormone internalization is necessary for the stimulation of steroid biosynthesis, because the compounds tested may inhibit these two processes in an independent fashion. To test this possibility, we examined the effects of these compounds on the ability of cAMP analogues to stimulate steroid biosynthesis (Ascoli, 1978). The data presented in Table IV show the results obtained with two of the compounds that inhibit internalization and one of the inhibitors of degradation of the internalized hormone (the same results were obtained with the other inhibitors of internalization and with chlorpromazine, quinacrine, and tosyl-Lys-CH$_2$Cl). The data show that these agents are also

**TABLE III**

**Inhibitors of hCG Internalization**[a]

| Inhibitor | Percentage of initial cpm bound | | | Progesterone (%) |
|---|---|---|---|---|
| | Surface | Internalized | Degraded | |
| None | 16 | 37 | 46 | 100 |
| NaN$_3$ (10 mM) | 69 | 19 | 11 | 8 |
| NaCN (10 mM) | 82 | 12 | 1 | 37 |
| None | 9 | 38 | 49 | 100 |
| Tosyl-Phe-CH$_2$Cl (150 μM) | 70 | 25 | 2 | 0.4 |
| CBZ-Phe-CH$_2$Cl[b] (150 μM) | 70 | 22 | 1 | 4 |
| p-Br-phenacyl bromide (100 μM) | 86 | 13 | 1 | 0.8 |
| Diphenylcarbamyl Chloride (100 μM) | 62 | 25 | 9 | 34 |
| Phenylglyoxal (1 mM) | 72 | 17 | 5 | 2 |

[a] Experimental details are the same as those given in the legend to Table II. The experiments using metabolic inhibitors were performed in medium devoid of glucose.
[b] CBZ, N-Carbobenzoxy.

TABLE IV

**Effect of Inhibitors of hCG Internalization and Degradation on Steroidogenesis**[a]

| | Progesterone (ng/μg DNA) | | |
|---|---|---|---|
| Inhibitor | Basal | hCG | 8-Br-cAMP |
| None | 0.08 | 24.1 | 29.4 |
| Monensin (25 μM) | 0.25 | 0.24 | 0.25 |
| p-Br-phenacyl bromide (100 μM) | 0.17 | 0.12 | 0.16 |
| Phenylglyoxal (1 mM) | 0.10 | 1.2 | 2.0 |

[a] Cells were incubated with the indicated inhibitors for 3 hr at 37°C in the presence or absence of hCG (40 ng/ml) or 8-Br-cAMP (1 mM) prior to the assay of progesterone.

effective in blocking cAMP-stimulated steroidogenesis, suggesting that they block hCG internalization and hCG-stimulated steroidogenesis by different mechanisms. Thus, they are not useful in the study of a possible relationship between these two processes.

A second approach to this problem was made possible by the development of a method that removes the surface-bound hCG without affecting cellular integrity or the subsequent ability of the cells to respond to freshly added hCG or cAMP (Ascoli, 1982b). Thus, it is possible to allow the cells to bind and internalize hCG, remove the surface-bound hormone, and test if this affects cAMP or steroid biosynthesis (Segaloff and Ascoli, 1981). The results of one such experiment are presented in Fig. 8 and show that, at any time after the addition of hCG, the removal of the surface-bound hormone results in the cessation of steroidogenesis (or of cAMP biosynthesis; see Segaloff and Ascoli, 1981). Inasmuch as this effect is obtained at any time after the addition of hCG, when increasing amounts of hormone have been internalized, it is reasonable to conclude that it is the surface-bound hCG, as opposed to the internalized and/or degraded hCG, that is responsible for the activation of cAMP and steroid biosynthesis.

Thus, neither the internalization nor degradation of hCG are required for its stimulatory effects on steroid biosynthesis. Since hormone internalization effectively removes the hormone from the cell surface, it is possible that this process is involved in the termination of cAMP and steroid biosynthesis.

## C. Internalization and Degradation of hCG and the Down-Regulation of hCG Receptors

The data discussed earlier show that, upon binding to its cell surface receptor, hCG is internalized by receptor-mediated endocytosis and degraded in the lysosomes and that these processes are not involved in the stimulation of cAMP and steroid biosynthesis. The results presented in this section deal with the

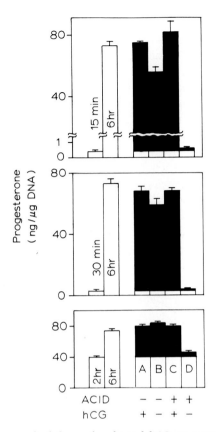

**Fig. 8.** Effect of removal of the surface-bound hCG on progesterone biosynthesis. Cells were incubated at 37°C with hCG (40 ng/ml) for 15 min (top panel), 30 min (middle panel), or 2 hr (bottom panel). At this time, the free hormone was removed by washing, and the cells were divided into four groups that received no further treatment (A or B) or were treated with acid (C and D) to remove the surface-bound hormone (Ascoli, 1982b). Then, groups A and C received warm medium containing hCG (40 ng/ml), while groups B and D received medium only. A second incubation (37°C) of variable length was then performed so that the total incubation time (first and second) was 6 hr. Progesterone was measured in the medium at the end of the first (open portion of bars) and second incubations (shaded portion of bars). The open bars on the left show progesterone biosynthesis by cells incubated in the continuous presence of hCG for the times indicated. (Reproduced with permission from Segaloff and Ascoli, 1981.)

involvement of receptor-mediated endocytosis in the regulation of hCG receptors.

Our initial observations on the internalization and degradation of the surface-bound hCG showed that during these processes there was a decrease in the number of functional hCG cell surface receptors (Ascoli and Puett, 1978a). This phenomenon can be easily demonstrated using two different experimental approaches. First, when cells are incubated with a saturating concentration of radio-labeled hCG at 37°C, the hormone is internalized and degraded, but in spite of the presence of an excess of free, biologically active hormone in the medium, the cells are unable to maintain an elevated level of surface-bound hormone (Fig. 9). Second, when cells are incubated with hCG at 37°C for increasing periods of time and then tested for their ability to bind $^{125}$I-hCG, there is a time-dependent loss of binding that cannot be accounted for by residual occupancy of the receptor (Fig. 10). This decrease in the binding of hCG has been shown to be due to a reduction in the number of functional cell surface receptors rather than to a change in the affinity of the receptor and will henceforth be referred to as "homologous down-regulation" (Freeman and Ascoli, 1981; Lloyd and Ascoli, 1983). The role of the receptor-mediated endocytosis of hCG in this process is stressed by the finding that, as discussed earlier, the

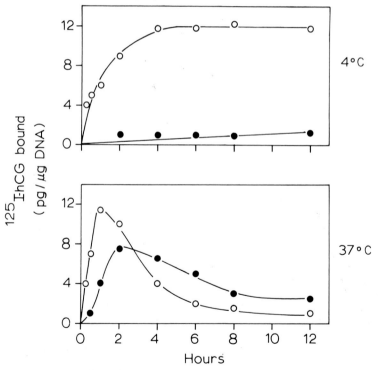

**Fig. 9.** Association of $^{125}$I-hCG with MA-10 cells at 4° and 37°C. Cells were incubated with $^{125}$I-hCG (20 ng/ml) at the indicated temperatures. The surface-bound (○) and internalized radioactivities (●) were measured at the times indicated. (Reproduced with permission from Ascoli, 1982b.)

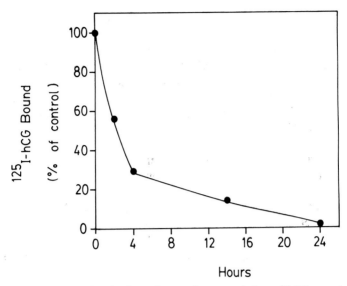

**Fig. 10.**   Time course for the homologous down-regulation of hCG receptors. Cells were incubated with hCG (40 ng/ml) at 37°C. At the times indicated, the cells were washed to remove the free hormone and treated with acid to remove the surface-bound hormone prior to determining $^{125}$I-hCG binding. (Reproduced with permission from Freeman and Ascoli, 1981.)

hormone is internalized while bound to its receptor and that this complex accumulates in the lysosomes, where the receptor can be degraded (see Section III,A).

Further support for this hypothesis includes the findings that (1) hCG increases the rate at which its receptors disappear from the surface and (2) the homologous down-regulation of hCG receptors can be measured not only in the cell surface but also in detergent extracts of the cells (Lloyd and Ascoli, 1983; see also Section V).

## IV. INTERACTION OF mEGF WITH MA-10 CELLS

The MA-10 cells have separate receptors for mEGF and hCG (Ascoli, 1981b). Evidence for the specificity of these receptors is provided by the findings that a high concentration of mEGF does not prevent the binding of $^{125}$I-hCG and a high concentration of hCG does not prevent the binding of $^{125}$I-mEGF. Moreover, the number of mEGF receptors present in MA-10 and MA-14 cells are nearly identical, while the number of hCG receptors in these two cells differs about 50-fold. The MA-10 cells have about 100,000 mEGF receptors per cell and bind mEGF with an affinity association constant (measured at 4°C) of $2-3 \times 10^8 \, M$ (Lloyd and Ascoli, 1983).

During the initial characterization of the several clonal lines of cultured Leydig tumor cells, it was noted that mEGF had little or no effect on cell multiplication but reduced the $^{125}$I-hCG binding activity of all the clonal lines of cultured Leydig tumor cells (Ascoli, 1981a,b). We subsequently showed that the maximal effect of mEGF on $^{125}$I-hCG binding was (1) observed after a 40–50 hr incubation, (2) dependent on the amount of mEGF used (half-maximal and maximal effects are observed at $8 \times 10^{-11} \, M$ and $3.2 \times 10^{-10} \, M$, respectively), and (3) due to a true reduction in the number of receptors rather than to a decrease in the affinity of the receptor (Ascoli, 1981b; Lloyd and Ascoli, 1983).

Thus, the number of surface hCG receptors in MA-10 cells can be down-regulated with two different hormones: hCG, the homologous hormone, and mEGF, a heterologous hormone (Table V). In addition, mEGF also down-regulates its own receptors, as expected from the results obtained in other cell types, but hCG has little or no effect on the number of mEGF receptors (Table V).

Inasmuch as the binding of mEGF to MA-10 cells in other cell types is followed by the internalization and degradation of mEGF and the down-regulation of mEGF receptors, it was possible that the reduction in hCG receptors resulted from the cointernalization of these receptors. This possibility, however, seems unlikely, since upon addition of mEGF, the binding and internalization

TABLE V

Effects of Prolonged Exposure to mEGF or hCG
on $^{125}$I-mEGF and $^{125}$I-hCG Binding in MA-10 Cells[a]

| Pretreatment | $^{125}$I-Labeled hormone bound (pg/μg DNA) | |
| --- | --- | --- |
| | $^{125}$I-mEGF | $^{125}$I-hCG |
| None | 5.24 | 21.40 |
| mEGF | 0.14 | 5.30 |
| hCG | 4.50 | 0.30 |

[a] Cells were incubated at 37°C for 48 hr with mEGF (5 ng/ml) or for 24 hr with hCG (40 ng/ml). The binding of $^{125}$I-mEGF and $^{125}$I-hCG was determined at 4°C, as described elsewhere (Lloyd and Ascoli, 1983). The conditions chosen for the incubations with mEGF and hCG have been shown to result in optimal reduction in the binding of $^{125}$I-hCG (Ascoli, 1981b; Freeman and Ascoli, 1981; Lloyd and Ascoli, 1983).

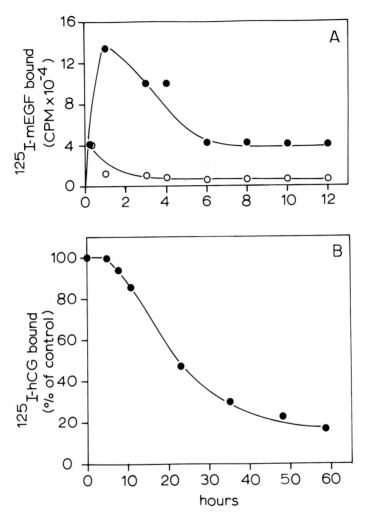

**Fig. 11.** Time course of the binding and internalization of $^{125}I$-mEGF (A) and the down-regulation of hCG receptors by mEGF (B). (A) Cells were incubated at 37°C with $^{125}I$-mEGF (5 ng/ml). The surface-bound and internalized ligands were measured as described by Ascoli (1982b). (B) Cells were incubated at 37°C with mEGF (5 ng/ml). $^{125}I$-hCG binding was determined, after washing the cells, at the times indicated. (The data shown in (B) are reproduced from Lloyd, C. E., and Ascoli, M., *The Journal of Cell Biology*, 1983, **96**, 521–526, by copyright permission of The Rockefeller University Press.)

attain a steady state within 6 hr, whereas the down-regulation of hCG receptors shows a lag of 6–8 hr and reaches a steady state within 40–50 hr (Fig. 11).

## V. MECHANISMS INVOLVED IN THE HOMOLOGOUS AND HETEROLOGOUS DOWN-REGULATION OF hCG RECEPTORS ARE DIFFERENT

In order to study the mechanisms by which mEGF and hCG down-regulate the surface hCG receptors, we (Lloyd and Ascoli, 1983) employed a model that was initially described by Wiley and Cunningham (1981, 1982). This analysis starts with the assumption that at any given steady state the number of surface hCG receptors $[R]_S$ is given by

$$[R]_S = V_R/kt \tag{1}$$

where $V_R$ is the rate of appearance of receptors at the cell surface and $kt$, the rate constant for the disappearance of the surface receptors. It follows then that $[R]_S$ can be decreased by decreasing $V_R$ and/or increasing $kt$.

By measuring these rates, it is possible to show that hCG reduces its own receptors by increasing the rate at which receptors disappear from the surface (Table VI), while mEGF decreases hCG receptors by decreasing the rate at which these receptors appear at the surface (Table VII). Clearly then, the homologous and heterologous down-regulation of hCG receptors occur by different mechanisms. It should also be pointed out that, in these and other cells, the homologous down-regulation of mEGF receptors also appears to occur as a result of an increase in the rate of receptor internalization (Lloyd and Ascoli, 1983; Wiley and Cunningham, 1981; McKanna et al., 1979; Stoscheck and Carpenter, 1984).

TABLE VI

**Homologous Down-Regulation of Surface hCG Receptors in MA-10 Cells**[a]

| Parameter | − hCG | + hCG |
|---|---|---|
| $[R_S]$ (receptors/cell) | 12,075 (100) | 600 (5) |
| $V_R$ (receptors/cell · min) | 21 (100) | 24 (114) |
| $kt$ (min$^{-1}$) | $1.7 \times 10^{-3}$ (100) | $4 \times 10^{-2}$ (2,352) |

[a] Cells were preincubated with or without hCG (40 ng/ml) for 24 hr at 37°C. The parameters shown were measured as described by Lloyd and Ascoli (1983). The numbers in parentheses show percentages relative to the control cells. (Reproduced from Lloyd, C. E., and Ascoli, M., *The Journal of Cell Biology*, 1983, **96**, 521–526, by copyright permission of The Rockefeller University Press.)

**TABLE VII**

**Heterologous Down-Regulation of Surface hCG Receptors by mEGF in MA-10 Cells[a]**

| Parameter | − mEGF | + mEGF |
|---|---|---|
| $[R_S]$ (receptors/cell) | 12,075 (100) | 2593 (20) |
| $V_R$ (receptors/cell · min) | 21 (100) | 4 (19) |
| $kt$ (min$^{-1}$) | $1.7 \times 10^{-3}$ (100) | $1.5 \times 10^{-3}$ (88) |

[a] The parameters shown were measured after a 48 hr incubation at 37°C in the presence or absence of mEGF (5 ng/ml). The numbers in parentheses show percentages relative to the control cells. (Reproduced from Lloyd, C. E., and Ascoli, M., *The Journal of Cell Biology*, 1983, **96**, 521–526, by copyright permission of The Rockefeller University Press.)

As used here $V_R$ and $kt$ are operational definitions that offer no insights into the biochemical processes involved in the appearance and disappearance of the surface hCG receptors. The changes observed may be due to changes in the transport of receptors to and from the surface with or without changes in the number of total (i.e., surface plus intracellular) cellular receptors. Thus, if the

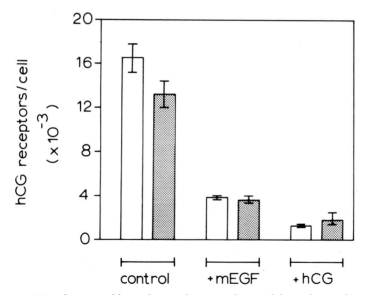

**Fig. 12.** Homologous and heterologous down-regulation of the surface and total hCG receptors. Cells were preincubated with mEGF (5 μg/ml) for 48 hr or hCG (40 ng/ml) for 24 hr and used to measure the number of hCG receptors in the intact cells (open bars) or in detergent extracts of the cells (shaded bars), as described by Lloyd and Ascoli (1983) and Ascoli (1983), respectively.

reduction in the surface hCG receptors is accompanied by a reduction in the total cellular hCG receptors, the data presented earlier could be interpreted to mean that $V_R$ and $kt$ are proportional to the rates of receptor synthesis and degradation, respectively. The results presented in Fig. 12 show that mEGF and hCG reduce the number of hCG receptors measurable in the cell surface and detergent extracts of the cells to about the same extent and together with the results shown earlier (Tables VI and VII) suggest that mEGF decreases the rate of synthesis and hCG increases the rate of degradation of the hCG receptor.

It should be stressed, however, that this is only a working hypothesis that will be tested in future experiments as new tools are developed to directly follow the synthesis, processing, and degradation of the hCG receptor.

## VI. REGULATION OF STEROIDOGENIC RESPONSES BY mEGF AND hCG

### A. Regulation of hCG-Stimulated Steroidogenesis

The results presented in Table VIII show that, when the surface hCG receptors are down-regulated with mEGF or hCG, there is a parallel loss of hCG-stimulated progesterone production. Two changes occur as a result of receptor down-regulation: (1) the maximal amount of progesterone produced is reduced to an extent similar to the reduction in hCG binding and (2) the amount of hCG required for half-maximal stimulation of progesterone biosynthesis ($ED_{50}$) increases 2- to 3-fold. These are the expected results for a cell type such as these that do not have a large excess of spare receptors (see Fig. 2).

### B. Regulation of Cholera Toxin- and cAMP-Stimulated Steroidogenesis

The results discussed earlier clearly show that in MA-10 cells, the down-regulation of hCG receptors leads to a decrease in the steroidogenic response to hCG. Inasmuch as this response can also be activated by compounds that do not bind to the hCG receptor, such as cholera toxin and cAMP analogues (see Section II), it was of interest to determine if the down-regulation of hCG receptors affects these responses. The results presented in Fig. 13 show that the steroidogenic response to these stimuli is reduced by about 50% in the hCG-treated cells but remains unchanged in the mEGF-treated cells. This hCG-induced loss of response to cholera toxin and cAMP analogues had been previously described in normal Leydig cells [reviewed in Catt *et al.* (1980) and Ascoli (1985)] and in freshly isolated Leydig tumor cells (Segaloff *et al.,* 1981a,b) and has been called desensitization. The results presented in Fig. 13 suggest that the

**TABLE VIII**

**Effects of mEGF and hCG on $^{125}I$-hCG Binding and $^{125}I$-hCG-Stimulated Progesterone Production[a]**

| Experiment number | Pretreatment | Binding | | Progesterone | |
|---|---|---|---|---|---|
| | | Maximum (molecules/cell) | $ED_{50}$ ($M \times 10^{-10}$) | Maximum (ng/$\mu$g DNA) | $ED_{50}$ ($M \times 10^{-10}$) |
| 1 | None | 10,440 (100) | 1.3 (100) | 29.4 (100) | 0.34 (100) |
| | hCG | 596 (5.7) | 1.2 (92) | 6.5 (22) | 0.90 (289) |
| 2 | None | 8799 (100) | 2.1 (100) | 28 (100) | 1.1 (100) |
| | mEGF | 1285 (15) | 1.7 (81) | 6 (21) | 2.3 (209) |

[a] Cells were preincubated with hCG (40 ng/ml) or mEGF (5 ng/ml) for 14 or 48 hr, respectively. After washing, $^{125}I$-hCG binding and $^{125}I$-hCG-stimulated progesterone production were measured during a 2 hr incubation at 37°C. The numbers in parentheses show percentages relative to the control cells. (Reproduced with permission from Ascoli, 1981b; Freeman and Ascoli, 1981.)

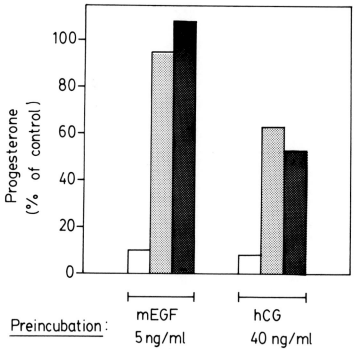

**Fig. 13.** Effects of hCG (open bars), cholera toxin (lightly shaded bars), and 8-Br-cAMP (heavily shaded bars) on progesterone production by hCG- and mEGF-treated cells. Cells were incubated (37°C) with the indicated concentrations of mEGF or hCG for 48 or 14 hr, respectively. The cells were washed and progesterone production was determined after a further incubation (4 hr, 37°C) in serum-free medium containing hCG (40 ng/ml), cholera toxin (1.2 nM), or 8-Br-cAMP (1 mM). (Redrawn by permission from Ascoli, 1981b; Freeman and Ascoli, 1981.)

effects of mEGF on the steroidogenic response of MA-10 cells is confined to the hCG receptors, while the effects of hCG are expressed at the level of the hCG receptor and at some other point(s) in the steroidogenic pathway. Inasmuch as the loss of response to hCG is more pronounced than the loss of response to cholera toxin and cAMP analogues, it can be concluded that the receptor defect is limiting.

An obvious difference between the effects of mEGF and hCG on MA-10 cells is that, while hCG has pronounced stimulatory effects on cAMP and steroid biosynthesis, mEGF, by itself, has little or no effect on these processes (Ascoli, 1981b; D. L. Segaloff, personal communication). Thus, it was of interest to determine if the hCG-induced desensitization (see Fig. 13) was due to the stimulation of steroidogenesis. This was done by exposing cells to either cholera toxin or cAMP analogues under conditions in which the stimulation of steroidogenesis is similar to that observed with hCG but there is no reduction in the number of hCG receptors (Freeman and Ascoli, 1981). As shown in Fig. 14, when cells that had been pretreated with cholera toxin or cAMP analogues are stimulated with hCG, cholera toxin, or cAMP analogues, their steroidogenic response is reduced by about 50%. It is important to note that, under these conditions, the loss of steroidogenic response to all three stimuli is very similar (Fig. 14), while in the hCG-treated cells there is a more drastic reduction in the response to hCG than in the response to cholera toxin or cAMP analogues (Fig. 13).

Taken together, these results show that the steroidogenic response of MA-10 cells can be regulated by changing the number of hCG receptors or by reducing the activity of one or more steps in the steroidogenic pathway. The reduction in the activity of the steroidogenic pathway is expressed independently of the reduction in hCG receptors and appears to be a direct result of the stimulation of steroid biosynthesis. As discussed in the following sections, this defect is entirely due to cholesterol depletion and can be corrected by adding low-density lipoprotein.

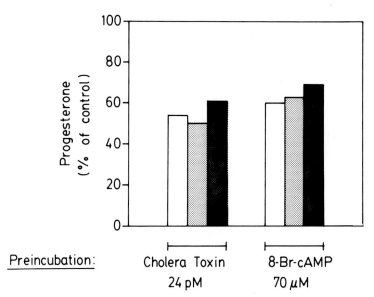

**Fig. 14.** Effects of hCG (open bars), cholera toxin (lightly shaded bars), and 8-Br-cAMP (heavily shaded bars) on progesterone production by cholera toxin- and 8-Br-cAMP-treated cells. Cells were incubated with the indicated concentrations of cholera toxin or 8-Br-cAMP for 14 hr at 37°C. After washing, the cells were incubated (37°C, 4 hr) in serum-free medium containing hCG (40 ng/ml), cholera toxin (1.2 nM), or 8-Br-cAMP (1 mM) prior to the determination of progesterone. (Reproduced with permission from Freeman and Ascoli, 1981.)

## VII. REGULATION OF STEROIDOGENIC RESPONSES BY CHOLESTEROL AVAILABILITY: THE ROLE OF LOW-DENSITY LIPOPROTEIN

### A. Sources of Cholesterol Used for Steroid Biosynthesis

Like other steroid-producing cells, MA-10 cells can derive the cholesterol needed for steroid biosynthesis from one or more of the following three sources: (1) intracellular stores, (2) de novo synthesis, and (3) lipoproteins. Our recent studies on this subject show that, when progesterone biosynthesis is stimulated with hCG, MA-10 cells utilize all of these sources to different extents, depending on their availability and on the duration of the hormonal stimulation (Ascoli, 1981c; Freeman and Ascoli, 1982a,b, 1983).

Under basal conditions, MA-10 cells have two to four times more free than esterified cholesterol (Albert *et al.*, 1980; Segaloff *et al.*, 1981b; Ascoli, 1981c; Freeman and Ascoli, 1982a,b, 1983) and the rate of progesterone biosynthesis is very low. The prevailing levels of intracellular cholesterol appear to be derived primarily from a relatively high level of de novo synthesis rather than from the uptake of lipoproteins (Morris and Chaikoff, 1959; Andersen and Dietschy,

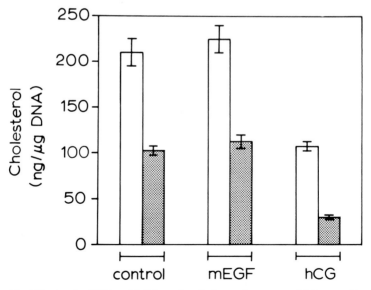

**Fig. 15.** Effects of mEGF and hCG on the cholesterol content of MA-10 cells. Cells were incubated at 37°C for 24 hr in lipoprotein-deficient medium containing mEGF (5 ng/ml) or hCG (40 ng/ml), prior to the determination of cholesterol. Symbols: open bars, free forms; shaded bars, esterified forms. (Reproduced with permission from Freeman and Ascoli, 1982b.)

1978; Ascoli, 1981c; Freeman and Ascoli, 1982a). Upon the addition of hCG, about 70% of the progesterone synthesized during the first 4 hr is derived from intracellular stores. Kinetic studies showed that the esterified cholesterol is used more rapidly than the free cholesterol. As a consequence of this, the preexisting stores of free and esterified cholesterol decline by about 50% and 70%, respectively, regardless of the presence of LDL.

If LDL is not present, there is a net decline in intracellular cholesterol and the rate of progesterone biosynthesis eventually falls because the rate of de novo synthesis of cholesterol is not high enough to keep up with the enhanced demand. Thus, under prolonged hormonal stimulation in the absence of LDL, about 95% of the progesterone synthesized by the cells after depletion of the intracellular stores is derived from newly synthesized cholesterol.

If LDL is present, there is no net decline in intracellular cholesterol because this pool is quickly replenished by the uptake of LDL. Under these conditions, the rate of progesterone biosynthesis remains elevated for long periods of time, and 50–60% of the cholesterol needed for steroid biosynthesis is derived from LDL (Freeman and Ascoli, 1982a, 1983).

It should also be noted that LDL is more active than HDL in supporting steroidogenesis in MA-10 cells (Freeman and Ascoli, 1982a,b).

## B. Cholesterol As a Determinant of the Steroidogenic Capacity of MA-10 Cells

As discussed earlier, cholesterol is an important determinant of the long-term steroidogenic response of MA-10 cells. Since our earlier studies (see Figs. 13 and 14) dealing with the effects of hCG and mEGF on the steroidogenic response of MA-10 cells to several steroidogenic stimuli had been done in media devoid of lipoproteins (Ascoli, 1981b; Segaloff et al., 1981a,b; Freeman and Ascoli, 1981), it seemed possible that the steroidogenic response of the cells was limited by the availability of cholesterol. This hypothesis was also supported by the finding that treatment of MA-10 cells with hCG, which results in a reduction in the subsequent steroidogenic response to cholera toxin and cAMP, depleted the intracellular stores of cholesterol; while treatment with mEGF, which does not affect the aforementioned responses, had no effect on intracellular cholesterol (Fig. 15).

We then reasoned that, if the lack of cholesterol is indeed responsible for desensitization, then our knowledge of cholesterol dynamics (see earlier) would predict that LDL should reverse this phenomenon. The results presented in Fig. 16A show the effects of LDL on the dibutyryl cAMP- (Bt$_2$-cAMP) stimulated steroidogenic response of cells that had been preincubated in lipoprotein-deficient medium with or without hCG or mEGF. The addition of LDL to control or mEGF-treated cells increases the Bt$_2$-cAMP-stimulated progesterone bio-

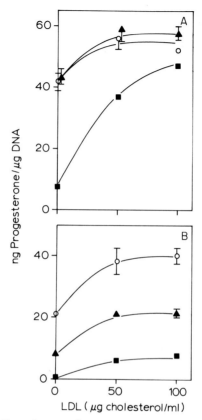

**Fig. 16.** Effects of LDL on the steroidogenic responses of hCG on mEGF-treated cells. Cells were preincubated (24 hr, 37°C) in lipoprotein-deficient medium containing buffer only (○), mEGF (5 ng/ml) (▲), or hCG (40 ng/ml) (■). After washing, the cells were placed in lipoprotein-deficient medium containing the indicated concentrations of LDL and Bt$_2$-cAMP (1 mM) (A) or hCG (40 ng/ml) (B). The amount of progesterone produced was determined at the end of an 8 hr incubation at 37°C. (Reproduced with permission from Freeman and Ascoli, 1982b.)

synthesis less than twofold. As expected, the response of these two groups of cells is very similar regardless of the presence of LDL. In contrast, the steroidogenic response of the hCG-treated cells to Bt$_2$-cAMP is drastically reduced if LDL is not present, but the addition of LDL restores this response to control levels. The results presented in Fig. 16B are from a similar experiment in which the control, mEGF-treated, or hCG-treated cells were restimulated with hCG. The addition of LDL again increased progesterone biosynthesis in all groups, as described earlier, but the maximal response of the mEGF- or hCG-treated cells remained well below control. Thus, it is concluded that (1) LDL restores the Bt$_2$-cAMP-stimulated progesterone biosynthesis to normal levels and

(2) the reduction of hCG receptors limits the steroidogenic response to hCG even when LDL is present.

The finding that LDL overcomes the hCG-induced desensitization is consistent with the hypothesis that cholesterol depletion is responsible for desensitization. If the effect of LDL is mediated by increasing the intracellular cholesterol stores, it should be possible to show that cells treated with hCG and LDL have greater cholesterol stores than cells treated with hCG in the absence of LDL. Moreover, the steroidogenic response of these cells to cAMP should correlate with their cholesterol content. The results presented in Table IX and Fig. 17 validate these predictions.

## C. LDL Pathway of MA-10 Cells

The effects of LDL on the steroidogenic response of MA-10 cells can be entirely explained on the basis of the delivery of cholesterol by the classical LDL pathway (Brown and Goldstein, 1979; Brown et al., 1979).

We have shown that MA-10 cells have specific receptors that mediate the binding, internalization, and subsequent degradation of LDL by a pathway with biochemical properties closely resembling those described for the LDL pathway in a number of other cell types (Freeman and Ascoli, 1983). As summarized earlier, LDL-derived cholesterol is readily used for steroid biosynthesis and in addition it exerts the expected effects that have been described in other cells (Brown and Goldstein, 1979). Thus, addition of LDL to MA-10 cells results in a decrease in HMG-CoA reductase activity, an increase in the rate of cholesterol esterification, and a decrease in the number of LDL receptors (Freeman and Ascoli, 1982a, 1983). On the other hand, the addition of hCG to MA-10 cells results in the expected effects of cholesterol depletion, an increase in HMG–CoA

TABLE IX

**Effects of LDL and hCG on Cholesterol Content of MA-10 Cells[a]**

| Additions | | Cellular cholesterol content (ng/µg DNA) | |
|---|---|---|---|
| hCG (40 ng/ml) | LDL (50 µg/ml) | Free | Esterified |
| − | − | 206 ± 12 | 92 ± 3 |
| + | + | 107 ± 5 | 31 ± 3 |
| − | + | 270 ± 6 | 286 ± 39 |
| + | + | 233 ± 22 | 169 ± 10 |

[a] Cells were incubated (37°C) in lipoprotein-deficient medium containing the indicated additions for 24 hr prior to the determination of cholesterol. (Reproduced with permission from Freeman and Ascoli, 1982b.)

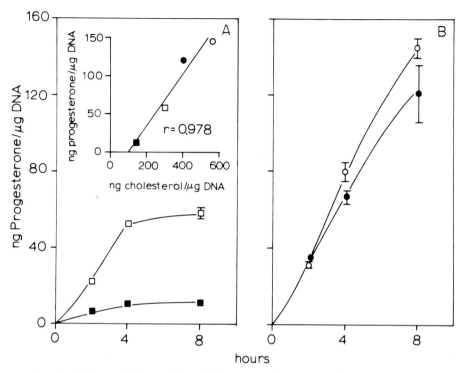

**Fig. 17.** Effects of LDL and Bt$_2$-cAMP on steroid production by hCG-treated cells. Cells were preincubated (24 hr, 37°C) in lipoprotein-deficient medium without (A) or with (B) LDL cholesterol (50 ng/ml) in the absence (O,□) or presence (●,■) of hCG (40 ng/ml). After washing, the cells were incubated (37°C) in lipoprotein-deficient medium containing Bt$_2$-cAMP (1 mM) without (A) or with (B) LDL cholesterol (50 μg/ml). The accumulation of progesterone in the medium was measured at the times indicated. (Reproduced with permission from Freeman and Ascoli, 1982b.)

reductase activity, a decrease in cholesterol esterification, and an increase in the number of LDL receptors (Freeman and Ascoli, 1982a, 1983). That these effects of hCG are mediated by the hCG-induced depletion of cholesterol is supported by the finding that when added with hCG, aminoglutethimide, an inhibitor of steroidogenesis, blocks steroidogenesis, cholesterol depletion, and the effects described earlier.

## VIII. SUMMARY

Based on the studies summarized here, we propose the following sequence of events to explain the effects of mEGF and hCG on the hCG receptors and the steroidogenic responses of MA-10 cells to hCG and cAMP.

Upon binding to its cell-surface receptors, hCG stimulates progesterone biosynthesis, while bound to the cell surface, by activating the adenylate cyclase–protein kinase pathway. The surface-bound hCG is quickly internalized (while bound to the receptor and without dissociating the subunits) into endocytic vesicles which deliver the intact hormone–receptor complex to the lysosomes. Once in the lysosomes, the hCG–receptor complex dissociates, and both subunits of the free hormone are eventually degraded to single amino acids which are released into the medium. This process also results in the accumulation of the hCG receptor in the lysosomes. Although it is not known with certainty if the receptor is degraded, the results discussed here suggest that this is the case and that the ability of hCG to down-regulate its own receptors is due to an increase in the rate of internalization and degradation of the receptor.

The stimulation of steroid biosynthesis by hCG results in an increased demand for cholesterol. During the early phase of stimulation, most of the cholesterol needed for steroid biosynthesis is derived from intracellular sources which are eventually depleted unless LDL is present. When LDL is present, the cellular content of cholesterol is maintained, and the rate of progesterone biosynthesis remains elevated for prolonged periods of time, because LDL delivers cholesterol to the cell via the classic LDL pathway. If LDL is not present, the elevated rate of progesterone biosynthesis declines because it becomes dependent on the rate of de novo synthesis of cholesterol. These metabolic changes have pronounced effects on the differentiated function of the cells. Thus, their steroidogenic response to compounds that do not act through the hCG receptor, such as cAMP (or analogues thereof), is absolutely dependent on the presence of LDL. If LDL is not present, the hCG-treated cells will fail to respond well with increased steroid biosynthesis, simply because they do not have enough cholesterol to support this response. On the other hand, if steroidogenesis is stimulated with freshly added hCG, the cells will fail to respond well even in the presence of LDL, because of the limitation imposed by the reduced number of hCG receptors.

A second ligand, mEGF, that binds to its own surface receptors, also affects the differentiated function of MA-10 cells. Upon binding to the cells, mEGF is rapidly internalized and degraded and its surface receptors are down-regulated. When the cells are in prolonged contact with mEGF, the surface hCG receptors are also down-regulated, because of a reduction in one or more of the processes (synthesis?) leading to the appearance of hCG receptors at the surface. In MA-10 cells, mEGF is not a growth factor, does not stimulate the adenylate cyclase–protein kinase pathway, and has little or no effect on the content of cellular cholesterol and steroid biosynthesis. The net result of the interaction of mEGF with MA-10 cells is a reduction in their steroidogenic response to hCG because of the reduction of hCG receptors. Cells pretreated with mEGF have a normal steroidogenic response to compounds that do not act through the hCG receptor, because their cholesterol stores are normal.

## ACKNOWLEDGMENTS

Unpublished observations from this laboratory were supported by a grant from the National Cancer Institute (CA-23603). It is a pleasure to acknowledge the contributions of Carolyn Lloyd and Drs. Dale A. Freeman and Deborah Segaloff to the work reviewed here.

## REFERENCES

Albert, D. A., Ascoli, M., Puett, D., and Coniglio, J. G. (1980). Lipid composition and gonadotropin-mediated lipid metabolism of the M5480 murine Leydig cell tumor. *J. Lipid Res.* **21,** 862–867.

Andersen, J. M., and Dietschy, J. M. (1978). Relative importance of high and low density lipoproteins in the regulation of cholesterol synthesis in the adrenal gland, ovary, and testis of the rat. *J. Biol. Chem.* **253,** 9024–9032.

Ascoli, M. (1978). Demonstration of a direct effect of inhibitors of the degradation of receptor-bound human choriogonadotropin on the steroidogenic pathway. *J. Biol. Chem.* **253,** 7839–7843.

Ascoli, M. (1979). Inhibition of the degradation of receptor-bound human choriogonadotropin by leupeptin. *Biochim. Biophys. Acta* **586,** 608–614.

Ascoli, M. (1980). Degradation of the subunits of receptor-bound human choriogonadotropin by Leydig tumor cells. *Biochim. Biophys. Acta* **629,** 409–417.

Ascoli, M. (1981a). Characterization of several clonal lines of cultured Leydig tumor cells: gonadotropin receptors and steroidogenic responses. *Endocrinology (Baltimore)* **108,** 88–95.

Ascoli, M. (1981b). Regulation of gonadotropin receptors and gonadotropin responses in a clonal strain of Leydig tumor cells by Epidermal Growth Factor. *J. Biol. Chem.* **256,** 179–183.

Ascoli, M. (1981c). Effects of hypocholesterolemia and chronic hormonal stimulation on sterol and steroid metabolism in a Leydig cell tumor. *J. Lipid Res.* **22,** 1247–1253.

Ascoli, M. (1982a). Regulation of steroid production in gonadal, adrenal, and placental tumor cells. *In* "Cellular Regulation of Secretion and Release" (P. M. Conn, ed.), pp. 409–445. Academic Press, New York.

Ascoli, M. (1982b). Internalization and degradation of receptor-bound human choriogonadotropin in Leydig tumor cells: fate of the hormone subunits. *J. Biol. Chem.* **257,** 13306–13311.

Ascoli, M. (1983). An improved method for the solubilization of stable gonadotropin receptors. *Endocrinology (Baltimore)* **113,** 2129–2134.

Ascoli, M. (1984a). Lysosomal accumulation of the hormone-receptor complex during receptor-mediated endocytosis of human choriogonadotropin. *J. Cell Biol.* **99,** 1242–1250.

Ascoli, M. (1985). Regulation of luteinizing receptors and actions. *In* "Luteinizing Hormone Receptors and Actions" (M. Ascoli, ed.), CRC Press, Boca Raton, Florida. In press.

Ascoli, M., and Puett, D. (1978a). Gonadotropin binding and stimulation of steroidogenesis in Leydig tumor cells. *Proc. Natl. Acad. Sci. U.S.A.* **75,** 99–102.

Ascoli, M., and Puett, D. (1978b). Degradation of receptor-bound human choriogonadotropin by murine Leydig tumor cells. *J. Biol. Chem.* **253,** 4892–4899.

Ascoli, M., and Puett, D. (1978c). Inhibition of the degradation of receptor-bound human choriogonadotropin by lysosomotropic agents, protease inhibitors, and metabolic inhibitors. *J. Biol. Chem.* **253,** 7832–7838.

Barret, A. J., and Heath, M. F. (1977). Lysosomal enzymes. *In* "Lysosomes. A Laboratory Handbook" (J. T. Dingle, ed.), pp. 19–146. Elsevier/North-Holland, New York.

Brown, M. S., and Goldstein, J. L. (1979). Receptor-mediated endocytosis: insights from the lipoprotein receptor system. *Proc. Natl. Acad. Sci. U.S.A.* **76,** 3330–3337.

Brown, M. S., Kovanen, P. T., and Goldstein, J. L. (1979). Receptor-mediated uptake of lipoprotein-cholesterol and its utilization for steroid biosynthesis in the adrenal cortex. *Rec. Prog. Horm. Res.* **35,** 215–257.

Brown, M. S., Anderson, R. G. W., and Goldstein, J. L. (1983). Recycling receptors: the round trip itinerary of migrant membrane proteins. *Cell* **32,** 663–667.

Catt, K. J., Harwood, J. P., Clayton, R. N., Davies, T. F., Chan, V., Katikineni, J., Nozu, K., and Dufau, M. L. (1980). Regulation of peptide hormone receptors and gonadal steroidogenesis. *Rec. Prog. Horm. Res.* **36,** 557–622.

De Duve, C., De Barsy, T., Poole, B., Trouet, A., Tulkeus, P., and Van Hoof, F. (1974). Lysosomotropic agents. *Biochem. Pharmacol.* **23,** 2495–2531.

Dufau, M. L., Charreau, E. H., and Catt, K. J. (1973). Characteristics of a soluble gonadotropin receptor from the rat testis. *J. Biol. Chem.* **248,** 6973–6982.

Freeman, D. A., and Ascoli, M. (1981). Desensitization to gonadotropins in cultured Leydig tumor cells involves loss of gonadotropin receptors and decreased capacity for steroidogenesis, *Proc. Natl. Acad. Sci. U.S.A.* **78,** 6309–6313.

Freeman, D. A., and Ascoli, M. (1982a). Studies on the source of cholesterol used for steroid biosynthesis in cultured Leydig tumor cells. *J. Biol. Chem.* **257,** 14231–14238.

Freeman, D. A., and Ascoli, M. (1982b). Desensitization of steroidogenesis in cultured Leydig tumor cells: role of cholesterol. *Proc. Natl. Acad. Sci. U.S.A.* **79,** 7796–7800.

Freeman, D. A., and Ascoli, M. (1983). The low density lipoprotein pathway of cultured Leydig tumor cells. Utilization of low density lipoprotein-derived cholesterol for steroidogenesis. *Biochim. Biophys. Acta* **754,** 72–81.

Goldstein, J. L., Anderson, R. G. W., and Brown, M. S. (1979). Coated pits, coated vesicles, and receptor-mediated endocytosis. *Nature (London)* **279,** 679–685.

Haigler, H. T., Willingham, M. C., and Pastan, I. (1980). Inhibitors of $^{125}$I-epidermal growth factor internalization. *Biochem. Biophys. Res. Commun.* **94,** 630–637.

Hunzicker-Dunn, M., and Birnbaumer, L. (1985). The involvement of adenylyl cyclase and cyclic AMP-dependent protein kinases in luteinizing hormone actions. *In* "Luteinizing Hormone Receptors and Actions." (M. Ascoli, ed.). CRC Press, Boca Raton, Florida, In press.

Lacroix, A., Ascoli, M., Puett, D., and McKenna, T. J. (1979). Steroidogenesis in hCG-responsive Leydig cell tumor variants. *J. Steroid Biochem.* **10,** 669–675.

Lloyd, C. E., and Ascoli, M. (1983). On the mechanisms involved in the regulation of the cell-surface receptors for human choriogonadotropin and mouse epidermal growth factor in cultured Leydig tumor cells. *J. Cell Biol.* **96,** 521–526.

McKanna, J. A., Haigler, H. T., and Cohen, S. (1979). Hormone receptor topology and dynamics: morphological analysis using ferritin-labeled epidermal growth factor. *Proc. Natl. Acad. Sci. U.S.A.* **76,** 5689–5693.

Maxfield, F. R. (1982). Weak bases and ionophores rapidly and reversibly raise the pH of endocytic vesicles in cultured mouse fibroblasts. *J. Cell Biol.* **95,** 676–681.

Mendelson, C., Dufau, M. L., and Catt, K. J. (1975). Gonadotropin binding and stimulation of cyclic adenosine 3':5'-monophosphate and testosterone production in isolated Leydig cells. *J. Biol. Chem.* **250,** 8818–8823.

Merion, M., and Sly, W. S. (1983). The role of intermediate vesicles in the adsorptive endocytosis and transport of ligand to lysosomes by human fibroblasts. *J. Cell Biol.* **96,** 363–371.

Morgan, F. J., Kaye, G. I., and Canfield, R. E. (1974). Characterization of preparations of radioiodinated human chorionic gonadotropin. *In* "Heterogeneity of Polypeptide Hormones" (D. Rabinowitz and J. Roth, eds.), pp. 81–89. Academic Press, New York.

Morris, M. D., and Chaikoff, L. L. (1959). The origin of cholesterol in liver, small intestine, adrenal gland, and testis of the rat: dietary *versus* endogenous contributions. *J. Biol. Chem.* **234,** 1095–1097.

Moyle, W. R., and Greep, R. O. (1974). Steroid-secreting tumors as models in endocrinology. *In* "Hormones and Cancer" (K. W. McKerns, ed.), pp. 329–361. Academic Press, New York.

Neaves, W. B. (1975). Growth and composition of a transplantable murine Leydig cell tumor. *J. Natl. Cancer Inst.* **55**, 623–631.

Ohkuma, S., and Poole, B. (1978). Fluorescence probe measurement of the intralysosomal pH in living cells and the perturbation of pH by various agents. *Proc. Natl. Acad. Sci. (U.S.A.)* **75**, 3327–3331.

Payne, A. H., Quinn, P. G., and Stalvey, J. R. D. (1985). The stimulation of steroid biosynthesis by luteinizing hormone. *In* "Luteinizing Hormone Receptors and Actions" (M. Ascoli, ed.). CRC Press, Boca Raton, Florida, In press.

Pressman, B. C. (1976). Biological applications of ionophores. *Annu. Rev. Biochem.* **45**, 501–530.

Schwartz, A. L., Fridovich, S. E., and Lodish, H. F. (1982). Kinetics of internalization and recycling of the asialoglycoprotein receptor in a hepatoma cell line. *J. Biol. Chem.* **257**, 4230–4237.

Segaloff, D. L., and Ascoli, M. (1981). Removal of the surface bound human choriogonadotropin results in the cessation of hormonal responses in cultured Leydig tumor cells. *J. Biol. Chem.* **256**, 11420–11423.

Segaloff, D. L., Puett, D., and Ascoli, M. (1981a). Dynamics of the steroidogenic response of Leydig tumor cells to ovine luteinizing hormone, human choriogonadotropin, cholera toxin, and adenosine 3′,5′-monophosphate. *Endocrinology (Baltimore)* **108**, 632–638.

Segaloff, D. L., Ascoli, M., and Puett, D. (1981b). Characterization of the desensitized state of Leydig tumor cells. *Biochim. Biophys. Acta* **675**, 351–358.

Steinman, R. M., Mellman, I. S., Muller, W. A., and Cohn, Z. A. (1983). Endocytosis and the recycling of plasma membrane. *J. Cell Biol.* **96**, 1–27.

Stoscheck, C. M., and Carpenter, G. (1984). Down-regulation of epidermal growth factor receptors: direct demonstration of receptor degradation in human fibroblasts. *J. Cell Biol.* **98**, 1048–1053.

Tsuruhara, T., Dufau, M. L., Cigorraga, S., and Catt, K. J. (1977). Hormonal regulation of testicular luteinizing hormone receptors. *J. Biol. Chem.* **252**, 9002–9009.

Wiley, H. S., and Cunningham, D. D. (1981). A steady state model for analyzing the cellular binding, internalization and degradation of polypeptide ligands. *Cell* **25**, 433–440.

Wiley, H. S., and Cunningham, D. D. (1982). The endocytotic rate constant. A cellular parameter for quantitating receptor-mediated endocytosis. *J. Biol. Chem.* **257**, 4222–4229.

# 12

# Somatostatin Receptors in Endocrine Cells

**BORIS DRAZNIN**
Research Service
Veterans Administration Medical Center and
Department of Medicine
University of Colorado Health Sciences Center
Denver, Colorado

## I. INTRODUCTION

In 1968 Krulich *et al.* (1) discovered a fraction of rat and sheep hypothalamus that inhibited growth hormone release *in vitro*. This provided the first evidence for a hypothalamic factor capable of inhibiting growth-hormone secretion from the anterior pituitary gland. Five years later, Brazeau, Vale, and colleagues isolated somatostatin (SRIF) from sheep hypothalamus (2). It was found to be a cyclic peptide containing 14 amino acid residues. Two cystine residues (in the third and fourteenth positions) were joined by two disulfide bonds. Later, a linear somatostatin was synthesized and found to have identical biological activity with the cyclic form (3,4).

Schally *et al.* (5,6) isolated and sequenced somatostatin from porcine hypothalamus in 1975. Porcine somatostatin was shown to have the same structure as ovine somatostatin. Similarly, analysis of pigeon and human somatostatins revealed that they were identical to the porcine and ovine peptides (7). In contrast, somatostatin isolated from the angler fish was found to be structurally different from the mammalian somatostatin (8).

**401**

Initially, somatostatin was discovered in high concentrations in the hypo-thalamus but soon it was localized to the cortex, midbrain, brain stem, spinal cord, sensory ganglia, delta cells of the islets of Langerhans, in gastric and intestinal epithelium, and in parafollicular cells in the thyroid (9).

The widespread and rather ubiquitous distribution of somatostatin suggested a diversity of physiological functions. Subsequent work has confirmed a variety of biological roles for somatostatin, both within the nervous system and in many tissues outside of the brain.

The biological actions of somatostatin can be divided into these three major categories: (1) somatostatin's action as neurotransmitter, (2) somatostatin's inhi-bition of hormone secretion, and (3) somatostatin's inhibition of nonendocrine secretion.

Endocrine and nonendocrine actions of somatostatin are summarized in Table I.

Although the precise mechanism of somatostatin action is still unknown, the initial step is believed to be an interaction of somatostatin with its specific plasma membrane receptors on target cells (10,11). Following the initial binding event, somatostatin has been suggested to lower cellular cyclic AMP levels (12–15) as well as to inhibit cyclic AMP-dependent protein kinase (16). In the nervous system, somatostatin has been shown to influence the uptake release of calcium by synaptosomes (17). Because the movement of calcium ions is crucial for both nerve terminal depolarization and hormone secretion, it was speculated that

**TABLE I**

**Effects of Somatostatin**[a]

| Hormonal secretion[b] | Nonhormonal secretion |
| --- | --- |
| GH | Gastric acid secretion |
| TSH | Gastric emptying |
| ACTH (Nelson's syndrome) | Pancreatic exocrine function |
| Gastrin | Splanchnic blood flow |
| CCK–Pancreozymin | Salivary amylase secretion |
| Secretin | |
| VIP | |
| GIP | |
| Motilin | |
| Pancreatic polypeptide | |
| Insulin | |
| Glucagon | |
| Renin | |

[a] Modified from Wass (9).

[b] CCK, cholecystokinin; VIP, vasoactive intestinal peptide; GIP, gastric inhibitory polypeptide.

somatostatin exerts its action by affecting calcium transport through the nerve cell and endocrine membrane (18).

Receptors for somatostatin have been discovered in a variety of endocrine and nonendocrine cells. The control mechanisms which allow somatostatin to act on one tissue or at a particular subcellular site or biochemical pathway must be very finely tuned and under very specific regulation. The magnitude of somatostatin action can be modulated by its ambient concentration, the number of receptor sites and their affinity for somatostatin, and the activity of yet unidentified intracellular postbinding events. Each of these variables may be a critical determinant of the level of somatostatin action. A central thesis which will be advanced in this chapter is that the plasma membrane somatostatin receptor concentration can undergo rapid change, which may be of paramount importance in determining the degree of specificity of somatostatin action in diverse tissues.

In this chapter, we will discuss somatostatin binding to its receptors in endocrine cells. Special attention will be given to the role of somatostatin receptor translocation from the cell interior to the plasma membrane in regulating target-cell sensitivity to somatostatin.

## II. CHARACTERISTICS OF SOMATOSTATIN BINDING IN ENDOCRINE CELLS

The initial description of somatostatin binding to rat pituitary tumor cell lines appeared in 1978 (10). In their article, Schonbrunn and Tashjian have reported that $^{125}I$-somatostatin binds to high-affinity binding sites on $GH_4C_1$ cells. The $GH_4C_1$ cells are a clonal strain of the rat pituitary tumor line known to secrete both prolactin and growth hormone. Although the main advantage in using these cells for studying somatostatin binding and action is their apparent homogeneity, they are transformed cells and as such may differ from normal cells in their binding ability and responsiveness to somatostatin. The experiments with these cells revealed one class of high-affinity binding sites for somatostatin. Scatchard analysis of binding data indicated that each cell contains approximately 13,000 receptor sites. The dissociation constant for somatostatin was found to be $6.3 \times 10^{-10} M$. Half-maximal binding of somatostatin to these cells occurred at a concentration of $6 \times 10^{-10} M$.

The specificity of somatostatin binding in this and other studies with different endocrine cells was proved by demonstrating reversibility of binding reaction, lack of ligand displacement by a host of peptides, and a positive correlation between somatostatin analogues' binding and their biological activity (10,19). Binding of $^{125}I$-SRIF was not inhibited by TRH, substance P, neurotensin, LRH, calcitonin, ACTH, or insulin, indicating specificity of binding sites of somatostatin.

When the ability of somatostatin to bind to its specific receptor in $GH_4C_1$ cells was correlated with its ability to inhibit growth hormone and prolactin release, a close relationship between somatostatin binding and action was found. Thus, the half-maximal inhibition of growth hormone and prolactin release was observed at approximately $5-10 \times 10^{-10}$ $M$ of somatostatin (10).

Somatostatin binding to rat anterior pituitary membranes was similar to that observed with $GH_4C_1$ cells (20–25). Specific binding of somatostatin was sensitive to pH, with maximum binding obtained at pH 7.5–8, as well as time and temperature dependent. At 37°C, the maximal somatostatin binding is usually reached at 20 min of incubation. At 4°C, somatostatin binding reaches a plateau at approximately 14 hr of incubation. Kinetics of SRIF binding in rat pituitary cells and plasma membranes is summarized in Table II.

Some discrepancies in somatostatin-binding characteristics outlined in this table could be accounted for by differences in the ligand used, its specific activity, the time and temperature of the binding reaction, the composition of binding buffers, variable rates of ligand degradation, and variations in performing Scatchard analysis.

Nevertheless, all investigators reported specificity of somatostatin binding and an excellent correlation of somatostatin analogues' binding with their biological potency.

Rapid degradation of $^{125}I$-labeled-tyrosine-somatostatin by intact cells, plasma membranes, and other subcellular organelles has been commonly observed (20). To minimize somatostatin degradation, bacitracin is usually added to the incubation media. In addition, it was discovered that specific somatostatin binding was inhibited by increasing concentrations of calcium in the incubation medium (20).

Somatostatin binding to a single cell population from normal pituitary has not been performed. It is possible that different pituitary cells may express a distinct binding capacity for somatostatin. However, demonstration of a single population of binding sites in membranes prepared from mixed anterior pituitary cells suggested that somatostatin receptors located on somatotroph, thyrotroph, and lactotroph cells are identical.

Somatostatin binding to bovine pituitary membranes differed from that observed in rat pituitary cells and membranes (24). In bovine membranes, the kinetics of somatostatin binding suggested the presence of binding sites of lower affinity ($1.9 \times 10^{-8}$ $M$) than those described in the experiments with rodent tissues. In contrast, the binding capacity of bovine membranes was much greater than that of rat. Again, differences in specific activity of the ligand, buffers used, time of binding reaction, and application of Scatchard analysis may be partially responsible for the differences observed.

Specific receptors for somatostatin have been found and characterized in the rat adrenal glomerulosa zone both *in vivo* and *in vitro* (25). In *in vivo* experi-

**TABLE II**

**Characteristics of Somatostatin Binding in Pituitary Cells and Membranes**

| $K_D$ (M) | $R_o$ (pmoles/mg protein) | Ligand used | Tissue | Reference |
|---|---|---|---|---|
| $6.3 \times 10^{-10}$ | 0.11 | $^{125}I\text{-}Tyr^I$ | $CH_4C_1$ | Schonbrunn and Tashjian (10) |
| $9.1 \times 10^{-10}$ | 0.10 | $^{125}I\text{-}Tyr^I$ | Rat anterior pituitary membranes | Enjalbert et al. (20) |
| $8.6 \times 10^{-9}$ | 0.054 | $^{125}I\text{-}Tyr^I$ ($D\text{-}Trp^8$) | Rat anterior pituitary membranes | Aguilera and Parker (21) |
| $2.43 \times 10^{-9}$ | 0.1 | $^{125}I\text{-}Tyr^{II}$ | Rat anterior pituitary membranes | Reubi et al. (23) |
| $2.12 \times 10^{-9}$ | 0.095 | $^{125}I\text{-}Tyr^{II}$ | Rat anterior pituitary membranes | Srikant and Patel (22) |

ments, $^{125}$I-somatostatin was rapidly injected into animals and the tissue uptake of somatostatin was studied. The highest uptake was in the adrenal capsule (with the tissue–blood ratio being 4.5) followed by kidney (ratio is 2), anterior pituitary (ratio is 1.8), and liver (ratio is 1.4*).

*In vitro* experiments demonstrated that specific binding of $^{125}$I-somatostatin to adrenal particles was temperature and time dependent. With increased temperature, the rate of association was more rapid but the maximal binding was decreased and the period of equilibrium was shorter. An excess of unlabeled somatostatin rapidly displaced the ligand from its binding sites, suggesting specificity of binding reaction.

Interestingly, increasing concentrations of calcium in incubation media did not affect somatostatin binding to adrenal capsule or particles, but an increase in monovalent cations (sodium and potassium) decreased somatostatin binding. Scatchard analysis of binding data revealed a single population of binding sites with an affinity constant of $1.5 \times 10^{-10} M$. The binding capacity for somatostatin was found to be $367 \pm 54$ fmoles/mg of protein. Several somatostatin analogues displaced native somatostatin from its binding sites but other peptides such as oxytocin, vasopressin, substance P, gonadotropin releasing hormone (GnRH), and angiotensin II showed little or no competition with native somatostatin.

Increasing concentrations of somatostatin reduced angiotensin II-induced aldosterone production. The half-maximal inhibition was observed at a somatostatin concentration of $2.6 \times 10^{-10} M$, suggesting good correlation between somatostatin binding and its biological activity. Furthermore, in studying somatostatin analogues, these authors have found a good correlation between their binding and action in terms of inhibition of steroidogenesis.

Another endocrine tissue that became the subject of intense investigation of somatostatin binding and action was the pancreatic islets and islet cells.

Leitner *et al.* (26) originally described somatostatin binding to isolated pancreatic islets in 1980. These observations have been subsequently confirmed in experiments with hamster insulinoma cells, primarily comprised of β cells (27). The affinity of somatostatin binding sites was lower in intact rat islets, but the binding capacity of islets was several fold greater than that of insulinoma cells. Table III demonstrates the results of somatostatin binding in pancreatic cells.

Biochemical observations of somatostatin binding in pancreatic islets were further supported by autoradiographic studies of Patel *et al.* (28). In these experiments, an association of $^{125}$I-somatostatin autoradiographic grains with β, α, and δ cells was clearly demonstrated. Quantitative assessment of somatostatin

---

*Somatostatin binding can be demonstrated in adipocytes for which there is no known action of this ligand. Furthermore, nonspecific somatostatin binding can be observed in hepatocytes and erythrocytes (26).

TABLE III

Characteristics of Somatostatin Binding in Pancreatic Islets and Insulinoma Cells

| $K_D$ (M) | $R_o$ (pmoles/mg) | Ligand | Tissue | Authors |
|---|---|---|---|---|
| $0.25 \times 10^{-9}$ | 0.068 | [Leu$^8$,D-Trp$^{22}$, Tyr$^{25}$]SS-28 | Hamster insulinoma cells | Reubi et al. (27) |
| $7.8 \times 10^{-8}$ | 5.67 | $^{125}$I-Tyr$^I$ | Rat pancreatic islets | Draznin et al. (33) |

binding demonstrated that α cells showed the highest density of labeling with $^{125}$I-somatostatin: 75% of α cells, 37% of β cells, and 33% of δ cells were labeled with radioactive ligand.

## III. MODULATION OF SOMATOSTATIN BINDING BY HORMONES AND SECRETAGOGUES

A most interesting development in studying somatostatin binding in endocrine cells was the discovery that several hormones and secretagogues can regulate the number of somatostatin receptors on the surface of these cells.

Schonbrunn and Tashjian (29) have demonstrated that an addition of 100 m$M$ TRH to cultures of GH$_4$C$_1$ cells for 60 min caused an increase in somatostatin binding to 150–200% (Fig. 1). The TRH effect was evident within 5 min of pretreatment. In these experiments, somatostatin-receptor affinity was unchanged and only receptor concentration was influenced by the TRH treatment. Moreover, the TRH effect was not affected by cycloheximide. The latter experiments suggested that TRH action involved a pool of receptors either in the plasma membrane or within the cell.

In contrast to the TRH effect, chronic treatment of GH$_4$C$_1$ cells with cortisol resulted in a 20–40% decrease in the specific binding of $^{125}$I-somatostatin (30). The maximal effect of cortisol was seen after 8 hr of treatment and persisted for at least 48 hr. Dexamethasone had a similar effect on somatostatin binding, but 17β-estradiol, 17α-methyltestosterone, testosterone, or progesterone had no effect. Similar to experiments with TRH, somatostatin-receptor affinity was the same in control and cortisol-treated cells ($K_D = 7.8 \times 10^{-10}$ $M$). The only variable was the number of somatostatin receptors on the cell surface.

A rapid modulation of somatostatin binding was also observed in pancreatic islets (31). Stimulatory concentrations of glucose or tolbutamide enhanced somatostatin binding to pancreatic islets within 10–30 min of incubation (Fig.

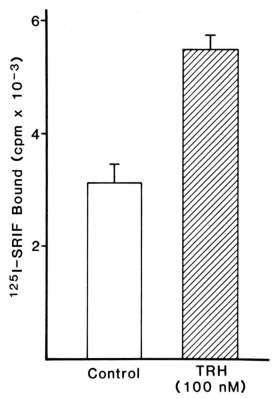

**Fig. 1.** Effect of TRH (100 nM) on $^{125}$I-somatostatin binding to $GH_4C_1$ pituitary cells. [Modified from Schonbrunn and Tashjian (29).]

2). In contrast, galactose, a sugar that does not evoke insulin release, did not affect somatostatin binding to pancreatic islets. The increase in somatostatin binding stimulated by the secretagogues (glucose or tolbutamide) was due to an increase in the number of somatostatin receptors on the cell surface. Receptor affinity remained unchanged.

The stimulatory effect of glucose was dose-dependent with the maximal stimulation of both insulin release and somatostatin binding at a glucose concentration of 200 mg/dl (Fig. 3). An increase in somatostatin binding evoked by the stimulatory levels of glucose was observed as early as within 10 min of incubation. This increase in plasma membrane receptor concentration was not prevented by the protein synthesis inhibitor, cycloheximide (32), suggesting that glucose, like TRH in $GH_4C_1$ cells, influenced an intracellular pool of somatostatin receptors.

Inhibitors of secretion vesicle migration (colchicine and $D_2O$) and the meta-

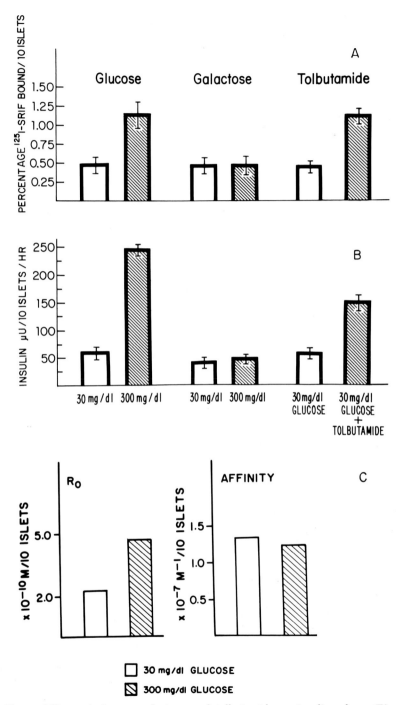

**Fig. 2.** Effects of glucose, galactose, and tolbutamide on insulin release (B) and somatostatin binding (A) in isolated pancreatic islets. $R_0$ represents receptor concentration. [From Sussman *et al.* (11).]

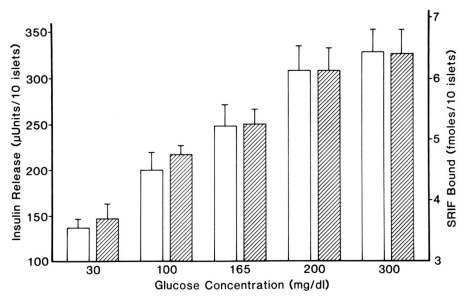

**Fig. 3.** Effect of increasing concentration of glucose on insulin release (open bars) and somatostatin binding (hatched bars) in isolated pancreatic islets. Results represent mean ± SEM of six experiments.

bolic inhibitor DNP inhibited both insulin release and glucose-induced enhancement in somatostatin binding (32). These experiments strongly suggested not only the existence of a somatostatin receptor pool but also a somatostatin receptor translocation from the cell interior to the plasma membrane during exocytosis. Furthermore, from the previous work of Leitner *et al.* (26), we knew that somatostatin receptors were located in association with secretion vesicles.

Based on this evidence and following this logic, we then attempted to study the size and mobility of the intracellular somatostatin receptor pool. To do this we have compared somatostatin binding in intact pancreatic islets with that in homogenized pancreatic islets. If homogenization exposes intracellular somatostatin receptors to the ligand, then the binding of [125]I-somatostatin in disrupted islets would be greater than somatostatin binding in intact islets. This approach revealed that the bulk of somatostatin receptors in pancreatic islets is located intracellularly (33). There was no difference in the affinity of SRIF receptors regardless of their cellular localization (Fig. 4).

In the presence of nonstimulatory glucose (30 mg/dl), only 8% of the total cellular receptor pool was present on the surface of pancreatic islets. Upon stimulation with higher glucose concentrations (300 mg/dl), an additional 10–12% of the intracellular somatostatin receptors were translocated to the plasma membrane (33).

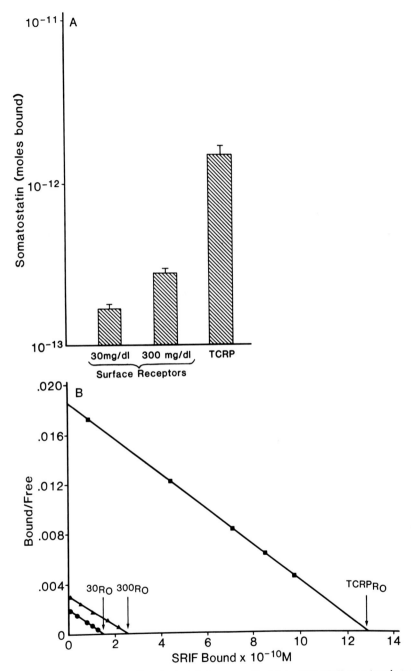

**Fig. 4.** Somatostatin binding in isolated pancreatic islets. [125]I-SRIF was incubated with either intact islets (in the presence of 30 or 300 mg/dl of glucose) or homogenized islets. The latter reveals the total cellular receptor pool (TCRP). Results of binding experiments are shown in (A) and Scatchard analysis in (B), for which $K_D$ values were TCRP = $7.4 \times 10^{-8}$ M; 300 mg/dl glucose = $8.3 \times 10^{-8}$ M; and 30 mg/dl glucose = $7.6 \times 10^{-8}$ M. [From Draznin et al., (33). Reproduced with permission from the American Diabetes Association, Inc.]

**TABLE IV**

**Specific $^{125}$I-SRIF Binding to Plasma Membranes and Secretion Vesicles**

| Organelle | $R_o$ |
|---|---|
| Pituitary (bovine) plasma membranes | $8.4 \times 10^{-14}$ moles/U 5'-nucleotidase |
| Pituitary (bovine) secretion vesicles | $4.4 \times 10^{-13}$ moles/U 5'-nucleotidase |
| Isolated rat islets | $4.8 \times 10^{-15}$ moles/$\mu$g protein |
| Islet secretion vesicles | $1.7 \times 10^{-14}$ moles/$\mu$g protein |

The presence of intracellular somatostatin receptors and strong evidence in favor of their translocation via secretion vesicles prompted us to compare somatostatin binding to a variety of subcellular organelles. In pancreatic islets, this comparison revealed the highest somatostatin binding in the fraction of secretion granules. Similarly, secretion granules from bovine anterior pituitary cells bound more somatostatin than any other subcellular organelle isolated from these cells (Table IV). Thus, these experiments further supported an association of somatostatin receptors with secretion vesicles in both pancreatic islets and anterior pituitary cells (11).

To further prove this association, we have examined binding of gold-conjugated somatostatin to isolated secretion vesicles. In these experiments, a 95% pure fraction of secretion vesicles was isolated from bovine anterior pituitary cells. Gold-conjugated somatostatin was prepared by the method of Jennes *et al.* (34). One particle of colloidal gold was associated with 60 molecules of somatostatin. Gold-conjugated somatostatin retained its biological activity and inhibited glucose-induced insulin release as native somatostatin. Gold-conjugated somatostatin was incubated with the isolated intact secretion vesicles for 30 min at 37°C and fixed for electron micsrocopy. Electron micrographs (Fig. 5) clearly demonstrated the existence of somatostatin receptors on the surface of secretion vesicles (35). An excess of "cold" somatostatin reduced binding of gold-conjugated ligand by approximately 10-fold.

The mechanism of translocation of somatostatin receptors from the secretion vesicle to the plasma membrane is not fully understood. It most certainly involves migration of the secretion vesicle to the cell surface and possibly its fusion with the plasma membrane. If translocation of somatostatin receptors is closely associated with secretion vesicle margination to the plasma membrane, then the recruitment of somatostatin receptors could be used as an index of secretion vesicle margination following secretagogue-induced insulin release.

We have recently designed a new experimental approach to separately study secretion vesicle margination to the plasma membrane and secretion vesicle lysis. The separation of the two events can be achieved by substituting sodium isethionate for sodium chloride in the incubation media. Sodium isethionate is an

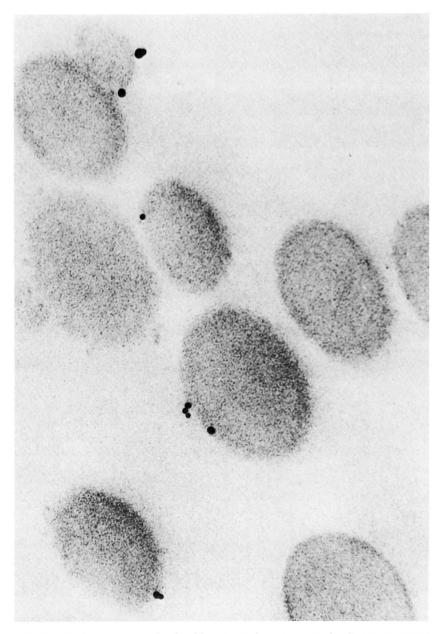

**Fig. 5.** Electron micrograph of gold-conjugated somatostatin binding in secretory granules isolated from bovine anterior pituitaries. Gold-conjugated somatostatin was prepared according to the method described by Jennes et al. (34).

impermeant anion that inhibits osmotic lysis of exocytotic vesicles (36,37). Utilizing sodium isethionate we were able to prevent secretagogue-induced insulin release without any change in the rate and magnitude of somatostatin receptor recruitment (37). The results of these experiments demonstrated that somatostatin binding in islets incubated with high glucose (300 mg/dl) and sodium isethionate (120 m$M$) rose similarly to that in control islets, despite an inhibition of insulin release (Fig. 6).

Since somatostatin receptors are located in the secretion vesicle fraction and become accessible to an extracellular somatostatin even without secretion vesicle lysis, one can draw the following conclusions: (1) Somatostatin receptors must be located on the outer surface of secretion vesicles. (2) Somatostatin binding may be occurring at the junction of the secretion vesicle with the plasma mem-

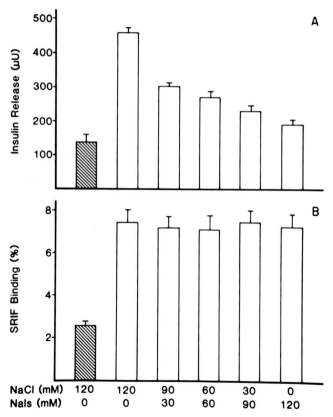

**Fig. 6.** Effect of sodium isethionate (NaIs) on glucose-induced insulin release (A) and recruitment of somatostatin receptors (B) in isolated pancreatic islets. Control experiments (glucose 30 mg/dl) are shown by the hatched bars.

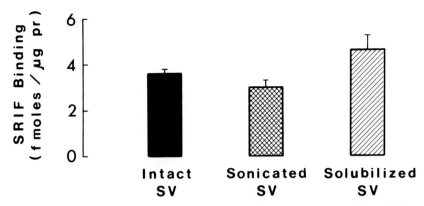

**Fig. 7.** Somatostatin binding to intact, sonicated (15 sec × 3), or solubilized (Triton X-1%) secretion vesicles (SV) isolated from bovine anterior pituitaries. Results represent mean ± SEM of five experiments.

brane. This site of somatostatin action might be important for a specific biological function of somatostatin: an inhibition of hormone release by acting at a particular anatomic locus rather than interference with all cyclic AMP-dependent intracellular processes.

The presence of somatostatin receptors on the cytoplasmic surface of secretion vesicles was so much unexpected that it necessitated a series of additional experiments to establish the orientation of these receptors. If secretion vesicles contain both cytoplasmic and intraluminal somatostatin receptors, then disruption of these vesicles or their solubilization would increase somatostatin binding. Following this logic, we have compared somatostatin binding to intact secretion vesicles with somatostatin binding to solubilized and sonicated secretion vesicles. We found that somatostatin binding to all secretion vesicles preparations remained the same and was not altered by either sonication or solubilization of secretion vesicles (Fig. 7).

In the next series of experiments, we have treated isolated intact secretion vesicles with pronase. Pronase treatment removes somatostatin receptors from the surface while pronase does not penetrate secretion vesicles. In these experiments, pronase treatment removed approximately 85% of specific somatostatin binding in intact secretion vesicles. When the secretion vesicles, previously treated with pronase, were subsequently disrupted and incubated with [125]I-somatostatin, there was no additional somatostatin binding to these disrupted pronase-treated secretion vesicles (Fig. 8). These experiments support our observation that the somatostatin receptor is located on the outer surface of the secretion vesicles.

At present, we have no explanation how somatostatin receptors originally located on the cytoplasmic surface of secretion vesicle become accessible to the extracellular somatostatin without lysis of the secretion vesicle.

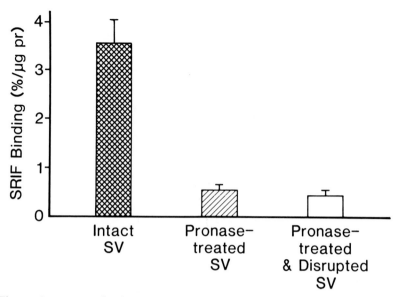

**Fig. 8.** Somatostatin binding to intact, pronase-treated, and pronase-treated and soni-cated secretion vesicles (SV). Secretion vesicles were incubated with pronase (13 μg/ml) for 60 min at 37°C. The reaction was stopped by diluting the incubation mixture 8-fold with an ice-cold buffer containing phenylmethylsulfonyl fluoride (PMSF) (1 mM). Secretion vesicles were then collected by centrifugation, and one group was placed into binding buffer containing [125]I-SRIF for 16 hr at 4°C. The second group was sonicated for 45 sec prior to binding studies. Results represent the mean ± SEM of five experiments.

## IV. BIOLOGICAL SIGNIFICANCE OF SOMATOSTATIN-RECEPTOR TRANSLOCATION

The biological action of a polypeptide hormone could be regarded as a function of ambient hormone concentration, binding to its target-cell receptor, and postbinding intracellular events. Alterations at any one of these levels could result in augmentation or diminution of the biological action of a hormone.

It is conceivable, therefore, that a secretagogue-induced increase in the somatostatin receptor concentration may represent a significant mechanism in enhancing somatostatin action, particularly when paracrine regulation of insulin secretion is concerned.

Although the mechanism of the translocation of somatostatin receptors from secretion vesicles to the plasma membrane is still unknown, we were able to study the biological significance of this translocation. We have hypothesized that an increased number of somatostatin receptors on the cell surface of endocrine cells would render these cells more sensitive to an inhibitory action of somatostatin.

Traditionally, the concept of paracrine regulation in the endocrine pancreas is based upon the assumption that increasing local concentrations of one pancreatic hormone affects the neighboring cells and, thus, either enhances or diminishes the release of other hormones. According to this model, glucose stimulates both β and δ cells to secrete insulin and somatostatin, respectively. The increased concentration of somatostatin then acts on the β cells and inhibits insulin release.

We theorized that an alternate model, based upon augmented cellular sensitivity to somatostatin as a result of a greater somatostatin receptor concentration, can also be operational and at least equally important. If our theory is correct, then by enhancing a number of receptors for a particular hormone, one may observe an increase in the biological action of this hormone, even without an increase in the ambient concentration of this hormone. Theoretically, it would be a much more efficient system to provide immediate feedback in paracrine regulation than the system based upon alterations in hormone concentration.

To prove this theory, we have incubated pancreatic islets in the presence of sodium isethionate and increasing concentrations of glucose. This phase of the experiment (phase A) was designed to stimulate the migration of the vesicles to the plasma membrane and translocate somatostatin receptors to the plasma membrane without lysing marginated vesicles. Indeed, insulin release in the presence of sodium isethionate was identical regardless of glucose concentration. Following short washout periods (phase B) designed to remove sodium isethionate and stimulatory concentrations of glucose from the perifusion media, islets were challenged with a fixed dose of isobutylmethylxanthine (IBMX) alone (400 $\mu M$) or IBMX (400 u$M$) with a fixed dose of somatostatin (5 $\mu M$), (period C).

Islets perifused during period A with a higher glucose concentration responded to IBMX stimulation with a greater insulin release (Fig. 9). Because the main action of IBMX is on lysis of marginated vesicles (38), these results would suggest that more secretion vesicles were translocated to the plasma membrane. Subsequently, when the islets were challenged with both IBMX and SRIF, the latter inhibited IBMX-induced insulin release (Fig. 10). The inhibitory action of somatostatin was greater in islets perifused with a higher glucose concentration during period A. As glucose concentration rose, the recruitment of somatostatin receptors was enhanced and an inhibitory effect of somatostatin was significantly augmented.

These experiments enabled us to establish a correlation between somatostatin binding and its biological activity. It became clear that an enhancement in somatostatin binding resulted in greater sensitivity of islets to subsequent inhibitory action of somatostatin. The maximum inhibition of insulin release was achieved when 5.4 fmoles of somatostatin were bound to 10 pancreatic islets.

These results do not negate the model of paracrine regulation based upon augmentation in release of both hormones. Nevertheless, they favor an alternate model being responsible for the paracrine regulation of insulin secretion.

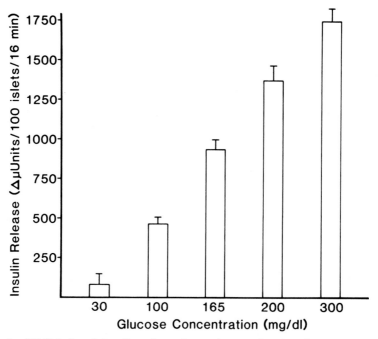

**Fig. 9.** IBMX-induced insulin release from islets perifused with increasing concentrations of glucose.

When glucose stimulates insulin release, it concomitantly enhances translocation of somatostatin receptors from the cell interior to the plasma membrane. This renders insulin-secreting cells more sensitive to somatostatin. This putative pathway in paracrine regulation would have certain advantages in the cellular control of hormone secretion (Fig. 11). Mainly, the effect of somatostatin in inhibiting hormone release would be closely integrated with the exocytotic secretory process.

The concentration of somatostatin receptors on the cell surface is a function of the combined processes of appearance of receptors at the plasma membrane and the rate of receptor internalization. Theoretically, in the scheme being proposed, the secretagogues regulate the appearance of receptors at the plasma membrane. The combining of somatostatin with its receptor would probably accelerate the rate of receptor internalization, thus giving rise to a well-integrated system.

Rapid modulation of the somatostatin-receptor concentration on the surface of endocrine cells provides an excellent means for the feedback regulation of hormone secretion. When the secretion of one hormone is being stimulated, the receptors for an inhibitory regulator are translocated to the surface of the cell, rendering this hormone-secreting cell more sensitive to the subsequent inhibition of its secretion.

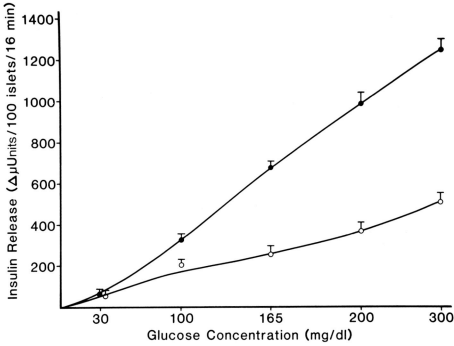

**Fig. 10.**   Effect of somatostatin (SRIF) on IBMX-induced insulin release from islets perifused with increasing concentrations of glucose. Symbols: IBMX (●); IBMX plus SRIF (○).

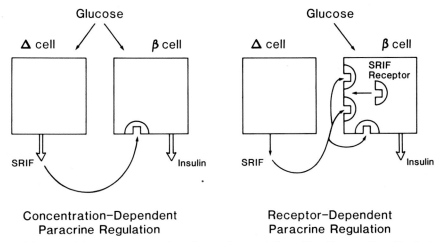

**Fig. 11.**   Schematic presentation of paracrine regulation of insulin secretion. (See text for details.)

It would seem rational to assume that some other regulatory systems may operate in a similar manner. To date, however, we know of no other example of such regulation. It is possible that some of the neurotransmitter interactions may be governed by their receptor translocation. Thus, a stimulatory signal would enhance secretion, secretion vesicle migration, and receptor concentration for an inhibitory substance. Another likely pair of candidates for this type of regulation are dopamine and prolactin release. According to the proposed model, an increase in prolactin secretion would be accompanied by an increase in dopamine-receptor concentration. This, in turn, would augment lactotroph sensitivity to the inhibitory action of dopamine. In a recent report, dopamine has been shown to localize in prolactin-containing secretion granules of the pituitary (39). It is conceivable, as we discussed earlier, that prolactin-containing secretion vesicles possess receptors capable of binding dopamine.

There exists some evidence that the secretion vesicles may indeed be actively involved in the regulation of hormone secretion. Sternberger and Petrali (40), using immunocytochemical staining with antisera to GnRH, have reported the presence of GnRH receptors on the membranes of large secretory granules of gonadotrophs. Hazum and co-workers (41) have demonstrated that approximately 20% of the total cellular pool of GnRH receptors is associated with secretion vesicles. Jennes and colleagues (34) have, however, reported a much lower figure.

Even if the translocation of somatostatin receptors is unique and no other similar system is found, one should bear in mind the existence of unique regulatory mechanisms. One example is an inhibition of PTH secretion by high concentrations of calcium, a cation known to stimulate hormone secretion.

In summary, specific binding sites for somatostatin are found in pituitary cells, adrenal cells, and pancreatic islet cells. Moreover, somatostatin receptors in endocrine cells (at least pituitary and islet cells) are being rapidly translocated from the cell interior to the plasma membrane during exocytosis. This translocation of receptors renders the target cells more sensitive to somatostatin, providing a unique and highly efficient system for regulating hormone secretion.

## REFERENCES

1. Krulich, L., Dhariwal, A. P. S., and McCunn, S. M. (1968). Stimulatory and inhibitory effects of purified hypothalamic extracts on growth hormone release from rat pituitary *in vitro*. *Endocrinology (Baltimore)* **83,** 783–790.
2. Brazeau, P., Vale, W., Burgus, R., Ling, N., Butcher, M., Rivier, J., and Guillemin, R. (1973). Hypothalamic polypeptide that inhibits the secretion of immunoreactive pituitary growth hormone. *Science (Washington, D.C.)* **178,** 77–79.
3. Coy, D. H., Coy, E. J., Arimura, A., and Schally, A. V. (1973). Solid phase synthesis of growth hormone release inhibiting factor. *Biochem. Biophys. Res. Commun.* **54,** 1267–1273.

4. Yamashiro, D., and Li, C. H. (1973). Synthesis of a peptide with full somatostatin activity. *Biochem. Biophys. Res. Commun.* **54**, 882–888.

5. Schally, A. V., Dupont, A., Arimura, A., Redding, T. W., and Lithincum, G. L. (1975). Isolation of porcine GH-release inhibiting hormone (GH RIH); the existence of 3 forms of GH-RIH. *Fed. Proc., Fed. Am Soc. Exp. Biol.* **34**, 584.

6. Schally, A. V., Dupont, A., Arimura, A., and Redding, T. W. (1976). Isolation and structure of somatostatin from porcine hypothalami. *Biochemistry* **15**, 509–514.

7. Spiess, J., Rivier, J. E., Rodkey, J. A., Bennett, C. D., and Vale, W. (1979). Isolation and characterization of somatostatin from pigeon pancreas. *Proc. Natl. Acad. Sci. U.S.A.* **76**, 2974–2978.

8. Reichlin, S. (1983). Somatostatin. *N. Engl. J. Med.* **309**, 1495–1501.

9. Wass, J. A. H. (1982). Somatostatin and its physiology in man in health and disease. *In* "Clinical Neuroendocrinology" G. M. Besser and L. Martini, eds.), Vol. 2, p. 359–395. Academic Press, New York.

10. Schonbrunn, A., and Tashjian, A. H., Jr. (1978). Characterization of functional receptors for somatostatin in rat pituitary cells in culture. *J. Biol. Chem.* **253**, 6473–6483.

11. Sussman, K. E., Draznin, B., Leitner, J. E., and Mehler, P. S. (1982). The endocrine secretion granule revisited—postulating new functions. *Metab., Clin. Exp.* **31**, 959–965.

12. Borgeat, P., Labrie, F., Drouin, J., Belanger, A., Immer, H., Sestanj, K., Nelson, V., Gatz, M., Schally, A. F., Coy, D. H., and Coy, E. J. (1974). Inhibition of adenosine $3',5'$-monophosphate accumulation in anterior pituitary gland *in vitro* by growth hormone-release-inhibiting hormone. *Biochem. Biophys. Res. Commun.* **56**, 1052–1059.

13. Labrie, F., Borgeat, P., Drouin, J., Beaulieu, M., Lagace, L., Ferland, L., and Raymond, V. (1979). Mechanism of action of hypothalamic hormones in the adenohypophysis. *Annu. Rev. Physiol.* **41**, 555.

14. Dorflinger, L. J., and Schonbrunn, A. (1983). Somatostatin inhibits vasoactive intestinal peptide-stimulated cyclic adenosine monophosphate accumulation if GH pituitary cells. *Endocrinology (Baltimore)* **113**, 1541–1550.

15. Bilezikjian, L. M., and Vale, W. W. (1983). Stimulation of adenosine $3',5'$-monophosphate production by growth hormone-releasing factor and its inhibition by somatostatin in anterior pituitary cell *in vitro*. *Endocrinology (Baltimore)* **113**, 1726–1731.

16. Sussman, K. E., Leitner, J. W., and Rifkin, R. M. (1978). Somatostatin: selective inhibition of cyclic AMP stimulated protein kinase. *Trans. Assoc. Am Physicians.* **91**, 129–143.

17. Tan, A. T., Tsang, D., Renaud, L. P., and Martin, J. B. (1977). Effect of somatostatin on calcium transport in guinea pig cortex synaptosomes. *Brain Res.* **123**, 193–196.

18. Luft, R., Efendic, S., and Hokfelt, T. (1978). Somatostatin: both hormone and neurotransmitter? *Diabetologia* **14**, 1–13.

19. Schonbrunn, A., Rorstad, O. P., Westendorf, J. M., and Martin, J. B. (1983). Somatostatin analogs: correlation between receptor binding affinity and biological potency in GH pituitary cells. *Endocrinology (Baltimore)* **113**, 1559–1567.

20. Enjalbert, A., Tapia-Arancibia, L., Rieutort, M., Brazeau, P., Kordon, C., and Epelbaum, J. (1982). Somatostatin receptors on rat anterior pituitary membranes. *Endocrinology (Baltimore)* **110**, 1634–1640.

21. Aguilera, G., and Parker, D. S. (1982). Pituitary somatostatin receptors: characterization by binding with a nondegradable peptide analogue. *J. Biol. Chem.* **257**, 1134–1137.

22. Srikant, C. B., and Patel, Y. C. (1982). Characterization of pituitary membrane receptors for somatostatin in the rat. *Endocrinology (Baltimore)* **110**, 2138–2144.

23. Reubi, J.-C., Perrin, M., Rivier, J., and Vale, W. (1982). High affinity binding sites for somatostatin to rat pituitary. *Biochem. Biophys. Res. Commun.* **105**, 1538–1545.

24. Leitner, J. W., Rifkin, R. M., Maman, A., and Sussman, K. E. (1979). Somatostatin binding to pituitary plasma membranes. *Biochem. Biophys. Res. Commun.* **87**, 919–927.
25. Aguilera, G., Parker, D. S., and Catt, K. J. (1982). Characterization of somatostatin receptors in the rat adrenal glomerulosa zone. *Endocrinology (Baltimore)* **111**, 1376–1384.
26. Leitner, J. W., Rifkin, R. M., Maman, A., and Sussman, K. E. (1980). The relationship between somatostatin binding and cyclic AMP-stimulated protein kinase inhibition. *Metab., Clin. Exp.* **29**, 1065–1073.
27. Reubi, J.-C., Rivier, J., Perrin, M., Brown, M., and Vale, W. (1982). Specific high affinity binding sites for somatostatin-28 on pancreatic B-cells: differences with brain somatostatin receptors. *Endocrinology (Baltimore)* **110**, 1049–1051.
28. Patel, Y. C., Amherdt, M., and Orci, L. (1982) Quantitative electron microscopic autoradiography of insulin, glucagon, and somatostatin binding sites on islets. *Science (Washington, D.C.)* **217**, 1155–1156.
29. Schonbrunn, A. and Tashjian, A. H. Jr. (1980). Modulation of somatostatin receptors by thyrotropin-releasing hormone in a clonal pituitary cell strain. *J. Biol. Chem.* **255**, 190–198.
30. Schonbrunn, A. (1982). Glucocorticoids down-regulate somatostatin receptors in pituitary cells in culture. *Endocrinology (Baltimore)* **110**, 1147–1154.
31. Mehler, P. S., Sussman, A. L., Maman, A., Leitner, J. W., and Sussman, K. E. (1980). Role of insulin secretagogue in the regulation of somatostatin binding by isolated rat islets. *J. Clin. Invest.* **66**, 1334–1338.
32. Sussman, K. E., Mehler, P. S., Leitner, J. W., and Draznin, B. (1982). Role of the secretion vesicle in the transport of receptors: modulation of somatostatin binding to pancreatic islets. *Endocrinology (Baltimore)* **111**, 316–323.
33. Draznin, B., Leitner, J. W., and Sussman, K. E. (1982). Kinetics of somatostatin receptor migration in isolated pancreatic islets. *Diabetes* **31**, 467–469.
34. Jennes, L., Stumpf, W. E., and Conn, P. M. (1983). Intracellular pathways of electron-opaque gonadotropin-releasing hormone derivatives bound by cultured gonadotropes. *Endocrinology (Baltimore)* **113**, 1683–1689.
35. Draznin, B., Mehler, P., Steinberg, J., Leitner, J. W., and Sussman, K. E. (1984). Cytoplasmic orientation of somatostatin receptors on secretion vesicles in anterior pituitary gland and pancreatic islets. *Clin. Res.* **32**, 393A.
36. Pollard, H. B., Pozoles, C. J., Creutz, C. E., and Zinder, O. (1979). The chromaffin granule and possible mechanism of exocytosis. *Int. Rev. Cytol.* **58**, 160–198.
37. Sussman, K. E., Pollard, H. B., Leitner, J. W., Nesher, R., Adler, J., and Cerasi, E. (1983). Differential control of insulin secretion and somatostatin-receptor recruitment in isolated pancreatic islets. *Biochem. J.* **214**, 225–230.
38. Steinberg, J. P., Leitner, J. W., Draznin, B., and Sussman, K. E. (1984). Calmodulin and cyclic AMP: possible different sites of action of these two regulatory agents in exocytotic hormone release. *Diabetes* **33**, 339–345.
39. Nansel, D. D., Gudelsky, G. A., and Porter, J. C. (1979). Subcellular localization of dopamine in the anterior pituitary gland of the rat: apparent association of dopamine with prolactin secretory granules. *Endocrinology (Baltimore)* **105**, 1073–1077.
40. Sternberger, L. A., and Petrali, J. P. (1975). Quantitative immunocytochemistry of pituitary receptors for luteinizing hormone-releasing hormone. *Cell Tissue Res.* **162**, 141–176.
41. Hazum, E., Meidan, R., Keinan, D., Okon, E., Koch, Y., Lindner, H. L., and Amsterdam, A. (1982). A novel method for localization of gonadotropin releasing hormone receptors. *Endocrinology (Baltimore)* **111**, 2135–2137.

# Index

## A

Acetylcholine receptors
role of microaggregation in antibody–receptor–effector system and, 68–69
studies of, 246
role of carbohydrate, 333–334
N-Acetylglucosaminylpyrophosphoryl polyisoprenol, tunicamycin and, 313–314
Actinomycin D, 1,25-dihydroxyvitamin $D_3$ action and, 4, 13, 14
Activation, of adenylate cyclase, persistence of, 45–46
Adenosine, lipolysis and, 295
Adenovirus(es), attachment proteins of, 145–147
high-affinity receptors for, 154–155
Adenovirus type 7, erythrocyte receptors for, 152
Adenylate cyclase
activation of catalytic component, kinetics of, 45
activity as function of agonist concentration, 74–77
antibodies to TSH and, 67
coupling factors, inhibitory and excitatory, 77–78
coupling to receptor by deglycosylated hormones, 326–331, 332–333
functional relationships between R, N and C, 46–47
association of N with C throughout activation, 51–53
association of N with R throughout stimulation, 47–51
transfer of N to C in activation, 53–54
inhibition by $\alpha_2$-adrenergic agents, possible mechanisms of, 292–294
receptor-mediated stimulation
general assumptions, definitions and nomenclature, 38–41
requirements for, 38
summary of experimental observations on different systems for, 48–49
regulation, TCD and, 111, 114
Adenylate cyclase receptor, catalyst or reactant, 41–42

kinetics of activation of catalytic component, 45
persistence of activation, 45–46
steady-state binding and response isotherms, 42–45
Adenylate cyclase system, properties of a hormone–receptor–coupling protein–effector system, 71–74
Adipocytes, insulin receptors on, 65
Adrenal, somatostatin uptake by, 404, 406
Adrenergic receptors, historical perspective, 281
$\alpha$-Adrenergic receptors, identification of, 283, 284
$\alpha_1$-Adrenergic receptors, target size analysis of, 260–266
$\alpha_2$-Adrenergic receptors
linked to inhibition of adenylate cyclase
mechanism of eliciting physiological effects, 294–299
mechanism of inhibition of adenylate cyclase, 292–294
properties of, 283, 285–292
multiple affinity states of, 286–287
target size analysis of, 267–269
$\beta_1$-Adrenergic receptor, 273
$\beta_2$-Adrenergic receptor, 273
Adrenocorticotropic hormone, inhibition of secretion, dexamethasone and, 343
Affinity chromatography, of insulin receptors, 222
Aggregation, *see also* Clustering; Cross-linking; Microaggregation
of deglycosylated gonadotropin, 336
of 1,25-dihydroxyvitamin $D_3$ receptor, 9
Agonist, activation of adenylate cyclase and, 53
*Ah* receptor
biochemical characterization of
assay methods, 95–97
physical properties of crude and partially purified receptors, 98–100
purification, 97–98
stereospecificity, 100–101
biology of
coordinate gene expression, 103–104

**423**